PRINCIPLES OF STATISTICAL INFERENCE

from a Neo-Fisherian Perspective

ADVANCED SERIES ON STATISTICAL SCIENCE & APPLIED PROBABILITY

Editor: Ole E. Barndorff-Nielsen

This Series comprises accounts, by leading experts, from those areas of statistical science and applied probability where advanced mathematical tools are essential. Both research monographs and textbooks will be included, and the series promotes purely theoretical works as well as studies that combine theory and applications.

*Published**

*To view the complete list of the published volumes in the series, please visit:
http://www.worldscientific.com/series/asssap

Advanced Series on

Statistical Science &

Applied Probability

Vol. 4

PRINCIPLES OF STATISTICAL INFERENCE

from a Neo-Fisherian Perspective

Luigi Pace

Department of Statistics
University of Udine, Italy

Alessandra Salvan

Department of Statistics
University of Padua, Italy

World Scientific

Singapore • New Jersey • London • Hong Kong

Published by

World Scientific Publishing Co. Pte. Ltd.

5 Toh Tuck Link, Singapore 596224

USA office: 27 Warren Street, Suite 401-402, Hackensack, NJ 07601

UK office: 57 Shelton Street, Covent Garden, London WC2H 9HE

Library of Congress Cataloging-in-Publication Data
Pace, Luigi.
  Principles of statistical inference from a neo-Fisherian perspective /
Luigi Pace, Alessandra Salvan.
     p. cm. (Advanced series on statistical science & applied probability ; vol. 4)
     Includes bibliographical references (pp. 489–520) and index.
     ISBN-13 978-981-02-3066-1 -- ISBN-10 981-02-3066-4
     ISBN-13 978-981-238-694-6 (pbk) -- ISBN-10 981-238-694-7 (pbk)
     1. Mathematical statistics. I. Salvan, Alessandra. II. Title. III. Series.
     QA276.P14 1997
     519.5-dc21
                                                                      97-10525
                                                                          CIP

British Library Cataloguing-in-Publication Data
A catalogue record for this book is available from the British Library.

To our families

CONTENTS

**3 SURVEY OF SOME BASIC CONCEPTS
   AND TECHNIQUES**   **71**

PREFACE

The aim of this book is to give an integrated account of statistical inference from a frequentist likelihood-based viewpoint. Classical concepts and results are presented together with more recent developments which have been taking place over the last twenty years. A large part of this research is built upon ideas of R.A. Fisher, hence the term neo-Fisherian. Throughout, emphasis is placed more on general concepts and methods than on mathematical details and regularity conditions.

The volume might be used as a textbook in graduate courses and in advanced undergraduate courses on statistical inference.

Prerequisites include a basic knowledge of inference techniques based on the likelihood function (e.g. at the level of the book by Azzalini, 1996) and some ideas about classical optimality results (e.g. at the level of Beaumont, 1980, or of Bickel and Doksum, 1977). Basic knowledge of probability theory is also required. Measure theory is not essential for understanding most of the topics treated: except for section 3.6.2, measure theory is only used in integrals as a unified notation for continuous and discrete distributions. The basic definitions and results of measure theory used here can all be found in section 1.2 of the book by Lehmann (1983).

Chapters 1-3 are mainly introductory and give a unified view of background material such as: parametric statistical models and likelihood; data and model reduction (sufficiency, conditionality, ancillarity, completeness); optimal procedures and their relation to the Fisherian approach. This preliminary review also considers: moments, cumulants and their generating functions; likelihood and first-order asymptotics; the empirical distribution function. Inference in the presence of nuisance parameters is examined separately in Chapter 4. This includes: extension of data and model reduction concepts; marginal and conditional likelihoods; profile likelihood and orthogonal parameters; basic ideas on partial, quasi- and empirical likelihoods. Exponential families, exponential dispersion models and generalized linear models, and group families are introduced in Chapters 5-7. The structure and properties of these classes of statistical models are considered together with specific data and model reduction and inference techniques. The last part of the book is devoted to higher-order asymptotic methods. Basic traditional ideas are first described, e.g. variance stabilization and skewness reducing transformations. Index notation and asymptotic expansions for statistics and distributions are then introduced. The final chapter gives an account of some of the major applications to likelihood inference.

Illustrative examples are developed for specific statistical models, while numerical examples with real data sets are not considered here and could be the subject of a parallel or subsequent course on statistical modelling. Each chapter is supplemented with problems and with bibliographic notes.

We thank Ole E. Barndorff-Nielsen for his encouragement to write this book and for his comments on preliminary drafts. During the four years it has taken us to complete the project, we have greatly benefited from a number of visits to the Department of Theoretical Statistics of Aarhus University. We are grateful to Adelchi Azzalini, Preben Blæsild, Stuart G. Coles, Anthony C. Davison, Jens Ledet Jensen, Ib M. Skovgaard and Andrew T.A. Wood for their valuable comments on parts of the manuscript. Errors undoubtedly remain and we take full responsibility for them. We will greatly appreciate any comments or suggestions from the readers. Special thanks go to Hilary Creek, who helped us a lot with the written English, and to Patrizia Piacentini for her advice in the LaTeX typesetting of the manuscript.

Udine and Padua   **L.P.**
January 1997   **A.S.**

This work was partly supported by EU HCM contract ERB CHRX CT94 0449 and by CNR research grants 93.00824.CT10 and 94.02041.CT10.

LIST OF SYMBOLS

$[a_r]$	The vector with generic element a_r
$[a_{rs}]$	The matrix with generic element a_{rs}
$Bi(n, \theta)$	The binomial distribution with index n and parameter θ
c.g.f.	Cumulant generating function
$\xrightarrow{d}$	Converges in distribution to
$|A|$	Determinant of the matrix A
d.f.	Distribution function
$x \cdot y$	The scalar product of two vectors
$E(Y)$ or $E_\theta(Y)$	Expectation of the random variable Y
$E(Y|\cdot)$	Conditional expectation
$ED_p(\mu, \sigma^2 V(\mu))$	Exponential dispersion family of order p with mean μ and covariance matrix $\sigma^2 V(\mu)$
$\mathcal{F}_e^p$	Exponential family of order p
$\mathcal{F}_{ed}^p$	Exponential dispersion family of order p
$\mathcal{F}_{ne}^p$	Natural exponential family of order p
$Ga(\nu, \phi)$	The gamma distribution with shape parameter ν and scale parameter ϕ
$\Gamma(\cdot)$	The gamma function
$I_B(\cdot)$	The indicator function of the set B
I_d	The $d \times d$ identity matrix
$IG(\phi, \lambda)$	The two-parameter inverse Gaussian distribution
i.i.d.	Independent and identically distributed
int B	The interior of the set B
$K_Y(t)$	The cumulant generating function of Y
κ_r, $\kappa_r(Y)$	Cumulant of order r of the univariate random variable Y

m.g.f.	Moment generating function
m.l.e.	Maximum likelihood estimate
$M_Y(t)$	The moment generating function of Y
μ^*	The dominating measure
$\mathbb{N}$	The set of natural numbers
$\|\cdot\|$	Norm of a vector
$N(\mu, \sigma^2)$	The univariate normal distribution with mean μ and variance σ^2
$N_d(\mu, \Sigma)$	The d-dimensional normal distribution with mean μ and covariance matrix Σ
$\xrightarrow{P}$	Converges in probability to
$p_Y(\cdot)$	Probability density function of Y
$p_{Y\|X=x}(y; x)$	Conditional probability density function of Y given x
p.d.f.	Probability density function
$\phi(\cdot)$	$N(0,1)$ density function
$\phi_d(\cdot; \kappa)$	$N_d(0, \kappa)$ density function
$\Phi(\cdot)$	$N(0,1)$ distribution function
$P(E)$	Probability of the event E
$P(E\|A)$	Conditional probability
$P(\theta)$	The Poisson distribution with mean θ
$\mathbb{R}$	The set of real numbers
$\mathbb{R}^d$	The d-dimensional Euclidean space
$\mathbb{R}^+$	The set of non-negative real numbers
r.v.	Random variable
$\text{sgn}(\cdot)$	The sign function
$\sim$	Distributed as
$\dot{\sim}$	Approximately distributed as
y^T, A^T	Transpose of a vector or a matrix
$\theta^r, \theta^s, \ldots$	Generic components of the parameter vector θ
$U(a, b)$	The uniform distribution on $[a, b]$
$\text{Var}(Y)$ or $\text{Var}_\theta(Y)$	Variance or covariance matrix of the random variable Y
$\mathbb{Z}$	The set of integers

PRINCIPLES OF STATISTICAL INFERENCE

from a Neo-Fisherian Perspective

CHAPTER 1

STATISTICAL MODELS

1.1 The Theory of Statistical Inference

Often the information collected regarding a phenomenon consists of data that are inherently variable. Thus, it is essential that schemes for interpretation should be available in order to distinguish, in the data, possible underlying regularities from the superficial patterns. The underlying regularities can be conjectured to be effective, the superficial patterns appear clear because of the common psychological distortion that leads to over-rationalization of experience. Both formal and informal tools for interpretation are provided by statistics, which could be depicted as a collection of ideas and methods which aim to describe and evaluate variability and its consequences.

The discipline has developed only rather recently and, encouraged by both internal and external pressures, is still evolving. Among the external pressures are the changing needs and resources of science, technology and society, whilst knowledge of previous progress, the elaboration of hitherto ill-defined analogies and links, the search for clarity, synthesis, speed and efficiency in disseminating important ideas outside of the specialist context are internal pressures.

Naturally, a method that has proved fruitful for one specific problem could, potentially, be of assistance when exploring other problems that are similar either in form or in content. However, analysis of a set of observations always requires a new interpretation and isolated ideas cannot provide a broad enough vision of the data and of the method appropriate for approaching them. The study of statistics, or rather parts of it, can usefully follow structures which tend to unify and prepare for not entirely predictable applications, hence, statistics is not seen as a rigid catalogue of cases and methods.

There are ways of approaching the discipline which aim directly towards

1

applications, and ways which offer the possibility of an overall view in which details fade and abstraction unites areas that would otherwise be very separate. Focussing on the details of a specific type of observations, and on the appropriate material for understanding them, is typical of applied statistics. Clarifying more general ideas, which are useful for analyzing broad classes of data and for suggesting methods for the analysis of more specific classes of data, is the aim of the theory of statistics.

This book deals in particular with the theory of statistical inference. In this context, the **fundamental assumption** is that the observed data y^{obs}, often of the form $y^{obs} = (y_1^{obs}, \ldots, y_n^{obs})$, with y_i^{obs} an observation on the i-th observed unit, are the realization of a random vector Y (or, more generally, of a stochastic process) whose probability distribution is unknown. The way in which this basic abstraction conforms to the observations being studied can vary considerably according to their nature and to the concept of probability that is considered to be the most suitable.

If, as in the frequency view of probability, one wishes to associate probability to a physical meaning, as a way of idealizing the long-run stability of the frequencies observed in a large number of homogeneous repetitions of the generation process, then this requires experimental interpretation of the genesis of the data. This is plausible in many areas of scientific and technological research, indeed, it can sometimes be justified on the basis of randomized allocation of units to treatments, or, in the area of human sciences, whenever a random sampling mechanism is explicitly set up. But, outside of these situations, if one thinks for instance of economic time series, the probability model is essentially an attempt to distinguish systematic and noisy aspects within the variability of the observations.

The context of inference can be summarized by thinking of y^{obs} as a realization of $Y \sim p^0(y)$, $y \in \mathcal{Y}$, where $p^0(y)$ represents the unknown probability density function (p.d.f.), with respect to a suitable measure, and where $\mathcal{Y}$ is the sample space. The aim of statistical analysis is to reconstruct $p^0(y)$ on the basis of both data and suitable assumptions and, possibly, on the grounds of previous information, in order to obtain a concise description of the phenomenon being studied which will permit both interpretation and prediction. Density $p^0(y)$ or, more accurately, the probability distribution it represents, will be referred to below by the expression **probability model**. Some of the assumptions that facilitate reconstruction of the probability model are usually expressed through a limitation of the possible forms of $p^0(y)$, that is, through the specification of a family $\mathcal{F}$ of probability distributions which are, at least qualitatively, compatible with y^{obs}. The family $\mathcal{F}$ will be called the statistical model. If the density function of the probability model which generates the data is an element of $\mathcal{F}$, i.e. if $p^0(\cdot) \in \mathcal{F}$, the statistical model is said to be correctly specified, otherwise the model is said to be misspecified.

The probability model has been defined above using the probability density function $p(y)$. In some contexts, such a specification would be better expressed in terms of other functions. For instance, if Y is a univariate random variable (r.v.), we could equivalently describe the distribution of Y through its distribution function (d.f.) $F(y) = P(Y \leq y)$, its moment generating function $M(t) = E\{\exp(tY)\}$ or its characteristic function $C(t) = E\{\exp(itY)\}$ (see section 3.2). With a continuous non-negative random variable which describes a lifetime, the failure rate $r(y) = p(y)/\{1 - F(y)\}$ might be more easily interpretable. The quantity $r(y)dy$ expresses the probability that a failure occurs in the interval $(y, y + dy)$, given that it did not occur before y. If Y has failure rate $r(y)$ the probability density of Y is

$$p(y) = r(y) \exp\{- \int_0^y r(t)dt\} \, . \tag{1.1}$$

In the following, when it is necessary to indicate, explicitly, the random variable to which the functions $p(\cdot)$, $F(\cdot)$, $r(\cdot)$, etc. pertain, the symbol of the random variable will be written as a subscript, i.e. notations such as $p_Y(\cdot)$, $F_Y(\cdot)$, $r_Y(\cdot)$ will be used.

1.2 Four Paradigms of Inference

From the theoretical point of view, the problem of looking for general principles for statistical inference arises immediately. Principles, that is, which can guide the statistician when seeking suitable techniques for reconstructing $p^0(y)$. This search is largely beyond the scope of statistics as it is linked to the more general problem of knowledge and, especially, to the epistemological rules of experimental sciences which are, in the broadest sense, inductive. The literature on this question, although fascinating, has so far proved incapable of offering a satisfactory clarification comparable to that which research on logic has provided for the meaning and the limits of axiomatic structures in mathematics. In other words, the philosophical clarity obtained for deduction has not yet been matched by that for induction. A discussion about foundations of statistical inference is outside the scope of this book. The bibliographic note, section 1.6, gives some references.

However, even in mathematics some scholars, independently of the research in logic, have proposed various conceptions of the foundations of the discipline. Similar investigations have been developed, with even greater urgency, alongside the growth of statistics in the 20th century and have favoured the emergence of some broad structures for interpreting inference. Four general views, or paradigms, for statistical inference are distinguished here, using schematization which is, of course, reductive. The essential differences relate to the interpretation of probability and to the objectives of statistical inference.

The first, which goes back to Bayes and Laplace, is the **personalistic Bayesian paradigm** (or subjectivist paradigm), according to which, interpretation of probability in terms of frequency is only a side-issue, while the fact that it is a description of the subject's state of knowledge is crucial. In the simplest formulation of this view, the subject is required to describe his or her initial state of knowledge in terms of a prior probability distribution on the elements of $\mathcal{F}$. Inference is the formalization of how the initial distribution changes in the light of empirical evidence acquired through the data available, according to the one scheme of up-dating that maintains internal consistency, given by Bayes' formula.

Often considered as being out of place in experimental science, subjectivity has been mitigated or disguised by the specification of prior distributions that represent ignorance (Laplace), and hence are somewhat inter-subjective, like the prior distributions deduced from the assumption of equiprobability of the elements of $\mathcal{F}$, provided $\mathcal{F}$ has a finite number of elements. The definition of ignorance or **uninformative prior distributions** for more complex statistical models was formalized later, from the 1940s on. Some uninformative prior distributions will be considered in sections 2.11.3 and 7.7.2. These later developments identify a non-personalistic **Bayesian paradigm**.

In the early 1920s, in opposition to the Laplace view and, in particular, arguing against the mingling of probability–frequency and probability–opinion, which have an epistemologically different status, Fisher claimed that statistical inference can, indeed must when possible, be *entirely* based on probabilities with experimental interpretation. Here, probability has a dual role: the first is descriptive, that is, modelling variability in a population; the second is epistemological: probability permits quantification of sampling variability of inductions by allowing the variability of samples to be modelled. According to the **Fisherian paradigm**, the variability of inductions should be modelled in accordance with the **principle of repeated sampling**, taking into consideration how the conclusions change with variations in the samples which can be obtained through the hypothetical repetition, under the same conditions, of the experiment which first generated the observations. This hypothetical element refers, in the first place, to the physical non-availability of further samples. It can also be ascribed to the statistical model, in the sense that the behaviour of the inductions is examined for each element of the statistical model. A leading role is played by the concept of likelihood, which is essentially the probability that, within a hypothetical re-running of the experiment, the various competing stochastic mechanisms assign to re-observation of the data that were produced in the actual experiment. A further crucial point in the Fisherian vision is that the probability which describes an event, in order to be relevant, must be considered as conditional on everything that is known. Thus, when judging the probability that a particular seventy year old person will live to the age of seventy-five, not just marginal probability should be used but an attempt should be made to take covariates into account such as sex, state of health, eating habits, family situation and anything else that could prove to be important. Consequently, even when probability is used epistemologically in inference, it must be as relevant as possible to the data y^{obs}: all the aspects which may be shown by the observations that can be considered as analogous to covariates must be taken into account.

During the 1930s and 1940s, Neyman and Egon Pearson and, later, Wald and Lehmann, contributed to offering a new paradigm for inference that, initially imperceptibly, but later more and more markedly, moved away from the Fisherian view. The starting point was the nucleus of ideas put forward by Fisher and in-depth mathematical study of some of his fundamental concepts, such as likelihood and sufficiency. The emphasis shifted from inference as a summary of data to inferential procedures (hypothesis testing, point and interval estimation) seen as decision problems, in the mathematical form of constrained optimization problems. Further elements which differentiate these new developments were dictated by the need for clarity in mathematical formulations. These required that optimum inference procedures should be identified before the observations were available so as to obey the principle of repeated sampling and leave no shadow of ambiguity about interpretation. This vision will be called the frequency-decision paradigm.

Most of the concepts and methods which will be dealt with here fall within the Fisherian paradigm. Many recent innovations have, directly or indirectly, drawn their inspiration from it. Some such innovations, which have proved as fruitful for theoretical research as they have for applications, are: robust methods, bootstrap, higher-order asymptotic methods and the various con-

cepts of pseudo-likelihood. This does not mean that the ideas of the other
three paradigms will be ignored. In particular, classical results of optimal theory of inference (reviewed in section 3.5) will be used; some close relations with
the non-personalistic Bayesian paradigm will also be stressed.

1.3   Model Specification

It is worth starting from the radically innovative ideas in Fisher (1922a), where
the Fisherian paradigm was sketched out for the first time. On the premise
that the aim of a statistical analysis is to summarize the data y^{obs} by means
of the reconstruction of $p^0(y)$, Fisher divided the problems encountered into
three classes:

- **problems of specification**, that are linked to the identification of a statistical model $\mathcal{F}$ which is appropriate for the observations y^{obs} being studied;
  ideally the probability model that generates data is exactly captured by
  $\mathcal{F}$ ($p^0(y) \in \mathcal{F}$) or, at least, its most essential aspects are captured;

- **problems of inference**, referred to by Fisher as *problems of estimation*,
  that is, in general, finding statistical procedures able to locate $p^0(y)$
  within $\mathcal{F}$ or, with the help of $\mathcal{F}$; if the statistical model $\mathcal{F}$ is correctly
  specified, the reconstruction of $p^0(y)$ will, usually, be all the easier the
  less mathematically complex $\mathcal{F}$ is; this class of problems also includes
  finding procedures which are appropriate for giving indications about
  model goodness of fit, that is, about the plausibility of the assumption
  $p^0(y) \in \mathcal{F}$;

- **problems of distribution**, that is the evaluation of how sensitive the reconstruction of $p^0(y)$ is to the fact that the data used are only a sample;
  with regard to this, we can presume, in general, that the reconstruction
  of $p^0(y)$ will be more effective the lower the extension of $\mathcal{F}$ is.

The three classes of problems (specification, inference and distribution) are
closely interlinked. In applications, use of a model $\mathcal{F}$ for which the available
inference methods are either difficult to use or not particularly effective is
discouraged. Vice versa, any substantial progress related to the problems of
the second and third classes, facilitates the solution of specification problems
and broadens the reserve of statistical models that can, reasonably, be drawn
upon. These interactions are complementary, in the sense that interest in the

application of statistical models which pose complex problems of inference is a stimulus to further research.

In applications, a statistical model is usually considered as an approximation, in the sense that no one expects it to capture the probability model accurately, rather, the idea is that the model can be considered adequate for the aims of the research. In this sense, the three classes of problems should not be understood as corresponding to successive phases in data analysis, rather, they should be understood as logical moments along a necessarily iterative path. This same specification phase is usually, to a greater or lesser extent, guided by the data through the feedback effects of the inference phases: as, for example, in the analysis of residuals in a linear regression model.

Although model specification is very important, and usually reflects more on the conclusions than does the inference paradigm adopted, the theory of statistical inference, traditionally, lacks explicit indications about specification problems. Fisher, who probably felt that the unique aspects of any single type of data were pre-eminent, assigned these problems to applied statistics. This vision seems to be an over-simplification because it introduces a split between theory and application. Theory puts some particular models in the spotlight and, even though it is true that the specification phase is hard to formalize, some guidelines, based on common sense, should be offered.

A starting point could be that of setting limitations which are based upon the nature of the data being examined. Different models are suitable for dealing with qualitative (nominal, ordinal) or quantitative (discrete, continuous) variables, and with functions, images, etc.. In addition, the observed variables could be subdivided into subsets, for example, into response and explanatory variables. In order to be specified reasonably accurately, the model must respect both the support and the role of variables.

A second point is that the information about the observation scheme is also important: for example, idealizations such as random sampling, randomization, censoring or other models for missing data, sequential sampling, time and space dependence.

Furthermore, attention should be focussed on aspects of the data that the model must be able to catch, for example, the centre of distribution, unimodality or bimodality, dependence on explanatory variables etc., and complementary aspects, such as dispersion, asymmetry, heteroschedasticity, etc., which should be taken into account in order to perfect the analysis. The statistical model must be able to succinctly describe the aspects that are of primary interest and must also be sufficiently flexible to allow a realistic description of additional aspects.

1.3.1 Levels of Specification

Depending on the information available it may be deemed appropriate to extend the statistical model $\mathcal{F}$ to a greater or a lesser degree. In increasing order of extension, hence in decreasing order of presumed information, the following three levels of specification can be outlined.

- **Parametric specification.** Previous knowledge and conjecture produce a rather restricted class $\mathcal{F}$; the elements that make it up can be indexed by a finite number p of real parameters, that is

$$\mathcal{F} = \{p(y; \theta), \ \theta \in \Theta \subseteq \mathbb{R}^p\}.$$

  If the model is correctly specified, we have $p^0(y) = p(y; \theta_0)$ for a value $\theta_0 \in \Theta$, called the **true parameter value**.

- **Semiparametric specification.** The elements of $\mathcal{F}$ can be identified through both a parametric and a nonparametric component, that is, $\mathcal{F} = \{p(y) = p(y; \theta), \ \theta \in \Theta\}$ where $\theta = (\tau, h(\cdot))$, with $\tau \in \mathrm{T} \subseteq \mathbb{R}^k$ whereas the set of possible specifications of the function $h(\cdot)$ cannot be indexed by a finite number of real parameters. In this specification, τ usually represents the aspects of primary interest of the distribution, while high flexibility is sought in the description of the aspects of secondary concern. Think, for example, of the class of continuous symmetrical distributions on $\mathbb{R}$, which are partially parameterized by their centre of symmetry, that is, have density of the form $p(y; \mu) = p_0(y - \mu)$, with $\mu \in \mathbb{R}$ and $p_0(\cdot)$ a probability density symmetric around the origin. Furthermore, think of the usual linear regression model of Y on X, $E(Y_i) = \alpha + \beta x_i$, where the Y_i, $i = 1, \ldots, n$, are independent with a common variance σ^2, the x_i are n known constants, $\tau = (\alpha, \beta, \sigma^2)$, and the distribution of Y is not further specified. One last, rather important, example is the **proportional hazards model** (Cox, 1972), which is suitable for expressing the dependence of the lifetime Y_i of the i-th unit on a k-dimensional explanatory variable x_i. This model is specified through the failure rate function as

$$r_{Y_i}(y_i) = r_0(y_i) \exp\{\beta \cdot x_i\}, \tag{1.2}$$

  where $r_0(\cdot)$ is an unknown **baseline hazard function**, $\beta = (\beta_1, \ldots, \beta_k)$ denotes a vector of regression coefficients, which are also unknown, and $\beta \cdot x_i$ denotes the scalar product of the two vectors.

- Nonparametric specification. The model $\mathcal{F}$ is a restriction of the family of all the probability distributions defined on a support which is suitable for the data under analysis. It is defined by means of global simplifying assumptions which do not expressly identify a finite number of parameters that are primary subject of inference. For example if the observations are $y = (y_1, \ldots, y_n)$, a possible nonparametric model is given by the restriction of the distributions on $\mathbb{R}^n$ to the subfamily $\mathcal{F}$ which is made up of distributions with independent and identically distributed components (random sampling of size n from a random variable with an entirely unknown distribution).

Because of the, usually unavoidable, element of approximation that is inherent in a statistical model, it is often a guide rather than a prescription. The choice of the level of specification depends, among other factors, both upon how much information one can reasonably expect to extract from the data and on the aim for which the model is used. When there is only a small amount of data a rather parsimonious parametric model can be useful. If the model is purely empirical, a black box that directs action, then its mathematical manageability is of prime importance. If, on the other hand, it must explain the subject being studied and represent an element for understanding, then a parametric model is usually required. This model should be carefully deduced from simple assumptions elicited through a fundamental examination of the mechanism that generates the data. Usually, in this type of model, each parameter represents one specific aspect endowed with a physical meaning. One strategy that is always useful when building a complex model is that of combining elementary blocks. For example, it is worth separating any consideration of the deterministic part from that of the stochastic part. Usually, the former can be more closely linked to the essential knowledge or to the aims of the analysis and then modelled parametrically, while, in the latter, it is better to avoid over-restrictive assumptions. Just think of the description of the effect of covariates carried out through regression analysis, that is, through modelling the connection between the parameters of the distribution of the response variable with the values of the covariates.

1.3.2 Notes on the Specification of a Parametric Model

The problem of specification is always present but is enhanced in the parametric case. Here, a direct comparison is required between the knowledge

available regarding the mechanism that has generated the data and the probabilistic genesis, exact or asymptotic, of the various families of distributions.
The binomial distribution describes the random variability of the number of
successes in a given number of trials *if* the trials are independent *and if* the
probability of success in each trial is constant. These assumptions give a suitable statistical model if they are met to a reasonable degree of approximation.
Such a statistical model could still be useful even if the approximation is not
accurate, so long as the usual inferential procedures are suitably modified, e.g.
so that they can take lack of independence or overdispersion produced by the
heterogeneity of trials into account. When fitting a continuous distribution to
time to failure, the exponential distribution is useful to describe populations
that do not age, in the sense that if an item works at a given time, the distribution of its additional lifetime is the same as that of new items (see for example,
Azzalini, 1996, section A.2.4). If essential knowledge suggests ageing for the
phenomenon being studied, the data should be modelled through specification
of the failure rate.

More generally, characterization results that express, in a simple form,
which aspects will emerge from the data under a particular model can also
be helpful. A characterization of a family of probability distribution $\mathcal{F}$ states
a necessary and sufficient condition for the density $p(y)$ to belong to $\mathcal{F}$. For
example, a constant failure rate characterizes the exponential distribution in
the class of continuous distributions. As a further example, the assumption
that $Y_1, \ldots, Y_n$ are independent and identically distributed (i.i.d.) and follow
a normal distribution with mean 0 and variance σ^2, $N(0, \sigma^2)$, is equivalent
to the assumption that their joint density $p_Y(y)$ can be factorized according
to stochastic independence and is spherically symmetrical with respect to the
origin of $\mathbb{R}^n$ (see e.g. Lehmann, 1990, Example 2.1).

Some parametric models can be justified on the basis of asymptotic considerations. Besides the well known case of the normal distribution, which is
tied to the central limit theorem, **extreme value distributions** are also defined
by limiting arguments. If the maximum of n independent observations from
a univariate distribution, suitably standardized, has a limiting distribution as
n tends to infinity, then this limiting distribution can only belong to one of
three families. These distributions are useful when describing phenomena that
could be interpreted in terms of a large value of either a detectable or a hidden
variable (maximum annual rainfall, failure time, etc.). The asymptotic results
relating to stable distributions (see section 3.2.6) have also attracted some interest in statistics. Stable distributions arise from generalizations of the central

limit theorem to the infinite variance case. In many applications a model that permits infinite variance in the observable variables is unsuitable. Exceptions can be found in models for economic and financial data, see Du Mouchel (1983). Basic convergence results for sums and extremes are collected in Appendices A and B, respectively.

The connection between the genesis of a parametric model or characterization results and the data may be weak or unclear. Asymptotic arguments require an evaluation of the adequacy of the approximation given. The assumptions of independence and identical distribution, which may have been made hastily, should be critically examined. Furthermore, particularly in the specification of complex models, it may happen that some parts of the model can be expressed with reasonable confidence, while others require closer examination in the light of the information provided by the data. All these elements make it clear that the specification of a model is usually the product of an iterative process. The choice between competing models can be made through informal and formal tools, such as plots, analysis of residuals, selection procedures and tests of goodness-of-fit.

The rest of this book will mainly deal with problems of inference and of distribution and will only offer the reader indirect help with specification problems. It will, possibly, broaden the range of models which could be considered familiar, both with their definition and the mastery of their salient inferential properties. Even though complex models will only rarely be expressly mentioned, in-depth understanding of the blocks from which they are built will facilitate their definition and study in the specific applications of statistics.

1.4 Parametric Statistical Models and Likelihood

Assume that the family $\mathcal{F}$ has been specified as a class of probability models compatible with y^{obs}. Of course, below, this assumption is understood as being provisional, it is a working hypothesis that is liable to be reformulated in the light of any new information which becomes available in later stages of statistical analysis.

1.4.1 General Formulation of a Statistical Model

The statistical model $\mathcal{F}$ is usually described as a family of density functions. A mathematically precise formulation requires notions of measure theory. Measure theory is not essential to most of the topics treated in this book; however,

the reader is assumed to be familiar with the basic definitions and results, at
the level, for instance, of section 1.2 in Lehmann (1983).

In general terms, the class $\mathcal{F}$ is specified once the triple

$$(\mathcal{Y}, P_\theta, \Theta)$$

has been assigned, where $\mathcal{Y}$ denotes the sample space, P_θ a probability measure
on a σ-algebra $\mathcal{B}$ of subsets of $\mathcal{Y}$, called events. The distribution P_θ is indexed
by the parameter θ, which takes values in the parameter space Θ. If Θ is,
abstractly, meant as a set of indices, then both nonparametric, semiparametric
and parametric specifications fall within this formulation. It is assumed that
the identifiability condition is met, $P_\theta \neq P_{\theta'}$ if $\theta \neq \theta'$. This means that there is
at least one event $B \in \mathcal{B}$ such that $P_\theta(B) \neq P_{\theta'}(B)$.

Even though some of the ideas presented below could be interpreted as
referring to this more general specification, attention will be concentrated on
classes $\mathcal{F}$ of parametric models for which the following further conditions are
satisfied.

- There exists a σ–finite measure $\mu^\bullet$ on $\mathcal{B}$ such that all the probability
  measures P_θ, $\theta \in \Theta$, are absolutely continuous with respect to $\mu^\bullet$, that
  is, whichever $\theta \in \Theta$, $P_\theta(B) = 0$ for every event B for which $\mu^\bullet(B) = 0$.
  The family $\mathcal{F}$ is then called dominated. Then, by the Radon–Nikodym
  theorem, the density function of P_θ with respect to the dominating mea-
  sure $\mu^\bullet$, $p_Y(y; \theta) = dP_\theta/d\mu^\bullet$, is defined. The probability of $B \in \mathcal{B}$ may
  be expressed as $P_\theta(Y \in B) = \int_B p_Y(y; \theta)\, d\mu^\bullet$. In most applications $\mathcal{Y}$
  is a subspace of a Euclidean space and $\mu^\bullet$ is the Lebesgue measure (absolutely continuous case) or a counting measure (discrete case), so that
  $P_\theta(Y \in B) = \int_B p_Y(y; \theta)\, dy$ or $P_\theta(Y \in B) = \sum_{y \in B} p_Y(y; \theta)$, respectively.

- The parameter space Θ is a subset, possibly the entire space, of the p-
  dimensional Euclidean space $\mathbb{R}^p$. In the simplest setting one assumes
  that Θ is an open non-empty subset of $\mathbb{R}^p$; on one hand, this makes the
  treatment of aspects of differential calculus for smooth functions defined
  on Θ easier, on the other hand, it gives p the meaning of an effective
  geometric dimension of Θ.

Thus, a dominated parametric statistical model $\mathcal{F}$ can be specified in the
form

$$(\mathcal{Y}, p_Y(y; \theta), \Theta), \tag{1.3}$$

where $p_Y(y;\theta)$ is a density with respect to μ^* and $\Theta \subseteq \mathbb{R}^p$.

Below, especially when theoretical considerations are presented, the components of θ are indicated by $\theta^1, \ldots, \theta^p$. It should be noted that with this notation, θ^2 is the second component of θ and not θ squared. Furthermore, the symbols θ^r, θ^s, etc., $r, s, \ldots = 1, \ldots, p$, are used to indicate generic components of θ. The notational convention of using upper indices for the components of θ may be new to the reader. The advantages of using this notation will become clearer later.

One fact which should be underlined is that, by attributing to θ the mere function of indicator of elements in $\mathcal{F}$, it makes no difference if it is replaced by a one-to-one smooth function (infinitely differentiable with infinitely differentiable inverse), considering a **reparameterization** for $\mathcal{F}$. In applications, it might be more suitable to choose a parameterization in which every component of the parameter describes an easy to interpret characteristic of distribution. On the other hand, inference could be simplified by the choice of a different parameterization, without changing the essence of the problem which remains that of locating $p^0(y)$ within $\mathcal{F}$.

If the data consist of n observations $y = (y_1, \ldots, y_n)$, both the sample space $\mathcal{Y}$ and the density $p_Y(\cdot)$ depend on n. As regards the parameter space, we will mainly deal with situations where the dimension of Θ does not depend on n, thus excluding, at least initially, having to deal with models with incidental parameters (see section 4.1). If the n observations are independent, the sample space $\mathcal{Y}_n$ is the Cartesian product of the sample spaces of the individual observations, the density is $p_Y(y;\theta) = \prod_{i=1}^{n} p_{Y_i}(y_i;\theta)$, with $p_{Y_i}(y_i;\theta)$ density of the i-th observation. If the n observations are dependent, it is usually convenient to write their joint density in product form, as

$$p_Y(y_1, \ldots, y_n; \theta) = p_{Y_1}(y_1; \theta)\, p_{Y_2 | Y_1 = y_1}(y_2; y_1, \theta)$$
$$\cdots p_{Y_n | Y_1 = y_1, \ldots, Y_{n-1} = y_{n-1}}(y_n; y_1, \ldots, y_{n-1}, \theta),$$

where $p_{X|Z=z}(x; z)$ denotes here, and in the following, the conditional density of X given $Z = z$.

1.4.2 Likelihood and Related Quantities

The concept of likelihood is crucial for the Fisherian view of parametric statistical inference.

Definition 1.1 Let $\mathcal{F}$ be a parametric statistical model for data y specified in the form (1.3). The likelihood function is

$$L = L(\theta) = L(\theta; y) = c(y)\, p_Y(y; \theta) , \tag{1.4}$$

where $\theta \in \Theta$ and $c(y) > 0$ is an arbitrary constant of proportionality.

The function $L(\theta; y)$ gives the natural information summary on θ, based on the parametric statistical model (1.3) and on the observed data y. The arguments for this statement will be discussed in Chapter 2. Even though likelihood was introduced into a strictly parametric context, recent developments have demonstrated that the concept of likelihood is productive in the context of semi-parametric and nonparametric models too (see sections 4.8-4.10). The term *likelihood* was first introduced in statistics by Fisher (1921). In Fisher (1922a) the following concise definition was given.

> Likelihood.–*The likelihood that any parameter (or set of parameters) should have any assigned value (or set of values) is proportional to the probability that if this were so, the totality of observations should be that observed.*

The definition of the likelihood function up to constant factors can be justified in various ways. At an intuitive level, the fact that likelihood serves as an indicator of coherence between probability models, which correspond to the possible values of the parameter, and observations, means that only relative comparisons are possible, where the constant is irrelevant. Moreover, the definition of $L(\theta)$ is independent of the dominating measure μ^*. This has two aspects. Firstly, as in the discrete case where $L(\theta; y)$ is proportional to the probability of re-observing y in a hypothetical repetition of the experiment, hence this interpretation is valid in the continuous case, according to a clear limiting procedure. Secondly, the likelihood function does not change under one-to-one transformations of the data; in the continuous case, the Jacobian determinant does not depend on the parameter and is incorporated in the constant of proportionality. For a further reason, see Example 2.4. One important consequence of the presence of $c(y)$ in the definition (1.4) is that the likelihood function is independent of the sampling rule provided that this rule depends only on the data and not on θ (see Problem 2.2).

When dealing with the likelihood function, it may in many cases prove more convenient to consider the **log-likelihood function**

$$l = l(\theta) = l(\theta; y) = \log L \,, \tag{1.5}$$

where $\log(\cdot)$ denotes the natural logarithm and $l(\theta) = -\infty$ if $L(\theta) = 0$. Since $L(\theta)$ is defined up to a multiplicative constant, the log-likelihood is, in its turn, determined up to an additive constant, which only depends on y. In the case of independent observations

$$l(\theta) = \sum_{i=1}^{n} \log p_{Y_i}(y_i; \theta) \,. \tag{1.6}$$

Adopting the repeated sampling principle as a criterion of evaluating inferential procedures requires study of the probability distribution of the random variable $l(\theta) = l(\theta; Y)$, or of related quantities, for θ fixed and as y varies in the sample space $\mathcal{Y}$ according to a density $p_Y(y; \tilde{\theta})$ in $\mathcal{F}$, where $\tilde{\theta} \in \Theta$ is a parameter value not necessarily equal to θ. Generally, we speak of a null distribution if $\tilde{\theta} = \theta$ and of a non-null distribution if $\tilde{\theta} \neq \theta$. Analogously, we refer to null moments, that is evaluated with respect to a null distribution, and to non-null moments, evaluated with respect to a non-null distribution. Symbols such as $E_\theta(\cdot)$, $\text{Var}_\theta(\cdot)$ are used to indicate expectation, variance (or covariance matrix), etc., calculated with reference to $p_Y(y; \theta)$. In some cases, it must be explicitly indicated which distribution has been used to calculate expectation, variance, etc.; in these cases, $E_{Y;\theta}(\cdot)$, $\text{Var}_{Y;\theta}(\cdot)$ and other analogous notations are used. It should be noted that while the likelihood function is not sensitive to the sampling rule, the distributions and the moments introduced above depend on hypothetical repetitions of the experiment.

Jensen's inequality offers a basic result for the expectation of log-likelihood which will be referred to as the Wald inequality:

$$E_\theta(l(\theta)) > E_\theta(l(\tilde{\theta})), \qquad \tilde{\theta} \neq \theta \,. \tag{1.7}$$

According to (1.7), the null expectation of log-likelihood is always greater than the non-null expectation.

In the following, it will be assumed that log-likelihood is a sufficiently smooth function of θ, that is, that it has partial derivatives with respect to the components of θ up to the required order. Furthermore, it is assumed that all the null moments of these derivatives are finite. If non-null moments are also referred to, then their existence is implicit. The partial derivatives of the log-likelihood function are indicated by

$$l_r = l_r(\theta; y) = \frac{\partial}{\partial \theta^r} l(\theta) \,,$$

$$l_{rs} = l_{rs}(\theta; y) = \frac{\partial^2}{\partial\theta^r \partial\theta^s} l(\theta) ,$$

$$l_{rst} = l_{rst}(\theta; y) = \frac{\partial^3}{\partial\theta^r \partial\theta^s \partial\theta^t} l(\theta) ,$$

etc.. Such derivatives, and more generally the quantities obtained from the likelihood function will be called likelihood quantities. The derivatives of $l(\theta)$ up to the second order, play a crucial role in inference. Derivatives up to the fourth order and their joint moments are important for refining the asymptotic theory of inference (see Chapters 9 and 11). Likelihood quantities that depend on the observed data are called observed likelihood quantities and functions of their moments are called expected likelihood quantities.

Definition 1.2 The score function l_* is the vector of the partial derivatives of $l(\theta)$ with respect to θ, i.e. $l_* = l_*(\theta; y) = (l_1, \ldots, l_p)$.

The notation $l_* = [l_r]$ will also be used, where $[a_r]$ denotes a vector with generic element a_r and the range of the index r is understood.

Assuming that the conditions which permit differentiation and integration to be interchanged are satisfied, from the identity

$$\int_{\mathcal{Y}} p_Y(y; \theta) \, d\mu^* = 1$$

it follows that

$$E_\theta(l_*) = E_\theta(l_*(\theta; Y)) = 0 , \tag{1.8}$$

that is, the null first moment of the score is zero. This is what one would expect from (1.7).

Definition 1.3 The observed information matrix, $j(\theta)$, is

$$j = j(\theta) = \begin{pmatrix} -l_{11} & \cdots & -l_{1p} \\ \vdots & \ddots & \vdots \\ -l_{p1} & \cdots & -l_{pp} \end{pmatrix} ,$$

or, in compact notation, $j = [-l_{rs}]$; here and below the matrix with elements a_{rs} is indicated by $[a_{rs}]$. Furthermore,

$$i = i(\theta) = E_\theta\{j(\theta)\} = [E_\theta\{-l_{rs}(\theta; Y)\}]$$

denotes the expected information or Fisher information matrix.

It is assumed here that the conditions which ensure the validity of the information identity

$$E_\theta\{-l_{rs}(\theta)\} = E_\theta\{l_r(\theta)l_s(\theta)\} , \qquad (1.9)$$

$r, s = 1, \ldots, p$, are satisfied (see e.g. Azzalini, 1996, section 3.2.4). In other words, the expected information matrix is the null second moment of the score and, as such, is a non-negative definite matrix. The elements of the inverse of a matrix $[a_{rs}]$ will be denoted by upper indices, that is, $[a_{rs}]^{-1} = [a^{rs}]$; in particular $j^{-1} = [j^{rs}]$ and $i^{-1} = [i^{rs}]$, when these inverses exist.

The sequence with three indices of third order partial derivatives of log-likelihood, l_{rst}, can be considered collectively as an aggregate of objects endowed with the relevant order structure, called array, indicated by $[l_{rst}]$. Analogous notations may be used for higher-order derivatives. Generally, in an array, the indices can appear both as upper and lower indices; a specific meaning is attributed to the position of an index in section 9.5. For the moment, attention should be drawn to the fact that an expression such as l_{rst} denotes a generic element of $[l_{rst}]$, that is, a generic function which appears in such an array; the range of the indices $(r, s, \ldots = 1, \ldots, p)$ is usually understood.

The likelihood function highlights some probability models included in the parametric statistical model $\mathcal{F}$ as being particularly suitable for interpreting the variability observed.

Definition 1.4 A value of θ that maximizes $L(\theta; y)$ over Θ, that is a value $\hat{\theta}$ such that $L(\hat{\theta}) = \sup_{\theta \in \Theta} L(\theta)$, is called **maximum likelihood estimate** (m.l.e.) of θ.

Of all the elements of $\mathcal{F}$, $\hat{\theta}$ denotes that (or those) for which data y offer the maximum empirical support. The notations $\hat{l} = l(\hat{\theta})$, $\hat{j} = j(\hat{\theta})$, $\hat{i} = i(\hat{\theta})$, etc., are used to denote likelihood quantities evaluated at $\hat{\theta}$. By slightly adjusting the terminology, if $\theta = (\tau, \zeta)$ and $\hat{\theta} = (\hat{\tau}, \hat{\zeta})$, we say that $\hat{\tau}$ is the maximum likelihood estimate of τ.

Maximizing $l(\theta)$ instead of $L(\theta)$ is often an advantage in applications, and, whenever $l(\theta)$ can be differentiated, it is worth seeking the maximum likelihood estimate from among the solutions of the likelihood equation

$$l_*(\theta) = 0 . \qquad (1.10)$$

In the following, unless otherwise stated, it will be assumed that the maximum likelihood estimate is unique and that it is the unique solution to (1.10).

Sufficient conditions for specific classes of parametric models will be given in Chapters 5, 6 and 7. Equation (1.10) is an estimating equation naturally associated with the parametric model $\mathcal{F}$.

Under a parametric statistical model $\mathcal{F}$ with data y^{obs}, inference about θ can be carried out in qualitative terms on the basis of the behaviour of the likelihood function. Intuitively, the true parameter value θ_0 is located close to the value $\hat{\theta}$ which has maximum empirical support in terms of likelihood. Furthermore, the rapidity of the decrease of log-likelihood around its maximum which is measured, for example, by $\hat{\jmath}$, is a source of information about the variations in the empirical support for θ values which are close to $\hat{\theta}$, in the sense that any drastic variations are a sign of efficacious localization. Lastly, further aspects such as the asymmetry of $l(\theta)$ around the global maximum, the presence of local maxima, slow decrease on one tail, and so on, give an articulated vision of the information gleaned from the data. The practise of examining the log-likelihood function is to be recommended and it is facilitated by the graphic options of many popular statistics packages. For an approach to inference which is based entirely on the likelihood function, see Edwards (1972).

1.4.3 Reparameterizations

The likelihood and the log-likelihood functions do not depend on the parameterization chosen for $\mathcal{F}$. In fact, let $\psi = \psi(\theta)$ be a one-to-one smooth function from $\Theta \subseteq \mathbb{R}^p$ to $\Psi \subseteq \mathbb{R}^p$, infinitely differentiable together with its inverse. Then ψ defines an alternative parameterization of the model. Since θ and $\psi(\theta)$ identify the same element of $\mathcal{F}$, we can write

$$L^{\Psi}(\psi) = L^{\Theta}(\theta(\psi)) \tag{1.11}$$

and

$$l^{\Psi}(\psi) = l^{\Theta}(\theta(\psi)) , \tag{1.12}$$

where likelihood and log-likelihood in the parameterization ψ are denoted by $L^{\Psi}(\cdot)$ and $l^{\Psi}(\cdot)$ and the same functions in the original parameterization θ are denoted by $L^{\Theta}(\cdot)$ and $l^{\Theta}(\cdot)$; the inverse function of $\psi(\theta)$ is denoted by $\theta(\psi)$.

The transformation law (1.11) excludes any interpretation of likelihood, multiplied by the suitable normalizing constant, as a probability distribution on Θ. In this case, for continuous parameters, $L^{\Theta}(\cdot)$ and $L^{\Psi}(\cdot)$ should differ through the Jacobian determinant of the transformation $\psi(\theta)$.

On the basis of relations (1.11) and (1.12), likelihood may be considered as an intrinsic function, i.e. one not dependent on the coordinate system expressed by the parameterization, defined on the collection of probability models $\mathcal{F}$. Indeed, it is possible to define the likelihood function as $L(p(\cdot))$, $p(\cdot) \in \mathcal{F}$. When a parameterization is assigned as a function $\theta : \mathcal{F} \to \mathbb{R}^p$, then the expression $L(\theta)$ should be understood as the abbreviated form of $L(\theta(p(\cdot)))$. Analogously, the expression $l(\theta)$ is the abbreviated form of $l(\theta(p(\cdot)))$. Even though the notations $L(\theta)$ and $l(\theta)$ will, henceforth, be adopted, it should be remembered that every function of $L(\theta)$ and $l(\theta)$ can be defined directly on $\mathcal{F}$ and only for convenience is expressed in terms of a given parameterization. One important example is the maximum likelihood estimate: $\hat{\theta}$ indicates the element of $\mathcal{F}$ for which the data offer maximum support. If such an element is indicated by $\hat{p}(\cdot)$ it will be $\hat{\theta} = \theta(\hat{p}(\cdot))$. From this follows the important property of **equivariance under reparameterization** of the maximum likelihood estimate. If ψ is an alternative parameterization of $\mathcal{F}$, we will have $\hat{\psi} = \psi(\hat{p}(\cdot)) = \psi(\hat{\theta})$ and $\hat{\theta} = \theta(\hat{\psi})$.

On the other hand, the log-likelihood derivatives l_r, l_{rs}, $\ldots$, and their expectations depend on the parameterization chosen for $\mathcal{F}$ and they transform according to regular patterns under a reparameterization. Let us denote by $\psi^a, \psi^b, \ldots (a, b = 1, \ldots, p)$, the generic components of ψ, whereas $\theta^r, \theta^s, \ldots$, denote the generic components of θ. According to the differentiation rule for composite functions, the relation between the scores in the two parameterizations can be expressed as

$$\bar{l}_a = \sum_{r=1}^{p} l_r \theta_a^r \, , \tag{1.13}$$

where

$$\bar{l}_a = \frac{\partial}{\partial \psi^a} l^\Psi(\psi) \, , \tag{1.14}$$

$l_r = l_r(\theta(\psi))$ and $\theta_a^r = (\partial/\partial \psi^a)\theta^r(\psi)$. Let

$$\bar{l}_{ab} = \frac{\partial^2}{\partial \psi^a \partial \psi^b} l^\Psi(\psi) \, , \tag{1.15}$$

$$\bar{l}_{abc} = \frac{\partial^3}{\partial \psi^a \partial \psi^b \partial \psi^c} l^\Psi(\psi) \, , \tag{1.16}$$

etc., denote partial log-likelihood derivatives in the parameterization ψ. By reapplying the differentiation rule for composite functions the following further relationships are obtained

$$\bar{l}_{ab} = \sum_{r,s=1}^{p} l_{rs}\, \theta_a^r\, \theta_b^s + \sum_{r=1}^{p} l_r\, \theta_{ab}^r \,, \tag{1.17}$$

$$\bar{l}_{abc} = \sum_{r,s,t=1}^{p} l_{rst}\, \theta_a^r\, \theta_b^s\, \theta_c^t + \sum_{r,s=1}^{p} l_{rs}\, (\theta_{ab}^r\, \theta_c^s + \theta_{ac}^r\, \theta_b^s + \theta_{bc}^r\, \theta_a^s) + \sum_{r=1}^{p} l_r\, \theta_{abc}^r \,, \tag{1.18}$$

etc., where $l_{rs} = l_{rs}(\theta(\psi))$, $l_{rst} = l_{rst}(\theta(\psi))$, $\theta_{ab}^r = (\partial^2/\partial\psi^a\,\partial\psi^b)\theta^r(\psi)$ and $\theta_{abc}^r = (\partial^3/\partial\psi^a\,\partial\psi^b\,\partial\psi^c)\theta^r(\psi)$.

Information in the parameterization θ and information in the parameterization ψ are linked by the identities

$$\bar{\jmath}_{ab} = \sum_{r,s=1}^{p} j_{rs}\, \theta_a^r\, \theta_b^s - \sum_{r=1}^{p} l_r\, \theta_{ab}^r \,, \tag{1.19}$$

$$\bar{\imath}_{ab} = \sum_{r,s=1}^{p} i_{rs}\, \theta_a^r\, \theta_b^s \,. \tag{1.20}$$

Perhaps the reader would be more familiar with the two expressions (1.13) and (1.20) in matrix notation, in the form

$$l_*^{\Psi}(\psi) = [\theta_a^r]^{\mathrm{T}}\, l_*^{\Theta}(\theta(\psi))$$

and

$$i^{\Psi}(\psi) = [\theta_a^r]^{\mathrm{T}}\, i^{\Theta}(\theta(\psi))\, [\theta_a^r] \,, \tag{1.21}$$

where Θ and Ψ denote the reference parameterization, $[\theta_a^r]$ denotes the $p \times p$ matrix with θ_a^r as element of position (r,a) and the symbol T is used to denote transposition. As will be discussed later in sections 9.1 and 9.2, matrix notation is not really suitable for expressing, in compact form, the rules of transformation of quantities which are linked to derivatives of order higher than two, as for example in (1.18); therefore, it will be convenient to use a different notation.

Example 1.1 *Poisson distribution*
Let $y_1, \ldots, y_n$ be independent and identically distributed observations drawn from a Poisson distribution with mean θ. The log-likelihood function is

$$l(\theta) = \log(\theta) \sum_{i=1}^{n} y_i - n\theta \,,$$

$\theta > 0$, $y_i = 0, 1, \ldots$. The score function and the expected and observed information are

$$l_*^{\Theta}(\theta) = l_1 = \frac{1}{\theta} \sum_{i=1}^{n} y_i - n,$$

$$j^{\Theta}(\theta) = \frac{1}{\theta^2} \sum_{i=1}^{n} y_i,$$

$$i^{\Theta}(\theta) = \frac{n}{\theta}.$$

Consider the new parameterization $\psi = \psi(\theta) = e^{-\theta} = p_{Y_i}(0; \theta)$, with inverse $\theta = \theta(\psi) = -\log \psi$. Instead of calculating the quantities $l_*^{\Psi}(\psi)$, $j^{\Psi}(\psi)$ and $i^{\Psi}(\psi)$ starting from the new parameterization, the relationships (1.13), (1.19) and (1.20) can be used, by taking into account that $(\partial/\partial\psi)\theta(\psi) = \theta_1^1 = -1/\psi$ and $(\partial^2/\partial\psi^2)\theta(\psi) = \theta_{11}^1 = 1/\psi^2$, thus directly obtaining,

$$l_*^{\Psi}(\psi) = \left(\frac{\sum y_i}{-\log \psi} - n \right) \left(-\frac{1}{\psi} \right) = \frac{1}{\psi} \left(\frac{\sum y_i}{\log \psi} + n \right),$$

$$j^{\Psi}(\psi) = \frac{\sum y_i}{(\log \psi)^2} \frac{1}{\psi^2} - \frac{1}{\psi^2} \left(\frac{\sum y_i}{-\log \psi} - n \right) = \frac{1}{\psi^2} \left(\frac{\sum y_i}{(\log \psi)^2} + \frac{\sum y_i}{\log \psi} + n \right),$$

$$i^{\Psi}(\psi) = \frac{n}{(-\log \psi)\psi^2}.$$

Note that $\hat{\psi} = e^{-\hat{\theta}}$. $\triangle$

1.5   Examples of Likelihood Functions

In this section some examples are given of likelihood functions for parametric statistical models which are different, in some way or another, from the more elementary cases of independent, identically distributed observations which, we presume, the reader is already familiar with.

Example 1.2 *Censored observations*
Suppose $T_1, \ldots, T_n$ are independent and identically distributed continuous r.v.'s with density $p_T(t; \theta)$, $t \in T \subseteq \mathbb{R}^+$, $\theta \in \Theta$, and distribution function $F_T(t; \theta)$. Let $c_1, \ldots, c_n$ be given positive constants. Suppose that the i-th observation ($i = 1, \ldots, n$) is a realization of the r.v. (Y_i, δ_i), with $Y_i = \min(T_i, c_i)$ and $\delta_i = I_{[0, c_i]}(T_i)$, where $I_A(\cdot)$ denotes the indicator function of the set A,

defined by $I_A(x) = 1$ if $x \in A$, $I_A(x) = 0$ if $x \notin A$. This observation scheme is called type I censoring (see e.g. Lawless, 1982, section 1.4) and is important in survival analysis. In this context, T_i denotes the survival time of the i-th unit and c_i is a preset censoring time for the same unit. Once the time c_i has elapsed, the unit is no longer observed and, if it is still alive its survival time is censored, that is, the only observation made is that T_i is greater than c_i. If $\delta_i = 1$, then Y_i is an uncensored lifetime; if $\delta_i = 0$, then the information on the i-th unit is that $T_i > c_i$. The r.v. (Y_i, δ_i) has a mixed distribution, that is, it has one continuous and one discrete component, with joint density

$$p_{Y_i, \delta_i}(y_i, \delta_i; \theta) = \{p_T(y_i; \theta)\}^{\delta_i} \{1 - F_T(y_i; \theta)\}^{1 - \delta_i} .$$

The likelihood function is

$$L(\theta) = \prod_{i=1}^{n} \{p_T(y_i; \theta)\}^{\delta_i} \{1 - F_T(y_i; \theta)\}^{1 - \delta_i} . \tag{1.22}$$

If, for example, T_i has an exponential distribution with mean θ, (1.22) becomes

$$L(\theta) = \theta^{-r} \exp\left(-\theta^{-1} \sum_{i=1}^{n} y_i\right),$$

where $r = \sum_{i=1}^{n} \delta_i$ is the observed number of uncensored lifetimes.

If the censoring times c_i are independent realizations of a continuous r.v. with density $p_C(c)$ and distribution function $F_C(c)$ and if T_i and C_i are independent, then (Y_i, δ_i) has density

$$p_{Y_i, \delta_i}(y_i, \delta_i; \theta) = \{p_T(y_i; \theta)(1 - F_C(y_i))\}^{\delta_i} \{p_C(y_i)(1 - F_T(y_i; \theta))\}^{1 - \delta_i} .$$

$$\tag{1.23}$$

Whenever the distribution of C_i does not depend on θ, and this is a special case of uninformative censoring, the likelihood function based on (y_i, δ_i), $i = 1, \ldots, n$, is still given by (1.22). The same likelihood (1.22) is also obtained under weaker assumptions on the dependence of the survival and censoring mechanisms (see, for example, Lawless, 1982, section 1.4.1d). $\triangle$

Example 1.3 *Two-state Markov chain*

Let $y = (y_1, \ldots, y_n)$ be a realization of $(Y_1, \ldots, Y_n)$ whose p.d.f. is factorized as

$$p_Y(y) = p_{Y_1}(y_1) \prod_{i=2}^{n} p_{Y_i | Y_{i-1} = y_{i-1}}(y_i; y_{i-1}) .$$

This holds if

$$p_{Y_i|Y_{i-1}=y_{i-1},\ldots,Y_1=y_1}\left(y_i;y_{i-1},\ldots,y_1\right) = p_{Y_i|Y_{i-1}=y_{i-1}}\left(y_i;y_{i-1}\right),$$

that is, if the dependence between the observations (which for ease of interpretation can be envisaged as a time series) is Markovian.

Consider the simplest case, where each Y_i can only take the values 0 and 1. In other words, we observe dependent binary variables which follow a first-order Markov chain. Assuming that the initial state y_1 is fixed, the distribution of the observations for $i = 2,\ldots,n$ is entirely specified by the one-step transition probabilities, i.e. by the conditional probabilities

$$P\left(Y_i = 1 \mid Y_{i-1} = 0\right) = \theta_{01}$$

and

$$P\left(Y_i = 1 \mid Y_{i-1} = 1\right) = \theta_{11}.$$

Thus

$$p_Y(y) = \prod_{i=2}^{n} p_{Y_i|Y_{i-1}=y_{i-1}}\left(y_i;y_{i-1}\right) = \theta_{00}^{n_{00}}\theta_{01}^{n_{01}}\theta_{10}^{n_{10}}\theta_{11}^{n_{11}},$$

where $\theta_{00} = 1 - \theta_{01}$, $\theta_{10} = 1 - \theta_{11}$ and n_{jk}, $(j,k = 0,1)$ denote the overall number of one-step transitions from state j to state k. Therefore, the likelihood function for $\theta = (\theta_{01},\theta_{11})$ is

$$L(\theta;y) = \prod_{j,k} \theta_{jk}^{n_{jk}}, \tag{1.24}$$

and this expression generalizes straightforwardly to Markov chains with more than two states. $\triangle$

Example 1.4 *Non-homogeneous Poisson process*

Poisson processes are special counting processes, i.e., continuous time stochastic processes with discrete state space. A counting process is a collection of random variables $\{N_t, t \geq 0\}$, where N_t can take the values $0,1,\ldots$ and will express the number of arrivals or events that take place within the interval $[0,t)$. Let $N_0 = 0$. Indicate with $N(t,t+h)$ the number of arrivals within the interval $[t,t+h)$, with $t \geq 0$ and $h > 0$; assume that N_t and $N(t,t+h)$ are independent, that is, that the counting process has independent increments. It is said that $\{N_t, t \geq 0\}$ is a non-homogeneous Poisson process if, as $h \to 0$,

$$P\left\{N(t,t+h) = 0\right\} = 1 - \lambda(t)h + o(h)$$

and
$$P\{N(t, t+h) = 1\} = \lambda(t)h + o(h) ,$$
where $\lambda(t)$ is a positive function called the **rate function** of the process, and $o(h) \to 0$ as $h \to 0$. It can be shown that $N(s, s+t)$ follows a Poisson distribution with mean $\mu(s, t) = \int_s^{s+t} \lambda(u)du$. The process is called **homogeneous** if $\lambda(t) = \lambda$; in this case the times between two successive arrivals are independent and exponentially distributed with mean $1/\lambda$.

Suppose that the data correspond to the observation of a non-homogeneous Poisson process in the interval $[0, t_0)$ and that the events take place at times $t_1, \ldots, t_n$. One simple procedure for defining the likelihood function associated with these data, consists of subdividing the observation interval into m subintervals each $h = t_0/m$ long, then, calculating the contribution (multiplicative, because of the independence of the increments) of each of the subintervals to the likelihood and, lastly, taking the limit of the factorization obtained as $h \to 0$. Denote these subintervals as $[u_j, u_j + h)$, $j = 1, \ldots, m$. The observation related to $[u_j, u_j + h)$ contributes to the likelihood with a factor $\lambda(u_j)h + o(h) = \lambda(t_i)h + o(h)$ if $u_j \le t_i < u_j + h$ for some i, that is, if there is one arrival in the interval, and with a factor of $1 - \lambda(u_j)h + o(h)$ if in this interval there has been no arrival. Thus the likelihood has to be obtained from

$$\prod_{i=1}^{n} \{\lambda(t_i)h + o(h)\} \prod_{j}^{*} \{1 - \lambda(u_j)h + o(h)\} , \qquad (1.25)$$

where the second product is over all values of j such that the interval $[u_j, u_j + h)$ does not contain any arrival time $t_1, \ldots, t_n$. Since

$$\prod_{j}^{*} \{1 - \lambda(u_j)h + o(h)\} = \exp\left\{ \sum_{j}^{*} \log\left(1 - \lambda(u_j)h + o(h)\right) \right\}$$

$$= \exp\left\{ -\sum_{j}^{*} (\lambda(u_j)h + o(h)) \right\} ,$$

the limit of this quantity as $h \to 0$ is

$$\exp\left\{ -\int_0^{t_0} \lambda(u)du \right\} .$$

Omitting the factor h^n in (1.25) we obtain

$$L(\lambda(t)) = \exp\left\{ -\int_0^{t_0} \lambda(u)du \right\} \prod_{i=1}^{n} \lambda(t_i) . \qquad (1.26)$$

The argument of $L(\cdot)$ in (1.26) is a function, which is usually specified in a parametric class. If the process is homogeneous, i.e. with $\lambda(t) = \lambda$, (1.26) becomes

$$L(\lambda) = \lambda^n e^{-\lambda t_0} . \tag{1.27}$$

This likelihood is equivalent to that obtained regarding n as a realization of a Poisson distribution with mean λt_0.

Extensions of the above considerations to spatial Poisson processes are particularly important in applications. Let A denote a bounded region of the plane and $N(A)$ the number of arrivals in A. Then the function

$$\lambda(\mathbf{y}) = \lim_{|d\mathbf{y}| \to 0} \left\{ \frac{E(N(d\mathbf{y}))}{|d\mathbf{y}|} \right\},$$

where $d\mathbf{y}$ denotes a neighbourhood of the point $\mathbf{y} \in \mathbb{R}^2$ with area $|d\mathbf{y}|$, is called the intensity function. The collection $\{N(A), A \subset \mathbb{R}^2, \text{bounded}\}$ is a non-homogeneous Poisson process in the plane if arrivals in disjoint regions are independent and if

$$P\{N(d\mathbf{y}) = 0\} = 1 - \lambda(\mathbf{y})|d\mathbf{y}| + o(|d\mathbf{y}|)$$

and

$$P\{N(d\mathbf{y}) = 1\} = \lambda(\mathbf{y})|d\mathbf{y}| + o(|d\mathbf{y}|) .$$

Let

$$\Lambda(A) = \int_A \lambda(\mathbf{y})d\mathbf{y}$$

denote the intensity measure of the process. By a repetition of the procedure leading to (1.26), the likelihood associated with the observation of points $\mathbf{y}_1, \ldots, \mathbf{y}_n$ in a fixed region A_0 is seen to be

$$L(\lambda(\cdot); \mathbf{y}_1, \ldots, \mathbf{y}_n) = \exp\{-\Lambda(A_0)\} \prod_{i=1}^{n} \lambda(\mathbf{y}_i) .$$

$$\triangle$$

Example 1.5 *Brownian motion*
Brownian motion offers the most important example of a continuous time stochastic process with continuous state space. Brownian motion with drift coefficient μ and unit variance is a collection of random variables $\{Y(t), 0 \leq t < +\infty\}$ such that

(i) $Y(0) = 0$;

(ii) $P_\mu\{Y(t) - Y(s) \le z\} = \Phi((z - (t - s)\mu)/(t - s)^{1/2})$, $0 \le s < t < +\infty$;

(iii) for every finite sequence of points $0 \le s_1 < t_1 \le s_2 < t_2 \le \ldots \le s_n < t_n < +\infty$, $n = 2, 3, \ldots$, the r.v.'s $Y(t_i) - Y(s_i)$, $i = 1, \ldots, n$, are independent;

(iv) $Y(t)$, $0 \le t < +\infty$, is a continuous function of t.

In (ii) $\Phi(\cdot)$ denotes the standard normal distribution function. Brownian motion is thus a process with normally distributed, independent increments.

Assume that the data correspond to the observation of Brownian motion in the interval $[0, t_0]$, that is, the partial trajectory $\{y(t), \ 0 \le t \le t_0\}$ is observed. As in the preceding Example 1.4, a simple procedure for constructing a likelihood function is that of subdividing the observation interval into m subintervals $h = t_0/m$ long, then calculating the multiplicative contribution (for independence of increments) of each subinterval to the likelihood and, lastly, by considering the limit as $h \to 0$ of the factorization obtained. Let $[u_j, u_j + h)$, $j = 1, \ldots, m$, denote the subintervals. The observation related to $[u_j, u_j + h)$ contributes to the likelihood with a factor

$$\exp\left\{-\frac{1}{2h}\left\{y(u_j + h) - y(u_j) - \mu h\right\}^2\right\},$$

proportional to

$$\exp\left\{-\frac{1}{2}\mu^2 h + \mu\{y(u_j + h) - y(u_j)\}\right\}.$$

Hence, the overall likelihood is

$$L(\mu) = \exp\left\{-\frac{1}{2}t_0\mu^2 + \mu y(t_0)\right\},\tag{1.28}$$

which is equivalent to the likelihood obtained regarding $y(t_0)$ as a realization of a normal distribution with mean μt_0 and variance t_0. Note that obtaining (1.28) does not require special limiting considerations. However, in more complicated cases, stochastic calculus is essential. From a general point of view, defining a likelihood requires it to be possible to express the probability measures associated with the process as a dominated family (cf. section 1.4.1). One technical difficulty that may arise is that the measures associated with processes with continuous sample paths could not be mutually absolutely continuous.   △

1.6 Bibliographic Note

A large part of the material in this chapter is usually dealt with in introductory texts to the theory of statistical inference (see for example, Chapters 1 and 2 in Cox and Hinkley, 1974). General discussions of models, principles and methods can be found, among others, in Dawid (1983a), Fraser (1983), Cox (1958, 1986).

Other approaches to data analysis either supplement or oppose that of the theory of statistical inference which is based on the fundamental assumption that the data y^{obs} are a realization of $Y \sim p^0(y)$. Some views emphasize exploratory elements, description or classification. The expression *exploratory data analysis* is confined to all the preliminary techniques of data analysis suggested by Tukey (1977); see also Tukey (1980) and Chatfield (1985). The *analyse des données*, which was initially developed in France in the early 1960s, rejects the interpretation of data through probability models (see, for example, Benzécri, 1973, section TIA no:2) in favour of algebraic-geometric techniques of reduction and classification. Other possible non-probabilistic models that can be used to describe variability or uncertainty should also be mentioned. The most recent, and outstanding, example is given by the theory of chaotic deterministic systems. These have been discussed by Bartlett (1990), Berliner (1992) and Chatterjee and Yilmaz (1992). One final example is that of the theory of fuzzy sets, which aims to explain uncertainty as possibility: see, for example, Kaufmann and Gupta (1991).

For a lively resumé of the debate on the foundations of statistical inference, see Barnett (1982). The idea that because even the foundations of mathematics embody intricacies and elements of incompleteness, far more difficult problems are encountered when seeking to define general rules for statistical inference, has been repeatedly stressed by Barnard (see e.g. Barnard, 1974a). The distinction according to four visions of section 1.2 follows Cox (1993, 1995).

A classic text on the personalistic Bayesian paradigm is de Finetti (1974, 1975). Accounts of inference from the personalistic and non-personalistic Bayesian points of view may be found in the volumes by Berger (1985) and Bernardo and Smith (1994). Gelman, Carlin, Stern and Rubin (1995) give an introduction to Bayesian modelling.

For a fuller account of the Fisherian paradigm, see the appreciation of Fisher's contribution in Fienberg and Hinkley (1980), and, naturally, Fisher's collected works, Fisher (1950, 1971), and his last book, Fisher (1956). Rao (1992) summarizes Fisher's contribution to the development of statistics. Formalization and extensions form the nucleus of the first part of Barndorff-Nielsen (1978). Cox and Hinkley (1974) offer a balanced introduction to the theory of statistical inference also with reference to other paradigms. The recent book by Lindsey (1996) stresses the central role of likelihood in frequentist as well as Bayesian approaches to statistical inference. The main references on the theory of inference, from the decisional point of view, are the classic books by Lehmann (1983, 1986) and Ferguson (1967). Efron (1986a) offers other points for reflection in favour of the adoption of frequentist paradigms within scientific research. McPherson (1989) and Hand (1994) discuss the interaction between statistics and its users in various scientific fields.

Useful hints on model specification are given in Lehmann (1990) and Cox (1990); see also Cox and Hinkley (1974, sections 1.2 and 1.3). Chatfield (1995) and Draper (1995) discuss model uncertainty. Introductions to statistical inference in the context of nonparametric models, with emphasis on different aspects, can be found in Randles and Wolfe (1979), Tarter and Lock (1994), Efron and Tibshirani (1993). Pfanzagl (1990) and Bickel, Klaassen, Ritov, Wellner (1993) are recent monographs on semiparametric models. As an example of parametric statistical models which explain the phenomenon being studied, see Barndorff-Nielsen, Blæsild, Jensen and Sørensen (1985) where a parametric model which aims to describe the variability of the mass of grains of sand is derived. See Tawn (1990) for one use of stable distributions as an element for modelling unobservable quantities. For a more in-depth examination of methods for selecting a parametric model, goodness-of-fit tests and analysis of residuals, see Linhart and Zucchini (1986), D'Agostino and Stephens (1986), Cook and Weisberg (1982).

For a detailed introduction to the notions of measure theory mentioned in section 1.4.1, a useful reference is for instance Billingsley (1986). The notation in sections 1.4.2 and 1.4.3 is introduced in Barndorff-Nielsen (1988). The concept of likelihood appears in the first scientific paper published by Fisher, in 1912, when he had just taken his honours degree in mathematics at Cambridge University. Fisher's early works emphasized likelihood as a tool for tracing estimators which, in many ways, are preferable to those based on the method of moments. He subsequently developed the theory of likelihood in papers published in 1921, 1922a, 1925 and 1934, and in the book published in 1956. From the historical point of view, the idea of estimates based on the maximization of $L(\theta)$ is usually attributed to Fisher (1912, 1922a, 1925, 1935), however, an embryonic version of the idea had already been put forward by Lambert, Daniel Bernoulli and Lagrange in the eighteenth century, see Edwards (1974) for a lively account. Recent extensions of the theory to semiparametric and nonparametric models are dealt with e.g. in Gill (1989), Wong and Severini (1991), Gill and van der Vaart (1993), Lindsay (1995), Murphy (1995a), van der Vaart (1996). There are many examples of the application of the method of maximum likelihood to problems of estimation in complex models, to cite just a few: Vardi and Lee (1993), Ying (1993), Coles and Tawn (1994).

The examples offered in section 1.5 are common. See Cox and Snell (1989, section 2.11) for a more in-depth treatment of two-state Markov chains and Bhat (1985) and the references given there for more complex Markov processes. The procedure of constructing the likelihood function used in Examples 1.4 and 1.5 is adopted in Cox and Hinkley (1974, p. 15), Barndorff-Nielsen (1991a, section 10.2.6), Barndorff-Nielsen and Cox (1984a). Further examples of the technique for calculating the likelihood function for a stochastic process with a continuous parameter can be found in Barndorff-Nielsen (1991a) and in Barndorff-Nielsen and Cox (1994, section 2.2.3). A rigorous formalization of the procedure adopted could be based on the notion of **product integral** (Gill and Johansen, 1990); see also Andersen, Borgan, Gill and Keiding (1993, Chapter 2). For a definition of the likelihood function related to a general class of processes, the **diffusion processes**, see Sørensen (1983) and Barndorff-Nielsen and Sørensen (1994) where various examples are presented.

1.7 Problems

1.1 Justify the statement that, for a discrete r.v. with finite support, any statistical model is a parametric one.

[Section 1.3.1]

1.2 Consider the parametric model $(\mathcal{Y}, P_\theta, \Theta)$, with $\mathcal{Y} = [0, 1]$, $\Theta = \{\theta_1, \theta_2\}$, where P_{θ_1} is the measure that corresponds to the uniform distribution on $[0, 1]$, while P_{θ_2} is the measure that corresponds to the discrete distribution taking values $1/2$ and 1 with equal probability. Show that the probability measures P_{θ_1} and P_{θ_2} are not mutually absolutely continuous. Show that the parametric family defined above is dominated by the measure $P_{\theta_1} + P_{\theta_2}$.

[Section 1.4.1]

1.3 A subject undergoes n tests of ability. Let Y_i, $i = 1, \ldots, n$, be the result of the i-th test; put $Y_i = 1$ if the answer is correct and $Y_i = 0$ if the answer is wrong. Assume that the probability of a correct answer at the i-th test is $\pi_i(\theta) = \exp(\theta - x_i)/\{1 + \exp(\theta - x_i)\}$, $\theta \in \mathbb{R}$, with $x_1, \ldots, x_n$ given real constants. Furthermore, assume that $Y_1, \ldots, Y_n$ are independent. Specify the statistical model for $Y_1, \ldots, Y_n$. Generally, the parameter θ and the constant x_i are referred to, respectively, as *ability of the subject* and *difficulty of the i-th test*; justify this interpretation. Write the likelihood equation for θ and say when it has a finite solution.

[Section 1.4]

1.4 Suppose $y_1, \ldots, y_n$ are i.i.d. observations from an exponential distribution with mean $1/\lambda$, $\lambda > 0$. Only the integer part is observed for the last $n - m$ $(m < n)$ observations. Specify the statistical model. Write the likelihood equation for λ and show that this has only one solution.

[Section 1.4]

1.5 Prove and interpret the relation $\hat{\jmath}_{ab} = \sum_{r,s} \hat{\jmath}_{rs} \, \hat{\theta}_a^r \, \hat{\theta}_b^s$.

[Section 1.4.3]

1.6 Let $y_1, \ldots, y_n$ be n i.i.d. observations from a normal distribution with mean μ and variance σ^2. Obtain l_*, i and $\hat{\jmath}$ in the parameterizations $\theta = (\mu, \sigma^2)$ and $\psi = (\mu, \sigma)$.

[Section 1.4.3]

1.7 Let $y_1, \ldots, y_n$ be n i.i.d. observations from a Poisson distribution with mean μ. Obtain l_*, i and $\hat{\jmath}$ in the parameterizations $\theta = \mu$ and $\psi = \mu^\delta$, with $\delta \neq 0$ fixed.

[Section 1.4.3]

1.8 Let $y_1, \ldots, y_n$ be n independent observations from a binomial distribution with index m and parameter θ, $Bi(m, \theta)$. Obtain l_*, i and $\hat{\jmath}$ in the parameterization $\psi = \Phi^{-1}(\theta)$, where $\Phi(\cdot)$ indicates the standard normal d.f., starting from the same quantities calculated in the parameterization θ, and using the transformation laws (1.13), (1.19) and (1.20).

[Section 1.4.3]

1.9 Let $y_1, \ldots, y_n$ be independent observations from a two-parameter gamma distribution, with p.d.f. $p(y; \nu, \lambda) = \lambda(\lambda y)^{\nu-1} e^{-\lambda y}/\Gamma(\nu)$, with $y > 0$, $\nu > 0$, $\lambda > 0$. Obtain l_*, i and $\hat{\jmath}$ in the parameterizations $\theta = (\nu, \lambda)$ and $\psi = (\nu, \nu/\lambda)$.

[Section 1.4.3]

1.10 Let $y_1, \ldots, y_n$ be independent observations from a two-parameter Weibull distribution with p.d.f.

$$p(y; \nu, \lambda) = \lambda\nu(\lambda y)^{\nu-1} e^{-(\lambda y)^\nu}, \tag{1.29}$$

with $y > 0$, $\nu > 0$, $\lambda > 0$. Obtain l_*, i and $\hat{\jmath}$ in parameterizations $\theta = (\nu, \lambda)$ and $\psi = (\nu, \lambda^\nu)$.

[Section 1.4.3]

1.11 Write the likelihood function (1.24) according to the alternative parameterization $\psi = (\beta_0, \beta_1)$, with $\theta_{01} = e^{\beta_0}/(1 + e^{\beta_0})$, $\theta_{11} = e^{\beta_0 + \beta_1}/(1 + e^{\beta_0 + \beta_1})$. Interpret the parameter β_1.

[Section 1.5]

1.12 Let $y_0, y_1, \ldots, y_n$ be realizations of $n + 1$ random variables $Y_0, Y_1, \ldots, Y_n$ such that:
(i) $Y_0 \sim N(\mu, \sigma^2)$;
(ii) The distribution of Y_i given $Y_0 = y_0, \ldots, Y_{i-1} = y_{i-1}$ is the same as that of Y_i given $Y_{i-1} = y_{i-1}$ and is normal with mean μ and variance $\sigma^2(1 + y_{i-1}^2)$, $i = 2, \ldots, n$.
Write the likelihood function for (μ, σ^2) and the corresponding likelihood equations. Obtain the maximum likelihood estimate for (μ, σ^2) and suggest an interpretation of the formula obtained for $\hat{\mu}$.

[Section 1.5]

1.13 Specialize the likelihood function (1.26) when $\lambda(t) = \alpha + \beta t$ and when $\lambda(t) = \alpha t^\beta$.

[Section 1.5]

1.14 Suppose $\{N_y, y \geq 0\}$ is a non-homogeneous Poisson process with rate function

$$\lambda(y; \xi, \sigma, \mu) = \sigma^{-1}\{1 + \xi(y - \mu)/\sigma\}^{-1/\xi - 1},$$

if $1 + \xi(y - \mu)/\sigma > 0$ and zero elsewhere, with $\xi, \mu \in \mathbb{R}$, $\sigma > 0$. For $\xi = 0$ the rate function $\lambda(y; 0, \sigma, \mu)$ is defined as $\lim_{\xi \to 0} \lambda(y; \xi, \sigma, \mu) = \exp\{-(y - \mu)/\sigma\}$. Arrivals are observed at $y_1, \ldots, y_m$. Using (1.26), write the likelihood function for (ξ, μ, σ). (*Hint*: the observations must fall within $I = \{y > 0 : 1 + \xi(y - \mu)/\sigma > 0\}$.) This model is used for the analysis of extremes.

[Section 1.5; Smith, 1989, section 4]

CHAPTER 2

DATA AND MODEL REDUCTION

2.1 Introduction

The likelihood function offers an immediately interpretable summary of data under a statistical model $\mathcal{F}$. The likelihood function is, however, conditional on the observed sample y^{obs}. To take sampling variability into account, the likelihood should be explicitly related to the principle of repeated sampling. The Wald inequality, the result on the first null moment of the score, the relation between observed information and expected information are all interesting, but still rather vague, links.

To justify the reduction to the likelihood function on the basis of the principle of repeated sampling, it is convenient to consider the problem of extracting information about θ from the sample in general terms. This extraction can be carried out using statistics or combinants. The former are transformations of the data, usually not one-to-one, which do not depend on the parameter, while the latter also depend on the parameter: the likelihood function is the first example of a combinant.

The main problem is to define criteria which establish under which conditions the reduction of data takes place, under $\mathcal{F}$, without losing information about the quantity of interest. Of all the possible reductions which do not entail loss of information, the choice usually falls on those which offer the maximum synthesis. Given this, the privileged position held by the likelihood function and likelihood quantities is soon recognized, as the discussion in this chapter will highlight.

This chapter will concentrate on the case where the interest is *global*, that is, where inference is to be carried out about the whole parameter θ. However, in many inferential problems interest is focused on some primary aspects of the data, described by a function of the parameter θ which is not one-to-one. In such cases we say that interest is *partial*. The extraction of information from a sample in this second situation is taken up in Chapter 4.

Basic definitions and concepts on point estimation, testing hypotheses and confidence regions are assumed to be known. For convenience, the standard terminology is briefly recalled in Appendix C.

2.2   Statistics

Summarizing quantities offer an immediate way of reducing data. Let $\mathcal{S}$ be a space endowed with a σ-algebra $\mathcal{B}_{\mathcal{S}}$ of subsets of $\mathcal{S}$; the elements of $\mathcal{B}_{\mathcal{S}}$ are called events in $\mathcal{S}$. A function s defined on the sample space $\mathcal{Y}$ with values in $\mathcal{S}$ is said to be measurable if for all $B \in \mathcal{B}_{\mathcal{S}}$ also $s^{-1}(B) = \{y \in \mathcal{Y} : s(y) \in B\}$ is an event in $\mathcal{Y}$, that is, if $s^{-1}(B) \in \mathcal{B}$.

Definition 2.1 Given the statistical model $\mathcal{F}$ as $(\mathcal{Y}, P_\theta, \Theta)$ and a space $\mathcal{S}$ with a σ-algebra $\mathcal{B}_{\mathcal{S}}$, a **statistic** s is a measurable function $s : \mathcal{Y} \to \mathcal{S}$.

A statistic s induces a statistical model $(\mathcal{S}, P_\theta^S, \Theta)$. This is the model of the transformed data $s^{obs} = s(y^{obs})$. It is said that P_θ^S is the image measure of P_θ under s; this is defined by $P_\theta^S(B) = P_\theta(s^{-1}(B))$ for $B \in \mathcal{B}_{\mathcal{S}}$.

In the following, it is implicitly assumed that all the data transformations mentioned are measurable functions and that all the subsets of a sample space considered belong to some suitable σ-algebra.

The usefulness of a statistic s for inference about θ depends on the model reduction that s performs, that is, on the relation between the statistical model of the transformed data and that of the original data. From this point of view, the case where s is a one-to-one transformation of the data is not that interesting because the inferential problem is reformulated in equivalent terms. This seemingly harmless observation has major consequences if it is taken to be a principle: for example, it implies that the method of moments is not acceptable (see Problem 2.1), while inferences based on the likelihood function are acceptable. Let us consider transformations which are not one-to-one. It is more important to consider which data y a statistic poses as equal within the sample space $\mathcal{Y}$, than it is to consider the values of s in themselves: $\sum y_i$ and $\exp\{\sum y_i\}$ are equivalent statistics, while $\sum y_i$ and $(\sum y_i)^2$ are not necessarily equivalent statistics. In other words, the summary provided by a given statistic is most clearly represented by the corresponding partition of the sample space.

The usefulness in inference of a given statistic, according to the principle of repeated sampling, is assessed on the basis of the induced statistical model. Thus one is faced with a problem of determining distributions. The solution is explicit or reasonably simple in very few cases. If, for example, $y_1, \ldots, y_n$ are n independent realizations of a $N(\mu, \sigma^2)$ distribution, the usual summary of data in terms of sample mean and of sample variance gives $s = (\bar{y}_n, s_n^2)$, where $\bar{y}_n = \sum y_i / n$ and $s_n^2 = \sum (y_i - \bar{y}_n)^2 / (n-1)$, which is associated with the statistical image model where $\bar{Y}_n \sim N(\mu, \sigma^2/n)$, $S_n^2 \sim \sigma^2 \chi_{n-1}^2 / (n-1)$, $\bar{Y}_n$ and S_n^2 are independent. As well as allowing a dimension reduction, from n to 2, the statistical model of the transformed data also separates the information about σ^2 from that about μ.

The problem remains of whether reduction to s takes place at the cost of sacrificing part of the information about θ which can be extracted from the original data and from their model.

2.3 Distribution Constant Statistics

It is convenient to begin by defining absence of information about θ in the transformed data.

Definition 2.2 If the statistical model associated with the statistic s has only one element, that is, if the distribution of S does not depend on θ, we call the statistic s distribution constant (with respect to $\mathcal{F}$).

In this section a distribution constant statistic is indicated by c and the corresponding random variable by C. The distribution of C depends on $\mathcal{F}$ but not on the particular element of $\mathcal{F}$ which generates the data. Thus, distribution constant statistics offer information regarding data that can be predicted on the basis of the model and which are independent of the specific position of $p^0(y)$ within $\mathcal{F}$. Clearly, the observation c^{obs} alone cannot be used for inference about θ, while it can be used to test the adequacy of the model $\mathcal{F}$. Note that if c is distribution constant with respect to $\mathcal{F}$ it is also distribution constant under any subfamily of $\mathcal{F}$, while if $\mathcal{F}$ is enlarged c is usually no longer distribution constant.

A random sample size n is a distribution constant statistic whenever its distribution does not depend on θ (this is not the case, for instance, in *sequential sampling*, see Problem 2.2). A less immediate example arises with random sampling of size n from a normal distribution. The statistic

$$c = \left(\frac{y_2 - y_1}{y_n - y_1}, \ldots, \frac{y_{n-1} - y_{n-2}}{y_n - y_1} \right) ,$$

called the configuration of the sample (Fisher, 1934), is distribution constant because of implicit standardization.

Example 2.1 *A nonparametric setting*
Consider a random sample of size n from a univariate continuous distribution with known median which, without loss of generality, may be taken as zero. Under this nonparametric model $\mathcal{F}$, the statistic $c = \sum I_{\mathbf{R}+}(y_i)$, is distribution constant because C has a binomial distribution with index n and probability $1/2$, $C \sim Bi(n, 1/2)$. Thus c cannot be used to locate $p^0(y)$ within $\mathcal{F}$, say to investigate the third quartile. However, it can be used to test the adequacy of $\mathcal{F}$, in particular to test the hypothesis of zero median (sign test, see, e.g. Gibbons, 1988). $\triangle$

Understanding aspects of the data that are irrelevant for inference about θ under $\mathcal{F}$ is usually difficult. There are no general criteria for finding distribution constant statistics and, indeed, in many models one does not know whether non-trivial instances of such statistics even exist. Think, for example, of the family associated with sampling from a gamma distribution with an unknown shape parameter. On the other hand, there can be more than one distribution constant statistic under $\mathcal{F}$. In this case it would be natural to turn one's attention to those statistics that separate out the largest amount of aspects that are irrelevant for inference about θ.

Definition 2.3 A distribution constant statistic c is called maximal distribution constant (with respect to $\mathcal{F}$) if there is no other distribution constant statistic of which c is a function, except for one-to-one functions.

Thus, a maximal distribution constant statistic indirectly highlights the important information in the most concise way. However, maximal distribution constant statistics are not always unique either.

Example 2.2 *Bivariate normal distribution*
Let (x_i, y_i), $i = 1, \ldots, n$, be n independent observations from a bivariate normal distribution with standardized marginals and correlation coefficient θ, $|\theta| < 1$. Both $c_X = (x_1, \ldots, x_n)$ and $c_Y = (y_1, \ldots, y_n)$ are maximal distribution constant but one cannot be expressed as a function of the other. Each distribution

constant statistic based on the x_i is a function of c_X and, analogously, each distribution constant statistic based on the y_i is a function of c_Y. Observe that $c_1 = (y_1, x_2, \ldots, x_n)$, $\ldots$, $c_{n-1} = (y_1, y_2, \ldots, y_{n-1}, x_n)$ are also maximal distribution constant. $\triangle$

Various extensions of the notion of distribution constancy are possible. For example, a statistic $u(y)$ is said to be first-order distribution constant if its expectation $E_\theta(u(Y))$ does not depend on θ. In a (semiparametric) linear regression model the residuals of least squares regression are first-order distribution constant.

Finding a distribution constant statistic does not only have the rather negative meaning of showing what inference *cannot* be directly based on. As Fisher (1934, 1935) argued, a distribution constant statistic permits a reduction by conditioning of the inferential problem. Suppose that c is distribution constant and t is an additional statistic such that (c, t) is a one-to-one transformation of y. In a dominated model the following factorization holds

$$p_{C,T}(c, t; \theta) = p_C(c)\, p_{T|C=c}(t; c, \theta) \qquad (2.1)$$

and the inferential equivalence between y and (c, t) suggests that inference about θ should refer to the conditional model $p_{T|C=c}(t; c, \theta)$. But while it is unquestionable that no inference about θ can be based only on c, the use of $p_{T|C=c}(t; c, \theta)$ as a reduced model for inference is not widely accepted. Above all, some practical problems have to be satisfactorily resolved in order to base inference on $p_{T|C=c}(t; c, \theta)$. Firstly, how to choose the statistic c when different options are offered; secondly, how to solve problems of inference and of distribution conditionally on $C = c$; in addition, when the sample space is discrete conditioning may lead to almost degenerate distributions. Moreover, reduction by conditioning raises the delicate problem of the choice of the relevant sample space for the principle of repeated sampling.

Observe that, because of (2.1), the definition of the likelihood function, of the observed information j and of other observed likelihood quantities could be thought to automatically refer to the conditional model. However, any study of the frequency properties of these quantities will require further analysis.

Since isolating information about the parameter in order to remove irrelevant aspects from the data raises non-trivial problems, further discussion on this issue is deferred to sections 2.6-2.8. The following two sections deal with the complementary notion of sufficiency reduction, that is, reductions of data which can concentrate the information about the parameter without any losses.

Elementary facts about sufficiency are, presumably, known to the reader; the following section reviews the basic definitions and results.

2.4   Sufficient Statistics

A statistic s whose distribution depends on θ is said to be an information carrier. Let $(\mathcal{S}, p_s(s; \theta), \Theta)$ be the image statistical model $\mathcal{F}_S$, which encapsulates the information about θ carried by s. Let u be an additional statistic such that (s, u) is a one-to-one transformation of y. Consider the factorization of the joint density of (s, u) as

$$p_{S,U}(s, u; \theta) = p_S(s; \theta) p_{U|S=s}(u; s, \theta) \ .$$

If the statistical model of U given $S = s$ has a single element, for any s, the reduction to s does not entail loss of information about θ, since, given s, U is distribution constant. It is important to define this property without explicitly considering the additional statistic u.

Definition 2.4 The statistic s is said to be sufficient for the statistical model $\mathcal{F}$ if the conditional distribution of Y given $S = s$ is constant within $\mathcal{F}$, whatever s is.

A statistic sufficient for $\mathcal{F}$ is equivalently said to be sufficient for θ; note, however, that sufficiency is a property of the model and does not depend on the parameterization. If s is sufficient and the model $\mathcal{F}$ is dominated, the factorization

$$p_Y(y; \theta) = p_S(s; \theta) p_{Y|S=s}(y; s) \tag{2.2}$$

holds for the densities in $\mathcal{F}$. Relation (2.2) expresses the fact that, within $\mathcal{F}$, it is possible to interpret the process of generation of y as an experiment which takes place in two stages:
i) first, s is generated according to $p_S(s; \theta)$;
ii) then y is generated according to $p_{Y|S=s}(y; s)$.
Given s, the second stage is governed by a distribution that does not depend on θ. Hence, this second stage could be omitted without any loss of information on the parameter. In fact, $p_{Y|S=s}(y; s)$ depends only on $\mathcal{F}$ and, given s, in principle, fictitious data y^* can be produced which are equivalent to the observed data y. This shows, convincingly, that so long as there is no doubt that $p^0(y)$ belongs to $\mathcal{F}$ a sufficient statistic is a summary of the data which is inferentially equivalent to the data themselves. The measures of information i

and j (see section 1.4.2) are in line with this interpretation. If s is sufficient, it follows from (2.2) that $l(\theta; y) = \log p_s(s(y); \theta)$; therefore, $j(\theta)$ and $i(\theta)$ do not change if they are referred to the model induced by s rather than to the original model.

Note that inferential equivalence of y and s does not require any reinterpretation of the principle of repeated sampling. In fact, reduction by sufficiency is supported in all the paradigms for inference which were mentioned in section 1.2, except for minor exceptions in the frequency-decision view (for example, in the theory of point estimation, if the loss function is not convex the optimum estimator might not be based on a sufficient statistic; in the theory of hypothesis testing, the notion of randomized tests conflicts with the obvious sufficiency of the data, see sections 3.5.2 and 3.5.3). This leads to the statement of the

> sufficiency principle: in the context of an assigned statistical model, if s is sufficient, two observations y_1 and y_2 for which $s(y_1) = s(y_2)$ must lead to the same inferential conclusions about $p^0(y)$.

It is immediately clear from (2.2) that the likelihood function, and the observed likelihood quantities, such as l_* and j, obey the sufficiency principle.

Furthermore, (2.2) expresses an inferential separation between aspects of the data that are useful for locating $p^0(y)$ within $\mathcal{F}$, given that $p^0(y) \in \mathcal{F}$, and aspects that are useful for checking whether $p^0(y) \in \mathcal{F}$. The former aspects are summarized by s. All other aspects of the data can be predicted on the basis of s, independently of the specific location of $p^0(y)$ within $\mathcal{F}$. For an example of the use of conditioning on a sufficient statistic in order to test the adequacy of a parametric model, see Quesenberry and Starbuck (1976). The discussion here will be limited to the following example, which is related to a nonparametric problem.

Example 2.3 *Order statistics*
Let $y = (y_1, \ldots, y_n)$ be a random sample of size n from a continuous univariate r.v. with unknown distribution function $F_0(\cdot)$. Let $s(y) = (y_{(1)}, \ldots, y_{(n)})$, with $y_{(i)} \le y_{(j)}$ for $i < j$ $(i, j = 1, \ldots, n)$, be the order statistics. In the statistical model

$$\mathcal{F} = \{F_Y(y): \quad F_Y(y) = \prod F(y_i), \quad F(\cdot) \text{ continuous}\}$$

the probability of observing ties is zero. Furthermore, given s, the vector y is one of the $n!$ permutations of the components of s, each of which are equally likely under $\mathcal{F}$. Consequently, the distribution of Y given $S = s$ does not depend on the parameter of $\mathcal{F}$, $F(\cdot)$, and s is a sufficient statistic for $\mathcal{F}$. The usual nonparametric estimate of $F_0(\cdot)$, given by the empirical distribution function $\hat{F}_n(z) = n^{-1} \sum I_{(-\infty, z]}(y_i)$, is based on s, indeed, it is equivalent to s. The distribution of Y given $S = s$ can be used to test the adequacy of $\mathcal{F}$ against deviations such as correlation between the observations or dependence of the marginal distribution functions $F_{Y_i}(\cdot)$ on covariates. For example, while maintaining the independence of the observations, consider the possibility that the first $m < n$ observations $(y_1, \ldots, y_m)$ come from a different population than that of the last $n - m$, in particular that the two populations have different means with $E(Y_n) < E(Y_1)$. It is then natural to investigate whether the observed value of $t = \sum_{i=1}^{m} y_i$ is extreme on the right-hand tail of its distribution given $S = s$, which is called the **permutation distribution** of t.

$$\triangle$$

Observe that if s is sufficient under $\mathcal{F}$, it is also sufficient under any subfamily of $\mathcal{F}$; on the contrary, sufficiency is not usually retained if $\mathcal{F}$ is enlarged.

Sufficiency would be of limited importance were there not operational criteria for tracing sufficient statistics. The following **Fisher–Neyman factorization theorem** offers a simple but very useful criterion.

Theorem 2.1 In a dominated statistical model $\mathcal{F} = \{p_Y(y; \theta), \theta \in \Theta\}$ a statistic s is sufficient if, and only if, every density $p_Y(y; \theta)$ of the model can be expressed in the form

$$p_Y(y; \theta) = g(s(y); \theta)\, h(y)\,, \qquad (2.3)$$

where $g(\cdot; \theta)$ and $h(\cdot)$ are measurable non-negative functions.

This is simple to prove in the discrete case, where no measure-theoretic considerations are required to define the conditional distribution of Y given $S = s$. Theorem 2.1 was formalized in the general measure-theoretic framework by Halmos and Savage (1949); for a brief account, see Lehmann (1986, section 2.6).

Example 2.4 *Sufficiency of the likelihood ratio*
The simplest statistical model is $\mathcal{F} = \{p(y; 0), p(y; 1)\}$, with two elements. Assume, for simplicity, that both the probability models in $\mathcal{F}$ have the same support. Under $\mathcal{F}$ the likelihood ratio $s(y) = p(y; 1)/p(y; 0)$ is a sufficient statistic. The factorization criterion (2.3) is satisfied with $h(y) = p(y; 0)$, $g(y; 0) = 1$, $g(y; 1) = p(y; 1)/p(y; 0)$, since, with clear identifications,

$$p(y; \theta) = \left(\frac{p(y; 1)}{p(y; 0)}\right)^{\theta} p(y; 0)\,, \qquad \theta = 0, 1\,.$$

This equation suggests that s is a sufficient statistic under the broader one-parameter family

$$\tilde{\mathcal{F}} = \left\{ p(y; \theta) = c(\theta) \left(\frac{p(y; 1)}{p(y; 0)} \right)^{\theta} p(y; 0), \quad \theta \in \Theta \right\},$$

where Θ is the set of values $\theta \in \mathbb{R}$ for which the normalizing constant $c(\theta)$ is finite. It is easy to see that $[0, 1] \subseteq \Theta$. $\triangle$

The sufficiency of the likelihood ratio rather than, for instance, of the likelihood differences, offers a further justification for the presence of a constant of proportionality in the definition of the likelihood function. Only ratios are meaningful in the light of the sufficiency principle.

Usually under a statistical model $\mathcal{F}$ there are many sufficient statistics. For example, when $\mathcal{F}$ is as in Example 2.4, the likelihood ratio $s(y)$, the order statistics $s' = (y_{(1)}, \ldots, y_{(n)})$, and, obviously, the data themselves $s'' = y$ are sufficient statistics. The statistic s' is more concise than s'' and s is more concise than s'. Clearly, under $\mathcal{F}$, the more concise a sufficient statistic is, the more valuable it is for condensing the information about θ given by the data.

Definition 2.5 A statistic s is said to be minimal sufficient for $\mathcal{F}$ if it is sufficient and is a function of every other sufficient statistic s', that is $s = s(s')$.

A minimal sufficient statistic always exists under $\mathcal{F}$ and is, under mild regularity conditions, supplied by the likelihood function (Barndorff-Nielsen, 1978, pp. 41-43). One constructive way of tracing a minimal sufficient statistic s is as follows. Because s is sufficient it must satisfy the Fisher–Neyman factorization theorem and, because it is minimal, it must provide the coarsest possible partition of the sample space. If s' is a generic sufficient statistic, the factorization theorem implies that

$$\frac{p_Y(y'; \theta)}{p_Y(y; \theta)} = \frac{g(s'(y'); \theta)}{g(s'(y); \theta)} \frac{h(y')}{h(y)},$$

for $y, y' \in \mathcal{Y}$. Therefore, if $s'(y') = s'(y)$, the ratio $p_Y(y'; \theta)/p_Y(y; \theta)$ does not depend on θ. For a fixed $y \in \mathcal{Y}$, the minimal sufficient statistic s reduces as many points $y' \in \mathcal{Y}$ as possible to the value $s(y)$, provided that (2.3) is satisfied, that is, all the points y' for which the likelihood ratio is independent of θ. Thus

$$s^{-1}(s(y)) = \left\{ y' \in \mathcal{Y} : \frac{p_Y(y'; \theta)}{p_Y(y; \theta)} = c(y, y') \right\}.$$

When y varies, a partition of $\mathcal{Y}$ is obtained which is called the likelihood partition. Any statistic s whose partition of the sample space coincides with the likelihood partition is minimal sufficient. The likelihood function itself is, in a sense, a minimal sufficient statistic: data points y which produce proportional likelihoods are unified for inference. From now on, unless explicitly stated otherwise, the term *sufficient statistic* will be used to indicate the minimal sufficient statistic.

Example 2.5 *Normal distribution*
Let y be a realization of n independent r.v.'s $Y_i \sim N(\mu, \sigma^2)$, $i = 1, \ldots, n$. Then

$$p_Y(y; \mu, \sigma^2) \quad \propto \quad \exp\left\{-\frac{1}{2\sigma^2}\sum(y_i - \mu)^2 - \frac{n}{2}\log\sigma^2\right\}$$

$$\propto \quad \exp\left\{-\frac{n}{2\sigma^2}(\bar{y}_n - \mu)^2 - \frac{1}{2\sigma^2}\sum(y_i - \bar{y}_n)^2 - \frac{n}{2}\log\sigma^2\right\}$$

and construction of the likelihood partition requires that points y' are found such that

$$\frac{n}{2\sigma^2}\left\{(\bar{y}_n - \mu)^2 - (\bar{y}_n' - \mu)^2\right\} + \frac{1}{2\sigma^2}\left\{\sum(y_i - \bar{y}_n)^2 - \sum(y_i' - \bar{y}_n')^2\right\} = c(y, y').$$

These points are such that $\bar{y}_n' = \bar{y}_n$ and $\sum(y_i' - \bar{y}_n')^2 = \sum(y_i - \bar{y}_n)^2$, and therefore $(\bar{y}_n, s_n^2)$ is minimal sufficient. $\triangle$

Example 2.6 *One-parameter exponential family*
Let y be a realization of n independent r.v.'s Y_i, $i = 1, \ldots, n$, with densities that belong to a one-parameter exponential family, that is, of the form

$$p_{Y_i}(y; \theta) = e^{\theta s(y) - K(\theta)} h(y) ,$$

where $h(\cdot)$ is a non-negative function, $s(\cdot)$ is non-constant and $\theta \in \Theta \subseteq \mathbb{R}$, with Θ containing a non-empty open set. Then

$$p_Y(y; \theta) \propto \exp\left\{\theta\sum s(y_i) - nK(\theta)\right\} \tag{2.4}$$

and to construct the likelihood partition requires finding the locus of points y' such that
$$\exp\left\{\theta\left(\sum s(y_i) - \sum s(y_i')\right)\right\} = c(y, y').$$

This leads to points such that $\sum s(y_i') = \sum s(y_i)$, which shows that $\sum s(y_i)$ is minimal sufficient. $\triangle$

Example 2.7 *Uniform distribution on* $[0, \theta]$
Let y be a realization of n independent r.v.'s with a uniform distribution on $[0, \theta]$, $Y_i \sim U(0, \theta)$, $i = 1, \ldots, n$. Then

$$p_Y(y; \theta) = \frac{1}{\theta^n} I_{\mathbb{R}^+}(y_{(1)}) I_{[0,\theta]}(y_{(n)})$$

and the likelihood partition criterion

$$\frac{p_Y(y'; \theta)}{p_Y(y; \theta)} = \frac{I_{[0,\theta]}(y'_{(n)})}{I_{[0,\theta]}(y_{(n)})} \frac{I_{\mathbb{R}^+}(y'_{(1)})}{I_{\mathbb{R}^+}(y_{(1)})} = h(y', y)$$

shows that $y_{(n)}$ is a minimal sufficient statistic. $\triangle$

Example 2.8 *A curved exponential family: the normal parabola*
Let y be a realization of n independent r.v.'s $Y_i \sim N(\sigma, \sigma^2)$, $\sigma > 0$, $i = 1, \ldots, n$.
Then

$$p_Y(y; \sigma) \propto \exp\left\{-\frac{1}{2\sigma^2} \sum y_i^2 + \frac{1}{\sigma} \sum y_i - n \log \sigma\right\},$$

thus $s = (\sum y_i, \sum y_i^2)$ is minimal sufficient for σ. Note that here, unlike in the preceding examples, the minimal sufficient statistic does not have the same dimension as the parameter. This is typical of curved exponential families (cf. section 5.10) of which $N(\sigma, \sigma^2)$ is an instance. $\triangle$

Under $\mathcal{F}$, the likelihood function is fully specified if the value of a sufficient statistic is known. The sufficiency principle, when referred to a minimal sufficient statistic, is the same as the

weak likelihood principle. in the context of an assigned parametric statistical model, two observations y and y' such that $L(\theta; y) = c(y, y') L(\theta; y')$ for every θ must lead to the same inferential conclusions about θ.

The likelihood function has the advantage over other expressions of a minimal sufficient statistic, in that it summarizes information about θ in an intuitively accessible form. There are many examples where the maximum likelihood estimate $\hat{\theta}$ is, itself, a sufficient statistic (see e.g. Examples 2.5 and 2.7). Nevertheless, this is not the general situation, as Example 2.8 shows. If the maximum likelihood estimate is unique, then $\hat{\theta} = \hat{\theta}(s)$ with s minimal sufficient, i.e. $\hat{\theta}$ is a function of the data only through s.

2.5 Completeness

After reduction by sufficiency to s, in the statistical model with density $p_s(s;\theta)$ there could still be, implicit, distribution constant aspects. Hence a further reduction by conditioning would be possible. This section describes situations where this possibility does not arise. Inference about θ is then based directly on the reduced model with density $p_s(s;\theta)$, which is also the reference model for repeated sampling evaluations.

A sufficient statistic s, whose only distribution constant functions are trivial functions (i.e., essentially, constant functions), is called an **inferentially complete statistic**. The statistic s is inferentially complete if, and only if, the only events in its sample space whose probability does not depend on θ have probability either 0 or 1.

There are several definitions of completeness and they are usually referred to a family $\mathcal{F}$ of probability distributions, such as the family $\mathcal{F}_S$ of probability distributions for s.

A family $\mathcal{F} = \{p_Y(y;\theta),\ \theta \in \Theta\}$ of probability distributions on $\mathcal{Y}$ is called **weakly complete** if, for any statistic $u(y)$ taking only two values

$$E_\theta(u(Y)) = 0 \quad \text{for each } \theta \in \Theta \quad \Longrightarrow \quad P_\theta\{u(Y) = 0\} = 1 \qquad (2.5)$$

for each $\theta \in \Theta$. A statistic whose induced model is weakly complete is, itself, called a weakly complete statistic. The definition of a weakly complete family can be re-formulated in an equivalent way using indicator functions. Thus, a weakly complete sufficient statistic is inferentially complete (see Lehmann and Scholz, 1992, pp. 35-38).

For mathematical convenience, the notion of completeness can take stronger forms. $\mathcal{F}$ is said to be **boundedly complete** if (2.5) holds for any bounded statistic $u(Y)$. $\mathcal{F}$ is said to be **complete** if (2.5) holds for any statistic $u(Y)$. Completeness implies bounded completeness which, in its turn, implies weak completeness.

Exponential families are the main example of statistical models which admit complete sufficient statistics, see section 5.5. In Example 2.6, the sufficient statistic $\sum s(y_i)$ is complete.

The notion of completeness, in its strongest sense, was introduced by Lehmann and Scheffé (1950), as a condition for the essential uniqueness of the unbiased estimator of θ based on s, whenever such an estimator exists (cf. section 3.5.2).

Completeness plays an important role in inferential reduction. If the statistic s is complete, the distribution of each statistic $u(s)$ which is first-order distribution constant must be degenerate. Consequently, the reduction to s is so efficacious that in the induced statistical model $\mathcal{F}_S$ even first-order distribution constant aspects no longer exist. See Lehmann (1981) for an extensive treatment of this comment.

It is not difficult to show that if a sufficient statistic u is boundedly complete then it is minimal sufficient. In fact, if s is minimal sufficient, $s = s(u)$. On the other hand, the function $s(u)$ must be one-to-one, i.e. the conditional distribution of U given $S = s$ must be degenerate. If this were not true, an event A would exist in the sample space of U such that, with positive probability,

$$I_A(U) \neq E\{I_A(U) \mid S = s\} = P\{U \in A \mid s(U) = s\} \ .$$

However, since

$$E_{U;\theta}\left\{I_A(U) - E\{I_A(U) \mid s(U) = s\}\right\} = 0 \ ,$$

this would be in conflict with the bounded completeness of u.

The most important result concerning boundedly complete sufficient statistics is that they are independent of every distribution constant statistic. This is known as **Basu's theorem** (Basu, 1955).

Theorem 2.2 Under a statistical model $\mathcal{F}$ specified in the form $(\mathcal{Y}, P_\theta, \Theta)$, let $c = c(y)$ be a distribution constant statistic and let $s = s(y)$ be a boundedly complete sufficient statistic. Then S and C are independent for each $\theta \in \Theta$.

Proof. The distribution of C does not depend on θ and, because s is sufficient, the distribution of C given $S = s$ does not depend on θ either. Therefore, for every event E in the sample space of C, the quantity

$$u(s) = P(C \in E \mid S = s) - P(C \in E)$$

is a statistic for which $E_\theta(u(S)) = 0$ for each $\theta \in \Theta$, because

$$E_{S;\theta}\left\{P(C \in E \mid S)\right\} = P(C \in E) \ .$$

Since s is boundedly complete, this implies that $P(C \in E \mid S = s) = P(C \in E)$.
□

Under the assumptions of Basu's theorem, the distribution of S given $C = c$ coincides with the marginal distribution of S. This demonstrates the irrelevance for inference of any distribution constant aspect if s is sufficient and boundedly complete.

There is also a converse to the theorem: if $s = s(y)$ is a boundedly complete sufficient statistic and $C = c(Y)$ is independent of S, then, under weak regularity conditions, C is distribution constant (see Koehn and Thomas, 1975).

Under a statistical model $\mathcal{F}$ with sample space $\mathcal{Y}$, an event $R \subseteq \mathcal{Y}$ is called a **similar region** (meaning: similar to the sample space) if $P_\theta(R)$ does not depend on θ. Trivially, $\emptyset$ and $\mathcal{Y}$ are similar regions. If the statistic s is inferentially complete and, hence, *a fortiori*, if s is boundedly complete, there are no similar regions in the sample space S of s that are essentially different from $\emptyset$ and $\mathcal{Y}$. However, if s is a boundedly complete sufficient statistic, there can be similar, but not trivial, regions based on s in the original sample space $\mathcal{Y}$. These similar regions can be easily characterized. For $R \subseteq \mathcal{Y}$, because of sufficiency, R is similar conditionally on $S = s$; in addition, if the conditional probability $P(Y \in R \mid S = s)$ does not depend on s, then R is also, unconditionally, similar. In this case, R is said to be a **region with Neyman structure**. One immediate consequence of Basu's theorem is that if a model admits a sufficient statistic which is boundedly complete, then *all* the similar regions in $\mathcal{Y}$ will have Neyman structure. An important application of these concepts will be considered in sections 3.5.3 and 5.8.2.

2.6 Conditioning on Distribution Constant Statistics

On the basis of the principle of repeated sampling, the uncertainty of inferences must be evaluated according to their behaviour in hypothetical repetitions of the experiment carried out under identical conditions. This evaluation requires the study of sampling distributions. It would be reasonable to demand that hypothetical repetitions should be as relevant as possible to the data being analyzed. In other words, sampling distributions should be calculated conditionally on aspects of the data that are uninformative on θ.

One classic example, already mentioned in section 2.3, is when the sample size n is not determined prior to the experiment, but is a realization of a random variable whose distribution is independent of θ. In this case, the additional variability which is due to the random nature of n, because it cannot be attributed to our ignorance about θ, should not be taken into account in evaluations based on the principle of repeated sampling for which n should, therefore, be considered as fixed. Despite the fact that n offers no information about θ, it does, nonetheless, express the *amount* of information about θ which is contained in the data: all things being equal, we would expect more accurate analyses on the basis of one hundred observations than we would from five observations. Furthermore, the larger the variability of n is, the more variable are the conditional evaluations based on the principle of repeated sampling, e.g. the mean squared error of an estimator of θ based on n observations.

In a more general context, consider a statistical model specified as a collection of statistical models

$$\mathcal{F} = \{\mathcal{F}_c,\, c \in \mathcal{C}\}, \tag{2.6}$$

where each element $\mathcal{F}_c$ of $\mathcal{F}$ is, in its turn, a parametric model $(\mathcal{Y}_c, P_{\theta,c}, \Theta)$. Once a value c has been obtained in $\mathcal{C}$, according to a known probability distribution $p_C(c)$ not dependent on θ, the observation y is generated by a probability model in $\mathcal{F}_c$. The statistic c is distribution constant and, by conditioning, evaluation of the accuracy of inferences about θ is referred to the conditional model $\mathcal{F}_c$. Furthermore if, as in the case of a random sample size, c can be associated to some conditional measure of precision of the inferences about θ, conditioning on c assumes the further meaning of restricting the hypothetical repetitions to those samples which contain a quantity of information about θ comparable to that contained in y. In this situation the more variable c is, the more the conditional evaluations based on $\mathcal{F}_c$ may differ from unconditional evaluations.

A model of the form (2.6) does not only describe special experimental situations. From factorization (2.1), any parametric model that admits a distribution constant statistic could be expressed as in (2.6). In other words, if (2.1) holds, the sample y, equivalent to (c,t), could be considered as the result of a two-stage experiment: the first stage would give the value c of C, according to $p_C(c)$, the second would provide the observation t according to $p_{T|C=c}(t; c, \theta)$. Only the second of these two stages provides information about θ, while the observed value c of C could be considered to be fixed within the hypothetical repetitions of the experiment. Thus, considerations regarding models of the form (2.6) can be generalized to models which have (non-trivial) distribution constant statistics. This leads to the statement of the

> conditionality principle: if c is a distribution constant statistic under $\mathcal{F}$, evaluations based on the principle of repeated sampling are to be carried out conditionally on the observed value of c.

The application of the conditionality principle will be all the richer in interpretations the more efficacious a conditioning statistic is for identifying a partition of the sample space such that the samples belonging to the same element of the partition contain (at least approximately) the same amount of information about θ.

2.7   Discussion and Examples

The conditionality principle is not as widely accepted as is the sufficiency principle. Discussion involves both foundational and practical issues. One first point is that inference should not only be relevant to the observed data but also be relevant to the target population, while maintaining a manageable form. The maximum degree of relevance to the data is achieved in a Bayesian framework by conditioning to the whole data; however this requires a reasonably firm prior distribution. Relevance to the population depends on the experimental situation and on the scope of the analysis. It is clear that if the experiment is to be performed many times (as may occur, for example, in quality control) and one wishes to check on a global precision measure, it is not possible to reason conditionally. If, on the other hand, the possibility of repetition is almost entirely hypothetical, and is, essentially, only an element of the interpretative model of the observed data, then the arguments in favour of applying the conditionality principle, that is, making inference as relevant as possible to the data available, are all the more convincing. However, even in this last context, the target population should suggest which degree of conditioning is appropriate. Consider for instance a laboratory situation, where data consist of repeated measurements made using several instruments randomly selected according to a known distribution; it is known which instrument gave each measurement. Conditional inference is appropriate to assess the accuracy of each instrument, while unconditional inference is appropriate to assess the accuracy of the laboratory. The appropriate degree of conditioning is seen to necessarily involve criteria that are external to the statistical model; see Helland (1995) for a detailed discussion. One additional viewpoint is introduced with the aid of the following examples.

Example 2.9 *Random sample size*
Let N be a r.v. taking the value 1 with probability 0.5 and the value 100 with probability 0.5. Conditionally on $N = n$, let $y_1, \ldots, y_n$ be independent observations of a normal distribution with mean θ and unit variance. The sample size is distribution constant. Consider the estimator $\bar{y}_N = N^{-1} \sum_{i=1}^{N} y_i$

of θ. Acceptance, or non-acceptance of the conditionality principle means that the evaluation of the precision of $\bar{y}_N$ either is or is not carried out considering the observed value of N as fixed. This estimator is unbiased both conditionally on $N = n$ and marginally. Its conditional variance is $\mathrm{Var}_\theta(\bar{Y}_N \mid N = n) = 1/n$, while its marginal variance is $\mathrm{Var}_\theta(\bar{Y}_N) = 101/200$. $\triangle$

Example 2.10   *Uniform distribution on* $[\theta, \theta + 1]$
Let $y_1, \ldots, y_n$ be independent realizations of a r.v. with uniform distribution on $[\theta, \theta + 1]$. The sample range $v = y_{(n)} - y_{(1)}$, with $y_{(1)}$ and $y_{(n)}$ the minimum and maximum sample observations, respectively, is distribution constant. The minimal sufficient statistic $(y_{(1)}, y_{(n)})$ is a one-to-one function of $(y_{(1)}, v)$. Therefore, on the basis of the sufficiency principle, inference about θ must be based on $(y_{(1)}, v)$ and, according to the conditionality principle, the accuracy of inferences must refer to the conditional distribution of $Y_{(1)}$ given v. The statistic $y_{(1)} - 1/(n + 1)$ is, marginally, an unbiased estimator of θ, while conditionally

$$E_\theta\{Y_{(1)} - 1/(n + 1) \mid V = v\} = \theta + (1 - v)/2 - 1/(n + 1)$$

and a conditionally unbiased estimator based on $y_{(1)}$ is given by $y_{(1)} - (1-v)/2$. This estimator is unbiased also unconditionally. A unique unbiased estimator based on the sufficient statistic $(y_{(1)}, y_{(n)})$ does not exist, because this statistic is not complete (see also section 3.5.2). Further analysis of this example is left for an exercise, see Problems 2.11–2.13. $\triangle$

Examples 2.9 and 2.10 represent markedly different situations. In Example 2.9 the experiment splits into two stages, of which only the second offers information about the parameter. In the other example, v is distribution constant because of the mathematical structure of the parametric model adopted. Some authors, (Basu, 1964; Kalbfleisch, 1975) feel that this distinction is important and classify distribution constant statistics as either experimental or mathematical, according to whether they fit into the first case or the second. Obviously, even in the second case, the model could always be re-formulated as in (2.6), however, this construction could appear to be too artificial and render the reasons for conditioning less strong and objective. Again, this distinction is based on aspects *outside* the statistical model (cf. Lehmann, 1986, p. 543). Provided no statistically valid interpretation suggests rejecting conditioning in a specific analysis, reduction to a conditional model entails, as a practical advantage, a dimension reduction. Moreover, conditional distributions are sometimes simpler to calculate (cf. section 7.6) and often simpler to approximate (cf. section 11.2).

When applying the conditionality principle some conceptual and practical difficulties may arise when identifying distribution constant statistics (cf. section 2.3). Above all, for a family $\mathcal{F}$, there might well not be any non-trivial statistic which is exactly distribution constant, or rather, one may only suspect that it does not exist in that no such statistic is known. There are no general criteria available for finding distribution constant statistics for a given $\mathcal{F}$ unless this is a group family (see Chapter 7). In other words, there is no criterion for finding distribution constant statistics analogous to the likelihood partition for minimal sufficient statistics.

When a distribution constant statistic is available it is desirable to recognize whether maximum simplification has been achieved, that is, whether the conditioning distribution constant statistic is maximal. Although there are no criteria for finding maximal distribution constant statistics, there is one to check whether maximality has been achieved, given by the following characterization of maximality (Lehmann and Scholz, 1992).

Theorem 2.3 A distribution constant statistic c is maximal if, and only if, the conditional model of Y given $C = c$ is weakly complete.

Example 2.11 *Random sample size (cont.)*
Under the same assumptions of Example 2.9, as a consequence of Theorem 2.3, the statistic $c = (N, y_2 - y_1, \ldots, y_N - y_1)$ is maximal distribution constant. Indeed, the data $(N, y_1, \ldots, y_N)$ are a one-to-one function of $(c, \bar{y}_N)$ and the conditional model of $\bar{Y}_N$ given $C = c$ is complete, as a consequence of the completeness of the sufficient statistic $\bar{y}_n$ (whose distribution belongs to a one-parameter exponential family) and of Basu's theorem. $\triangle$

However, there is no guarantee of the uniqueness of a maximal distribution constant statistic. Consequently, unlike for minimal sufficient statistics, where the direct opposite is true, a distribution constant statistic is not necessarily a function of a given maximal distribution constant statistic (cf. Example 2.2).

One additional problem is the possible conflict between the conditionality principle and the sufficiency principle. Distribution constant statistics may exist which are not functions of the minimal sufficient statistic. In the model of Example 2.2 a preliminary sufficiency reduction leads to $s = (\sum(x_i^2 + y_i^2), \sum x_i y_i)$; however, there is no distribution constant statistic which is a function of s (see Amari, 1982). One way of resolving possible conflicts with

the sufficiency principle and, at least partly, the problem of non-uniqueness is
to require that sufficiency reduction is carried out first and that conditioning
concerns distribution constant statistics which are functions of the minimal
sufficient statistic. However, not even this restriction will be able, generally,
to avoid possible non-uniqueness of a maximal distribution constant statistic.
One classic example of this is taken from Cox (1971).

Example 2.12 *Multinomial distribution*
Let (y_1, y_2, y_3, y_4) be an observation from a multinomial distribution with index
n and parameter

$$\theta = \left(\frac{1-\phi}{6}, \frac{1+\phi}{6}, \frac{2-\phi}{6}, \frac{2+\phi}{6} \right) ,$$

with $\mid \phi \mid < 1$. The minimal sufficient statistic is (y_1, y_2, y_3, y_4) with $y_1 + y_2 + y_3 + y_4 = n$. It is not difficult to check that both

$$a = (a_1, a_2) = (y_1 + y_2, y_3 + y_4)$$

and

$$b = (b_1, b_2) = (y_1 + y_4, y_2 + y_3)$$

are distribution constant statistics. $\triangle$

In cases analogous to the above, it would be useful to have some criterion
available for choosing between alternative distribution constant statistics. Cox
(1971) proposes basing this choice on the variability of expected information
conditionally on the distribution constant statistic. For simplicity, restricting
the discussion to a scalar parameter and to variance as a measure of variability, Cox suggests choosing the distribution constant statistic where variance
of the conditional information is maximum. This suggestion appears to be
motivated by the arguments in favour of conditioning discussed in section 2.6.
Among the possible partitions of the sample space determined by competing distribution constant statistics, that which is closest to the *ideal* partition
is usually preferred, that is, the distribution constant statistic for which all
the samples belonging to the same element of the partition contain the same
amount of information about the parameter. In other words, Cox's criterion
means choosing the distribution constant statistic which, more than any other,
accurately represents the information effectively contained in the sample.

Example 2.13 *Multinomial distribution (cont.)*
Simple calculations (Cox, 1971, p. 253) give

$$\text{Var}_\phi(l_*(\phi) \mid a) = i_a(\phi) = \frac{3a_1/n + (1 - \phi^2)}{(1 - \phi^2)(4 - \phi^2)}$$

and

$$\text{Var}_\phi(l_*(\phi) \mid b) = i_b(\phi) = \frac{2\phi b_1/n + (1 - \phi)(2 + \phi)}{(1 - \phi^2)(4 - \phi^2)} .$$

Since a_1 and b_1 have both a binomial distribution with index n and parameters, respectively, 1/3 and 1/2, then

$$\text{Var}_\phi(i_a(\phi)) = \frac{2}{n\{(1 - \phi^2)(4 - \phi^2)\}^2}$$

and

$$\text{Var}_\phi(i_b(\phi)) = \frac{\phi^2}{n\{(1 - \phi^2)(4 - \phi^2)\}^2} ,$$

so that $\text{Var}_\phi(i_a(\phi)) > \text{Var}_\phi(i_b(\phi))$ for each ϕ. Consequently, on the basis of Cox's criterion, a is preferable to b. $\triangle$

Becker and Gordon (1983) show that Cox's criterion always leads to conditioning on a maximal distribution constant statistic and that, under suitable assumptions, this is a function of the minimal sufficient statistic.

However, possible non-uniqueness of the maximal distribution constant statistic need not undermine the conditionality principle. Barndorff-Nielsen (1995a) argues that this possibility is just an instance of the natural and unavoidable multiplicity of interpretations of data in scientific inference.

2.8 Ancillary Statistics

In this section a weaker version of the conditionality principle is considered which is closer to the Fisherian view where conditioning is seen, above all, as a strategy for *recovering the information* lost when reducing a sufficient statistic s to the maximum likelihood estimate $\hat\theta = \hat\theta(s)$ which, usually, does not maintain sufficiency.

Definition 2.6 Under a parametric model $\mathcal{F}$ with minimal sufficient statistic s, a statistic a is called auxiliary if $(\hat\theta, a)$ is a one-to-one function of s. If a is auxiliary and distribution constant, a is said to be **ancillary**.

If a is ancillary, the following factorization holds:

$$p_{\hat{\theta},A}(\hat{\theta}, a; \theta) = p_A(a)\, p_{\hat{\theta}|A=a}(\hat{\theta}; a, \theta)\,,$$

thus, $\hat{\theta}$ is minimal sufficient in the model conditional on the observed value of a (Fisher, 1934). Therefore, an ancillary statistic can recover the information lost in the reduction from s to $\hat{\theta}$. This is particularly true for information relating to the accuracy with which the maximum likelihood estimate locates θ_0 within Θ. The expected information i behaves according to this interpretation (j is clearly, in itself, a conditional measure of information). Let $i_Y(\theta)$ denote the expected information under $\mathcal{F}$ and let $i_{\hat{\theta}}(\theta)$ indicate the expected information in the model induced by $\hat{\theta}(y)$. Generally, the difference $i_Y(\theta) - i_{\hat{\theta}}(\theta)$ is a non-negative definite matrix, whereas, because of sufficiency, $i_Y(\theta) = i_s(\theta)$, where $i_s(\theta)$ is the expected information in the model induced by s. The conditional information contained in $\hat{\theta}$ is

$$
\begin{aligned}
i_{\hat{\theta}|a}(\theta) &= -E_\theta\left\{\frac{\partial^2}{\partial\theta^2}\log p_{\hat{\theta}|A=a}(\hat{\theta}; a, \theta)\,|A = a\right\} \\
&= -E_\theta\left\{\frac{\partial^2}{\partial\theta^2}\log p_{\hat{\theta},A}(\hat{\theta}, a; \theta)\,|A = a\right\} \\
&= -E_\theta\left\{\frac{\partial^2}{\partial\theta^2}\log p_s(s; \theta)|A = a\right\}\,.
\end{aligned}
$$

Thus, taking the expectation of both sides with respect to the distribution of A, we get

$$E_A\{i_{\hat{\theta}|a}(\theta)\} = i_s(\theta)\,.$$

Consequently, conditioning on the ancillary statistic a permits total information recovery about θ: the expectation of the conditional information coincides with the overall information.

From now on the conditionality principle will mainly refer to ancillary statistics.

Example 2.14 *The von Mises distribution*

Suppose that $y_1,\ldots,y_n$ are a random sample from a distribution with density

$$p_Y(y; \theta, \lambda) = \frac{1}{2\pi I_0(\lambda)}\,\exp\{\lambda\cos(y - \theta)\}\,, \tag{2.7}$$

with $\lambda > 0$, $y \in (0, 2\pi)$, $\theta \in (0, 2\pi)$. The function $I_0(\cdot)$ which appears as a normalizing constant is a **Bessel** function of zero order and imaginary argument

$$I_0(\lambda) = \frac{1}{2\pi}\int_0^{2\pi} \exp\{\lambda\cos t\}\,dt\,. \tag{2.8}$$

The von Mises distribution is one of the basic models for the analysis of **directional data**, that is, of observations which correspond to measurements of angles (Mardia, 1972). For example wind speed in a given location can be recorded jointly with the angle of incidence. Density (2.7) is an analogy of the normal distribution on the unit circle; the parameter θ is the mode of the distribution and λ represents the **precision** (as λ increases the dispersion of the distribution around θ diminishes). If λ is known and equal to λ_0, the log-likelihood is given by

$$l(\theta) = \lambda_0 \sum_{i=1}^{n} \cos(y_i - \theta) = \lambda_0 \left(\cos\theta \sum_{i=1}^{n} \cos y_i + \sin\theta \sum_{i=1}^{n} \sin y_i \right). \qquad (2.9)$$

From (2.9) it immediately follows that $(\sum_{i=1}^{n} \cos y_i, \sum_{i=1}^{n} \sin y_i)$ is minimal sufficient. Since

$$l_* = \lambda_0 \left(\cos\theta \sum_{i=1}^{n} \sin y_i - \sin\theta \sum_{i=1}^{n} \cos y_i \right),$$

the maximum likelihood estimate of θ satisfies

$$\tan\hat\theta = \frac{\displaystyle\sum_{i=1}^{n} \sin y_i}{\displaystyle\sum_{i=1}^{n} \cos y_i} \quad \text{and} \quad \text{sgn}(\sin\hat\theta) = \text{sgn}(\sum_{i=1}^{n} \sin y_i),$$

where $\text{sgn}(z)$ is the **sign function** which is equal to 1 if $z > 0$, 0 if $z = 0$ and -1 if $z < 0$. Let

$$a = \left\{ \left(\sum_{i=1}^{n} \sin y_i \right)^2 + \left(\sum_{i=1}^{n} \cos y_i \right)^2 \right\}^{1/2};$$

the minimal sufficient statistic can be expressed as

$$\left(\sum_{i=1}^{n} \cos y_i, \sum_{i=1}^{n} \sin y_i \right) = (a\cos\hat\theta, \ a\sin\hat\theta)$$

so that a is auxiliary. The joint density of $y_1, \ldots, y_n$ can be written as

$$p_Y(y_1, \ldots, y_n; \theta, \lambda_0) = (2\pi I_0(\lambda_0))^{-n} \ \exp\{\lambda_0 a \cos(\hat\theta - \theta)\},$$

and the joint density of $(\hat\theta, a)$ is

$$p_{\hat\theta, a}(\hat\theta, a; \theta, \lambda_0) = (2\pi I_0(\lambda_0)^n)^{-1} \ \exp\{\lambda_0 a \cos(\hat\theta - \theta)\} h_n(a),$$

for $\hat{\theta} \in (0, 2\pi)$, $a \in (0, n)$; $h_n(a)$ is a function of a which does not depend on $\hat{\theta}$ (see Mardia, 1972, section 4.5.3). Therefore, a has marginal density

$$p_A(a; \lambda_0) = I_0(\lambda_0)^{-n} \, h_n(a) \, I_0(\lambda_0 a) \,,$$

which shows that a is distribution constant and, since it is auxiliary, it is also ancillary. The conditional density of $\hat{\theta}$ given a is

$$p_{\hat{\theta}|A=a}(\hat{\theta}; a, \theta) = \frac{1}{2\pi I_0(\lambda_0 a)} \, \exp\{\lambda_0 a \cos(\hat{\theta} - \theta)\} \,.$$

Again this is a von Mises distribution with precision $\lambda_0 a$. Note that n does not appear in this conditional model. The ancillary statistic a clearly assumes the role of a measure of the information contained in the observed sample, through the new precision parameter $\lambda_0 a$.   △

One practical difficulty is that a decomposition of the form $s = (\hat{\theta}, a)$ may be very difficult to trace, so that the reduction of s to $\hat{\theta}$ usually entails a loss of information which cannot be recovered.

Example 2.15 *Bivariate normal distribution (cont.)*
Let $(x_1, y_1), \dots, (x_n, y_n)$ be independent observations from a bivariate normal distribution with zero means, unit variances and correlation coefficient $\theta \in (-1, 1)$. The m.l.e. of θ is a solution of

$$n\theta^3 - \theta^2 \sum_{i=1}^n x_i y_i + \theta \left\{ \sum_{i=1}^n (x_i^2 + y_i^2) - n \right\} - \sum_{i=1}^n x_i y_i = 0 \,,$$

while the minimal sufficient statistic $s = (\sum(x_i^2 + y_i^2), \sum x_i y_i)$ is two-dimensional and there is no exact ancillary statistic (cf. Amari, 1982).   △

In cases where either there is no exact ancillary, or where one is unable to identify it, the natural response is to turn to approximate solutions, i.e. to auxiliary statistics which are asymptotically distribution constant (Cox, 1980; McCullagh, 1984a). One important example of approximate ancillary statistic is given in section 5.10. A *continuity of principles* is sought through approximate solutions. Indeed, it would seem rather unsatisfactory if the conditionality principle were to be applied only in situations where there is an exact ancillary available and in all other cases marginal procedures had to be adopted.

The following qualitative observations are rather important.

Above all, calculating conditional distributions is not usually easy even when simulations or numerical methods are used. Thus, it is particularly interesting to look for transformations which render, either exactly or, at least, approximately, the statistic of interest independent of the conditioning statistic. When this is possible, while still obeying the conditionality principle, we can refer to the marginal distribution and ignore the observed value of the conditioning statistic. Procedures which follow this logic are sometimes called *conditional inference without tears* or *conditional inference without conditioning*. Theorem 2.2 offers an important general result for exact conditioning. Under the assumptions of this theorem there are no non-trivial ancillary statistics and conditioning on distribution constant statistics is automatic.

Barndorff-Nielsen and Cox (1994, section 7.2) use the term **stable** to qualify procedures, and the corresponding statistics or combinants, which have approximately the same frequency properties both conditionally and unconditionally for any reasonable choice of the ancillary statistic. Stable procedures have the further advantage of avoiding the choice of the ancillary. In this context, the main result, which will be illustrated in greater detail in section 11.3, concerns the **log-likelihood ratio** statistic $2\{l(\hat{\theta}) - l(\theta)\}$: in one sense, which will be explained more fully, this statistic is independent, to a high order of accuracy, of any exact ancillary, or of any statistic which is ancillary to a suitable order.

2.9   Relevant Subsets

The long-run perspective, or **reference set**, in relation to which the information offered by y^{obs} is evaluated, is usually entirely conceptual. Therefore it can be made to depend on aspects expressed by the data themselves. However, as was seen in section 2.7, there are several problems in the frequency theory of inference for satisfying, through conditioning on distribution constant statistics, the Fisherian requirement of relevance of probability calculations. Even greater logical difficulties arise if the class of possible conditioning events is extended beyond similar regions.

While restricting the reference set from $\mathcal{Y}$ to a similar region R, with $y^{obs} \in R$, can be justified fairly convincingly, there are no specific reasons for choosing non-similar conditioning regions. However, even these latter may have an indirect role in inference, as was highlighted by Fisher (1956, p. 96). To appreciate this point, consider the following example, drawn from Brown (1967).

Example 2.16 *The conditional level of t intervals*
Let y_1, y_2 be two observations from a $N(\mu, \sigma^2)$ distribution, and let $C_{1/2}(y_1, y_2)$ be the usual confidence interval for μ based on the Student's t statistic, with confidence level 1/2. Then it can be shown that

$$P_{\mu, \sigma^2}\left(\mu \in C_{1/2}(Y_1, Y_2) \;\middle|\; \frac{|Y_1 + Y_2|}{|Y_1 - Y_2|} \leq 1 + \sqrt{2}\right) > \frac{2}{3}$$

for each value of (μ, σ^2). The conditioning event does not define a similar region. △

In a confidence region setting, let $C_{1-\alpha}(Y) \subseteq \Theta$ be a random region where $P_\theta(\theta \in C_{1-\alpha}(Y)) = 1 - \alpha$ for every $\theta \in \Theta$. An event Q in the sample space $\mathcal{Y}$ is called a **relevant subset** if, for a value $\varepsilon > 0$, either $P_\theta(\theta \in C_{1-\alpha}(Y) \mid y \in Q) \geq 1 - \alpha + \varepsilon$ or $P_\theta(\theta \in C_{1-\alpha}(Y) \mid y \in Q) \leq 1 - \alpha - \varepsilon$, for each $\theta \in \Theta$. Therefore, if Q is a relevant subset, and $y^{obs} \in Q$, the conditional confidence level of $C_{1-\alpha}(Y)$ is uniformly greater or less than the marginal level $1 - \alpha$ and a conflict arises between the pre-experimental and post-experimental attribution of confidence. Fisher (1956) offered an example of a relevant subset in Welch's (1947) solution of the Behrens–Fisher problem (test of equality of the means of two normal populations with unknown variances, see Example 11.9). Many other examples of relevant subsets were identified in the following period and the discussion about these pathologies led to the conclusion that only Bayesian procedures do not admit relevant subsets. See Pierce (1973) for the technical details and Casella (1992) for later developments.

The impact of the existence of relevant subsets on the Fisherian paradigm is not as important as Example 2.16 would lead one to believe. Certainly, their very existence demonstrates that frequentist procedures, even when they obey the conditionality principle, are not able to extract all the information in the data y under the statistical model $\mathcal{F}$. This statement is, however, somewhat mitigated by a variety of considerations. Firstly, $\mathcal{F}$ is, above all, a tool designed to summarize information about the important features of the phenomenon being studied. Secondly, when there is a large amount of information in the data, for example, when the sample size is large, one could plausibly expect that the variations in significance induced by the relevant subsets will be minor, comparable to those due to numerical error or other sources. Lastly, comparison between frequency and Bayesian methods is not carried out between homogeneous subjects: there is no way of comparing the extraction of *all* the information in the data and $\mathcal{F}$ using prior information too, with the extraction of information *only* present in the data according to $\mathcal{F}$.

2.10   Combinants and Pivotal Quantities

Statistical procedures are often described by quantities that depend on the data y, and on the statistical model through the parameter θ. Observed likelihood quantities are the most obvious examples of this. The definition below follows the terminology introduced in Barndorff-Nielsen (1988, p. 19).

Definition 2.7 Given a statistical model $\mathcal{F}$ specified as $(\mathcal{Y}, P_\theta, \Theta)$, a function $q = q(y; \theta)$, which for every fixed $\theta \in \Theta$ is a statistic, is called a **combinant**.

The probability distribution of $q = q(Y; \theta)$ can be evaluated for θ fixed as y varies in $\mathcal{Y}$ according to $p_Y(y; \tilde{\theta})$, with $\tilde{\theta} \in \Theta$. By slightly extending the analogous notions set out in section 1.4.2, we can speak of a null distribution of q when $\tilde{\theta} = \theta$ and, otherwise, of a non-null distribution. Hence, the statistical model $\mathcal{F}$ for y induces a statistical model $\mathcal{F}_Q$ for q, with density $p_Q(q; \theta, \tilde{\theta})$, where $\theta \in \Theta$ is considered to be fixed and $\tilde{\theta} \in \tilde{\Theta}$ with $\tilde{\Theta}$ a copy of Θ. A null distribution is $p_Q(q; \theta, \theta)$.

Example 2.17 *Normal distribution with known variance*
Consider random sampling of size n from a $N(\theta, 1)$ distribution. The quantity $q = \bar{y}_n - \theta$ is a combinant, which depends on the data through a complete sufficient statistic. The induced statistical model $\mathcal{F}_Q$ has density

$$p_Q(q; \theta, \tilde{\theta}) = (\sqrt{n}/\sqrt{2\pi}) \exp\{-(n/2)(q - \tilde{\theta} + \theta)^2\}.$$

The null distribution of q is $N(0, 1/n)$, independent of θ. $\triangle$

The combinants whose null distribution is independent of θ for every $\theta \in \Theta$ are particularly important for inference. Their distribution can depend on $\mathcal{F}$ and is denoted by $p_Q(q)$.

Definition 2.8 A combinant whose null distribution does not vary within a statistical model $\mathcal{F}$ is called a **pivotal quantity** (under $\mathcal{F}$).

If θ is a scalar and the pivotal quantity $q(y; \theta)$ is a monotone function of θ, it is possible to obtain one-sided or two-sided confidence intervals for θ, with an assigned confidence level, by simply solving, with respect to θ, the equation which connects $q(y^{obs}; \theta)$ to the suitable quantiles of the null distribution of q.

A theory of statistical inference based on pivotal quantities was first out-lined by Barnard (1974b, 1977a,b) and, later, developed by Chamberlin (1989), Chamberlin and Sprott (1989), Sprott (1990) and Barndorff-Nielsen (1994). Sufficiency of a statistic was generalized (Barnard, 1977b) into the property of full efficiency for a pivotal quantity. A pivotal quantity is said to be fully effi-cient if the likelihood function for θ based on q, $L(\theta; q) \propto p_Q(q) = p_Q(q(y; \theta))$, is equivalent to that based on y, $L(\theta; y)$. The pivotal quantity in Example 2.17 is fully efficient. However, the definition of the likelihood function based on a pivotal quantity is not unique. Indeed, the density of $q(y; \theta)$ and that of a one-to-one function of q do not give equivalent likelihoods, because the Jaco-bian determinant of the transformation usually depends on θ. To overcome this drawback, for a scalar θ, Sprott (1990) proposed only paying attention to linear pivotal quantities, of the form $q(y; \theta) = (t_1(y) - \theta)/t_2(y)$, with $t_2(y) > 0$; however, this restriction would seem to be rather unnatural. A pivotal quan-tity that is useful for inference about θ will always be based on the minimal sufficient statistic $(\hat{\theta}, a)$. If $q = q(\hat{\theta}, a; \theta)$ is a pivotal quantity with the same dimension as θ and if, for a fixed, q is a one-to-one function of $\hat{\theta}$, the original $L(\theta; y)$ can be recovered by defining the likelihood function based on q as

$$L(\theta; q(\hat{\theta}, a; \theta)) = p_Q(q(\hat{\theta}, a; \theta)) \left| \frac{\partial q}{\partial \hat{\theta}} \right|,$$

where $\partial q / \partial \hat{\theta}$ is the Jacobian of the transformation of $\hat{\theta}$ into q, considering θ and a to be fixed. See Barndorff-Nielsen (1994), where this basic idea is extended to the nuisance parameter case and q is not a one-to-one function of $\hat{\theta}$.

A natural weakening of pivotality is given by the notion of first-order piv-otality. A combinant q is a first-order pivotal quantity if its null expectation is independent of θ. The score is a first-order pivotal quantity; further examples are unbiased estimating functions, that is, combinants for which $E_\theta(q(Y; \theta)) = 0$ for every $\theta \in \Theta$. These combinants are called estimating functions because of the analogy between the maximum likelihood estimate $\hat{\theta}$, for which $l_*(\hat{\theta}) = 0$, and the estimate $\hat{\theta}^q$, for which $q(y; \hat{\theta}^q) = 0$. Godambe (1960) shows that the score is optimal among the unbiased estimating functions, in the sense that, if θ is a scalar and $q(y; \theta)$ is an unbiased estimating function, then, for every $\theta \in \Theta$,

$$\frac{\text{Var}_\theta(l_*(\theta))}{\{E_\theta(\partial l_*(\theta)/\partial \theta)\}^2} \leq \frac{\text{Var}_\theta(q(Y; \theta))}{\{E_\theta(\partial q(Y; \theta)/\partial \theta)\}^2} .$$

For extensions to vector parameters, see Godambe and Kale (1991, section 1.7). However, Godambe's result does not attribute an exclusive role to the score function: other requirements could arise, for example robustness (see section 3.6.3). Inference based on unbiased estimating functions is analyzed further in sections 3.4.3 and 4.9.

A second-order pivotal quantity q is a combinant whose null mean and variance do not depend on θ. If, because of the central limit theorem, the null distribution of q is asymptotically normal, second-order pivotality induces approximate pivotality and q is called an **asymptotically pivotal quantity**. Many widely used statistical procedures are based on this property. Think for example, in a multiple linear regression model under second-order assumptions, of the significance test on one regression coefficient, which is based on the asymptotic normal distribution of the least squares estimator divided by its estimated standard deviation.

2.11   The Principle of Parameterization Invariance

It could prove useful to consider a statistical model $\mathcal{F}$, from a geometrical point of view, as a space or a differentiable manifold; this latter term refers to a subspace which is *regular* in a geometrical sense. Each point in this space represents one probability distribution belonging to $\mathcal{F}$. A parametric model is characterized by the possibility of placing the points in $\mathcal{F}$ in one-to-one correspondence with a p-dimensional Euclidean space, according to a mapping

$$\theta : \mathcal{F} \to \Theta \subseteq \mathbb{R}^p \ ,$$

that is, according to a parameterization. Thus, fixing a parameterization is the same as choosing a coordinate system for the manifold $\mathcal{F}$.

If θ is a parameterization of $\mathcal{F}$, any transformation $\psi = \psi(\theta)$, where $\psi(\cdot)$ is a one-to-one continuous function, could substitute as an alternative coordinate system. One-to-one correspondence ensures that identifiability is maintained and, because of continuity, the transformed parameter space $\Psi = \psi(\Theta)$ is still a non-empty open set in $\mathbb{R}^p$. Furthermore, it is worth focussing attention only on **smooth functions** $\psi(\theta)$, i.e. infinitely differentiable functions with infinitely differentiable inverse. In this way, differentiability up to a certain order of a function of the parameter does not depend on the coordinate system and this is important because, in applications, we are often interested in differentiating combinants with respect to the components of θ. Any one-to-one and smooth mapping $\psi : \Theta \to \Psi \subseteq \mathbb{R}^p$ gives a reparameterization of $\mathcal{F}$ (see section 1.4.3).

Reparameterization leaves the inferential problem virtually untouched. Furthermore, smoothness of the transformation ensures that the most common inferential procedures (for example, those based on likelihood with the related first-order asymptotic approximations) can be applied under any reparameterization of the model.

It is natural to require that inferential conclusions should not depend on the parameterization. This leads to placing the following principle alongside the guiding principles of sufficiency and of conditionality for data and model reduction.

> Principle of parameterization invariance: if θ and ψ are two alternative parameterizations for the parametric model $\mathcal{F}$, $\pi(y)$ is an inferential procedure, C^Θ and C^Ψ are the conclusions that $\pi(\cdot)$ leads to, expressed, respectively, in the parameterizations θ and ψ, then the same conclusions C^Ψ should be reached both by the application of $\pi(\cdot)$ in the parameterization ψ and by the translation into the parameterization ψ of the conclusions C^Θ.

This principle is equivalent to the requirement that the diagram in Figure 2.1 is commutative.

2.11.1 Consequences for Parameter Estimation

If the procedure $\pi(\cdot)$ produces a point estimate, according to the principle of parameterization invariance, the estimate must be equivariant. This property holds for the maximum likelihood estimate, as seen in section 1.4.3. Fisher (1922a, section 6), considered this property of the maximum likelihood estimate as being particularly important and, even though not explicitly stated by Fisher, the principle of parameterization invariance may be considered as a fundamental constituent of the Fisherian paradigm.

Example 2.18 *Equivariance under reparameterization of estimators based on the method of moments*
If $t(y)$ is a p-dimensional statistic, used for inference about θ, with a finite mean vector, whose components are non-constant functions of θ, then an estimator based on the method of moments is given by the solution $\tilde{\theta}$ (which is assumed to exist and be unique) to the equation in θ

$$E_\theta\{t(Y)\} = t^{obs} \, ,$$

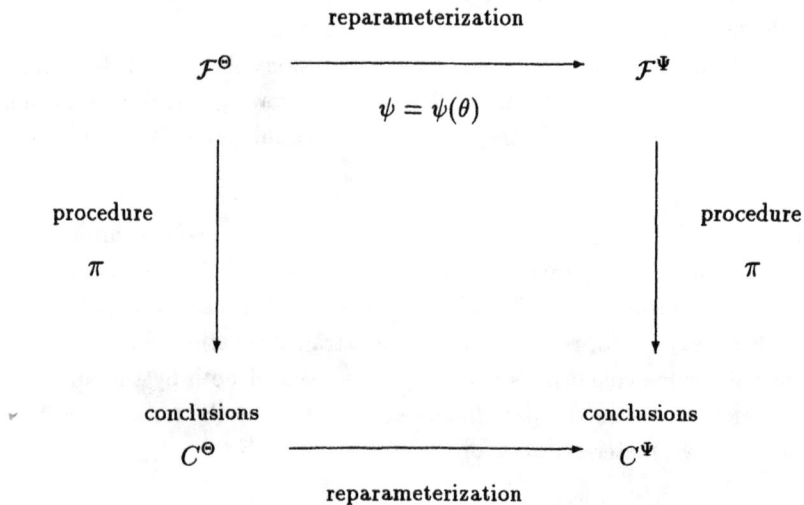

Figure 2.1: *The statistical procedure π is invariant under reparameterization if the conclusions C^Θ and C^Ψ, in the parameterizations θ and ψ, respectively, differ only through reparameterization.*

with $t^{obs} = t(y^{obs})$. If ψ is a reparameterization of $\mathcal{F}$

$$E_\theta\{t(Y)\} = E_{\psi(\theta)}\{t(Y)\}$$

and, hence, the estimator $\tilde\psi$, based on the method of moments in the parameterization ψ, which solves the equation $E_\psi\{t(Y)\} = t^{obs}$, is equivariant, i.e. such that $\tilde\psi = \psi(\tilde\theta)$. On the other hand, the estimators based on the method of moments do not behave in an invariant manner with respect to one-to-one transformations of $t(\cdot)$ (see Problem 2.1). $\triangle$

Only some of the criteria used for evaluating estimators according to the principle of repeated sampling are coherent with the principle of parameterization invariance. For instance, the properties of unbiasedness and of minimum variance are not invariant to parameterization. Hence, the estimators chosen on the basis of such criteria are not equivariant. When θ is a scalar, median

unbiasedness may be considered as a centring property alternative to mean un-biasedness. If $\tilde{\theta} = \tilde{\theta}(y)$ has a continuous distribution, it is median unbiased if

$$P_\theta\{\tilde{\theta}(Y) \leq \theta\} = 1/2, \qquad \theta \in \Theta. \tag{2.10}$$

If $\tilde{\theta}$ is median unbiased for θ then $\tilde{\psi} = \psi(\tilde{\theta})$ is also median unbiased for ψ. For a brief introduction to median unbiased estimators, see Read (1985).

Equivariance of estimators based on unbiased estimating equations is close-ly connected to equivariance of estimators based on the method of moments. The former are defined by an equation of the form $q(y^{obs}; \theta) = 0$, where $q(y; \theta)$ is a combinant such that $E_\theta(q(Y; \theta)) = 0$, $\theta \in \Theta$.

As regards the construction of confidence intervals or confidence regions, satisfying the principle of parameterization invariance translates into a demand for equivariance. If $C^\Theta(y)$ is a confidence region for θ with level $1 - \alpha$, that is if

$$P_\theta\{\theta \in C^\Theta(Y)\} = 1 - \alpha, \qquad \theta \in \Theta,$$

then the same procedure should, in the parameterization ψ, lead to the region

$$C^\Psi(y) = \{\psi \in \Psi : \theta(\psi) \in C^\Theta(y)\}, \qquad y \in \mathcal{Y}.$$

Likelihood based confidence regions

$$C^\Theta(y) = \{\theta \in \Theta : l(\hat{\theta}) - l(\theta) < k\},$$

with k a fixed constant, satisfy this requirement.

2.11.2 Consequences for Hypothesis Testing

When testing statistical hypotheses, reparameterization entails translating the null hypothesis $H_0 : \theta \in \Theta_0$ ($\Theta_0 \subset \Theta$) into $H_0 : \psi \in \Psi_0 = \{\psi \in \Psi : \theta(\psi) \in \Theta_0\}$. The principle of parameterization invariance requires that the test statistic should be invariant. More specifically, it requires that the observed significance level of a test statistic should be the same whether the statistic is calculated with reference to the parameterization θ or to the parameterization ψ, for each $y \in \mathcal{Y}$.

Parameterization invariance of the likelihood function ensures invariance of tests based on the likelihood ratio. Denote by $\hat{\theta}_0$ the value of θ which maximizes $l(\theta)$ for $\theta \in \Theta_0$. According to the likelihood ratio test, large values of $l(\hat{\theta}) - l(\hat{\theta}_0)$ offer evidence against H_0. The likelihood ratio test, when referred to the parameterization ψ, is based on $l^\Psi(\hat{\psi}) - l^\Psi(\hat{\psi}_0)$. The value of the test calculated in the two parameterizations is the same, because $\hat{\psi} = \psi(\hat{\theta})$, $\hat{\psi}_0 = \psi(\hat{\theta}_0)$ and bearing in mind (1.12).

Example 2.19 *Poisson distribution*
Let $y_1, \ldots, y_n$ be independent observations from a Poisson distribution with mean θ. Consider the log-likelihood ratio statistic for $H_0 : \theta = 1$. Because $\hat{\theta} = \bar{y}_n$, then $l(\hat{\theta}) = n\bar{y}_n(\log \bar{y}_n - 1)$ and $l(1) = -n$. In the parameterization $\psi = e^{-\theta}$ the null hypothesis becomes $H_0 : \psi = e^{-1}$, the log-likelihood is $l^\Psi(\psi) = n(\bar{y}_n \log(-\log \psi) + \log \psi)$ and the m.l.e. of ψ is $\hat{\psi} = e^{-\bar{y}_n}$. Hence, $l^\Psi(\hat{\psi}) = l(\hat{\theta}) = n\bar{y}_n(\log \bar{y}_n - 1)$ and $l^\Psi(e^{-1}) = l(1) = -n$. $\quad\triangle$

Not all tests based on likelihood quantities are parameterization invariant. Taking, for simplicity, the one-parameter case $(p = 1)$, one example of this is the **Wald statistic** for $H_0 : \theta = \theta_0$,

$$W_e = (\hat{\theta} - \theta_0)^2 i(\hat{\theta}) \, .$$

If we consider a reparameterization $\psi = \psi(\theta)$ and, referring to it, we calculate the Wald statistic W_e^Ψ, taking into account the relation

$$i^\Psi(\psi) = i^\Theta(\theta(\psi)) \left(\frac{\partial}{\partial \psi} \theta(\psi) \right)^2 ,$$

we obtain

$$W_e^\Psi = (\psi(\hat{\theta}) - \psi(\theta_0))^2 i^\Theta(\theta(\hat{\psi})) \left(\frac{\partial}{\partial \psi} \theta(\psi) \Big|_{\psi = \hat{\psi}} \right)^2 ,$$

which usually does not coincide with W_e calculated in the parameterization θ.

Example 2.20 *Poisson distribution (cont.)*
The Wald statistic for $H_0 : \theta = 1$ is $W_e = n(\bar{y}_n - 1)^2/\bar{y}_n$. In the parameterization $\psi = e^{-\theta}$ we obtain $W_e^\Psi = n(e^{-\bar{y}_n} - e^{-1})^2 e^{2\bar{y}_n}/\bar{y}_n$ which is not the same as W_e. If, for example, $n = 50$ and $\bar{y}_n = 1.3$, then $W_e \doteq 3.45$ and $W_e^\Psi \doteq 4.70$. Hence, when referring to the asymptotic chi-squared distribution with one degree of freedom the observed significance level is 0.06 for W_e and 0.03 for W_e^Ψ. $\quad\triangle$

2.11.3 Uninformative Prior Distributions and Invariance

Let $\mathcal{F}$ be a parametric statistical model specified in the form (1.3) and suppose that a prior probability distribution on Θ is given, with density $p_\theta(\theta)$. In the

Bayesian vision of inference, the prior distribution is to be combined through Bayes' formula with the density of the observations (given θ), $p_Y(y;\theta)$, to give the posterior distribution of θ given $Y = y$ with density

$$p_{\theta|Y=y}(\theta;y) \propto p_\theta(\theta)\, p_Y(y;\theta). \qquad (2.11)$$

For density (2.11) to be well defined, it is sufficient that the integral with respect to θ of the right-hand side of (2.11) is finite. The prior distribution might also be improper, that is the integral over Θ of $p_\theta(\theta)$ might not be finite. In the non-personalistic Bayesian paradigm (see section 1.2) the specification of uninformative prior distributions is crucial. The principle of parameterization invariance plays a role in this.

In the case of parameters with a continuous distribution, the main criticism of the representation of prior ignorance through, possibly improper, uniform distributions is lack of invariance of the representation under reparameterizations (see, e.g., Fisher, 1956, pp. 16-17). If, for example, $\theta \in \Theta = (0,1)$ is assigned a uniform prior distribution on $(0,1)$ the reparameterization $\psi = \theta^2$ neither changes the parameter space nor should it modify the state of lack of knowledge about θ; however, it does change the uniform prior density into $p_\psi(\psi) = 1/(2\sqrt{\psi})$, $\psi \in (0,1)$. Jeffreys (1946) proposed resolving this difficulty by choosing an invariant prior density, i.e. such that, under reparameterization of $\mathcal{F}$ from θ to $\psi = \psi(\theta)$, the form of the prior distribution does not change. This requirement is satisfied by the choice

$$p_\theta(\theta) \propto |i(\theta)|^{1/2}, \qquad \theta \in \Theta. \qquad (2.12)$$

In fact, in the parameterization $\psi = \psi(\theta)$, density (2.12) is transformed into

$$p_\psi(\psi) \propto |i(\theta(\psi))|^{1/2} |[\theta_a^r]|, \qquad \psi \in \Psi, \qquad (2.13)$$

where $|[\theta_a^r]|$ represents the Jacobian determinant of the inverse transformation $\theta = \theta(\psi)$ and $\Psi = \psi(\Theta)$. On the basis of (1.21), (2.13) could be rewritten as

$$p_\psi(\psi) \propto |i^\Psi(\psi)|^{1/2}, \qquad \psi \in \Psi, \qquad (2.14)$$

which has the same structure as (2.12) in the reparameterized model. In the one-parameter case, Jeffreys' rule has the same effect as assigning a uniform prior distribution in the parameterization for which Fisher information is constant. However, Jeffreys' proposal is anything but conclusive within the debate about the representation of prior ignorance in Bayesian inference. One point

is that (2.12) is based on a quantity, expected information, whose definition
is related to the principle of repeated sampling. For further details of the
controversy, see Dawid (1983b). In some cases (see e.g. section 7.7.2 and Example 9.10) there is a close relation between inferential procedures based on
uninformative prior distributions and non-Bayesian inferential procedures.

2.12 Bibliographic Note

Data and model reduction by sufficiency or by conditioning on either distribution constant statistics or ancillary statistics were introduced by Fisher (1920,
1922a, 1934, 1935). These topics are treated in many textbooks, e.g. in Cox
and Hinkley (1974, Chapter 2) and in Azzalini (1996, sections 2.3 and 3.3.7).
A rigorous and complete exposition of the concepts and main results on sufficiency and ancillarity can be found in Barndorff-Nielsen (1978, sections 4.1 and
4.2). A recent survey of theoretical developments in the concept of sufficiency
is in Yamada and Morimoto (1992).

Usually completeness is introduced within the theory of point estimation
(see, for example, Beaumont, 1980, sections 2.4 and 2.5). The role of completeness in data and model reduction is highlighted in Barndorff-Nielsen (1978,
section 4.3), in Lehmann (1981) and in Lehmann and Scholz (1992).

The term ancillary is used here as in Fisher (1956, section 6.7), Barndorff-
Nielsen (1988) and Barndorff-Nielsen and Cox (1994). Many authors, however,
use the term *ancillary statistic* merely in the sense of distribution constant
statistic. Reid (1995a) gives an account of the various purposes of conditioning in inference. McCullagh (1987, Chapter 8) treats distribution constant
statistics in detail and, in particular, concentrates upon approximate conditioning. See Lloyd (1992) also for some arguments in favour of approximate
conditioning. Examples 2.9 and 2.10 in this chapter are taken from section
8.1 in McCullagh (1987). McCullagh (1992) studies an interesting instance of
non-uniqueness of the ancillary statistic under random sampling from a two-
parameter Cauchy distribution. A further important reference is Lehmann
(1986, Chapter 10), where section 10.4 deals with relevant subsets.

One monograph on estimating equations is by McLeish and Small (1988).
See the volume edited by Godambe (1991) too.

The statement of the principle of parameterization invariance and the
schematic representation in Figure 2.1 are taken from Barndorff-Nielsen (1988,
section 1.3). Parameterization invariance has played a prime role in the recent
interest in differential geometry in statistics. The importance of the relation be-

tween the two disciplines was first pointed out by Rao (1945) and the links have drawn closer since Efron (1975). For an introduction, see Barndorff-Nielsen, Cox and Reid (1986) and Kass (1989); for further study, see the monographs by Amari (1985) and Murray and Rice (1993). Differential geometry is particularly suitable for dealing with intrinsic properties, that is, properties that are invariant with respect to the coordinate system: think for example of its use in physics, especially in relativity theory. Tools from differential geometry are suitable for developing procedures that are *automatically* coherent with the principle of parameterization invariance (see section 9.5). Recent developments in asymptotic theory have highlighted further points of contact between the two disciplines and research in the area is currently evolving (see, for example, Critchley, Marriot and Salmon, 1993, 1994).

Bernardo and Smith (1994, sections 5.4 and 5.6.2) give an updated account on uninformative prior distributions.

2.13 Problems

2.1 Prove that estimators based on the method of moments are not invariant under one-to-one transformations of the data.

[Section 2.2]

2.2 Consider **sequential sampling** where observations are independent realizations of a r.v. with p.d.f. $p_Y(y; \theta)$. At the i-th step one decides whether or not to take a further observation on the basis of whether there has been a *success* in a random experiment with a probability of success $\pi(y_1, \ldots, y_i)$, where $\pi(\cdot)$ is a known function. Suppose that sampling stops at the N-th step. Write the likelihood function for θ based on $y_1, \ldots, y_N$ and show that N is not distribution constant.

[Section 2.3]

2.3 Consider the family $\mathcal{F} = \{p(y; 0), \ldots, p(y; k)\}$, where the $k + 1$ elements have common support. Show that the statistic

$$s(y) = \left(\frac{p(y; 1)}{p(y; 0)}, \ldots, \frac{p(y; k)}{p(y; 0)} \right)$$

is sufficient.

[Section 2.4]

2.4 Determine the minimal sufficient statistic under the same assumptions as in Problem 1.3 and in Problem 1.4, respectively.

[Section 2.4]

2.5 Referring to Example 1.2, consider n independent realizations of an exponential r.v. with mean θ subject to type I censoring. Determine the minimal sufficient statistic.

[Section 2.4]

2.6 With reference to observation of a homogeneous Poisson process which gives the likelihood (1.27), find the minimal sufficient statistic when: (i) t_0 is fixed; (ii) t_0 is a realization of a r.v. with distribution independent of λ.

[Section 2.4]

2.7 Justify, in detail, the equivalence between weak completeness and inferential completeness.

[Section 2.5]

2.8 Show that the minimal sufficient statistic obtained in Example 2.15 is not complete.

[Section 2.5]

2.9 Let y be an observation of $Y \sim p_Y(y; \theta)$, $\theta \in \Theta$. Let s be a sufficient and complete statistic for θ. Furthermore, let c^{obs} be the observed value of a distribution constant statistic c such that $y = g(s, c)$, with $g(\cdot)$ a one-to-one function. The density of s is indicated by $p_s(s; \theta)$. Let s^* be a realization of the distribution with density $p_s(s; \theta_0)$. Show that $y^* = g^{-1}(s^*, c^{obs})$ has marginal density $p_Y(y; \theta_0)$ and illustrate this construction when $y = (y_1, \ldots, y_n)$ is a random sample from $N(\mu, \sigma^2)$. Consider the application of the result for defining a goodness-of-fit test for a specified parametric family.

[Section 2.5; Durbin, 1961]

2.10 Under the same assumptions as in Problem 2.9, let s^{obs} be the observed value of s and let $p_c(c)$ be the p.d.f. of c. Show that $Y^* = g^{-1}(s^{obs}, C)$ has the same distribution as Y given $S = s^{obs}$. Use this result to define an algorithm for generating pseudo-random observations of Y conditionally on $S = s^{obs}$. Specialize this algorithm to obtain n realizations of an $N(\mu, \sigma^2)$ with fixed values of the sample mean and variance.

[Section 2.5; Pace and Salvan, 1990]

2.11 With reference to Example 2.10, do detailed calculations in order to obtain the expectations given. Compare the two estimators in terms of their efficiency. (*Hint*: show that v has a beta distribution with parameters $n-1$ and 2, and that $Y_{(1)}$ given $V = v$ has a uniform distribution on $[\theta, \theta + 1 - v], \ldots$)

[Section 2.7]

2.12 With reference to Example 2.10, show that

$$(y_{(1)} - (1 - v)(1 - \alpha/2),\ y_{(1)} - (1 - v)\ \alpha/2)$$

is a confidence interval for θ whose conditional level given v is $1 - \alpha$. Interpret v as a measure of the information contained in the sample.

[Section 2.7]

2.13 With reference to Example 2.10, show that, for $n = 2$, the interval $(y_{(1)} - 0.7764, y_{(1)} - 0.0253)$ represents a confidence interval with unconditional level $1 - \alpha = 0.9$. Calculate the coverage probability (that is, the actual level) of such an interval conditionally on $v = 0.1, 0.2, \ldots, 0.9$ and on $v = 0.95$. Comment on the results.

[Section 2.7]

2.14 Let y_1, y_2 be independent observations from a Cauchy distribution with density

$$p_Y(y; \theta) = \frac{1}{\pi \{1 + (y - \theta)^2\}},$$

$-\infty < y < +\infty$, $-\infty < \theta < +\infty$. Show that $v = y_{(2)} - y_{(1)}$ is distribution constant. Show that the likelihood function is unimodal if $v \leq 2$ and bimodal if $v > 2$ and that only in the first case the maximum likelihood estimate of θ is $\hat{\theta} = (y_1 + y_2)/2$.

[Section 2.7]

2.15 Under the same assumptions as in Example 2.12, obtain the likelihood equation for ϕ and show that it has a unique solution in $(-1, 1)$, which corresponds to the m.l.e.. Check whether the distribution constant statistics a and b defined in the example are also ancillary.

[Sections 2.7 and 2.8]

2.16 Consider Examples 2.9 and 2.10 and check whether the distribution constant statistics identified there are ancillary.

[Section 2.8]

2.17 Using the same assumptions as in Example 2.8, obtain the maximum likelihood estimate of σ and prove its uniqueness. Determine an ancillary statistic a such that the minimal sufficient statistic may be written in the form $(\hat{\sigma}, a)$.

[Section 2.8]

2.18 Pivotal quantities are often used to construct confidence regions. Check whether such an approach is coherent with the principle of parameterization invariance.

[Sections 2.10 and 2.11]

2.19 Let $y_1, \ldots, y_n$ be independent observations from a distribution with density $p_Y(y; \theta)$, $y \in \mathcal{Y}$, $\theta \in \Theta \subseteq \mathbb{R}$. Denote by $l_*(\theta; y_i)$ the score for a single observation and by $i_1(\theta)$ the expected information for a single observation (assume that the usual regularity conditions are satisfied, see section 1.4.2). Prove that the quantity

$$\delta(\theta) = \frac{\sum_{i=1}^n (l_*(\theta; y_i))^2}{n i_1(\theta)}$$

is parameterization invariant (see also section 3.4.3).

[Section 2.11]

CHAPTER 3

SURVEY OF SOME BASIC CONCEPTS AND
TECHNIQUES

3.1 Introduction

In the previous chapter general principles for data and model reduction were
reviewed. Separating the informative portion of data from the uninformative
is the first step for solving inference problems. The next step is that of seeking
appropriate inference procedures and solving distribution problems (see section 1.3). This chapter reviews some basic notions that are useful when facing
these problems. First, the definitions and main properties of moments, cumulants and their generating functions are given in section 3.2. These tools make
it possible to deal with distribution problems in both exact and approximate
terms. The elementary asymptotic results and techniques reviewed in section
3.3 and in Appendix A are the basis for many approximations used in statistical inference, particularly for distributions of likelihood quantities: these are
set out in section 3.4. Section 3.5 offers a brief description of some of the main
results of the theory of inference according to the frequency-decision paradigm.
Some comments on the relations between the frequency-decision paradigm and
the Fisherian paradigm are in Appendix D. The empirical distribution function
plays a leading role in nonparametric problems, numerous inference procedures
refer to it as does the solution of distribution problems by means of the bootstrap: a brief description of this is offered in section 3.6. The contents of this
chapter should already be somewhat familiar to the reader. Here, the aim is
both to stimulate critical reappraisal and to provide a compendium of results
required later. Some readers may prefer to skip entire sections and just test
their knowledge on the final problems.

3.2   Moments, Cumulants and their Generating Functions

In many situations, obtaining the moments of a distribution may be easier than determining the distribution itself; think, for example, of the simple expressions of the mean and variance of the distribution of the sum of independent random variables in terms of the mean and variance of the basic random variables. Studying a distribution through its moments has the further advantage of supplying direct information about easily interpretable characteristics of the distribution. However, some moments may not be finite and, even when a distribution does have finite moments of any order, generally these do not uniquely identify the distribution (this is the classic moment problem, cf. Grimmet and Stirzaker, 1992, p. 166). One example of this is the log-normal distribution, see Problem 3.1.

There are some quantities related to the moments of a distribution which are more convenient to use. The most important of these are the characteristic function, the moment generating function, or Laplace transform, and the cumulant generating function. There are other generating functions, see Smith (1983). Characteristic functions and moment generating functions are one traditional chapter of probability theory and are used in most proofs of central limit theorems. Some notions about these topics are necessary for the theory of exponential families in Chapter 5 and for asymptotic methods, Chapters 8-11.

Some parametric models can be described more directly in terms of the characteristic function than by the probability density function. This is true for stable distributions, mentioned in section 1.3.2, which will be briefly discussed again here in section 3.2.6, together with some important generalizations.

3.2.1   The Moment Generating Function

Let Y be a one-dimensional random variable with density $p_Y(y)$ and let

$$\mu_r = E(Y^r), \qquad r = 1, 2, \ldots,$$

denote the moments of Y (about the origin). Writing, for simplicity, $\mu = \mu_1$, the central moments of Y are denoted by $\bar{\mu}_r = E(Y - \mu)^r$, $r = 2, 3, \ldots$. The moment generating function (m.g.f.) of Y is defined as

$$M_Y(t) = E(e^{tY}), \qquad t \in \mathbb{R}. \tag{3.1}$$

Observe that $M_Y(0) = 1$. The function $M_Y(t)$ is also called the Laplace transform of the measure on $\mathbb{R}$ which corresponds to the probability distribution of Y. The moment generating function takes positive values and is defined on an interval containing the origin. In fact $E\left(I_{[0,+\infty)}(Y)e^{tY}\right)$ is finite for every $t \leq 0$ and, if it is finite for $t = t_0 > 0$ then it is also finite for each $t < t_0$; a specular argument applies to $E\left(I_{(-\infty,0)}(Y)e^{tY}\right)$. It is not true, in general, that the moment generating function is finite for values of t in an *open* interval containing the origin; for instance, if Y has a log-normal distribution $M_Y(t)$ is not finite for $t > 0$ (see Problem 3.1). It may also be that $M_Y(t)$ is finite only for $t = 0$.

Example 3.1   *A distribution with $M_Y(t) < +\infty$ only for $t = 0$*
Let Y be a discrete random variable with values $\{\ldots, -2, -1, 1, 2, \ldots\}$, probability function $p_Y(j) = k/j^2$, where $k = 12/\pi^2$. Then

$$M_Y(t) = \sum_{j=-\infty}^{-1} e^{tj}k/j^2 + \sum_{j=1}^{+\infty} e^{tj}k/j^2 .$$

The first series on the right-hand side converges only for $t \geq 0$, while the second converges only for $t \leq 0$.   △

In the following, when we say that the moment generating function of Y exists, or that Y has moment generating function, we mean that $M_Y(t)$ is finite in an open interval containing the origin.

If Y is such that $M_Y(t) < +\infty$ for $|t| < t_0$, with $t_0 > 0$, then (see, for example, Billingsley, 1986, p. 285)

i) Y has finite moments of each order;

ii) the moment generating function of Y may be expanded as a power series with a radius of convergence $R \geq t_0$:

$$M_Y(t) = 1 + \mu_1 t + \mu_2 \frac{t^2}{2!} + \mu_3 \frac{t^3}{3!} + \mu_4 \frac{t^4}{4!} + \ldots . \qquad (3.2)$$

All the moments from the origin of Y can be deduced from the moment generating function $M_Y(t)$ through the relation

$$\mu_r = \frac{d^r}{dt^r} M_Y(t)\Big|_{t=0} .$$

The moment generating function characterizes the distribution of Y, in the sense that if $M_Y(t)$ is finite for $|t| < t_0$, with $t_0 > 0$, then both $M_Y(t)$ and the sequence of moments μ_r, $r = 1, 2, \ldots$, identify, uniquely, the distribution. Hence, the moment problem can be solved if the moment generating function of Y exists. See Table 3.1 for a summary of the moment generating functions of the main univariate distributions.

3.2.2   The Characteristic Function

If the real argument t in the definition of the moment generating function is replaced by the imaginary quantity it, where $i = \sqrt{-1}$, we have the **characteristic function** (c.f.) of Y

$$
\begin{aligned}
C_Y(t) &= E(e^{itY}) = \int_{-\infty}^{+\infty} e^{ity} p_Y(y) d\mu^* \\
&= \int_{-\infty}^{+\infty} \cos(ty) p_Y(y) d\mu^* + i \int_{-\infty}^{+\infty} \sin(ty) p_Y(y) d\mu^* .
\end{aligned}
\tag{3.3}
$$

If the density $p_Y(y)$ is symmetrical around zero, the second integral on the right-hand side of (3.3) is zero, so that $C_Y(t)$ is real-valued. Since e^{ity} is bounded, the characteristic function, unlike the moment generating function, is always finite and $C_Y(t)$ characterizes the distribution of any random variable Y. This is formalized in the following theorem, called the Lévy's inversion theorem.

Theorem 3.1 If the random variable Y has distribution function $F_Y(y)$ and characteristic function $C_Y(t)$ and if y_1 and y_2 are continuity points for $F_Y(y)$ then

$$
F_Y(y_2) - F_Y(y_1) = \lim_{T \to +\infty} \frac{1}{2\pi} \int_{-T}^{T} \frac{e^{-ity_1} - e^{-ity_2}}{it} C_Y(t) dt .
$$

Different distributions on $\mathbb{R}$ cannot have the same characteristic function.

Proof. See Billingsley (1986, p. 356).   □

Since there is a one-to-one correspondence between distribution functions and characteristic functions, it is important to have criteria which establish under which conditions a complex-valued function defined on $\mathbb{R}$ is a characteristic function. To this end, the following definitions are important. A complex-valued function, $h(\cdot)$, of a real variable is said to be **positive definite** if for every positive integer n, for every finite sequence of real numbers $t_1, \ldots, t_n$

Table 3.1: *Moment generating functions of some univariate distributions.*

Distribution	Probability density	$M_Y(t)$
Binomial: $Bi(n, \theta)$ $\theta \in (0, 1)$; $y = 0, 1, \ldots, n$	$\binom{n}{y} \theta^y (1 - \theta)^{n-y}$	$(\theta e^t + (1 - \theta))^n$, $t \in \mathbb{R}$
Poisson: $P(\theta)$ $\theta > 0$; $y = 0, 1, \ldots$	$e^{-\theta} \theta^y / y!$	$e^{\theta(e^t - 1)}$, $t \in \mathbb{R}$
Negative Binomial: $NB(\theta, r)$ $\theta \in (0, 1), r > 0$; $y = 0, 1, \ldots$	$\dfrac{\Gamma(r + y)}{\Gamma(r)\Gamma(y + 1)} \theta^r (1 - \theta)^y$	$\left\{ \dfrac{\theta}{1 - (1 - \theta)e^t} \right\}^r$, $1 - \theta < e^{-t}$
Normal: $N(\mu, \sigma^2)$ $\mu \in \mathbb{R}, \sigma^2 > 0$; $y \in \mathbb{R}$	$\dfrac{1}{\sqrt{2\pi}\,\sigma} e^{-\frac{1}{2\sigma^2}(y-\mu)^2}$	$e^{t\mu + \frac{t^2}{2}\sigma^2}$, $t \in \mathbb{R}$
Uniform: $U(a, b)$ $a, b \in \mathbb{R}, a < b$; $y \in [a, b]$	$\dfrac{1}{b - a}$	$\dfrac{e^{tb} - e^{ta}}{t(b - a)}$ if $t \neq 0$ 1 \quad\quad if $t = 0$
Gamma: $Ga(\nu, \phi)$ $\nu, \phi > 0$; $y > 0$	$\dfrac{y^{\nu-1} e^{-y/\phi}}{\phi^\nu \Gamma(\nu)}$	$(1 - \phi t)^{-\nu}$, $t < \dfrac{1}{\phi}$
Laplace: $La(\mu, \sigma)$ $\mu \in \mathbb{R}, \sigma > 0$; $y \subset \mathbb{R}$	$\dfrac{1}{\sigma} e^{-|y-\mu|/\sigma}$	$\dfrac{e^{t\mu}}{1 - (\sigma t)^2}$, $|t| < \dfrac{1}{\sigma}$
Logistic: $Lo(\mu, \sigma)$ $\mu \in \mathbb{R}, \sigma > 0$; $y \in \mathbb{R}$	$\dfrac{e^{-(y-\mu)/\sigma}}{\sigma \left\{1 + e^{-(y-\mu)/\sigma}\right\}^2}$	$e^{t\mu} \Gamma(1 - \sigma t) \Gamma(1 + \sigma t)$, $|t| < \dfrac{1}{\sigma}$
Inverse Gaussian: $IG(\phi, \lambda)$ $\phi, \lambda > 0$; $y > 0$	$\dfrac{\sqrt{\lambda}\, e^{-\frac{1}{2}\left(\frac{\lambda}{y} + \phi y\right) + \sqrt{\lambda\phi}}}{\sqrt{2\pi y^3}}$	$e^{\sqrt{\phi\lambda} - \sqrt{\lambda(\phi - 2t)}}$, $t < \dfrac{\phi}{2}$

and for every choice of n complex numbers $z_1, \ldots, z_n$,

$$\sum_{j=1}^{n} \sum_{k=1}^{n} z_j \bar{z}_k h(t_j - t_k) \geq 0 \,,$$

where $\bar{z}_k$ is the conjugate of z_k. A positive definite function $h(t)$ is said to be normalized if $h(0) = 1$.

Theorem 3.2 Let $h(t)$ be a complex-valued function on $\mathbb{R}$. Then $h(t)$ is a characteristic function if, and only if, it is continuous, positive definite and normalized.

Proof. See Laha and Rohatgi (1979, p. 162). □

For other characterization results for characteristic functions, see Laha and Rohatgi (1979, section 3.6).

The inversion Theorem 3.1 holds for both continuous and discrete random variables. If, however, the characteristic function is absolutely integrable, that is if

$$\int_{-\infty}^{+\infty} |C_Y(t)| dt < +\infty \,, \tag{3.4}$$

then Fourier's inversion theorem (see, for example, Billingsley, 1986, formula (26.60)) states that Y is continuous with density

$$p_Y(y) = \frac{1}{2\pi} \int_{-\infty}^{+\infty} e^{-ity} C_Y(t) dt \,. \tag{3.5}$$

Naturally, if Y has moment generating function, the preceding inversion results can be directly applied to the moment generating function calculated in it and, hence, the distribution with an assigned moment generating function is uniquely determined. Indeed, $M_Y(t)$ uniquely determines the distribution of Y provided it is finite for $t \in [0, t_0)$ or for $t \in (-t_0, 0]$ with $t_0 > 0$ (see e.g. Billingsley, 1986, Problem 26.7).

3.2.3 Convergence Results

There is a correspondence between convergence in distribution of a sequence of random variables and pointwise convergence of the sequence of their moment generating functions. More precisely, if $\{Y_n\}$ is a sequence of random variables with a corresponding sequence of moment generating functions $\{M_{Y_n}(t)\}$, and if Y is a random variable with moment generating function $M_Y(t)$, then:

i) if $M_{Y_n}(t) < c < +\infty$ for $|t| < t_0$, where c is a constant not dependent on n, $t_0 > 0$, and if the sequence $\{Y_n\}$ converges in distribution to Y then, $\lim_{n \to +\infty} M_{Y_n}(t) = M_Y(t)$ for $|t| < t_0$;

ii) if for $n \to +\infty$, $M_{Y_n}(t)$ converges to a moment generating function $M_Y(t)$ for $|t| < t_0$, $t_0 > 0$, then the sequence $\{Y_n\}$ converges in distribution to the random variable Y having $M_Y(t)$ as its moment generating function;

iii) under the same assumptions either as in i) or as in ii), the r-th moment of Y_n converges to the r-th moment of Y, $r = 1, 2, \ldots$

(see, e.g. Grimmet and Stirzaker, 1992, sections 5.9 and 7.10; Billingsley, 1986, section 30).

3.2.4 Generating Functions for Sums

The moment generating function $M_{S_n}(t)$ of $S_n = \Sigma Y_i$, where the Y_i are n independent copies of Y, can be fairly simply expressed as a function of $M_Y(t)$. Since the components of $(Y_1, \ldots, Y_n)$ are independent and identically distributed we immediately obtain

$$M_{S_n}(t) = (M_Y(t))^n \tag{3.6}$$

and

$$C_{S_n}(t) = (C_Y(t))^n . \tag{3.7}$$

This, together with the limiting results mentioned above, renders the moment generating function and the characteristic function useful tools for studying exact and asymptotic distributions of sums of independent random variables; see Chapter 10.

From (3.6) it is possible to obtain the moments of S_n in terms of the moments of Y. For example, the first four moments of S_n can be obtained from the expansion

$$
\begin{aligned}
M_{S_n}(t) &= \left(1 + \mu_1 t + \mu_2 \frac{t^2}{2!} + \mu_3 \frac{t^3}{3!} + \mu_4 \frac{t^4}{4!} + \ldots\right)^n \\
&= 1 + n\left(\mu_1 t + \mu_2 \frac{t^2}{2!} + \mu_3 \frac{t^3}{3!} + \mu_4 \frac{t^4}{4!}\right) \\
&\quad + \binom{n}{2}\left(\mu_1 t + \mu_2 \frac{t^2}{2!} + \mu_3 \frac{t^3}{3!}\right)^2
\end{aligned}
$$

$$+ \binom{n}{3} \left(\mu_1 t + \mu_2 \frac{t^2}{2!} \right)^3 + \binom{n}{4} \mu_1^4 t^4 + \dots \,, \qquad (3.8)$$

as the coefficients of $t^j/j!$, $j = 1, \dots, 4$ (Problem 3.4).

3.2.5   The Cumulant Generating Function

Let $M_Y(t)$ be finite for $|t| < t_0$; the cumulant **generating function** (c.g.f.) of Y is defined as

$$K_Y(t) = \log M_Y(t) \,. \qquad (3.9)$$

Like $M_Y(t)$, $K_Y(t)$ uniquely determines the distribution of Y. The function $K_Y(t)$ can be expanded into a power series, with the same radius of convergence as in (3.2), as

$$K_Y(t) = \kappa_1 t + \kappa_2 \frac{t^2}{2!} + \kappa_3 \frac{t^3}{3!} + \kappa_4 \frac{t^4}{4!} + \dots \,.$$

The coefficient κ_r of $t^r/r!$ in this expansion is called the **cumulant of order r** of Y. Clearly,

$$\kappa_r = \kappa_r(Y) = \left. \frac{d^r}{dt^r} K_Y(t) \right|_{t=0} . \qquad (3.10)$$

From (3.6), the cumulant generating function of S_n is

$$K_{S_n}(t) = n K_Y(t) \,, \qquad (3.11)$$

giving the simple relation

$$\kappa_r(S_n) = n \kappa_r(Y) = n \kappa_r \,. \qquad (3.12)$$

This motivates the usefulness of cumulants when dealing with sums of independent random variables.

Example 3.2 *Normal distribution*

If $Y \sim N(\mu, \sigma^2)$, then $M_Y(t) = e^{\mu t + \sigma^2 \frac{t^2}{2}}$, and the c.g.f. of $N(\mu, \sigma^2)$ is

$$K_Y(t) = \mu t + \sigma^2 \frac{t^2}{2} \,,$$

hence, $\kappa_1 = \mu$, $\kappa_2 = \sigma^2$, $\kappa_r = 0$ for $r = 3, 4, \dots$. Cumulants of order higher than two are all zero if, and only if, Y has a normal distribution. $\triangle$

A location change from Y to $Y + a$, affects the cumulant generating function of Y simply through the addition of a linear term, since

$$M_{Y+a}(t) = E\left(e^{t(Y+a)}\right) = e^{at} M_Y(t)$$

so that

$$K_{Y+a}(t) = at + K_Y(t) . \tag{3.13}$$

Only the first cumulant is affected by a location change: $\kappa_1(Y + a) = a + \kappa_1$. As a consequence of (3.13), the first cumulant is a function of the moments from the origin, while higher-order cumulants can be expressed in terms of central moments. This is done explicitly below for the first four cumulants. Using the expansion

$$\log(1 + z) = z - \frac{z^2}{2} + \frac{z^3}{3} - \frac{z^4}{4} + \cdots \qquad |z| < 1$$

in (3.2) and ignoring powers of t higher than four, we get

$$
\begin{aligned}
K_Y(t) &= \mu_1 t + \mu_2 \frac{t^2}{2!} + \mu_3 \frac{t^3}{3!} + \mu_4 \frac{t^4}{4!} - \frac{1}{2}\left(\mu_1 t + \mu_2 \frac{t^2}{2!} + \mu_3 \frac{t^3}{3!}\right)^2 \\
&\quad + \frac{1}{3}\left(\mu_1 t + \mu_2 \frac{t^2}{2!}\right)^3 - \frac{1}{4}\mu_1^4 t^4 + \cdots \\
&= \mu_1 t + \mu_2 \frac{t^2}{2!} + \mu_3 \frac{t^3}{3!} + \mu_4 \frac{t^4}{4!} \\
&\quad - \frac{1}{2}\left(\mu_1^2 t^2 + \frac{1}{4}\mu_2^2 t^4 + \mu_1 \mu_2 t^3 + \frac{1}{3}\mu_1 \mu_3 t^4\right) \\
&\quad + \frac{1}{3}\left(\mu_1^3 t^3 + \frac{3}{2}\mu_1^2 \mu_2 t^4\right) - \frac{1}{4}\mu_1^4 t^4 + \cdots \\
&= \mu_1 t + (\mu_2 - \mu_1^2)\frac{t^2}{2!} + (\mu_3 - 3\mu_1 \mu_2 + 2\mu_1^3)\frac{t^3}{3!} \\
&\quad + (\mu_4 - 3\mu_2^2 - 4\mu_1 \mu_3 + 12\mu_1^2 \mu_2 - 6\mu_1^4)\frac{t^4}{4!} + \cdots .
\end{aligned}
$$

Therefore

$$
\begin{aligned}
\kappa_1 &= \mu_1 & (3.14) \\
\kappa_2 &= \mu_2 - \mu_1^2 = \bar{\mu}_2 = \text{Var}(Y) & (3.15) \\
\kappa_3 &= \mu_3 - 3\mu_1 \mu_2 + 2\mu_1^3 = E(Y - \mu_1)^3 = \bar{\mu}_3 & (3.16) \\
\kappa_4 &= \mu_4 - 3\mu_2^2 - 4\mu_1 \mu_3 + 12\mu_1^2 \mu_2 - 6\mu_1^4 \\
&= E(Y - \mu_1)^4 - 3(E(Y - \mu_1)^2)^2 = \bar{\mu}_4 - 3\bar{\mu}_2^2 . & (3.17)
\end{aligned}
$$

The expression of κ_3 and κ_4 in terms of central moments is obtained exploiting location invariance and supposing $\mu_1 = 0$.

Now let us consider a scale change Y/b, with $b > 0$. Since

$$M_{Y/b}(t) = E\left(e^{tY/b}\right) = M_Y(t/b),$$

it follows that

$$K_{Y/b}(t) = K_Y(t/b), \tag{3.18}$$

and

$$\kappa_r(Y/b) = \kappa_r(Y)/b^r = \kappa_r/b^r. \tag{3.19}$$

All the cumulants are affected by a scale change (with $b \neq 1$), the greater the change the higher the order of the cumulant is. The first four cumulants are descriptive measures, of location, of variability, of asymmetry, or skewness, and of kurtosis, respectively; see (3.14)-(3.17). Higher-order cumulants are further measures of non-normality.

It is useful to define the **standardized cumulants**

$$\rho_r = \kappa_r/\kappa_2^{r/2}, \qquad r = 3, 4, \ldots, \tag{3.20}$$

which are invariant under linear transformations of Y. The quantities ρ_3 and ρ_4 are often found in the literature with the alternative notation γ_1 and γ_2, the well-known adimensional indices of skewness and kurtosis. The range of ρ_3 and ρ_4 is constrained by the inequality $\rho_4 \geq \rho_3^2 - 2$ (see, e.g. Barndorff-Nielsen and Cox, 1989, Exercise 1.2).

3.2.6 Infinitely Divisible, Stable, Selfdecomposable Laws

If the characteristic function of a random variable Y can be represented in the form

$$C_Y(t) = (C_n(t))^n \qquad \text{for every } n \in \mathbb{N} \setminus \{0\},$$

where $C_n(t)$ is a characteristic function, the distribution of Y is said to be infinitely divisible and, for each $n \in \mathbb{N} \setminus \{0\}$, Y can be represented as the sum of n identically distributed independent random variables. The Poisson, the normal and the gamma distributions are all examples of infinitely divisible distributions.

One important subclass of the infinitely divisible distributions is that of stable distributions. A non-degenerate random variable Y has a **stable distribution** if, for $Y_1, \ldots, Y_n$ independent copies of Y and $S_n = \sum_{i=1}^{n} Y_i$, two sequences of constants a_n and b_n exist such that $S_n/b_n - a_n$ has the same distribution as Y. It can be shown that b_n must be of form $b_n = n^{1/\theta}b_0(n)$, with θ called the characteristic exponent, $0 < \theta \le 2$, and $b_0(n)$ a slowly varying function. A positive function $b_0(\cdot)$ defined on $[0, +\infty)$ is slowly varying if, for each $k > 0$, $\lim_{x \to +\infty} b_0(kx)/b_0(x) = 1$. A normal distribution is stable with $\theta = 2$, a Cauchy distribution is stable with $\theta = 1$. A stable law with $\theta < 2$ can only have finite moments of order less than θ. Stable laws make up a class of possible limit laws for the distribution of suitably standardized sums of identically distributed and independent random variables. If $\theta < 2$ the distribution has infinite variance (see Theorem A.5). Stable laws have characteristic functions of the form

$$C_Y(t) = \exp\{ita - b|t|^\theta(1 + i\gamma\,\mathrm{sgn}(t)\omega(t, \theta))\}\,,$$

with $|\gamma| \le 1$ and $\omega(t, \theta) = \tan(\pi\theta/2)$ if $\theta \ne 1$, $\omega(t, \theta) = (2/\pi)\log|t|$ if $\theta = 1$. The quantities a and b are location and scale parameters, while γ is an asymmetry parameter; when $\gamma = 0$ the distribution is symmetric. If the distribution of Y is symmetric around zero, then $a = 0$ and

$$C_Y(t) = e^{-b|t|^\theta}\,.$$

The density of a stable law can be expressed in an explicit form only for θ equal to 1, 1/2 and 2. When $a = \gamma = 0$ and $b = 1$ the value $\theta = 1/2$ corresponds to the density

$$p_Y(y) = \frac{1}{\sqrt{2\pi}}y^{-3/2}e^{-\frac{1}{2y}}\,. \tag{3.21}$$

A stable distribution with a characteristic exponent $0 < \theta < 2$ satisfies (B.6), consequently all the stable distributions with $0 < \theta < 2$ are in the domain of attraction of the extreme value distribution Λ_1 (see Resnick, 1987, p. 13). For further information about stable laws and their applications in statistics, see Monrad and Stout (1988).

A generalization of stable laws, still in the class infinitely divisible distributions, which suggests models that are particularly interesting for applications, is the subclass of **selfdecomposable laws** (Feller, 1971, p. 588). These arise as the possible limit distributions of $S_n/b_n - a_n$ when $Y_1, \ldots, Y_n$ are independent but not necessarily identically distributed and $b_n \to +\infty$, $b_{n+1}/b_n \to 1$. The term *selfdecomposable* is used because of the following characterization property. The distribution of the random variable Y is selfdecomposable if, and only if, for each $\rho \in (0, 1)$ there exists a characteristic function $C_\rho(t)$ such that

$$C_Y(t) = C_Y(\rho t)C_\rho(t)\,.$$

As an example, consider a stationary first-order autoregressive process $Y_n = \rho Y_{n-1} + U_n$, with $|\rho| < 1$ and the U_i independent and identically distributed. Since $\{Y_n\}$ is stationary, $C_{Y_n}(t) = C_{Y_{n-1}}(t)$. From the independence of Y_{n-1} and U_n, it follows that $C_{Y_n}(t) = C_{Y_n}(\rho t) C_{U_n}(t)$. Therefore, if the one-dimensional marginal distribution of $\{Y_n\}$ does not depend on ρ it is selfdecomposable. This property is useful in applications because it enables one to select an appropriate parametric model that fits the one-dimensional marginal distribution of the observations. See Barndorff-Nielsen, Jensen and Sørensen (1995) for further applications of selfdecomposability to stationary processes.

3.2.7 Multivariate Extensions

The above definitions and results can be extended to d-dimensional random variables, with $d > 1$. Let $Y = (Y^1, \ldots, Y^d)$ be a d-dimensional random vector. A generic moment (about the origin) of order r of Y is given by

$$\mu^{i_1 \cdots i_r} = E(Y^{i_1} \cdots Y^{i_r}) \tag{3.22}$$

for $i_1, \ldots, i_r = 1, \ldots, d$. In particular, $\mu^i = E(Y^i)$, $\mu^{ij} = E(Y^i Y^j)$. The moments of order r form an array, which is symmetrical with respect to permutations of the indices.

The moment generating function of Y is defined as

$$M_Y(t) = E(e^{t \cdot Y}) = E(e^{t_1 Y^1 + \cdots + t_d Y^d}), \qquad t = (t_1, \ldots, t_d) \in \mathbb{R}^d . \tag{3.23}$$

The moment generating function of Y will be said to exist if (3.23) is finite in a neighbourhood of the origin of $\mathbb{R}^d$, $\|t\| < t_0$, $t_0 > 0$. The moment generating function can thus be expanded into a multivariate power series with convergence radius $R \geq t_0$

$$M_Y(t) = 1 + \sum_{i=1}^{d} \mu^i t_i + \frac{1}{2} \sum_{i,j=1}^{d} \mu^{ij} t_i t_j + \ldots$$

and

$$\mu^{i_1 \cdots i_r} = \left. \frac{\partial^r M_Y(t)}{\partial t_{i_1} \cdots \partial t_{i_r}} \right|_{t=0} . \tag{3.24}$$

If the moment generating function exists, it characterizes a multivariate distribution in the same way as it does a univariate distribution. Thus the factorization

$$M_Y(t) = M_{Y^1}(t_1) \cdots M_{Y^d}(t_d) \tag{3.25}$$

holds if, and only if, the components of Y are independent.

As in the univariate case, the cumulant generating function is defined as

$$K_Y(t) = \log M_Y(t) \ .$$

If $M_Y(t)$ exists, the cumulant generating function can be expanded into a multivariate Taylor series in a neighbourhood of the origin and the coefficients of this expansion will define the cumulants of Y. The generic joint cumulant of order r is

$$\kappa^{i_1, i_2, \ldots, i_r} = \left. \frac{\partial^r K_Y(t)}{\partial t_{i_1} \partial t_{i_2} \cdots \partial t_{i_r}} \right|_{t=0} . \tag{3.26}$$

In the notation for cumulants notice the use of commas which separate indices: commas are absent in the symbol for moments. Multivariate notations (3.22) and (3.26) for moments and cumulants are not consistent with those used in the univariate case but they will make dealing with certain points in the following far simpler (for example, in Chapter 10). However, the multivariate notation would be rather clumsy for the univariate case; for instance, the third and fourth cumulant of a scalar Y would be denoted as $\kappa^{1,1,1}$ and $\kappa^{1,1,1,1}$.

The relations (3.14)-(3.17), which link moments and cumulants, can be extended to the multivariate case. In particular, $\kappa^i = \mu^i$ and

$$\kappa^{i,j} = \mu^{ij} - \mu^i \mu^j = E(Y^i Y^j) - E(Y^i)E(Y^j) \ . \tag{3.27}$$

Below, the covariance matrix $[\kappa^{i,j}]$ will be indicated by κ. Furthermore, the notation $[\kappa_{ij}]$ will be used to denote the inverse matrix $[\kappa^{i,j}]^{-1}$. The identities that link the arrays of third and fourth order cumulants to central moments are

$$\begin{aligned}
\kappa^{i,j,k} &= \mu^{ijk} - \mu^i \mu^{jk}[3] + 2\mu^i \mu^j \mu^k \\
&= E\left\{ (Y^i - \mu^i)(Y^j - \mu^j)(Y^k - \mu^k) \right\} ,
\end{aligned} \tag{3.28}$$

$$\begin{aligned}
\kappa^{i,j,k,l} &= \mu^{ijkl} - \mu^i \mu^{jkl}[4] - \mu^{ij} \mu^{kl}[3] \\
&\quad + 2\mu^i \mu^j \mu^{kl}[6] - 6\mu^i \mu^j \mu^k \mu^l \\
&= E\left\{ (Y^i - \mu^i)(Y^j - \mu^j)(Y^k - \mu^k)(Y^l - \mu^l) \right\} \\
&\quad - E\left\{ (Y^i - \mu^i)(Y^j - \mu^j) \right\} E\left\{ (Y^k - \mu^k)(Y^l - \mu^l) \right\} [3] , \tag{3.29}
\end{aligned}$$

where the symbol $[k]$ indicates the sum of k similar terms obtained by all suitable permutations of the indices (except for the obvious symmetry relations). For example,

$$\mu^i \mu^{jk}[3] = \mu^i \mu^{jk} + \mu^j \mu^{ik} + \mu^k \mu^{ij} .$$

For the relation that links a cumulant of a generic order to the moments from the origin, see formula (9.29).

Example 3.3 *Multivariate normal distribution*
If Y has a d-variate normal distribution, $N_d([\kappa^i], [\kappa^{i,j}])$, with p.d.f.

$$p_Y(y; [\kappa^i], \kappa) = (2\pi)^{-d/2} \left| [\kappa^{i,j}] \right|^{-1/2} \exp\left\{ -\frac{1}{2} \sum_{i,j=1}^{d} (Y^i - \kappa^i)(Y^j - \kappa^j)\kappa_{ij} \right\},$$

the c.g.f. is

$$K_Y(t) = \sum_{i=1}^{d} t_i \kappa^i + \frac{1}{2} \sum_{i,j=1}^{d} t_i t_j \kappa^{i,j}$$

and, consequently, all the cumulants of order $r > 2$ are zero. This property characterizes the multivariate normal distribution.   △

In the following, the density of $Y \sim N_d(0, \kappa)$ will be denoted by $\phi_d(y; \kappa)$.

3.3 Basic Notions of Asymptotic Methods

Many inference procedures are based on asymptotic arguments. For instance, asymptotic theory can give useful indications for model specification (see section 1.3.2). The reader is probably, at least to some extent, already familiar with the use of asymptotic methods for finding optimal statistical procedures (e.g. locally most powerful tests, see section 3.5.3) and, also, for dealing with distribution problems.

Often a mathematical problem becomes simpler when a variable or a parameter takes a large value or is close to a particular value, and the limiting result can be used as an approximate solution to the original problem. In some cases it is possible to express analytically, through inequalities, how close the approximate solution is to the real situation; more often, only a qualitative study is possible. For example, when x is close to zero the function $\sin x$ is well approximated by x, the error being negligible when compared with x, as can be deduced from $\lim_{x \to 0} \sin x/x = 1$; the explicit upper bound of the error given by $|x^3|/6$ is well-known in this case. As another example, the binomial distribution $Bi(n,p)$, can be approximated by the normal distribution $N(np, np(1-p))$, in the sense that the corresponding distribution functions are close when n is sufficiently large. However finding a useful upper bound for the error is not simple here (see section 8.2.1).

3.3.1 Orders of Magnitude of Sequences

A qualitative study of approximation results is made easier by the notation $O(\cdot)$ and $o(\cdot)$ for orders of magnitude of errors. These symbols are used to denote asymptotic relationships between the quantity being studied and reference quantities. First, consider the definition of the symbols $o(n^\alpha)$, $O(n^\alpha)$ for sequences of real numbers.

A sequence of real numbers a_n is said to be **asymptotically of order o(n^α)** if

$$\lim_{n\to+\infty} \frac{a_n}{n^\alpha} = 0 .$$

When referring to this and other notions of asymptotic order, the adverb *asymptotically* is often omitted. If a_n converges to zero, as does, for example, the sequence n^{-1}, it is asymptotically of order $o(1)$. Note that a_n does not have to converge if it is $o(n^\alpha)$ with $\alpha > 0$. Furthermore, the definition of $o(n^\alpha)$ establishes an ordering which is not necessarily the *closest* possible: we could equally well say that n^{-1} is either of order $o(n)$ or of order $o(n^{-1/2})$. The most informative order $o(n^\alpha)$ is given by the smallest power α for which a_n is asymptotically of order $o(n^\alpha)$. Reference sequences often considered are integer powers of $\sqrt{n}$. For instance, $o(n^{-1/2})$ is the most accurate order in terms of integer powers of $\sqrt{n}$ for $\sin(1/n)$, which is not of order $o(n^{-1})$.

A sequence of real numbers a_n is said to be **asymptotically of order O(n^α)** if the sequence a_n/n^α is bounded, that is, if there exists a real number A such that for every $n \in \mathbb{N}$

$$\left| \frac{a_n}{n^\alpha} \right| < A .$$

Note that this notion requires neither the convergence of a_n nor that of a_n/n^α. For example the sequence $(-1)^n$ is of order $O(1)$; $\log(1 + 1/n)$ is of order $O(n^{-1})$. If a sequence a_n converges to a non-zero finite limit, it is asymptotically of order $O(1)$ but not of order $o(1)$; furthermore, such a sequence a_n is both of order $O(n)$ and of order $o(n)$. Consequently, $O(1)$ is the most informative asymptotic order for converging sequences that do not converge to zero. Lastly, note that if a_n is of order $o(n^\alpha)$ it is also, obviously, of order $O(n^\alpha)$.

When considering the order of a sequence of random variables, first of all we have to specify which notion of convergence is used. In most statistical applications, comparisons are done with reference to convergence in probability. It is also possible to define almost sure orders of magnitude.

A sequence of random variables Y_n is said to be **asymptotically of order o(n^α) in probability**, and this is indicated with the symbol $o_p(n^\alpha)$, if Y_n/n^α converges in probability to zero, that is, if for every $\delta > 0$,

$$\lim_{n \to +\infty} P\left\{\left|\frac{Y_n}{n^\alpha}\right| > \delta\right\} = 0 \ .$$

For example, $Y_n \sim N(0, 1/n)$ is of order $o_p(1)$, while it is not of order $o_p(n^{-1/2})$. If $Y_n \sim \chi_n^2$ (chi-squared with n degrees of freedom), Y_n is of order $o_p(n^{3/2})$ but not $o_p(n)$.

A sequence of random variables Y_n is said to be **asymptotically of order** $O(n^\alpha)$ **in probability**, and this is indicated by the symbol $O_p(n^\alpha)$, if, for each $\varepsilon > 0$ there exist a real number $A = A_\varepsilon > 0$ and a natural number $\bar{n} = \bar{n}_\varepsilon$ such that for every $n \geq \bar{n}$

$$P\left\{\left|\frac{Y_n}{n^\alpha}\right| < A\right\} > 1 - \varepsilon \ .$$

A sequence Y_n of order $O_p(1)$ is said to be **bounded in probability**. If $Y_n \xrightarrow{d} Y$, then Y_n is bounded in probability. Therefore in full agreement with the situation for sequences of real numbers, we say that Y_n is $O_p(n^\alpha)$ if Y_n/n^α is bounded in probability. For example, $Y_n \sim N(0, n^{-1})$ is $O_p(n^{-1/2})$ and $Y_n \sim \chi_n^2$ is $O_p(n)$.

A sequence of random variables Y_n is said to be **asymptotically of order** $o(n^\alpha)$ **almost surely**, and this is denoted by the symbol $o_{as}(n^\alpha)$, if Y_n/n^α converges to zero almost surely, that is if

$$\frac{Y_n}{n^\alpha} \xrightarrow{as} 0 \ .$$

Analogously, a sequence Y_n is said to be **asymptotically of order** $O(n^\alpha)$ **almost surely**, and this is denoted by $O_{as}(n^\alpha)$, if Y_n/n^α is almost surely bounded, that is, if for every $\varepsilon > 0$ there exist a constant $A = A_\varepsilon > 0$ and a natural $\bar{n} = \bar{n}_\varepsilon$ such that

$$P\left\{\left|\frac{Y_n}{n^\alpha}\right| > A, \quad \text{for every } n \geq \bar{n}\right\} < \varepsilon \ .$$

A sequence of random variables Y_n whose supports $\mathcal{Y}_n$ are all included in a given bounded set is of order $O_{as}(1)$; if $\mathcal{Y}_m \subseteq \mathcal{Y}_n$ for every $m < n$, and $\bigcap \mathcal{Y}_n = \{0\}$, then Y_n is of order $o_{as}(1)$.

3.3.2   Convergence of Sums and Extremes

Many inference procedures are based upon statistics that can be expressed, either exactly or approximately, as the sum of n random variables, which are independent and identically distributed in the simplest cases. Limit theorems, as $n \to +\infty$, of probability theory are a rich source of asymptotic results on the distribution of appropriately standardized sums of random variables. Frequently, in statistical inference, such results are used to obtain an approximate solution for distribution problems. From this point of view, the most important limit theorems are the laws of large numbers and central limit theorems. Sometimes the laws of the iterated logarithm are useful too. These topics are usually covered in probability textbooks (see the bibliographic note in section 3.7). For ease of reference, the main results are listed in Appendix A.

It may prove useful in some situations to have approximations for the distribution of either statistics or combinants based on the maximum (or minimum) value of a vector of independent and identically distributed observations. The main results, already mentioned in section 1.3.2, are collected in Appendix B.

3.3.3 Orders in Probability: Examples

In the following it will often be useful to be able to recognize asymptotic orders of magnitude at first glance, especially those of functions of sums and of sample means, summarized in the following examples.

Example 3.4 *Sum of n i.i.d. random variables: orders in probability*
Let $S_n = \sum_{i=1}^{n} Y_i$ with the Y_i independent and identically distributed.

i) If $0 < \mathrm{Var}(Y_i) < +\infty$ and $E(Y_i) = 0$, then S_n is of order $O_p(n^{1/2})$, that is, of the order of the standard deviation of S_n.

ii) If $\mathrm{Var}(Y_i)$ is finite and $E(Y_i) \neq 0$, then S_n is $O_p(n)$, that is, of the order of the expectation of S_n.

iii) If $\mathrm{Var}(Y_i)$ is not finite and Y_i is in the domain of attraction of a symmetric stable law with characteristic exponent θ, then S_n is of order $O_p(n^{1/\theta})$. For example, if Y_i has a Cauchy distribution, then S_n is of order $O_p(n)$.

$\triangle$

Example 3.5 *Mean of n i.i.d. random variables: orders in probability*
Let $Y_i, i = 1, \ldots, n$, be independent r.v.'s with $E(Y_i) = 0$, $\mathrm{Var}(Y_i) = 1$; consider the sample mean $\bar{Y}_n = n^{-1} \sum Y_i$. By the weak law of large numbers $\bar{Y}_n \xrightarrow{P} 0$, so that $\bar{Y}_n$ is bounded in probability, that is, $\bar{Y}_n$ is of order $O_p(1)$. On the other hand, $\sqrt{n}\, \bar{Y}_n$ does not converge in probability to a constant; however, by the central limit theorem it does converge in distribution, thus it is bounded in probability, i.e. $\bar{Y}_n$ is of order $O_p(n^{-1/2})$. $\triangle$

Example 3.6 *Mean of n i.i.d. random variables: a.s. orders of magnitude*
With the same assumptions as in Example 3.5, by the strong law of large numbers $\bar{Y}_n \overset{as}{\to} 0$, so that $\bar{Y}_n$ is of order $o_{as}(1)$; it is also, therefore, of order $O_{as}(1)$, that is, it is bounded almost surely. On the other hand, $\sqrt{n}\,\bar{Y}_n$ converges in distribution to a non-degenerate law, and it is not almost surely bounded, $\bar{Y}_n$ is not $O_{as}(n^{-1/2})$. The most informative almost sure order for $\bar{Y}_n$ will, therefore, fall somewhere between $O_{as}(1)$ and $O_{as}(n^{-1/2})$. Since according to the law of the iterated logarithm (Theorem A.7) $\sqrt{n/(2\log\log n)}\,\bar{Y}_n$ is almost surely bounded, $\bar{Y}_n$ is of order $O_{as}\left(\sqrt{(\log\log n)/n}\right)$. $\triangle$

3.4   Likelihood and First-Order Asymptotic Theory

One of the reasons for the popularity of likelihood methods is that simple and general approximations for sampling distributions of likelihood quantities are available. These approximations are based on limit theorems of probability theory, which provide valid results as the sample size n diverges, that is, as $n \to +\infty$. The asymptotic parameter n should be regarded as an index of the *quantity of information* supplied by the sample. Clearly, the accumulation of information renders inference easier. For the statistician, n going to infinity is simply a device used to obtain an approximation which will, however, be applied for the value of n that is really available. The adequacy of the approximation should be judged for this specific sample size, for example, through simulation.

Below we will shortly review the basic limiting results for the usual likelihood quantities. These make up what is usually called first-order asymptotic likelihood theory so as to distinguish it from the body of higher-order results which will be dealt with in the final chapters of this book. The likelihood quantities considered below are: the score function $l_*(\theta)$, the maximum likelihood estimator, $\hat{\theta}$, the log-likelihood ratio combinant,

$$W = 2(l(\hat{\theta}) - l(\theta)) \tag{3.30}$$

and its two other versions which are asymptotically equivalent to the first order (see, for example, Azzalini, 1996, section 4.2.2), i.e. the score test

$$W_u = l_*(\theta)^{\mathrm{T}} i(\theta)^{-1} l_*(\theta) \tag{3.31}$$

and the Wald test

$$W_e = (\hat{\theta} - \theta)^{\mathrm{T}} i(\theta)(\hat{\theta} - \theta). \tag{3.32}$$

The score test W_u is also known as the Rao test or the Lagrange multipliers test. Both above and in the following, l_*, $\hat{\theta} - \theta$, etc. will be treated as column vectors when involved in matrix operations.

In the following, we will assume that the reference model is a regular parametric statistical model, $\mathcal{F} = \{p_Y(y; \theta), \theta \in \Theta \subseteq \mathbb{R}^p\}$, that is that $\mathcal{F}$ satisfies the following assumptions. The support of Y does not depend on θ. The identifiability condition is satisfied and $p^0(y) = p(y; \theta_0)$, with θ_0 an inner point of Θ. The log likelihood function can be expanded in a Taylor series, in a neighbourhood of θ_0, up to the second order and with a remainder term whose absolute value is uniformly bounded (hence, $l(\cdot)$ has derivatives up to the third order with the absolute value of the third derivatives absolutely bounded). The null expectation of $l(\cdot)$ and of its derivatives up to the third order is finite. In particular, it is equal to 0 for the score and the information identity (1.9) holds with $i(\theta_0)$ positive definite; furthermore, the function of Y which bounds the absolute value of the third derivative of $l(\cdot)$ has finite null expectation.

Lastly, we assume throughout that $\hat{\theta}$ is well defined and consistent. Consistency of the maximum likelihood estimator (even strong consistency) can be obtained without assuming differentiability of the log-likelihood function. If the parameter space Θ is compact and the log-likelihood function is continuous, consistency is proved through the Wald inequality (1.7), see Wald (1949). The case of unbounded parameter space can be dealt with by using further assumptions about the behaviour of the log-likelihood as the norm of θ diverges.

3.4.1 Null Asymptotic Distributions

The asymptotic results given below generally require that there should be a *large* amount of information available, in the sense that $i(\theta) = O(n)$. In particular, this condition is satisfied if the data y are made up of n independent observations, hence, $i(\theta) = n i_1(\theta)$, where $i_1(\theta)$ indicates Fisher information for a single observation. When the observations are independent, but not identically distributed, then $i_1(\theta)$ is defined as the average information in a limiting sense, as $i_1(\theta) = \lim_{n \to +\infty} i(\theta)/n$.

Under random sampling of size n, by the central limit theorem,

$$\frac{l_*(\theta)}{\sqrt{n}} \xrightarrow{d} N_p(0, i_1(\theta)) , \qquad (3.33)$$

so that the null distribution of the score is approximately normal.

For simplicity, consider the one-parameter case $p = 1$ and assume that $\hat{\theta}$ is a consistent solution of the likelihood equation $l_*(\theta) = 0$, i.e. $\hat{\theta}$ is such that $l_*(\hat{\theta}) = 0$ and $\hat{\theta} - \theta = o_p(1)$. The starting point to establish first-order asymptotic properties for $\hat{\theta}$ and other likelihood quantities is the expansion of the score function

$$0 = l_*(\hat{\theta}) = l_*(\theta) + (\hat{\theta} - \theta)l_2(\theta) + \frac{1}{2}(\hat{\theta} - \theta)^2 l_3(\tilde{\theta}) , \qquad (3.34)$$

where $|\tilde{\theta} - \theta| < |\hat{\theta} - \theta|$ and, with a notation which is more convenient for the one-parameter case, $l_m = (\partial/\partial\theta^m)l(\theta)$, $m = 2, 3, \ldots$; note that $l_2(\theta) = -j(\theta)$. Expansion (3.34) gives

$$\frac{l_*(\theta)}{\sqrt{n}} = \sqrt{n}(\hat{\theta} - \theta)\left(\frac{j(\theta)}{n} - \frac{1}{2}(\hat{\theta} - \theta)\frac{l_3(\tilde{\theta})}{n}\right) . \qquad (3.35)$$

If, by a law of large numbers,

$$\frac{j(\theta)}{n} = i_1(\theta) + o_p(1) , \qquad (3.36)$$

and if, in addition, $n^{-1}l_3(\tilde{\theta}) = O_p(1)$, so that the remainder term $(\hat{\theta} - \theta)l_3(\tilde{\theta})/n$ in (3.35) is of order $o_p(1)$, then

$$\sqrt{n}(\hat{\theta} - \theta) = i_1(\theta)^{-1}l_*(\theta)/\sqrt{n} + o_p(1) . \qquad (3.37)$$

If the limiting null distribution of the score is normal according to (3.33), expansion (3.37) gives

$$\sqrt{n}(\hat{\theta} - \theta) \xrightarrow{d} N(0, i_1(\theta)^{-1}) . \qquad (3.38)$$

The same result holds when $p > 1$, with a p-dimensional normal limit distribution. Note that the maximum likelihood estimator $\hat{\theta}$ is asymptotically unbiased and efficient (cf. section 3.5.2). From (3.37)

$$\hat{\theta} - \theta = i_1(\theta)^{-1}\frac{l_*(\theta)}{n} + o_p(n^{-1/2}) , \qquad (3.39)$$

so that $\hat{\theta} - \theta = O_p(n^{-1/2})$.

If $\hat{\theta}$ is the unique root of the likelihood equation it corresponds to the global maximum of the likelihood function, with probability one for $n \to +\infty$. In fact, the second derivative $l_2(\hat{\theta}) = -ni_1(\theta) + o_p(n)$ is negative with probability close to 1 for a sufficiently large n.

Consider the further expansion

$$l(\hat{\theta}) = l(\theta) + (\hat{\theta} - \theta)l_*(\theta) + \frac{1}{2}(\hat{\theta} - \theta)^2 l_2(\theta) + o_p(1) . \qquad (3.40)$$

By substituting the expansion of $l_*(\theta)$ obtained from (3.34) into (3.40), we get

$$l(\hat{\theta}) - l(\theta) = (\hat{\theta} - \theta)\{-(\hat{\theta} - \theta)l_2(\theta)\} + \frac{1}{2}(\hat{\theta} - \theta)^2 l_2(\theta) + o_p(1)$$

so that, because of (3.36),

$$l(\hat{\theta}) - l(\theta) = -\frac{1}{2}(\hat{\theta} - \theta)^2 l_2(\theta) + o_p(1) = \frac{1}{2}(\hat{\theta} - \theta)^2 i(\theta) + o_p(1) , \qquad (3.41)$$

Consequently, using (3.38), the asymptotic null distribution result for the log-likelihood ratio statistic is

$$W = 2\{l(\hat{\theta}) - l(\theta)\} \xrightarrow{d} \chi_1^2 .$$

If θ is p-dimensional, the asymptotic null distribution of W is χ_p^2.

These results hold not only for independent and identically distributed observations: they can be immediately extended to the case where independence is maintained but not identical distribution (for example, to regression models), using Theorems A.8–A.11 in Appendix A. There are also extensions to dependent observations and to stochastic processes, however the condition $i(\theta) = O(n)$ remains essential. Extensions to dependent observations are usually obtained from a martingale representation of the process associated to the score function, using martingale central limit theorems mentioned at the end of Appendix A. For an accessible introduction to these more advanced topics, see Barndorff-Nielsen (1991a, section 10.2.3) and Barndorff-Nielsen and Cox (1994, section 3.3); for a more in-depth account, see Barndorff-Nielsen and Sørensen (1994).

The results on the asymptotic distribution of the score and of the maximum likelihood estimator (3.33) and (3.38) imply that also the combinants W_u and W_e, given by (3.31) and (3.32), respectively, are asymptotically pivotal with null distribution χ_p^2. This still holds if Fisher information $i(\theta)$ is estimated using $i(\hat{\theta})$ or $j(\hat{\theta})$. Asymptotic expansions (3.41) and (3.37), when extended to the multiparameter case, show that W_u and W_e are asymptotically equivalent to W to the first order:

$$W = W_u + o_p(1), \qquad (3.42)$$

$$W = W_e + o_p(1). \qquad (3.43)$$

Tests for H_0: $\theta = \theta_0$ versus H_1: $\theta \neq \theta_0$ based on W, W_u and W_e are, therefore, asymptotically equivalent to the first order. In these, the expected null information $i(\theta_0)$ can be replaced with $i(\hat{\theta})$, $j(\theta_0)$, $j(\hat{\theta})$. Confidence regions based on the asymptotic null distribution of W, W_u or W_e have approximately the shape of homothetic ellipsoids centred at $\hat{\theta}$ with axes parallel to the eigenvectors of $j(\hat{\theta})$; this is exactly the shape of the regions based on $(\hat{\theta} - \theta)^{\mathrm{T}} j(\hat{\theta})(\hat{\theta} - \theta)$.

For a scalar parameter, $p = 1$, it is often useful to refer to the one-sided signed versions of W, W_u and W_e

$$Z = \mathrm{sgn}(\hat{\theta} - \theta)\sqrt{W} , \qquad (3.44)$$

$$Z_u = l_*(\theta)/\sqrt{i(\theta)} , \qquad (3.45)$$

$$Z_e = (\hat{\theta} - \theta)\sqrt{i(\theta)} , \qquad (3.46)$$

which usually have null asymptotic $N(0, 1)$ distribution (see also section 11.5). With $\theta = \theta_0$, these statistics are suitable for testing H_0: $\theta = \theta_0$ against the one-sided alternative H_1: $\theta > \theta_0$ or H_1: $\theta < \theta_0$, with P-values based on the standard normal distribution.

Given the first order equivalence between the three versions of the likelihood ratio test, and between the corresponding signed versions, we should compare the three on the basis of some further criterion. The Wald test is based on an easily understandable comparison between estimated value and hypothetical value, taking estimation error into account. It is widely used in applications but does not satisfy the principle of parameterization invariance (see Example 2.20). Tests based on W_u and W are, on the other hand, parameterization invariant. One last aspect which distinguishes W from the other two versions is that the acceptance and confidence regions associated with it are qualitatively more suitable when the log-likelihood function is not almost quadratic in the region around its maximum. Marked asymmetry of the log-likelihood is captured by W, even if the nominal asymptotic level can turn out to be a poor approximation of the true level, for the given θ_0 and n. Furthermore, W does not place parameter values which do not belong to Θ in the confidence regions, which could happen with the other two versions. The advantage of W_u is that it is simple to calculate and it has maximum local power (see section 3.5.3). However, for non-local alternatives, the power of W is usually greater (Amari, 1985, p. 200). Lastly, Cox (1980) and McCullagh (1984a; 1987, section 8.4) have shown that, in the absence of nuisance parameters, W is asymptotically independent to a higher order of any asymptotically ancillary statistic, such as, for example, the Efron–Hinkley ancillary (see section 5.10). Thus, inference based on W conforms, asymptotically, to the conditionality principle, and leaves no doubt about the choice of ancillary (see also section 11.3). The same is not true of W_u. However, in this respect using the observed information $\hat{j}$ when defining the statistic is preferable to using expected information (Efron and Hinkley, 1978). One last point in favour of W is the possibility of making this combinant asymptotically pivotal to a higher order, simply by correcting its null expectation so that it matches, to a higher order, the mean of the relevant chi-squared distribution (see section 11.4). Analogous corrections for W_u are rather more complex.

3.4.2 Non-null Asymptotic Distributions

Now, we consider first-order asymptotic results concerning the non-null distribution of the main likelihood quantities, i.e. the asymptotic distributions of likelihood quantities evaluated at θ when θ_0 is the true value of the parameter. Consider for simplicity $p = 1$ and random sampling of size n.

Because of the central limit theorem, for finite n the non-null distribution of the score $l_*(\theta)$ is approximately normal,

$$l_* \sim N(E_{\theta_0}(l_*(\theta)), \mathrm{Var}_{\theta_0}(l_*(\theta))) \ .$$

However, the non-null limiting distribution of $l_*(\theta)/\sqrt{n}$ is usually degenerate at $+\infty$ or $-\infty$. A non-degenerate asymptotic distribution is obtained only for values of θ in a shrinking neighbourhood of θ_0. Let θ be such that $\theta_0 = \theta + \delta/\sqrt{n}$. The expansion

$$l_*(\theta) = l_*(\theta_0) + (\theta - \theta_0)l_2(\theta_0) + \frac{1}{2}(\theta - \theta_0)^2 l_3(\tilde{\theta}) \ ,$$

where $|\tilde{\theta} - \theta_0| < |\theta - \theta_u|$, gives

$$\frac{l_*(\theta)}{\sqrt{n}} = \frac{l_*(\theta_0)}{\sqrt{n}} - \delta \frac{l_2(\theta_0)}{n} + o_p(1) \ ,$$

since $l_3(\tilde{\theta})(\theta - \theta_0)^2 = O_p(n)O(n^{-1}) = O_p(1)$. Because

$$\frac{l_*(\theta_0)}{\sqrt{n}} \xrightarrow{d} N(0, i_1(\theta_0))$$

and

$$-\frac{l_2(\theta_0)}{n} = i_1(\theta_0) + o_p(1) \ ,$$

when θ_0 is the true parameter value we have

$$\frac{l_*(\theta)}{\sqrt{n}} \xrightarrow{d} N(\delta i_1(\theta_0), i_1(\theta_0)). \tag{3.47}$$

Even W has a non-degenerate limiting non-null distribution only locally; if $\theta \neq \theta_0$ is a fixed value

$$\lim_{n \to +\infty} P_{\theta_0}(2(l(\hat{\theta}) - l(\theta)) > c) = 1 \tag{3.48}$$

for any $c > 0$.

When $\theta - \theta_0 = O(n^{-1/2})$, through elementary Taylor expansions, the asymptotic equivalence

$$W = W_u + o_p(1) = \left(\frac{l_*(\theta)}{\sqrt{i(\theta)}}\right)^2 + o_p(1)$$

is again seen to hold. Therefore, if $\theta_0 = \theta + \delta/\sqrt{n}$,

$$W \xrightarrow{d} \chi_1^2(\delta^2 i_1(\theta_0)),$$

where $\chi_1^2(c^2)$ indicates a non-central chi-squared distribution with one degree of freedom and non-centrality parameter c^2, i.e. the distribution of $(c + U)^2$, where $U \sim N(0, 1)$.

3.4.3 Robustness of Likelihood Methods

A further problem concerns the robustness of likelihood quantities when the data generating model, $p^0(y)$, is not included in the family $\mathcal{F}$ wherein likelihood based inference is being carried out. Two concepts of robustness can be distinguished: *i)* robustness with respect to contamination; *ii)* robustness with respect to model misspecification.

The first aims to take the possible presence of outliers into account, in both the statistical model and in inference. Outliers are observations that do not effectively represent the population being studied, which is well captured by $\mathcal{F}$. The most common sources of outliers are errors in data collection and codification.

The second concept of robustness aims, to some degree, to protect inference from the possible inadequacies of $\mathcal{F}$ as a statistical model. Even the problems which arise from discretization can be dealt within this framework.

Inference procedures which are robust with respect to contamination are based on the assumption that $p^0(y)$ is in some neighbourhood of a distribution $p(y; \theta_0) \in \mathcal{F}$. The aim of inference remains to locate θ_0. For example, under random sampling, the generic sample component Y_i could be a mixture of a normal distribution $N(\mu, \sigma^2)$ and of a degenerate distribution at y_0, $D(y_0)$, with weights $1 - \varepsilon$ and ε, respectively. Only the parameters of the normal distribution are of interest, while the quantities y_0 and ε merely serve to take into account the possible presence of contamination into the model. A general approach to the study of robustness with respect to contamination is based on the notion of the influence function (see section 3.6.3 for some basic ideas and references).

Robustness with respect to misspecification may sometimes be achieved by means of suitable corrections to the usual likelihood inference procedures. As an example of a misspecified model, think, with count data, of the wrong specification of a Poisson model when there is overdispersion, that is, when the probability model that generates the data does not respect the equality between mean and variance which holds under a Poisson distribution.

Let $p^0(y)$ be the true probability model and let $\mathcal{F} = \{p(y; \theta), \theta \in \Theta \subseteq \mathbb{R}^p\}$ be an incorrectly specified statistical model, such that, with positive probability, $p^0(y) \neq p(y; \theta)$ for every $\theta \in \Theta$. Let us denote by $E_0(\cdot)$ expectation with respect to $p^0(y)$. If the problem is sufficiently regular, the maximum likelihood estimator $\hat{\theta}$ converges in probability to the value θ_{p^0} such that

$$E_0\left(\log p(Y; \theta_{p^0})\right) > E_0\left(\log p(Y; \theta)\right) \qquad (3.49)$$

for every $\theta \in \Theta$, $\theta \neq \theta_{p^0}$ (Huber, 1967). The parameter value θ_{p^0} corresponds to the element of $\mathcal{F}$ that satisfies the Wald inequality (1.7) when expectation is taken with respect to the true distribution.

Suppose that θ parameterizes $\mathcal{F}$ through quantities, such as moments and quantiles, which are also meaningful under $p^0(y)$, so that we can denote by $\theta(p^0(y))$ the value of these quantities for the true probability model. Suppose in addition that $\theta(p^0(y))$ is the quantity of primary interest, while the other aspects of distribution are of secondary interest. If

$$\theta_{p^0} = \theta(p^0(y)),$$

with θ_{p^0} defined by (3.49), then the maximum likelihood estimator $\hat{\theta}$ based on $\mathcal{F}$ is a consistent estimator of $\theta(p^0(y))$. Under regularity conditions $\hat{\theta}$ is the unique solution of $l_*(\theta) = 0$ and

$$E_0\left\{l_*(\theta(p^0(y)))\right\} = 0, \qquad (3.50)$$

so that the score associated to the misspecified model is still an unbiased estimating function.

Example 3.7 *Correlated observations*

Suppose that the working model for observations $y_1, \ldots, y_n$ is random sampling with marginal density $p(y; \theta)$ for one observation. If this marginal model is specified correctly, the score function

$$l_*(\theta) = \sum_{i=1}^{n} \frac{\partial}{\partial \theta} \log p(y_i; \theta)$$

remains an unbiased estimating function whatever the dependence structure between the observations is. $\triangle$

For simplicity, let us denote $\theta(p^0(y))$ by θ_0 and consider $p = 1$. The expansion

$$0 = l_*(\theta_0) + (\hat{\theta} - \theta_0) l_2(\theta_0) + O_p(1)$$

gives

$$\sqrt{n}(\hat{\theta} - \theta_0) = \frac{l_*(\theta_0)/\sqrt{n}}{-l_2(\theta_0)/n} + O_p\left(\frac{1}{\sqrt{n}}\right) .$$

Suppose that, as in a correctly specified model,

$$\frac{1}{n} l_2(\theta_0) \xrightarrow{p} E_0\left(\frac{1}{n} l_2(\theta_0)\right) = -i_1(\theta_0)$$

and that $l_*(\theta_0)/\sqrt{n}$ is asymptotically normal with zero mean. In general, however, the information identity does not hold if the model is misspecified, so that

$$\frac{1}{\sqrt{n}} l_*(\theta_0) \xrightarrow{d} N(0, \zeta) ,$$

where $\zeta > 0$ does not usually coincide with $i_1(\theta_0)$. A consistent estimator for the asymptotic variance ζ is

$$\hat{\zeta}_n = \frac{1}{n} \sum_{i=1}^{n} (l_*(\theta_0; y_i))^2 ,$$

where $l_*(\theta; y_i)$ denotes the contribution to the score by the i-th component of y, that is, $(\partial/\partial\theta) \log p(y_i; \theta)$. In addition, the same technique based on Taylor expansions used in section 3.4.1 shows that $\hat{\theta}$ is asymptotically normal with mean θ_0 and variance that can be estimated on the basis of the misspecified model. Specifically, for large n,

$$\sqrt{n}(\hat{\theta} - \theta_0) \sim N\left(0, \frac{1}{i_1(\theta_0)} \frac{\sum_{i=1}^{n} (l_*(\theta_0; y_i))^2}{n\, i_1(\theta_0)}\right). \tag{3.51}$$

Note that the usual asymptotic variance is corrected by the factor

$$\delta(\theta_0) = \frac{\sum_{i=1}^{n} (l_*(\theta_0; y_i))^2}{n\, i_1(\theta_0)},$$

which is parameterization invariant (see Problem 2.19). We may consider the further approximation, still with a parameterization invariant correction factor,

$$\sqrt{n}(\hat{\theta} - \theta_0) \,\dot{\sim}\, N\left(0, \frac{1}{i_1(\hat{\theta})} \delta(\hat{\theta})\right). \tag{3.52}$$

Detailed examples of these corrections can be found in Royall (1986).

From the Taylor expansion

$$2\{l(\hat{\theta}) - l(\theta_0)\} = i(\theta_0)(\hat{\theta} - \theta_0)^2 + o_p(1)$$

for large n the log-likelihood ratio statistic has a null approximate distribution

$$2\{l(\hat{\theta}) - l(\theta_0)\} \,\dot{\sim}\, \delta(\theta_0)\chi_1^2.$$

Except for a scale correction, the usual chi-squared asymptotic distribution is obtained. The correction factor itself can provide a test for correct specification (see White, 1982).

When the parameter θ is p-dimensional, the asymptotic result on the log-likelihood ratio statistic is slightly more complicated. Using arguments analogous to those above,

$$\frac{1}{\sqrt{n}} l_*(\theta_0) \xrightarrow{d} N_p(0, V_*(\theta_0)),$$

where

$$nV_*(\theta_0) = \left\{\sum_{i=1}^{n} l_*(\theta_0; y_i)(l_*(\theta_0; y_i))^{\mathrm{T}}\right\} + o_p(n).$$

Furthermore,

$$\sqrt{n}(\hat{\theta} - \theta_0) \xrightarrow{d} N_p\left(0, i_1(\theta_0)^{-1} V_*(\theta_0) i_1(\theta_0)^{-1}\right) \tag{3.53}$$

and

$$2\{l(\hat{\theta}) - l(\theta_0)\} \xrightarrow{d} \sum_{i=1}^{p} \lambda_i(\theta_0)U_i^2 \,, \tag{3.54}$$

where the U_i^2 are independent chi-squared random variables with one degree of freedom; the coefficients $\lambda_i(\theta_0)$ are the eigenvalues of the matrix $V_*(\theta_0)i_1(\theta_0)^{-1}$ (Kent, 1982). The matrix $V_*(\theta_0)$ and the $\lambda_i(\theta_0)$ can be consistently estimated in an obvious way.

The above considerations can be extended quite generally to unbiased estimating functions that are not necessarily connected with the score function of a given parametric family. Let $\theta = \theta(p(y))$ indicate the aspects of interest of the underlying distribution and let $q(y; \theta)$ be an unbiased estimating function, i.e. such that $E_p(q(Y; \theta)) = 0$, with $E_p(\cdot)$ denoting expectation under $p(y)$. If, by some law of large numbers, $n^{-1}q(Y; \theta) \xrightarrow{P} 0$ as $n \rightarrow +\infty$, when θ is the true value of the parameter, and if $q(\cdot; \theta)$ is a one-to-one function of θ in a neighbourhood of the true value of the parameter, then estimators defined as solutions with respect to θ of $q(y; \theta) = 0$ are consistent (see e.g. Liang and Zeger, 1995, section 2.3). In addition, one would expect that, under analogous conditions to those required by the score, the results (3.51) and (3.52) would generalize to estimates based on $q(\cdot)$. Suppose that θ is a scalar and denote by $\tilde{\theta}$ the solution of $q(y; \theta) = 0$. Consider the local linearization

$$\tilde{\theta} - \theta = -\left(\frac{\partial}{\partial \theta}q(y; \theta)\right)^{-1}q(y; \theta) + \dots \,.$$

Suppose that as n diverges, $q(y; \theta)/\sqrt{n}$ converges to a $N(0, V_q(\theta))$ distribution and that by a law of large numbers, $-(\partial q(y; \theta)/\partial \theta)/n$ converges to a function $G_q(\theta)$, which depends only on θ and not on other aspects of the distribution. Then,

$$\sqrt{n}(\tilde{\theta} - \theta) \xrightarrow{d} N\left(0, \frac{V_q(\theta)}{G_q(\theta)^2}\right) \,.$$

In the multiparameter case, the asymptotic covariance matrix is

$$(G_q(\theta))^{-1}V_q(\theta)(G_q(\theta))^{-1} \,.$$

The above conclusions are still true if the estimating equation is not exactly unbiased, but $E_p(q(Y; \theta)) = O(1)$, with a law of large numbers still applying both to $n^{-1}q(y; \theta)$ and to $-(\partial q(y; \theta)/\partial \theta)/n$, and with $q(y; \theta)/\sqrt{n}$ asymptotically normal. The whole argument breaks down if $E_p(q(Y; \theta)) = O(n)$.

If $q(\cdot)$ is not the score of a parametric statistical model, then, generally, there will be no log-likelihood function where a result which corresponds to (3.54) applies (see, however, quasi-likelihood in section 4.9).

3.5 Inference in the Frequency-Decision Paradigm

3.5.1 General Framework

The principle of repeated sampling may play an important role not only when evaluating uncertainty in inferences, but also when deciding which of the possible statistical procedures to choose for a given inference problem. The problem of choosing an inference procedure can sometimes be formulated as an optimization problem, under suitable constraints. The objective function, usually a moment or the probability of a certain event, depends on the distribution of a specific information carrier. Formalization may ensue in a simple and elegant manner if the inference problem can be reformulated as a **decision problem**. As was mentioned section 1.2, this viewpoint is the basis of the frequency-decision paradigm.

As regards the Fisherian paradigm, the main change in emphasis concerns the role of an inference procedure $\pi(y)$: from being a summary of information this becomes a **decision function**. This implies that the inference process leads to an **action**, i.e. that $\pi(y)$ is a mapping from $\mathcal{Y}$ to a set $\mathcal{A}$ of possible actions.

Each action corresponds to consequences that one assumes can be expressed by a **loss function**

$$D(\theta, \pi(y)): \quad \Theta \times \mathcal{A} \rightarrow \mathbb{R}^+, \qquad (3.55)$$

where $D(\cdot, \cdot)$ is a measurable function of y; the function $D(\cdot, \cdot)$ expresses the loss that occurs when applying $\pi(y)$ if θ is the true value of the parameter. Note that $D(\theta, \pi(y))$ is a combinant. Generally, one assumes that a procedure $\pi_\theta(y)$ exists for every θ such that $D(\theta, \pi_\theta(y)) = 0$. In this way any loss which may occur is evaluated with reference to a situation where there are no errors in the description of the population, at least none related to the aims of the decision.

The triple $(\mathcal{F}, \mathcal{A}, D(\cdot, \cdot))$ is the **statistical decision problem**. For example, formulating a point estimation problem as a decision problem identifies the set $\mathcal{A}$ with Θ, the statistical procedure $\pi(y)$ with an estimator $\tilde{\theta}(y)$ and, typically, the loss function with a convex function of $\tilde{\theta}(y) - \theta$ with a minimum at zero. In this context a significance test becomes merely a procedure which leads to acceptance or rejection of H_0, that is, a decision function in a problem with only two possible actions.

If the principle of repeated sampling is accepted, the performance of a procedure $\pi(y)$ may be summarized by the null expectation of the combinant (3.55), i.e. by the risk function,

$$R_\pi(\theta) = E_\theta(D(\theta, \pi(Y))) . \qquad (3.56)$$

In general, for a given statistical decision problem, there is no decision function which minimizes the risk for all θ. There are two main approaches to a more precise mathematical delineation of the problem.

The first introduces a preference ordering of the risk functions associated with the competing procedures. This ordering is based on the reduction of the function $R_\pi(\theta)$ to a real-valued summarizing quantity. The best known reduction criteria are the minimax and the **Bayesian**: as a summary of the function $R_\pi(\theta)$ the former selects its maximum for $\theta \in \Theta$, the latter requires the specification of a prior distribution on Θ, and bases comparisons on the **Bayes** risk, that is on the expectation of $R_\pi(\theta)$ with respect to the prior distribution. The two criteria are closely connected, see for example Ferguson (1967, section 2.11).

The second approach remains true to a frequency approach and to the global nature of the loss function (3.55) and imposes suitable constraints on the class of statistical procedures. The best known constraint criteria are **unbiasedness** and, in the case of group families, **invariance** or **equivariance under classes of transformations of the sample space** (see sections 7.6 and 7.7). See also Lehmann (1986, section 1.5) for a succinct discussion of such criteria in a unified framework.

Data reduction (or perhaps better *decision reduction*) is also important in statistical decision problems and here, too, sufficiency plays a key role. The main result is the Rao–Blackwell theorem which establishes that, if $\mathcal{A}$ is a convex subset in a finite-dimensional Euclidean space and if $D(\theta, a)$ is a convex function of $a \in \mathcal{A}$ for every θ, then the solution to the optimality problem does not change when attention is limited to procedures $\pi(y)$ which are functions of the data only through a sufficient statistic (see, for example, Ferguson, 1967, p. 121).

Optimality theory for point estimation and for hypothesis testing, together with the dual results on optimal confidence intervals, is one of the most traditional chapters within the theory of statistics. Here the aim is to offer a quick overview of the main results, where the reader may locate arguments already familiar and also find references to aspects dealt with in depth in other textbooks or, in some cases, in later parts of this book.

3.5.2 Point Estimation

For simplicity, suppose that θ is a scalar parameter and let $\tilde{\theta}(y)$ be an estimate of θ. The loss function most commonly adopted in point estimation is the quadratic loss

$$D(\theta, \tilde{\theta}(y)) = (\tilde{\theta}(y) - \theta)^2 \, .$$

Thus the problem of minimizing risk becomes that of minimizing the mean squared error $E_\theta(\tilde{\theta}(Y) - \theta)^2$. No estimator has minimum mean squared error uniformly in θ. Indeed, the mean squared error can be made equal to zero for every given value θ_0 of θ by choosing $\tilde{\theta}(y) = \theta_0$. Thus, the search for the optimum estimator should be restricted to a suitable class.

One first kind of constraint requires a *centring* property for the estimator, i.e. that a given location index of the distribution of $\tilde{\theta}(Y)$ coincides with θ when θ is the true parameter value. If this index is the mean, the problem becomes that of finding the uniformly minimum variance unbiased (UMVU) estimator. In this context, the main results are the Cramér–Rao lower bound and the Rao–Blackwell–Lehmann–Scheffé theorem.

The Cramér–Rao lower bound establishes that for regular models (see, for example Azzalini, 1996, sections 3.2.3-3.2.5), the variance of an unbiased estimator cannot be smaller than the reciprocal of Fisher information, i.e. that if $E_\theta(\tilde{\theta}(Y)) = \theta$, for every $\theta \in \Theta$, then

$$\text{Var}_\theta(\tilde{\theta}(Y)) \geq \frac{1}{i(\theta)} \, . \tag{3.57}$$

Indeed, differentiating the unbiasedness condition with respect to θ, it is not difficult to see that $\text{Cov}_\theta(l_*, \tilde{\theta}) = 1$; the inequality (3.57) is one immediate consequence. It follows that an estimator will reach the lower bound in (3.57) if, and only if, it is a linear function of the score.

The Rao–Blackwell–Lehmann–Scheffé theorem states that, if s is a complete sufficient statistic for θ under $\mathcal{F}$, and $\tilde{\theta}(y)$ is an unbiased estimator of θ, the UMVU estimator is essentially unique and is given by $E(\tilde{\theta}(Y) \mid S = s)$, which is a statistic based on s. This statistic can sometimes be identified directly among the functions of s, as that, essentially unique, which is an unbiased estimator of θ. For more details and extensions to the multi-parameter case, see Lehmann (1983, Chapter 2) and the references given there. An asymptotic approximation of the UMVU estimator will be given in Example 10.13.

If $y = (y_1, \ldots, y_n)$ is a random sample of size n from a distribution in a regular one-parameter model, the lower bound in (3.57) also holds for the asymptotic variance $v(\theta)/n$ of any estimator $\tilde{\theta}_n(y)$ such that

$$\sqrt{n}\left(\tilde{\theta}_n(Y) - \theta\right) \xrightarrow{d} N(0, v(\theta)),$$

except on a set of θ values having Lebesgue measure zero (Lehmann, 1983, Theorem 6.1.1). If $v(\theta)/n = 1/i(\theta)$, $\tilde{\theta}_n(y)$ is said to be **asymptotically efficient**. Under regularity conditions, the maximum likelihood estimator is asymptotically efficient (see (3.38)). The concept of asymptotic efficiency of an estimator is central also in the theory of point estimation of finite-dimensional functions of θ in semiparametric and nonparametric models (see section 3.6.3).

Sometimes unbiasedness is accompanied by a constraint on the analytical form of the estimator, typically (but not exclusively) of linearity in the observations. Generally, this further constraint allows the assumptions regarding the family $\mathcal{F}$ to be weakened, as for example in the Gauss–Markov theorem in the context of linear models (see e.g. Azzalini, 1996, section 5.2.3). Linear estimators play an important role in finite population sampling theory. In robust estimation, the requirement to give less weight to extreme observations sometimes leads to considering estimators which are a linear combination of the order statistics, such as **L-estimators**, see Hampel, Ronchetti, Rousseeuw and Stahel (1986, section 2.3b).

In the one-parameter case a perhaps more satisfying alternative to mean unbiasedness is median unbiasedness (see section 2.11.1, formula (2.10)). A median unbiased estimator has the same probability of underestimating or overestimating the parameter. This centring criterion is accompanied by the optimality requirement of maximum concentration around the true value of the parameter. If, for simplicity, we consider an estimator with a continuous distribution, the optimal estimator of θ among all the median unbiased estimators is that which minimizes

$$P_\theta\left(\tilde{\theta}(Y) < \theta'\right), \qquad \text{for all} \quad \theta' \; < \; \theta,$$

$$\text{(3.58)}$$

$$P_\theta\left(\tilde{\theta}(Y) > \theta''\right), \qquad \text{for all} \quad \theta'' \; > \; \theta.$$

A theory for optimal median unbiased estimators has been developed in close connection with the optimal theory for confidence intervals, by Lehmann (1959, section 3.5) and by Pfanzagl (1970a, 1970b, 1971, 1979); see Read (1985) for a schematic summary.

The second main criterion for restricting the class of competing estimators is that of equivariance under certain data transformations. This criterion is usually applied in group families (see Chapter 7), characterized by symmetry structures. This means that there is a class of transformations of the observations which give rise to distributions that still belong to the family, with a different value of the parameter. The criterion of equivariance requires that whenever data are subjected to such special transformations the estimate will also transform in a coherent manner. Essentially, this happens in situations such as the following: if the quantity that must be estimated is the weight of an object, then one seeks to obtain the same result, up to a change in the units of measurement, both when estimation is based on n measurements expressed in grams and when these are expressed in kilograms. Formalizing the criterion requires some basic notions about group families; discussion is deferred to section 7.6 (see in particular (7.30) for a precise definition). The constraint of equivariance is a weakening of the requirement of invariance of inferential conclusions under one-to-one transformations of the sample space (see section 2.2). In particular, the maximum likelihood estimator is always within the restricted class.

Lastly, we review one criterion for evaluating the relative accuracy of two estimators originally proposed by Pitman (1937) which has, of late, been the subject of renewed interest. This is the Pitman closeness criterion on the basis of which, if $\tilde{\theta}_1(y)$ and $\tilde{\theta}_2(y)$ are two estimators of θ, $\tilde{\theta}_2(y)$ is closer to θ than $\tilde{\theta}_1(y)$ if

$$P_\theta \left(|\tilde{\theta}_1(Y) - \theta| \geq |\tilde{\theta}_2(Y) - \theta| \right) \geq \frac{1}{2}, \qquad \text{for every } \theta \in \Theta .$$

Unlike criteria based on risk comparison, the Pitman closeness criterion does not refer to marginal distributions, but rather to joint distributions of estimators. Comparisons based on such a criterion are not transitive and this generates difficulties in the search for an optimum estimator within a suitable class. Ghosh and Sen (1989) made some interesting connections between median unbiasedness and Pitman closeness: in location or scale models, and limiting attention to the class of equivariant estimators, if an estimator is median unbiased then it is Pitman closest. See Sen (1992) for further comments.

3.5.3   Testing Hypotheses

In decision theory, a problem of hypothesis testing corresponds to a situation in which the statistical procedure leads to one of two possible decisions: θ belongs to the subset Θ_0 of Θ or θ belongs to $\Theta \setminus \Theta_0$. The test is thus a function $\pi(y)$, called the critical function, with only two values: $\pi(y) = 0$ if observation of y leads to the acceptance of $H_0 : \theta \in \Theta_0$ and $\pi(y) = 1$ if the alternative hypothesis $H_1 : \theta \in \Theta \setminus \Theta_0$ is accepted. (Obviously, the choice of values 0 and 1 for $\pi(y)$ is arbitrary, what is essential is that $\pi(\cdot)$ takes only two values.) The set $A_{\Theta_0} = \{y \in \mathcal{Y} : \pi(y) = 0\}$ is the acceptance region and the complementary set $\bar{A}_{\Theta_0} = \mathcal{Y} \setminus A_{\Theta_0}$ is the rejection region. A hypothesis is said to be simple if the corresponding subset of Θ contains only one element, otherwise, the hypothesis is said to be composite. Sometimes, for mathematical convenience, randomized tests are considered, i.e. critical functions which may take values in $[0, 1]$. In this case, the sample space is divided into three regions: the acceptance region, where $\pi(y) = 0$; the rejection region, where $\pi(y) = 1$; and the region where $0 < \pi(y) < 1$. Here, the final decision is to reject H_0 if an auxiliary random experiment, where the probability of success is equal to $\pi(y)$, gives a success.

The most commonly adopted loss function is the zero-one loss: $D(\theta, \pi(y)) = 1$ if $\theta \in \Theta_0$ and $\pi(y) = 1$, or if $\theta \in \Theta \setminus \Theta_0$ and $\pi(y) = 0$; otherwise $D(\theta, \pi(y)) = 0$, since the two other possible situations correspond to a correct decision. The function $\beta(\theta) = E_\theta(\pi(Y))$ is the power function of the test $\pi(\cdot)$. For a non-randomized test, risk is thus equal to $E_\theta(\pi(Y)) = P_\theta(Y \in \bar{A}_{\Theta_0}) = \beta(\theta)$ if $\theta \in \Theta_0$ and to $E_\theta(1 - \pi(Y)) = P_\theta(Y \in A_{\Theta_0}) = 1 - \beta(\theta)$ if $\theta \in \Theta \setminus \Theta_0$. In a randomized test the risk still represents the pre-experimental probability of taking a wrong decision.

A test uniformly minimizes risk if it makes $\beta(\theta)$ minimum for $\theta \in \Theta_0$ and maximum for $\theta \in \Theta \setminus \Theta_0$. As for point estimation, uniform minimization of the risk cannot be achieved, unless, in the case of a simple hypothesis versus a simple alternative, sequential sampling is adopted (Wald and Wolfowitz, 1948). The usual strategy is thus to constrain $\sup_{\theta \in \Theta_0} \beta(\theta)$ to be smaller than or equal to an assigned value α, the level or size of the test, and investigate the existence of a uniformly most powerful (UMP) test, that maximizes $\beta(\theta)$ uniformly for $\theta \in \Theta \setminus \Theta_0$. As is natural to expect, in general this problem can be solved when the alternative hypothesis is simple: in this case the test is called most powerful (MP). A test has exact size α if $\sup_{\theta \in \Theta_0} \beta(\theta) = \alpha$. The most important result is the following, known as the Neyman–Pearson lemma.

Theorem 3.3 Let $\Theta = \{\theta_0, \theta_1\}$ and consider the problem of testing $H_0 : \theta = \theta_0$ versus $H_1 : \theta = \theta_1$.

(*Existence.*) A critical function $\pi(y)$ and a constant c exist for this problem such that the test has exact size α and

$$
\pi(y) = \begin{cases} 1 & \text{if } \dfrac{p_Y(y; \theta_1)}{p_Y(y; \theta_0)} > c \\[2mm] 0 & \text{if } \dfrac{p_Y(y; \theta_1)}{p_Y(y; \theta_0)} < c. \end{cases} \tag{3.59}
$$

(*Sufficiency.*) If a test with exact size α satisfies (3.59), then this is the most powerful among the level α tests for H_0 versus H_1.

(*Necessity.*) If $\pi(y)$ is the critical function of the most powerful test with level α for H_0 versus H_1, then it will satisfy (3.59) for some c, with probability one. The test also has exact size α, provided there is no test with a significance level lower than α and with power equal to one.

Proof. See Lehmann (1986, section 3.2). $\qquad\qquad\qquad\qquad\qquad\qquad\qquad\square$

Note that the test statistic in (3.59) is the likelihood ratio, which is sufficient for a two-element family. It is easy to prove that the most powerful level α test is always unbiased, that is, such that $\beta(\theta_1) \geq \alpha$.

The most powerful level α test is uniquely defined by (3.59) except for the set of y values for which $p_Y(y; \theta_1)/p_Y(y; \theta_0) = c$. If this set has zero probability the test is essentially uniquely defined, otherwise the optimum test is randomized. Let

$$
\alpha(z) = P_{\theta_0} \left\{ \frac{p_Y(Y; \theta_1)}{p_{Y'}(Y; \theta_0)} > z \right\}
$$

and $\alpha(c^-) = \lim_{z \to c^-} \alpha(z)$. If in (3.59) the value c is chosen such that $\alpha(c) \leq \alpha \leq \alpha(c^-)$ and if, in addition, the critical function is defined by $\pi(y) = (\alpha - \alpha(c))/(\alpha(c^-) - \alpha(c))$ when $p_Y(y; \theta_1)/p_Y(y; \theta_0) = c$, then $E_{\theta_0}(\pi(Y)) = \alpha$.

A randomized test can be criticized from many points of view, in particular, because it does not respect the sufficiency principle. Thus, in applications involving discrete distributions, it is better to choose the level α from within the set of exactly attainable levels.

There are generalizations of the Neyman–Pearson lemma to problems where H_0 is composite while H_1 remains simple; see Lehmann (1986, section 3.2) for the main references. With composite alternative hypotheses, it is possible to obtain a uniformly most powerful level α test (without imposing additional constraints) essentially only in the one-parameter case, for problems of the form

$$H_0 : \theta \leq \theta_0 \qquad \text{versus} \qquad H_1 : \theta > \theta_0 . \qquad (3.60)$$

A parametric family $\mathcal{F}$ has a **monotone likelihood ratio**, if there exists a scalar statistic $t(y)$ such that for every pair of values θ and $\tilde{\theta}$, with $\theta < \tilde{\theta}$, the ratio $p_Y(y; \tilde{\theta})/p_Y(y; \theta)$ is a monotone function of $t(y)$. If $\mathcal{F}$ has a monotone likelihood ratio, then the optimum level α test still has a critical function given by (3.59) with c such that $\beta(\theta_0) = \alpha$. In this case in (3.59) the conditions can equally well be expressed in terms of $t(y)$. If for example the ratio $p_Y(y; \tilde{\theta})/p_Y(y; \theta)$ is a non-decreasing function of $t(y)$, then $\pi(y) = 0$ if $t(y) < c$ and $\pi(y) = 1$ if $t(y) > c$, with c such that the constraint on the level of the test is satisfied. The main class of distributions with a monotone likelihood ratio is that of one-parameter exponential families (see section 5.8.1). Other important examples are given by the hypergeometric distribution and by the non-central t, chi-square and F distributions with given degrees of freedom (and, more generally, by **totally positive families of order 2**; see Kim and Proschan, 1988). The condition that the elements of a one-parameter family are stochastically ordered with respect to θ (that is, in the case of stochastically increasing order, $\theta < \tilde{\theta}$ implies $P_\theta(Y \leq y) \geq P_{\tilde{\theta}}(Y \leq y)$, for every y) is necessary but not sufficient for the family to have a monotone likelihood ratio (Lehmann, 1986, Lemma 3.2).

Whenever $\mathcal{F}$ does not have a monotone likelihood ratio it would be reasonable to seek a level α test for (3.60) that is, at least, **locally most powerful**. Such a test maximizes power for alternatives close to H_0 among all level α tests. Suppose that $\beta(\theta_0) = \alpha$ and that, in a neighbourhood of θ_0, $\beta(\theta)$ can be expanded in a Taylor series

$$\beta(\theta) = \alpha + (\theta - \theta_0)\beta'(\theta_0) + \frac{(\theta - \theta_0)^2}{2}\beta''(\theta_0) + \dots . \qquad (3.61)$$

A level α test is called locally most powerful if $\beta'(\theta_0)$ is maximum. Assuming that differentiation and integration may be interchanged, the function which has to be maximized is

$$\beta'(\theta_0) = \int_{\bar{A}_{\Theta_0}} p'_Y(y; \theta_0) \, d\mu^* ,$$

where $p'_Y(y; \theta) = \partial p_Y(y; \theta)/\partial\theta$. The problem is thus re-formulated in such a way as the Neyman–Pearson lemma can still be applied, with $p_Y(y; \theta_1)$ substituted by $p'_Y(y; \theta_0)$. Therefore the locally most powerful test has the critical function

$$\pi(y) = \begin{cases} 1 & \text{if } \dfrac{p'_Y(y;\theta_0)}{p_Y(y;\theta_0)} > c \\[2ex] 0 & \text{if } \dfrac{p'_Y(y;\theta_0)}{p_Y(y;\theta_0)} < c \,. \end{cases} \tag{3.62}$$

Note that the ratio $p'_Y(y;\theta_0)/p_Y(y;\theta_0)$ coincides with the score statistic $l_*(\theta_0)$. This is one further, interesting aspect of the score function. See Cox and Hinkley (1974, section 4.9) for the extension of locally most powerful tests to the multiparameter case.

When the optimum level α test rejects the null hypothesis for large values of a scalar statistic $t(y)$ it is usually preferable to summarize the information carried by a test statistic by means of the observed significance level, or P-value,

$$\alpha^{obs} = \sup_{\theta \in \Theta_0} P_\theta \left(t(Y) \geq t(y^{obs}) \right) \,.$$

For some comments on this change of perspective, see Appendix D.

In the scalar parameter case, if the testing problem is two-sided,

$$H_0 : \theta = \theta_0 \qquad \text{versus} \qquad H_1 : \theta \neq \theta_0 \,, \tag{3.63}$$

then a uniformly most powerful test will only exist in certain specific instances (see for example, Ferguson, 1967, Exercise 5.2.7); it may be shown that a uniformly most powerful test does not exist even for one-parameter exponential families (Lehmann, 1986, Problem 3.31). If $\mathcal{F}$ is an exponential family an optimum test may be found in the class of unbiased level α tests, that is, tests such that $\beta(\theta) \geq \alpha$ for every $\theta \neq \theta_0$ (see also section 5.8.1), so that $\beta(\theta)$ has a global minimum equal to α at $\theta = \theta_0$. See Karlin (1952) and Lehmann (1986, Problems 3.27-3.30) for an extension of the results on the existence of unbiased uniformly most powerful tests to a class of distributions broader than the exponential family.

Even the idea of maximizing local power can be extended to two-sided problems so long as the constraint of unbiasedness is imposed. If $\beta(\theta)$ is smooth, unbiasedness implies that $\beta'(\theta_0) = 0$ and, on the basis of (3.61), the problem then translates into the maximization of $\beta''(\theta_0)$. It can be shown (see, for example, Cox and Hinkley, 1974, p. 119) that the locally most powerful unbiased test for problem (3.63) rejects H_0 if

$$l_*(\theta_0)^2 - j(\theta_0) > c\, l_*(\theta_0) + d \,, \tag{3.64}$$

where the constants c and d are chosen as to satisfy the constraints. The rejection regions of the locally most powerful unbiased test (3.64) are not nested, that is, included one into the other as α increases. Thus it is advisable to use the score function as a test statistic, with two-sided rejection regions (see Cox and Hinkley, 1974, p. 119).

Unbiasedness is complemented by a second type of constraint. A level α test for $H_0 : \theta \in \Theta_0$ is said to be similar on the boundary if the corresponding critical function satisfies the equation

$$E_\theta(\pi(Y)) = \alpha \qquad \text{for every } \theta \in \Upsilon , \tag{3.65}$$

where Υ denotes the common boundary of Θ_0 and $\Theta \setminus \Theta_0$. For example in the case of a scalar parameter with $H_0 : \theta = \theta_0$ and $H_1 : \theta \neq \theta_0$, the common boundary Υ is $\{\theta_0\}$. If the parametric family $\mathcal{F}$ is such that the power function of any test is continuous in θ, then the constraint (3.65) is a weakening of unbiasedness: an unbiased test with a continuous power function is also similar on the boundary (see Lehmann, 1986, section 4.1). Tests which are similar on the boundary are immediately characterized whenever the parametric subfamily $\mathcal{F}_\Upsilon = \{p_Y(y; \theta), \theta \in \Upsilon\}$ allows a boundedly complete sufficient statistic s (see section 2.5). In this case, the conditional distribution of any critical function $\pi(y)$ given s is independent of θ, for $\theta \in \Upsilon$. If $\pi(y)$ is a critical function such that

$$E(\pi(Y) \mid S = s) = \alpha \tag{3.66}$$

for every value of s (except for sets with zero probability), then also marginally $E_\theta(\pi(Y)) = \alpha$, for $\theta \in \Upsilon$; hence the test is similar on the boundary. A test is said to have Neyman structure if it satisfies (3.66). If s is boundedly complete, then any similar test must have Neyman structure (see section 2.5 and Lehmann, 1986, Theorem 4.2). The most important application of this result is in multiparameter exponential families and will be dealt with in detail in section 5.8.2.

As for point estimation so, too, for hypothesis testing, invariance under suitable classes of transformations of the sample space offers a further type of constraint. Whenever transformations of the observations exist, which leave the testing problem unaltered, then it is reasonable to consider only tests whose critical function is unaffected by those transformations. The criterion of invariance can be applied in the context of composite group families and will be illustrated in section 7.7.

In substance, the constraints of unbiasedness, similarity and invariance allow an optimum test to be found only in the cases where the constraint effectively reduces the original hypothesis testing problem to a more simple problem that the Neyman–Pearson lemma can be applied to.

3.5.4 Confidence Regions

Optimality in interval estimation is dual to optimality in hypothesis testing. The link between the two theories is given by the following considerations. Let A_{θ_0} be the acceptance region of a level α test for $H_0 : \theta = \theta_0$, then, for each $y \in \mathcal{Y}$,

$$C(y) = \{\theta \in \Theta : y \in A_\theta\} \tag{3.67}$$

is a confidence region with level $1 - \alpha$. If, for each $\theta_0 \in \Theta$, A_{θ_0} is the acceptance region of a uniformly most powerful test for $H_0 : \theta = \theta_0$ versus $H_1 : \theta \in \Theta_1$ ($\Theta_1 \subset \Theta$), then, among all the confidence regions with level $1 - \alpha$, $C(y)$ minimizes the non-null coverage probability

$$P_\theta \{\theta' \in C(Y)\} \qquad \text{for every} \quad \theta' \in \Theta, \ \theta \in \Theta_1 \,.$$

For a proof, see Lehmann (1986, Theorem 3.4). Thus, power maximization for tests translates for confidence regions into minimization of the probability of inclusion of parameter values different from the true one, that is, minimization of the non-null coverage probability

$$P_\theta \{\theta' \in C(Y)\} \,, \qquad \theta \neq \theta' \,. \tag{3.68}$$

A $1 - \alpha$ level confidence region which minimizes (3.68) is said to be uniformly most accurate. The preceding definitions could be thought to refer to the risk based on a loss function taking value 0 if $C(y)$ contains the true value of the parameter and 1, otherwise.

The region (3.67) is uniformly most accurate if $\Theta_1 = \Theta \setminus \{\theta_0\}$ and the corresponding test is optimal. However we can find an optimum test only when the parameter of interest is a scalar and suitable constraints are put on the class of tests. The same is true for confidence regions where the constraints usually adopted are translations of constraints in hypothesis testing. For example, a confidence region $C(y)$, with level $1 - \alpha$, is said to be unbiased if $P_\theta \{\theta' \in C(Y)\} \leq 1 - \alpha$, for each $\theta \neq \theta'$. When the most powerful unbiased level α test exists for problem (3.63), inversion of the acceptance region produces a uniformly most accurate unbiased confidence region. Analogously, if an optimal similar or invariant test exists, it is possible to define from (3.67) uniformly most accurate similar or unbiased confidence regions. For more details, see Lehmann (1986, sections 5.7 and 6.11).

Besides minimization of the non-null coverage probability, there are other criteria for optimality in confidence regions. One such criterion is minimization of the volume measure of the confidence region; however, this does not respect the principle of parameterization invariance.

3.5.5 Comments on the Relations with Likelihood Inference

Likelihood procedures may be defined in a rather broad range of settings, while optimum inference procedures only exist in very specific situations. Furthermore, when there is an optimum inference procedure, it often matches the corresponding likelihood procedure, or it gets very close to it when the sample size n diverges (cf. Problems 5.37 and 5.38). Indeed, in large samples likelihood methods have a broad range of features which it would be difficult to improve upon and, with a shrewdly chosen method, are almost entirely immune from the difficulties that frequentist inference meets with when establishing the proper reference sample space (see the discussion on the conditionality principle, sections 2.6-2.8, and the results of sections 11.3 and 11.5). Some more detailed comments on the relation between the frequency-decision and the Fisherian point of view are given in Appendix D.

3.6 The Empirical Distribution Function

Let $y = (y_1, \ldots, y_n)$ be a random sample of size n from a univariate random variable with unknown distribution function $F_0(\cdot)$. For the statistical model specified in the nonparametric form

$$\mathcal{F} = \{F_Y(y) : \quad F_Y(y) = \prod_{i=1}^{n} F(y_i), \quad F(\cdot) \text{ d.f. on } \mathbb{R}\} \qquad (3.69)$$

the empirical distribution function

$$\hat{F}_n(u) = \frac{1}{n} \sum_{i=1}^{n} I_{(-\infty, u]}(y_i) \qquad (3.70)$$

is a minimal sufficient statistic (see also Example 2.3).

3.6.1 Basic Properties

Since $\hat{F}_n(u)$ is the proportion of sample observations that have a value less than or equal to u, then $n\hat{F}_n(u) \sim Bi(n, F_0(u))$. Hence,

$$E_0(\hat{F}_n(u)) = F_0(u)$$

and

$$\text{Var}_0(\hat{F}_n(u)) = \frac{1}{n}F_0(u)(1 - F_0(u)) .$$

The subscript in $E_0(\cdot)$, $\text{Var}_0(\cdot)$ and in similar expressions indicates evaluation with respect to $F_0(\cdot)$. Furthermore,

$$\text{Cov}_0\left(\hat{F}_n(u), \hat{F}_n(v)\right) = \frac{1}{n}\left\{\min\left(F_0(u), F_0(v)\right) - F_0(u)F_0(v)\right\} . \qquad (3.71)$$

By the strong law of large numbers, $\hat{F}_n(u)$ will converge almost surely (hence also in probability) to $F_0(u)$ for every fixed value of u, as $n \to +\infty$. The following stronger result holds. Let $D_n = \sup_{u \in \mathbb{R}} |\hat{F}_n(u) - F_0(u)|$, then

$$P_0\left(\lim_{n \to +\infty} D_n = 0\right) = 1 .$$

If, furthermore, $F_0(u)$ is continuous, then for every fixed real values $u_1, \ldots, u_k$, the k-dimensional random variable with components $\sqrt{n}(\hat{F}_n(u_j) - F_0(u_j))$, $j = 1, \ldots, k$, will converge in distribution to a k-dimensional normal distribution $N_k(0, [\sigma_{jh}])$, with $\sigma_{jh} = \min(F_0(u_j), F_0(u_h)) - F_0(u_j)F_0(u_h)$, $j, h = 1, \ldots, k$. Note that this variable is non-degenerate provided $u_1, \ldots, u_k$ are inner points of the support of $F_0(\cdot)$.

3.6.2 Nonparametric Maximum Likelihood Estimate

Model (3.69) is not dominated. Hence, the likelihood function cannot be written according to the definition given in section 1.4.2. Kiefer and Wolfowitz (1956) suggested an extension of maximum likelihood estimation to non-dominated families. Let $P(\cdot)$ be the probability measure on $\mathbb{R}$ associated to a distribution function $F_Y(y)$ in (3.69). Denote by $\mathcal{P}$ the class of probability measures which correspond to $\mathcal{F}$. If P_1 and P_2 are in $\mathcal{P}$, P_1 is absolutely continuous with respect to $P_1 + P_2$ (cf. Problem 1.2). Therefore, the density function

$$p_{P_1+P_2}(y; P_1) = \frac{dP_1}{d(P_1 + P_2)}$$

can be defined using the Radon–Nikodym theorem. A probability measure $\hat{P}$ is a generalized maximum likelihood estimate if, for every $P \in \mathcal{P}$,

$$p_{\hat{P}+P}(y^{obs}; \hat{P}) \geq p_{\hat{P}+P}(y^{obs}; P) . \qquad (3.72)$$

In the dominated case, this definition will give the usual maximum likelihood estimate: one only has to identify the dominating measure, $P_1 + P_2$ with the common dominating measure μ^* and $P \in \mathcal{P}$ with the parameter θ that indexes P within $\mathcal{P}$. Note that the technical device behind definition (3.72) consists of only considering comparisons of likelihood between pairs of elements of the family.

It is possible to show that the empirical distribution function is the generalized maximum likelihood estimate of $F_0(\cdot)$. Essentially, the line of argument is as follows. Suppose, for simplicity, that there are no ties in y. If a measure $\hat{P}$ assigns positive probability to y^{obs}, then $p_{\hat{P}+P}(y^{obs}; P) = 0$ if P does not assign positive probability to y^{obs}. Determining $\hat{P}$ on the basis of (3.72) is, thus, the same as identifying the measure $\hat{P}$ on $\mathbb{R}$ such that

$$\hat{P}(y^{obs}) \geq P(y^{obs}), \qquad \text{for every } P \in \mathcal{P} , \tag{3.73}$$

where, in terms of the distribution function,

$$P(y) = P_F(y) = \prod_{i=1}^{n} \left\{ F(y_i) - F(y_i^-) \right\} ,$$

with $F(\cdot)$ a distribution function on $\mathbb{R}$. Thus, we should determine the vector of weights $(w_1, \ldots, w_n)$, with $w_i > 0$ and $\sum_{i=1}^{n} w_i \leq 1$, that maximizes the product

$$\prod_{i=1}^{n} w_i .$$

The weights w_i represent probabilities assigned to the components of y^{obs}. Using the well-known inequality between the geometric mean and the arithmetic mean we immediately conclude that the maximum is attained with $\hat{w}_i = n^{-1}$, $i = 1, \ldots, n$, that is, with the uniform distribution on $\{y_1^{obs}, \ldots, y_n^{obs}\}$; the corresponding distribution function is $\hat{F}_n(u)$. Intuitively, the result is obtained by interpreting the likelihood associated to a given distribution function as the probability which that distribution function assigns to a re-observation of y^{obs} in a repetition of the experiment.

3.6.3 Statistical Functionals

In the setting of the model (3.69), inference is often not focussed on the overall behaviour of the distribution function $F_0(\cdot)$, but is restricted to specific aspects, usually moments, or, more generally linear functionals

$$T(F_0) = \int g(y) \, dF_0(y) , \tag{3.74}$$

where $g(\cdot)$ is a real function. Since $\hat{F}_n$ is the generalized maximum likelihood estimate of $F(\cdot)$, by analogy, it is natural to estimate $T(F)$ using the linear statistical functional

$$T_n = T(\hat{F}_n) = \int g(y) \, d\hat{F}_n(y) = \frac{1}{n} \sum_{i=1}^{n} g(y_i) \, . \tag{3.75}$$

For example, if $g(y) = y$, the functional $T(F)$ is the expectation under $F(\cdot)$ and $T(\hat{F}_n)$ will be the sample mean $\bar{y}_n$. If $g(y) = I_{(-\infty, y_0]}(y)$, the corresponding linear functional is $F(y_0)$ and $T(\hat{F}_n) = \hat{F}_n(y_0)$.

Of course, inference can also concern non-linear functionals: for example, a generic M-estimator is defined as the solution of an estimating equation of the form

$$\sum_{i=1}^{n} q(y_i; \theta) = 0 \, , \tag{3.76}$$

where $q(y_i; \theta)$ is a real-valued combinant (or, more generally, a combinant with values in $\mathbb{R}^p$). Equation (3.76) implicitly defines a statistical functional $T(\hat{F}_n)$, which is the solution in θ of the equation

$$\int q(y; \theta) \, d\hat{F}_n(y) = 0 \, .$$

$T(\hat{F}_n)$ estimates the functional $T(F)$, solution in θ of the equation

$$\int q(y; \theta) \, dF(y) = 0 \, .$$

The maximum likelihood estimator under random sampling from a regular parametric model is an instance of a statistical functional. Indeed, it is the solution of (3.76) with $q(y_i; \theta) = l_*(\theta; y_i)$, where $l_*(\theta; y_i)$ is the contribution to the score of the i-th observation. An M-estimator of a location parameter is usually defined by using a combinant of the form $q(y; \theta) = q(y - \theta)$ in (3.76).

Under regularity conditions, the estimator $T(\hat{F}_n)$ is both strongly and weakly consistent. Furthermore, whenever the empirical distribution function exactly matches the population distribution function $F(\cdot)$ (clearly this is possible with a finite sample only if the random variable is discrete), then the estimate will coincide with the parameter $T(F)$. This property is known as Fisher consistency (Fisher, 1922a). This concept of consistency, which is not necessarily asymptotic, somehow fits the aims of inference better than it does the notion of convergence of a sequence of random variables. The latter in some way goes beyond the principle of repeated sampling.

The asymptotic behaviour of a linear statistical functional can be deduced directly from the central limit theorem. For a non-linear functional $T(\hat{F}_n)$, it may prove convenient to consider a local linearization. This technique, which was developed by von Mises (1947), is a generalization of the delta method (see section A.3) to a function space, in particular to a space of distribution functions. Here, firstly, a definition of the derivative of a functional must be given.

Let $T(\cdot)$ be a functional defined on a family $\mathcal{F}$ of distribution functions and assume that $\mathcal{F}$ is closed under convex linear combinations. The Gâteaux derivative of $T(F)$ at the point F_0 is defined as

$$T'_{F_0}(F - F_0) = \lim_{\varepsilon \to 0+} \frac{T((1 - \varepsilon)F_0 + \varepsilon F) - T(F_0)}{\varepsilon} , \qquad (3.77)$$

so long as there exists a function $H(y; T, F_0)$, independent of F, such that

$$T'_{F_0}(F - F_0) = \int H(u; T, F_0) d(F - F_0)(u) .$$

The function $H(u; T, F_0)$ is defined up to an additive constant and is usually normalized so as to satisfy $\int H(u; T, F_0)\, dF_0(u) = 0$. Since $H(\cdot)$ must be independent of F, it can be calculated on the basis of (3.77) using, for example, $F(u) = F_{D(y)}(u)$, the distribution function of a random variable degenerate at y. Thus,

$$H(y; T, F_0) = \lim_{\varepsilon \to 0+} \frac{T((1 - \varepsilon)F_0 + \varepsilon F_{D(y)}) - T(F_0)}{\varepsilon} . \qquad (3.78)$$

The function $H(\cdot)$ defined by (3.78) is called the influence function and is usually indicated by $IF(y; T, F_0)$. It describes the effect of an infinitesimal contamination of the point y on the functional $T(F_0)$, an effect which is standardized according to the weight of the contamination.

The above interpretation (Hampel, 1974) is useful in the study of estimators which are robust under contamination. Robustness is defined in terms of properties of the influence function. The most obvious and important condition is that the gross-error sensitivity

$$\gamma = \sup_y |IF(y; T, F_0)|$$

should be finite. It is not difficult to show, (see, e.g., Hampel, Ronchetti, Rousseeuw and Stahel, 1986, p. 101) that for an M-estimator, defined by an equation of the form (3.76), the influence function is

$$IF(y; q, F_0) = \frac{q(y; T(F_0))}{- \int (\partial/\partial\theta) q(u; \theta)|_{T(F_0)} \, dF_0(u)} \, ,$$

where $|_{T(F_0)}$ indicates evaluation at $\theta = T(F_0)$. Therefore, if the denominator of the preceding expression does not vanish, γ is finite if, and only if, $q(y; \theta)$ is bounded.

Example 3.8 *Huber estimator*
Let $y_1, \ldots, y_n$ be independent observations from a $N(\theta, 1)$ distribution. The maximum likelihood estimator $\bar{y}_n$ is the solution of (3.76) with $q(y_i; \theta) = y_i - \theta$. The corresponding influence function is not bounded because it is proportional to $y - \theta$. An estimator which is robust with respect to infinitesimal contamination can, thus, be defined by substituting the combinant $q(y; \theta) = y - \theta$ in the estimating equation (3.76) with the bounded combinant

$$q_b(y; \theta) = (y - \theta) \min(1, b/|y - \theta|) \, ,$$

where b is an assigned positive constant. This estimator was introduced by Huber (1964). It has minimum asymptotic variance in the class of Fisher consistent estimators with a bounded γ index (see Hampel, Ronchetti, Rousseeuw and Stahel, 1986, sections 2.4a-2.4b). In particular, γ satisfies the inequality $\gamma < b/\{2\Phi(b) - 1\}$. $\triangle$

Generalization of the delta method to differentiable statistical functionals $T(\hat{F}_n)$, is based on the local linearization obtained by expanding $T((1-\varepsilon)F_0 + \varepsilon\hat{F}_n)$ in a Taylor series around $\varepsilon = 0$. With $\varepsilon = 1$, we obtain the von Mises expansion

$$
\begin{aligned}
T(\hat{F}_n) &= T(F_0) + T'_{F_0}(\hat{F}_n - F_0) + R(\hat{F}_n, F_0) \\
&= T(F_0) + \int IF(y; T, F_0) d\hat{F}_n(y) + R(\hat{F}_n, F_0) \, ,
\end{aligned} \qquad (3.79)
$$

where $R(\hat{F}_n, F_0)$ is a remainder term. Thus we have the asymptotic representation

$$\sqrt{n}(T(\hat{F}_n) - T(F_0)) = \frac{1}{\sqrt{n}} \sum_{i=1}^{n} IF(y_i; T, F_0) + \sqrt{n} \, R(\hat{F}_n, F_0) \, , \qquad (3.80)$$

on the basis of which, provided $\sqrt{n}\, R(\hat{F}_n, F_0) = o_p(1)$ and

$$0 < E_0(IF(Y; T, F_0))^2 = \sigma_T^2 < +\infty\,, \qquad (3.81)$$

we get

$$\sqrt{n}(T(\hat{F}_n) - T(F_0)) \overset{d}{\to} N(0, \sigma_T^2)\,. \qquad (3.82)$$

The asymptotic linearization (3.80) is also crucial in the study of asymptotic efficiency of estimators in nonparametric and semiparametric settings (Pfanzagl, 1990, section A.2; Bickel, Klaassen, Ritov, Wellner, 1993, section 3.3). Any 'regular' asymptotically efficient estimator $T(\hat{F}_n)$ of $T(F_0)$ is *linear*, i.e. it has the representation (3.80) with $\sqrt{n}\, R(\hat{F}_n, F_0) = o_p(1)$, with a specific form of the influence function.

Substituting $\hat{F}_n$ for F in the functional $T(F)$ provides a general solution to inference problems in nonparametric models. The first-order asymptotic result (3.82) gives an approximate solution to the distribution problem for $T(\hat{F}_n)$.

One other approach to the problem of distribution for statistical functionals uses the empirical distribution function. The distribution function of $T(\hat{F}_n)$ for finite n can be represented in an abstract form as $F_{T(\hat{F}_n)}(t; F)$, for $F \in \mathcal{F}$, where the nature of a parameter of F in $\mathcal{F}$ is highlighted. One natural idea for eliminating the unknown parameter F in $F_{T(\hat{F}_n)}(t; F)$ is to estimate it using its generalized maximum likelihood estimate. Substituting F with $\hat{F}_n$ is the key idea of the **bootstrap** method. In most applications, this method also includes a second step of approximation: the distribution $F_{T(\hat{F}_n)}(t; \hat{F}_n)$, which is usually difficult to determine analytically, is estimated through simulation, that is, on the basis of a large number of random samples of size n from the distribution $\hat{F}_n$; this is the same as carrying out sampling with replacement from $\{y_1^{obs}, \ldots, y_n^{obs}\}$. Under regularity conditions, without further corrections, the asymptotic error of bootstrap approximations is of the same order in n as the error for the central limit approximation. Note that the bootstrap distribution can be obtained completely automatically, in particular, it does not require any evaluation of asymptotic variance.

3.7   Bibliographic Note

The notions about moments, cumulants and their generating functions, mentioned in section 3.2, can be found e.g. in Barndorff-Nielsen and Cox (1989, Chapters 1 and 5) and in McCullagh (1987, Chapter 2). Lukács (1970) is an important reference for characteristic functions. Some applications of stable laws in statistics were mentioned in Chapter 1. Hougaard (1986) highlights a parametric family of positive distributions, derived from stable laws, useful in survival analysis. Janicki and Weron (1994) illustrate the usefulness of stable distributions and processes in various contexts. The relations between moments and cumulants (formulae (3.14)-(3.17), (3.27), (3.28) and (3.29)) can be easily generalized to moments and cumulants of any order (see formulae (9.28) and (9.29)).

The notations $o_p(\cdot)$ and $O_p(\cdot)$, introduced by Mann and Wald (1943), are illustrated by detailed examples in Bishop, Fienberg and Holland (1975, Chapter 14). In addition to the list of asymptotic results given in Appendix A, it might be useful to see Rao (1973, sections 2c and 6a.2) and Serfling (1980). Azzalini (1996, sections 3.3, 4.2 and 4.3) gives an introduction to first-order asymptotic theory for likelihood quantities; see also Cox and Hinkley (1974, Chapter 9) and Barndorff-Nielsen and Cox (1994, sections 3.1-3.3) for a more in-depth treatment. Tsou and Royall (1995) introduce various notions of likelihood robustness. In this chapter, there has been no discussion of non-regular problems where the usual asymptotic results for the null distribution of the maximum likelihood estimator and related quantities do not hold; a recent reference is the discussion paper by Cheng and Traylor (1995).

The basic references for the classical theory of statistical inference from the decision viewpoint are the books by Lehmann (1983, 1986) and Ferguson (1967). See, in particular, Chapter 10 in Lehmann (1986) for a lucid discussion of the implications of the conditionality principle in testing hypotheses. One useful reference for a concise description of basic results is Fraser (1957, Chapter 2).

Ideas about the empirical distribution function, statistical functionals and the bootstrap in section 3.6, have been given for simplicity in the one-dimensional case with independent, identically distributed observations. Kaplan and Meier (1958) obtain a nonparametric generalized maximum likelihood estimate of the distribution function with censored data; see Lawless (1982, section 2.3.1) for an introduction and Andersen, Borgan, Gill and Keiding (1993) for a treatment that is based on the theory of the product integral and also for multivariate extensions (*ibidem*, section 10.3). For extensions of the definition of the influence function and M-estimators to multiparameter models, to regression problems and to dependent observations, see Hampel, Ronchetti, Rousseeuw and Stahel (1986) and the references given there; for regression problems in particular, see also, Rousseeuw and Leroy (1987), and Koul (1992). The bootstrap was introduced by Efron (1979, 1982). Brief introductions are given by Efron and Tibshirani (1986), DiCiccio and Romano (1988), Hinkley (1988), Davison (1990). Efron and Tibshirani (1993) and Hall (1992) are recent monographs on the subject; the former is mainly introductory while the latter is more specialized and gives a theoretical justification of asymptotic results. For a discussion of bootstrap methods, see Young (1994).

3.8 Problems

3.1 Show that if Y has a log-normal distribution with density

$$p(y; \mu, \sigma) = \frac{1}{\sqrt{2\pi}\,\sigma y} e^{-\frac{1}{2\sigma^2}(\log y - \mu)^2}, \quad y > 0, \mu \in \mathbb{R}, \sigma > 0,$$

then $M_Y(t)$ is not finite for $t > 0$.

[Section 3.2.1]

3.2 Show that, if X is an exponential r.v. with density $p_X(x) = e^{-x}$, $x > 0$, the m.g.f. of $Y = X^2$ does not exist.

[Section 3.2.1]

3.3 A parametric family is closed under convolution if the sum of two independent elements has a distribution that still belongs to the same family. Using (3.6) show that the following univariate families are closed under convolution: Poisson, binomial (with fixed probability), normal and gamma (with fixed scale).

[Section 3.2.4]

3.4 Rearrange the expression of $M_{S_n}(t)$ given by (3.8) collecting $t^r/r!$, $r = 1, \ldots, 4$. In this way obtain the first four moments of S_n as a function of the moments of Y.

[Section 3.2.4]

3.5 Obtain expressions for cumulants up to the fourth order for the distributions considered in Table 3.1.

[Section 3.2.5]

3.6 Show that the standardized cumulants defined by (3.20) are invariant under scale and location changes of Y.

[Section 3.2.5]

3.7 Show that all stable laws are continuous.

[Section 3.2.6]

3.8 Show that the p.d.f. (3.21) matches that of the reciprocal of a gamma random variable. Find the values of the parameters of this gamma distribution.

[Section 3.2.6]

3.9 Let (Y_1, Y_2) be a bivariate r.v. where the conditional distribution of Y_2 given $Y_1 = y_1$ is Poisson with mean μy_1 and where Y_1 has a marginal gamma distribution with unit mean and variance σ^2. Show that Y_2 has a negative binomial marginal distribution.

[Section 3.2]

3.10 Express the m.g.f. of the generic k-dimensional marginal distribution of the d-dimensional r.v. Y ($k < d$) as a function of the joint m.g.f. of Y.

[Section 3.2.6]

3.11 Check that for $d = 1$ the relations (3.27), (3.28) and (3.29) give (3.15)-(3.17). Adapt the expressions of $\kappa^{i,j}$ and $\kappa^{i,j,k}$ to the case $d = 2$.

[Section 3.2.6]

3.12 Let $(x_1, y_1), \ldots, (x_n, y_n)$ be n i.i.d. observations from a bivariate distribution with finite marginal variances. Let $(\bar{X}_n, \bar{Y}_n)$ be the bivariate sample mean. Use the delta method to obtain an approximation (up to order $O(n^{-1})$) for the mean and for the variance of $\bar{X}_n/\bar{Y}_n$.

[Sections 3.3.2 and A.3]

3.13 Let $y_1, \ldots, y_n$ be n observations whose joint distribution depends on a p-dimensional parameter θ. Let $q_n = q_n(y_1, \ldots, y_n; \theta)$ be a combinant having a limiting normal null distribution with zero mean. Let $\hat{\theta}_n = \hat{\theta}_n(y_1, \ldots, y_n)$ be an asymptotically normal and efficient sequence of estimators of θ and let us write $\hat{q}_n = q_n(y_1, \ldots, y_n; \hat{\theta}_n)$. Assume in addition that

$$\begin{pmatrix} \sqrt{n}q_n \\ \sqrt{n}(\hat{\theta}_n - \theta) \end{pmatrix} \xrightarrow{d} N_{p+1}\left(0, \begin{pmatrix} V_{11} & V_{12} \\ V_{21} & V_{22} \end{pmatrix} \right),$$

with a nonsingular covariance matrix, and that an expansion of the form

$$\sqrt{n}\hat{q}_n = \sqrt{n}q_n + B\sqrt{n}(\hat{\theta}_n - \theta) + o_p(1)$$

holds. Show that

$$\sqrt{n}\hat{q}_n \xrightarrow{d} N(0, V_{11} - BV_{22}B^{\mathrm{T}}).$$

[Sections 3.3.2 and A.3; Pierce, 1982; Barndorff-Nielsen and Cox, 1994, p. 104]

3.14 Let $y_1, \ldots, y_n$ be independent observations from an exponential distribution with mean θ. Use the results summarized in Appendix A to obtain an asymptotic approximation for the distribution of $\sum y_i^2 / (\sum y_i)^2$.

[Section 3.3]

3.15 Let $y_1, \ldots, y_n$ be independent observations from a $N(\mu, \sigma^2)$ distribution. Use the results in section A.3 to show that

$$\left(\frac{\bar{m}_3}{(\bar{m}_2)^{3/2}}, \frac{\bar{m}_4}{(\bar{m}_2)^2} \right) ,$$

with $\bar{m}_j = n^{-1} \sum_{i=1}^{n} (y_i - \bar{y}_n)^j$, $j = 2, 3, 4$, has an approximate bivariate normal distribution with mean vector $(0, 3)$, uncorrelated components and marginal variances $6/n$ and $24/n$.

[Sections 3.3.2 and A.3]

3.16 Let $a_n(\tau)$ be a sequence depending on a parameter τ; define the following property: $a_n(\tau)$ is of order $O(n^\alpha)$ uniformly in τ.

[Section 3.3.3]

3.17 Establish the orders of magnitude of the following sequences of random variables:
i) maximum of n independent observations from a uniform distribution on $[0, \theta]$, $\theta > 0$;
ii) minimum of n independent observations from a distribution with density

$$p_Y(y; \mu, \sigma) = (1/\sigma) \exp(-(y - \mu)/\sigma), \quad y \geq \mu, \ \mu \in \mathbb{R}, \ \sigma > 0 .$$

[Section 3.3.3]

3.18 Establish the order in probability of the minimum of n i.i.d. observations from a two-parameter Weibull distribution with p.d.f. (1.29).

[Section 3.3.3]

3.19 Show that in the local non-null asymptotic distribution (3.47) the information $i_1(\theta_0)$ can be replaced by $i_1(\theta)$.

[Section 3.4.2]

3.20 Show the consistency result (3.48) for W.

[Section 3.4.2]

3.21 Obtain the robust estimator of the asymptotic variance of $\hat{\theta}$ indicated in (3.51) when the working model is *i*) Poisson with mean θ; *ii*) $Bi(k, \theta)$.

[Section 3.4.3]

3.22 Let $y_1, \ldots, y_n$ be independent observations from a $N(\mu, \sigma^2)$ distribution. Obtain a correction for the maximum likelihood estimator of σ^2 which will make it median unbiased.

[Section 3.5.2]

3.23 Let y be one observation from a binomial distribution with index $n = 3$ and parameter θ. Consider the problem of testing $H_0 : \theta = 1/4$ versus $H_1 : \theta = 1/2$. Find the levels α which can exactly be reached by the likelihood ratio test.

[Section 3.5.3]

3.24 Show that the locally most powerful unbiased test (3.64) for problem (3.63) is parameterization invariant.

[Section 3.5.3]

3.25 Give a detailed proof of (3.71).

[Section 3.6.1]

CHAPTER 4

NUISANCE PARAMETERS AND PSEUDO-LIKELIHOODS

4.1 Nuisance Parameters

A statistical model $\mathcal{F}$ is a collection of possible distributions for the observed data. There are usually distributional aspects of primary concern for inference and further aspects, that the model must also take into account in order to adequately describe variability in the population.

Suppose that the aspects that are of primary interest are expressed through a k-dimensional parameter τ. Aspects of secondary interest can, in their turn, be described either parametrically, or nonparametrically: let us denote these aspects by ζ. The term **parameter of interest** is used to refer to τ and **nuisance parameter** to refer to ζ. This terminology refers to *nuisance* only in a conventional meaning even though it should be said that the statistical methods available for inference about τ, if ζ is known are, generally, simpler and more efficacious. If ζ is not known, then inferential procedures about τ which have satisfactory properties in a wider range of situations must be used.

The simplest setting is when a parametric model $\mathcal{F} = \{p_Y(y; \theta), y \in \mathcal{Y}, \theta \in \Theta\}$, $\Theta \subseteq \mathbb{R}^p$, is assumed, and θ is partitioned as $\theta = (\tau, \zeta)$, with τ of dimension $k < p$ and ζ of dimension $p - k$; furthermore, suppose that the parameters $\tau \in T$ and $\zeta \in Z$ are variation independent, that is, $\Theta = T \times Z$. More generally, $\tau = \tau(\theta)$, where $\tau(\cdot)$ is not a one-to-one function.

More complex examples can be found by looking at semiparametric models, such as those described in section 1.3.1. In the proportional hazards model, where the failure rate of the i-th item is given by (1.2), the parameter is $\theta = (\beta_1, \ldots, \beta_k, r_0(t))$ and, typically, the baseline hazard function $r_0(t)$ is treated as a nuisance parameter. Analogously, in the class of density functions $p_0(y - \mu)$, which are symmetric around $\mu \in \mathbb{R}$, μ often is the parameter of interest, while $p_0(\cdot)$ is treated as a nuisance parameter. Lastly, under nonparametric models, the quantity of interest can be a k-dimensional functional $\tau(F)$.

The definition of nuisance parameters has been given in general terms. It is convenient to distinguish some particular situations. Suppose that the data y are n observations $(y_1, \ldots, y_n)$, assumed, for simplicity, to be independent and with joint density

$$p_Y(y; \theta) = \prod_{i=1}^{n} p_{Y_i}(y_i; \theta_i) \,.$$

Two extreme possibilities arise:
i) $\theta_i = (\tau, \zeta)$, $i = 1, \ldots, n$,
ii) $\theta_i = (\tau, \zeta_i)$, $i = 1, \ldots, n$,
where we keep the assumption that the parameter of interest and the nuisance parameter are variation independent. In the first case, the nuisance parameter has the same value for each observation and the dimension of θ does not depend on n. In the second, the whole parameter is $\theta = (\tau, \zeta_1, \ldots, \zeta_n)$ and, while the parameter of interest remains the same for each observation, a nuisance component ζ_i is linked to every y_i, and takes into account the heterogeneity of the observations. Thus, the dimension of $\zeta = (\zeta_1, \ldots, \zeta_n)$ does depend on n. Nuisance parameters as in ii) are called **incidental parameters** (Neyman and Scott, 1948); in this context the parameter τ which expresses the common structure of the observations is called the **structural parameter**. Intermediate situations arise when different values of the nuisance parameter take block factors into account.

Data and model reduction and inference procedures examined in Chapters 2 and 3 can, to some extent, be extended to deal with inference problems when nuisance parameters are present. Incidental parameters, however, pose entirely new problems.

Example 4.1 *Neyman–Scott problem, normal distribution*
Let (y_{i1}, y_{i2}), $i = 1, \ldots, n$, be independent observations from $N(\mu_i, \sigma^2)$ distributions. The overall parameter is $\theta = (\sigma^2, \mu_1, \ldots, \mu_n)$. The maximum likelihood estimator of σ^2, $\hat{\sigma}^2$, is not consistent. Furthermore, the log-likelihood ratio statistic

$$W = 2\left\{ l(\hat{\sigma}^2, \hat{\mu}_1, \ldots, \hat{\mu}_n) - l(\sigma^2, \hat{\mu}_1, \ldots, \hat{\mu}_n) \right\}$$

is

$$W = -2n \log \left\{ 1 + \left(\frac{\hat{\sigma}^2}{\sigma^2} - 1 \right) \right\} + 2n \left(\frac{\hat{\sigma}^2}{\sigma^2} - 1 \right)$$

and its null expectation is of order $O(n)$, but not $o(n)$. Details are left to the reader, see Problem 4.1.   △

Example 4.2 *Rasch model*
A group of n subjects undergo k ability tests. Let y_{ij}, $i = 1, \ldots, n$, $j = 1, \ldots, k$, be the outcome of the j-th test on the i-th subject; $y_{ij} = 1$ if the answer is correct and $y_{ij} = 0$ if it is wrong. Assume that for the i-th subject, the probability of a correct answer to the j-th test is

$$\pi_{ij} = \frac{\exp(\zeta_i - \tau_j)}{1 + \exp(\zeta_i - \tau_j)} \, ,$$

where ζ_i and τ_j represent, respectively, the ability of the i-th subject and the difficulty of the j-th test. For identifiability, assume that $\sum \tau_j = 0$. If inference aims at evaluating the relative difficulty of tests, $\zeta = (\zeta_1, \ldots, \zeta_n)$ is considered to be a nuisance parameter. Andersen (1980, p. 224) showed that the maximum likelihood estimator of τ is not consistent.   △

The anomalies highlighted in Examples 4.1 and 4.2 are called **Neyman–Scott problems**. Solutions to deal with some particular cases will be given in Examples 4.7, 4.12 and 4.13. In this chapter we will deal mainly with the simplest setting, that is when the dimension of ζ does not depend on n. References to general methods for dealing with incidental parameters are given in section 4.11.

4.2 Data and Model Reduction with Nuisance Parameters

When the whole parameter θ is of interest, data and model reduction is relatively straightforward, even though with some controversial aspects related to the conditionality principle. Rather more complicated problems arise when reduction has to take into account nuisance components. There is no definitive treatment of this problem in the Fisherian paradigm: theoretical investigation has resulted in a multiplicity of definitions to justify specific reductions. Only some basic concepts will be introduced here; for a detailed account the reader should refer to Jørgensen (1993).

4.2.1   Lack of Information: No Nuisance Parameters

Firstly, we will again consider the special case where there are no nuisance
parameters. Reduction to a marginal model (carried out most efficaciously by
a minimal sufficient statistic) and reduction to a conditional model, given a
distribution constant statistic (carried out most efficaciously if that statistic is
maximal), are formally unified by the factorization

$$p_Y(y;\theta) = p_V(v;\theta)p_{U|V=v}(u;v,\theta)p_{Y|U=u,V=v}(y;u,v,\theta) , \qquad (4.1)$$

where $u = u(y)$ and $v = v(y)$ are suitable statistics. Equation (4.1) interprets
the data y as the outcome of a three-stage experiment and gives the following
factorization of the likelihood function

$$L_Y(\theta) = L_V(\theta)L_{U|V=v}(\theta)L_{Y|U=u,V=v}(\theta) , \qquad (4.2)$$

where $L_V(\theta)$, $L_{U|V=v}(\theta)$, $L_{Y|U=u,V=v}(\theta)$ are the likelihood factors associated to
the corresponding stages of the whole experiment.

Reductions via marginalizing and via conditioning can be treated in a unified way with reference to (4.1). Indeed, both reductions may be seen as the
consequence of the independence from θ of the factor $p_{U|V}(\cdot)$ in (4.1), so that
the likelihood factor $L_{U|V=v}(\cdot)$ can be neglected in (4.2). In particular, if y is
in a one-to-one correspondence with (u,v) and $p_{U|V}(\cdot)$ does not depend on θ,
the statistic v is sufficient and inference can be based on its marginal distri-
bution. If, on the other hand, V is degenerate and $p_{U|V}(\cdot) = p_U(\cdot)$ does not
depend on θ, the statistic u is distribution constant and, by the conditionality
principle, inference should be based on the distribution of Y conditionally on
the observed value of u. Of course, we must assume that the parameter θ
is identifiable both in the statistical model for Y and in reduced marginal or
conditional models. Note that reduction by conditioning on ancillary statistics can easily be fitted into the general scheme (4.1) by interpreting y as an
observation from a minimal sufficient statistic.

The completeness properties defined in section 2.5 are useful for establishing
whether further reduction is possible in a reduced marginal or conditional
model. However, completeness of a family of distributions may not be easy to
check and it is convenient to introduce a new definition of an exhaustive model
reduction, which corresponds somehow to a pragmatic notion of completeness.

Definition 4.1 A statistical model $\mathcal{F} = \{p_Y(y;\theta),\ \theta \in \Theta\}$ is called **saturated**
if for every $y \in \mathcal{Y}$ the maximum likelihood estimate $\hat{\theta} = \hat{\theta}(y)$ is unique and is
a one-to-one function of y.

This definition corresponds to the idea of *one parameter for every observation*. Note that saturation is a property of a statistical model that respects both parameterization invariance and invariance under one-to-one transformations of data.

Intuitively, following Jørgensen (1993), we could say that inference has no further possibility for reduction in a saturated model. Indeed, since interpretation of the data appears to be perfect in the light of the model, then there are no further degrees of freedom left to test the model. The maximum reduction to a marginal model, induced by a sufficient statistic, or to a conditional model, given a distribution constant statistic, must then correspond to saturation of the reduced model.

4.2.2 Lack of Information in the Presence of Nuisance Parameters

Now, let us move on to consider the extension of the notions of data and model reduction to the more general case where $\theta = (\tau, \zeta)$ and only τ is of interest. Here, the aim is to define absence of information about the parameter of interest.

Given two statistics u and v, we can say that, given $V = v$, u is not informative on τ if the distribution of U, conditionally on $V = v$, does not depend on τ (it can however depend on ζ). This notion can be termed partial non-information, meaning that there is no information about the component of interest of the parameter, and is the basis for the concepts of partially sufficient and partially distribution constant statistics, for suitable choices of u and v.

Consider factorization (4.1) written as

$$p_Y(y; \tau, \zeta) = p_V(v; \tau, \zeta) p_{U|V=v}(u; v, \tau, \zeta) p_{Y|U=u, V=v}(y; u, v, \tau, \zeta) . \qquad (4.3)$$

Whenever global reduction for inference about $\theta = (\tau, \zeta)$ is possible, then it is convenient to think that $p_Y(\cdot)$ in (4.3) is the corresponding reduced model. This has to be assumed in particular when a reduction by sufficiency is available.

Definition 4.2 With reference to (4.3), if y is a one-to-one function of (u, v), then v is partially sufficient for τ if $p_{U|V=v}(u; v, \tau, \zeta) = p_{U|V=v}(u; v, \zeta)$.

Definition 4.3 With reference to (4.3) with V degenerate, u is partially distribution constant for τ if $p_{U|V=v}(u; v, \tau, \zeta) = p_U(u; \zeta)$.

Partial sufficiency and partial distribution constancy are matched by the corresponding principles of data and model reduction: the partial sufficiency principle and the **partial conditionality principle**. If v is partially sufficient for τ, the contribution to the likelihood of $p_{U|V=v}(u; v, \zeta)$ in (4.3) can be neglected for inference about τ. If u is partially distribution constant for τ, one can neglect the contribution of $p_U(u; \zeta)$. In both cases, the remaining likelihood factor generally depends both on the parameter of interest τ and on the nuisance parameter ζ. This does not happen in the two special cases below.

(a) The statistic v is both partially sufficient for τ and partially distribution constant for ζ, i.e.

$$p_{U,V}(u, v; \tau, \zeta) = p_V(v; \tau)p_{U|V=v}(u; v, \zeta) . \tag{4.4}$$

(b) The statistic u is both partially distribution constant for τ and partially sufficient for ζ, i.e.

$$p_Y(y; \tau, \zeta) = p_U(u; \zeta)p_{Y|U=u}(y; u, \tau) . \tag{4.5}$$

An exhaustive model reduction in (4.4) and (4.5) is when the models with densities $p_V(v; \tau)$ and $p_{Y|U=u}(y; u, \tau)$, respectively, are saturated.

Under (4.4) or (4.5) the likelihood function $L(\theta)$ factorizes as

$$L(\theta) = L_1(\tau)L_2(\zeta) . \tag{4.6}$$

If a likelihood function factorizes as in (4.6), it is said to have **separable parameters**. Observed likelihood quantities for inference about τ only depend on $L_1(\tau)$ and inference about τ can take place separately from that about ζ, as if the value of ζ were known. Examples of likelihood functions with separable parameters are relatively rare.

Example 4.3 *Linear regression model*
Let $(y_i, x_{i1}, \ldots, x_{ik})$, $i = 1, \ldots, n$, be independent observations from a $(k+1)$-dimensional random variable $(Y, X_1, \ldots, X_k)$ such that

$$Y = \beta_1 X_1 + \ldots + \beta_k X_k + \sigma\varepsilon ,$$

where $\beta = (\beta_1, \ldots, \beta_k)$ is a vector of unknown regression coefficients and ε is a random error independent of $(X_1, \ldots, X_k)$ with known distribution $p_\varepsilon(\varepsilon)$. If the marginal distribution of $(X_1, \ldots, X_k)$ does not depend on $\tau = (\beta, \sigma)$, but only on a parameter ζ, which is variation independent of τ, the joint density of $(y, x_1, \ldots, x_k)$ can be factorized in the form

$$p_{Y, X_1, \ldots, X_k}(y, x_1, \ldots, x_k; \tau, \zeta) = p_{X_1, \ldots, X_k}(x_1, \ldots, x_k; \zeta)$$
$$\times\, p_{Y|X_1 = x_1, \ldots, X_k = x_k}(y; x_1, \ldots, x_k, \tau)$$
$$= p_{X_1, \ldots, X_k}(x_1, \ldots, x_k; \zeta)\, p_\epsilon((y - \beta_1 x_1 - \ldots - \beta_k x_k)/\sigma)\,.$$

The corresponding likelihood has separable parameters and $(x_{i1}, \ldots, x_{ik})$, $i = 1, \ldots, n$, is partially distribution constant for τ and partially sufficient for ζ. On the basis of the partial conditionality principle, inference about (β, σ) should be carried out as if the covariates $(x_{i1}, \ldots, x_{ik})$, $i = 1, \ldots, n$, were fixed. This is the usual procedure in applications, where covariates are treated as known experimental conditions. $\triangle$

Example 4.4 *Uninformative censoring*
Consider Example 1.2 with $\theta = \tau$ and with the censoring times c_i independent observations from a continuous random variable with density $p_c(c; \zeta)$, $\zeta \in Z$. The likelihood function based on (1.23) has separable parameters. Censoring times are partially distribution constant for τ and partially sufficient for ζ. According to the partial conditionality principle, inference about τ can still be based on the likelihood function (1.22). $\triangle$

4.2.3 Weaker Concepts of Lack of Information with Nuisance Parameters

Usually, the likelihood function does not have separable parameters and in most applications the first objective of data and model reduction in the presence of nuisance parameters is to single out a likelihood factor which depends only on τ. This can be obtained, for instance, by conditioning if a partially sufficient statistic for ζ is available, or by marginalizing if a partially distribution constant statistic for ζ is known.

Since a likelihood factor which depends both on τ and on ζ is neglected by this route, the key problem is now to define the **absence of relevant information about the parameter of interest** in a likelihood factor $L_{U|V=v}(\tau, \zeta)$. Jørgensen (1993) observed that the notion of a saturated model could be particularly helpful in this. Indeed, consider the conditional model $p_{U|V=v}(u; v, \tau, \zeta)$: if for every fixed v in the support $\mathcal{V}$ of V and for every fixed $\tau \in T$ this is a model for u saturated with respect to ζ, the model $p_{U|V=v}(u; v, \tau, \hat{\zeta}_\tau(u))$ *explains* the observation u equally well for every τ, in the sense that it always interprets it perfectly. If, given τ, the statistic u fulfills its informative role in determining the maximum likelihood estimate $\hat{\zeta}_\tau(u)$, we could say that there is no more useful information about τ in the likelihood factor being examined.

If the data y are a one-to-one function of (u, v) and v is partially distribution constant for ζ, then

$$p_{U,V}(u, v; \tau, \zeta) = p_V(v; \tau)p_{U|V=v}(u; v, \tau, \zeta) . \qquad (4.7)$$

If the model $p_{U|V=v}(u; v, \tau, \zeta)$, given τ, is saturated, all the relevant information about τ is contained in the marginal density $p_V(v; \tau)$. According to Jørgensen's (1993) definition, v is I-sufficient for τ. We shall see in section 7.7.1 that composite group families express a factorization of the form (4.7).

Now, let us consider the case where u is partially sufficient for ζ, i.e.

$$p_Y(y; \tau, \zeta) = p_U(u; \tau, \zeta)p_{Y|U=u}(y; u, \tau) , \qquad (4.8)$$

and the marginal model $p_U(u; \tau, \zeta)$, with τ fixed, is saturated. All the relevant information about τ is contained in the conditional density $p_{Y|U=u}(y; u, \tau)$. Following Jørgensen (1993), u is called I-distribution constant for τ. We shall see in section 5.4.2 that exponential families express a factorization of the form (4.8). Analogous examples can be found in exponential dispersion models (see Chapter 6).

However, saturation with respect to ζ of the model $p_{U|V=v}(u; \tau, \zeta)$, with τ fixed, does not exclude the possibility that u may still be useful to attribute different levels of plausibility to the various values of τ, under the model $p_{U|V=v}(u; \tau, \hat{\zeta}_\tau(u))$. One possible requirement for tightening up the argument for neglecting the likelihood factor $L_{U|V=v}(\tau, \zeta)$ is to require that τ should not be identifiable within the model $p_{U|V=v}(u; \tau, \hat{\zeta}_\tau(u))$, or, more pragmatically, that the maximum likelihood estimate of τ, based on the conditional model $p_{U|V=v}(u; \tau, \hat{\zeta}_\tau(u))$ with data u, is not defined, or does not depend on u. This corresponds to yet another notion of absence of relevant information about the parameter of interest, which will be referred to as partial non-information in the extended sense.

Example 4.5 *Normal distribution*
Consider n independent observations from a $N(\mu, \sigma^2)$ distribution, with parameter of interest σ^2 and nuisance parameter μ. Let $u = \bar{y}_n$ and $v = \sum(y_i - \bar{y}_n)^2$. The statistic (u, v) is sufficient for $\theta = (\mu, \sigma^2)$. The model for v depends only on the parameter of interest; hence, v is partially distribution constant for μ. Since U and V are independent, the model for u conditionally on v is

given by the marginal distribution of u, which is $N(\mu, \sigma^2/n)$. This is saturated with respect to μ, thus, v is I-sufficient for σ^2. In addition, whatever u is, $\hat{\mu}_{\sigma^2}(u) = u$, so that in the model $p_U(u; \sigma^2, \hat{\mu}_{\sigma^2}(u))$, which is $N(u, \sigma^2/n)$, with observation u, the maximum likelihood estimate for σ^2 is not defined. Therefore, the likelihood factor corresponding to the marginal density of u is partially uninformative on σ^2 in the extended sense as well. Note that if the parameter space for μ is not $\mathbb{R}$, but is, for example, $\mathbb{R}^+$, the maximum likelihood estimate of μ is not necessarily u, and the likelihood factor which corresponds to $p_U(u; \sigma^2, \hat{\mu}_{\sigma^2}(u))$ is not partially uninformative in the extended sense (see also Barnard, 1963). $\triangle$

The concept of I-ancillarity is particularly useful in multiparameter exponential families (cf. section 5.7); for further illustrations, see Examples 2.2 and 2.3 in Kalbfleisch and Sprott (1974).

4.2.4 *Nuisance Parameters and Reparameterizations*

Even when nuisance parameters are present parameterization invariance is an important property of a statistical procedure. However, a global reparameterization does not usually maintain the distinction between interest and nuisance components. Hence, the requirement for parameterization invariance must be restricted to **interest-respecting reparameterizations**. This means a reparameterization of the form $\psi = \psi(\theta) = \psi(\tau, \zeta)$, with $\psi = (\varphi, \chi)$, such that

$$\varphi = \varphi(\tau) \qquad \text{and} \qquad \chi = \chi(\tau, \zeta),$$

with $\varphi(\cdot)$ one-to-one. With $\mathcal{F}$ written as a collection of submodels, $\mathcal{F} = \cup_\tau \mathcal{F}_\tau$, where $\mathcal{F}_\tau = \{p_Y(y; \theta) \in \mathcal{F} : \tau(\theta) = \tau, \ \theta \in \Theta\}$, an interest-respecting reparameterization preserves each submodel $\mathcal{F}_\tau$. Thus, inference about τ should obey the principle of invariance under interest-respecting reparameterizations.

4.3 The Notion of a Pseudo-Likelihood

Consider a statistical model $\mathcal{F}$ with parameter space Θ, not necessarily finite-dimensional. Let $\tau = \tau(\theta)$, with $\tau \in \mathrm{T} \subseteq \mathbb{R}^k$, be the parameter of interest. The more complex the structure of the component complementary to τ in θ is, the more attractive the possibility of basing inference on a likelihood function which only depends on τ appears. It is of course desirable that this reduction in complexity takes place at the price of a negligible loss of information about τ.

The term **pseudo-likelihood** is used here to indicate any function of the data which depends only on the parameter of interest and which behaves, in some respects, as if it were a genuine likelihood (score with zero null expectation, maximum likelihood estimator with asymptotic normal distribution, likelihood ratio test with χ^2 null asymptotic distribution, etc.).

Two different situations must be distinguished.

a) The pseudo-likelihood is based on a statistical model defined as a reduction of the original model $\mathcal{F}$. If the density functions of this model only depend on τ, dependence on the nuisance parameter is also eliminated in sampling distributions of likelihood quantities defined on the pseudo-likelihood. The likelihood obtained in this way is a genuine likelihood. From the point of view of asymptotic considerations, the crucial assumption is that the information in the reduced model remains of order $O(n)$.

b) A reduced model from which a given pseudo-likelihood can be thought to be directly deduced is not defined. In other words, the pseudo-likelihood is not based on a factorization of the form (4.3). All the properties have to be specifically investigated.

From the discussion in the previous section, the first step when looking for a pseudo-likelihood, which has to be a genuine likelihood, is to identify factorizations of the form (4.3) where at least one factor depends only on τ. The following definitions consider the main possibilities.

Definition 4.4 Let y be in a one-to-one correspondence with (u, v), or, more generally, let (u, v) be a sufficient statistic for θ for which the factorization

$$p_{U,V}(u, v; \tau, \zeta) = p_V(v; \tau) p_{U|V=v}(u; v, \tau, \zeta) \qquad (4.9)$$

holds. Provided that the likelihood factor which corresponds to $p_{U|V}(\cdot)$ can be neglected, inference about τ can be based on the marginal model for V, with density $p_V(v; \tau)$. The corresponding likelihood function

$$L_M(\tau) = L_M(\tau; v) = p_V(v; \tau) \qquad (4.10)$$

is called the **marginal likelihood** based on v.

Definition 4.5 Let u be a statistic such that the factorization

$$p_Y(y; \tau, \zeta) = p_U(u; \tau, \zeta) p_{Y|U=u}(y; u, \tau) \qquad (4.11)$$

holds. Provided that the likelihood factor which corresponds to $p_U(\cdot)$ can be neglected, inference about τ can be based on the conditional model with density $p_{Y|U}(\cdot)$. The corresponding likelihood function

$$L_c(\tau) = L_c(\tau; y|u) = p_{Y|U=u}(y; u, \tau) \qquad (4.12)$$

is called the conditional likelihood based on conditioning on u.

Factorizations such as (4.9) or (4.11) arise, however, essentially only in exponential families (Chapter 5) and in group families (Chapter 7). When such an inferential separation cannot be obtained, useful pseudo-likelihoods have to be found outside the class of genuine likelihoods.

The rest of this chapter introduces and illustrates some concepts of pseudo-likelihood. Sections 4.4 and 4.5 deal with examples of marginal and conditional likelihoods. Outside of these cases, one simple, and general way of obtaining a pseudo-likelihood function in a parametric model is to replace the nuisance parameter ζ with its maximum likelihood for fixed τ, $\hat{\zeta}_\tau$, in the full likelihood $L(\tau, \zeta)$. The function $L(\tau, \hat{\zeta}_\tau)$ is called the *profile likelihood* and its properties are illustrated in section 4.6. It is not a genuine likelihood. However, when there is a marginal or conditional likelihood, the profile likelihood does provide a first-order approximation of such a likelihood (see Example 9.9 and section 10.10.2). Its inferential accuracy may not be entirely satisfactory, especially when the number of nuisance parameters is large. Neyman–Scott problems are extreme examples of this. Various modifications, aiming to improve profile likelihood, have been proposed in recent years. Section 4.7 introduces Cox and Reid's (1987) *approximate conditional likelihood*, which is based on the choice of an orthogonal parameterization. Further proposals for modifications have been put forward and described in, for example, Barndorff-Nielsen (1994); since the argument requires notions about higher-order asymptotic methods which will be dealt with in Chapter 11, discussion is deferred until section 11.6. Section 4.8 looks briefly at *partial likelihood*, which was introduced by Cox (1975) for inference about the regression coefficients in the proportional hazards model with censored observations. Section 4.9 describes *quasi-likelihood*, which is a pseudo-likelihood function associated to a semiparametric model specified in terms of first (and sometimes second) order moments of particular combinants of the form $q(y; \tau)$. Lastly, section 4.10 considers *empirical likelihood*. This pseudo-likelihood was introduced by Owen (1988) to deal with inference problems on k-dimensional smooth functionals in nonparametric models.

The above list does not cover all the notions of pseudo-likelihood currently available in the literature. In particular, the removal of nuisance parameters is not the only reason for introducing pseudo-likelihoods. Another reason could be the difficulty of expressing the true likelihood as, for example, in the case of some time series or spatial models. See section 4.11 for further indications.

Usually, once a pseudo-likelihood has been defined, the problems that have to be faced concern: *i)* specifying the information content of the pseudo-likelihood, that is, quantifying the information loss due to neglecting some likelihood factors; *ii)* studying its main distributional features, such as the expectation of the score, the information identity, etc.; *iii)* defining inference procedures on the parameter of interest and studying the corresponding distribution problems in exact, or at least approximate, terms.

4.4 Marginal Likelihood: Examples

The marginal likelihood defined in (4.10) is a genuine likelihood under the reduced marginal model. Thus, under regularity conditions, the marginal score has zero null expectation and the information identity holds. The usual first-order asymptotic properties continue to hold provided that the expected information under the reduced model, $i_M(\tau)$, is of order $O(n)$.

Example 4.6 *Normal distribution with known variance, inference about the absolute value of the mean*
Let $y_1, \ldots, y_n$ be independent observations from a $N(\theta, 1)$ distribution. Let $\tau = |\theta|$ be the parameter of interest and $\zeta = \text{sgn}(\theta)$ the nuisance component. Except in the case where $\theta = 0$, the parameters θ and τ are variation independent. The log-likelihood function is

$$l(\tau, \zeta; y) = -\frac{n}{2}(\bar{y}_n - \zeta\tau)^2 = -\frac{n}{2}\bar{y}_n^2 - \frac{n}{2}\tau^2 + n\zeta\tau\bar{y}_n .$$

Note that this does not have separable parameters. Let $V = |\bar{Y}_n|$. Since

$$P_{\tau,\zeta}\{V \leq v\} \quad = \quad P_{\tau,\zeta}\{|\bar{Y}_n| \leq v\} = P_{\tau,\zeta}\{-v \leq \bar{Y}_n \leq v\}$$

$$= \quad \Phi(\sqrt{n}(v - \zeta\tau)) - \Phi(\sqrt{n}(-v - \zeta\tau)) ,$$

then

$$l_M(\tau) = \log p_V(v; \tau, \zeta) = -\frac{n}{2}(v^2 + \tau^2) + \log(e^{nv\tau} + e^{-nv\tau}) . \tag{4.13}$$

Thus, we are now in the situation expressed by the factorization (4.9), with $u = \text{sgn}(\bar{y}_n)$. The neglected likelihood factor gives the log-likelihood summand

$$l_{U|V=v}(\tau, \zeta) = -\frac{n}{2}(\bar{y}_n - \zeta\tau)^2 + \frac{n}{2}(v^2 + \tau^2) - \log(e^{nv\tau} + e^{-nv\tau})$$

from which we obtain that the maximum likelihood estimate of ζ, for fixed τ, is $\hat{\zeta}_\tau(u) = uv/\tau$. Substituting $\hat{\zeta}_\tau(u)$ in $l_{U|V}(\cdot)$ for ζ shows that this log-likelihood does not depend on u and that it is thus partially non-informative, in the extended sense, for inference about τ (see section 4.2.3). The following expressions for the marginal score and for the observed marginal information can be obtained from (4.13)

$$l_M(\tau)_* = \frac{\partial l_M(\tau)}{\partial \tau} = -n\tau + \frac{nv(e^{nv\tau} - e^{-nv\tau})}{e^{nv\tau} + e^{-nv\tau}}, \qquad (4.14)$$

$$j_M(\tau) = n - (nv)^2 \left\{ 1 - \frac{(e^{nv\tau} - e^{-nv\tau})^2}{(e^{nv\tau} + e^{-nv\tau})^2} \right\} = n - \frac{4(nv)^2 e^{-2nv\tau}}{(1 + e^{-2nv\tau})^2}.$$

It is not difficult to check that the score (4.14) has null expectation equal to zero. In addition, the observed information is equal to n, plus a term that tends to zero exponentially. $\triangle$

Example 4.7 *Neyman–Scott problem*
As in Example 4.1, let (y_{i1}, y_{i2}), $i = 1, \ldots, n$, be independent observations from $N(\mu_i, \sigma^2)$. The parameter is $\theta = (\sigma^2, \mu_1, \ldots, \mu_n)$. Let σ^2 be the component of interest and let $(\mu_1, \ldots, \mu_n)$ be nuisance components. Under the usual assumptions, these parameters are variation independent. Hence

$$
\begin{aligned}
l(\theta; y) &= -\frac{1}{2\sigma^2} \sum_{i=1}^{n} \sum_{j=1}^{2} (y_{ij} - \mu_i)^2 - n\log\sigma^2 \\
&= -\frac{1}{\sigma^2} \sum_{i=1}^{n} (\bar{y}_i - \mu_i)^2 - \frac{1}{2\sigma^2} \sum_{i=1}^{n} \sum_{j=1}^{2} (y_{ij} - \bar{y}_i)^2 - n\log\sigma^2,
\end{aligned}
$$

with $\bar{y}_i = (y_{i1} + y_{i2})/2$. In this case too, the likelihood does not have separable parameters. It would seem reasonable to say that most of the information about σ^2 is contained in the second summand, $-\frac{1}{2\sigma^2}SS_w$, with

$$SS_w = \sum_{i=1}^{n} \sum_{j=1}^{2} (y_{ij} - \bar{y}_i)^2 = \sum_{i=1}^{n} \left(\frac{y_{i1} - y_{i2}}{\sqrt{2}} \right)^2.$$

The transformation

$$v_i = \frac{y_{i1} - y_{i2}}{\sqrt{2}}, \qquad u_i = \frac{y_{i1} + y_{i2}}{\sqrt{2}}, \qquad i = 1, \dots, n,$$

is orthogonal, so that $v = (v_1, \dots, v_n)$ is a vector of independent observations from a $N(0, \sigma^2)$ distribution; moreover, v is independent of $u_1, \dots, u_n$. Consequently,

$$l_M(\sigma^2; v) = -\frac{1}{2\sigma^2} SS_w - \frac{n}{2} \log \sigma^2. \tag{4.15}$$

The neglected likelihood factor corresponds to

$$l_U(\mu_1, \dots, \mu_n, \sigma^2) = -\frac{1}{2\sigma^2} \sum_{i=1}^{n} (u_i - \sqrt{2}\mu_i)^2 - \frac{n}{2} \log \sigma^2.$$

We can check that, whatever σ^2 is, $\hat{\mu}_i = u_i / \sqrt{2}$; therefore

$$l_U(\hat{\mu}_1, \dots, \hat{\mu}_n, \sigma^2) = -(n/2) \log \sigma^2$$

does not offer relevant information about σ^2. $\triangle$

Example 4.8 *Correlation coefficient in a bivariate normal distribution*
Let $(x_1, y_1), \dots, (x_n, y_n)$ be n independent observations from a bivariate normal distribution with mean vector $\mu = (\mu_X, \mu_Y)$ and covariance matrix

$$\Sigma = \begin{pmatrix} \sigma_X^2 & \rho \, \sigma_X \sigma_Y \\ \rho \, \sigma_X \sigma_Y & \sigma_Y^2 \end{pmatrix}.$$

Let ρ be the parameter of interest and consider $\zeta = (\mu_X, \mu_Y, \sigma_X, \sigma_Y)$ as a nuisance parameter. The statistic $(\bar{x}_n, \bar{y}_n, s_X^2, s_Y^2, r)$ is minimal sufficient, with s_X^2 and s_Y^2 the sample variances of $x_1, \dots, x_n$ and $y_1, \dots, y_n$, respectively, and

$$r = \frac{\sum_{i=1}^{n} (x_i - \bar{x}_n)(y_i - \bar{y}_n)}{\sqrt{\sum_{i=1}^{n} (x_i - \bar{x}_n)^2 \sum_{i=1}^{n} (y_i - \bar{y}_n)^2}}$$

the sample correlation coefficient. It is clear that the marginal density of r does not depend on ζ; moreover, r and $(\bar{x}_n, \bar{y}_n, s_X^2, s_Y^2)$ are independent as a consequence of Basu's theorem (Theorem 2.2). The marginal density of r is (Fisher, 1915)

$$p(r; \rho) = \frac{1}{\pi}(n-2)(1-\rho^2)^{(n-1)/2}(1-r^2)^{(n-4)/2} \int_0^\infty \frac{1}{(\cosh w - \rho r)^{n-1}} dw,$$
$$\tag{4.16}$$

where $\cosh x = (e^x + e^{-x})/2$ is the hyperbolic cosine function. The likelihood function can therefore be factorized according to (4.9). The marginal score is

$$l_M(\rho)_* = -(n-1)\frac{\rho}{1-\rho^2} + \frac{\displaystyle\int_0^\infty \frac{(n-1)r}{(\cosh w - \rho r)^n}\,dw}{\displaystyle\int_0^\infty \frac{1}{(\cosh w - \rho r)^{n-1}}\,dw}$$

so that the likelihood equation is

$$\int_0^\infty \frac{1}{(\cosh w - \rho r)^n}(\rho \cosh w - r)\,dw = 0 \ .$$

As n diverges, only values of w close to zero offer a non-negligible contribution to the above integral. Using the expansion $\cosh x = 1 + x^2/2 + \ldots$, the likelihood equation takes on the approximate form

$$\int_0^\infty \left(\rho(1 + \frac{w^2}{2}) - r\right)\exp\left\{-\frac{n}{2}\frac{w^2}{1-\rho r}\right\}dw = 0 \ ,$$

which gives the approximation

$$\hat\rho_M \doteq \frac{r}{1 + \dfrac{1 - \hat\rho_M r}{2n}} \doteq r\left(1 - \frac{1-r^2}{2n}\right) \tag{4.17}$$

for the marginal maximum likelihood estimate. $\triangle$

Example 4.9 *The proportional hazards model*
Using the same notation as in section 1.3.1, consider n independent and uncensored observations with proportional hazards, $r_{T_i}(t) = r_0(t)e^{\beta \cdot x_i}$. The data are n survival times $(t_1, \ldots, t_n)$ together with the corresponding covariate vectors $(x_1, \ldots, x_n)$. From (1.2) and (1.1), the density of each observation is

$$p(t_i; \beta, r_0(\cdot), x_i) = r_0(t_i)e^{\beta \cdot x_i}\exp\left\{-R_0(t_i)e^{\beta \cdot x_i}\right\} , \quad i = 1, \ldots, n \ ,$$

where $R_0(t) = \int_0^t r_0(u)du$ and, with the usual notation, $\beta = (\beta_1, \ldots, \beta_k)$, $x_i = (x_{i1}, \ldots, x_{ik})$. Therefore, the likelihood function for $\theta = (\beta, r_0(t))$ does not have separable parameters.

Let $w = (w_1, \ldots, w_n) = ((1), \ldots, (n))$ be the vector of inverse ranks of the observations. The symbol (i) denotes the index of the item which corresponds to the i-th ordered survival time, that is,

$$y_{w_i} = y_{(i)} , \quad i = 1, \ldots, n \ . \tag{4.18}$$

If, for example, with $n = 3$, we have $(t_1, t_2, t_3) = (15, 5, 10)$, then $t_{(1)} = 5$, $t_{(2)} = 10$, $t_{(3)} = 15$ and, hence, $w = ((1), (2), (3)) = (2, 3, 1)$. The name *inverse ranks* can be justified when compared with the definition of ranks. In fact, the set of ranks is the statistic $v = (v_1, \ldots, v_n)$ such that

$$y_{(v_i)} = y_i, \qquad i = 1, \ldots, n. \tag{4.19}$$

We have the one-to-one correspondence

$$(t_1, \ldots, t_n) \longleftrightarrow (t_{(1)}, \ldots, t_{(n)}, w).$$

The statistic w has a discrete distribution. Ignoring ties, which have zero probability if the T_i have continuous distributions, w takes $n!$ distinct values. It would seem reasonable that the distribution of $T_{(1)}, \ldots, T_{(n)}$ is governed by the baseline failure rate $r_0(t)$ and that the distribution of w is governed by the coefficients β that express the effect of the covariates. Let $\boldsymbol{x}_{(i)}$ be the vector of the covariates for the item with index (i); then

$$P_{\beta, r_0}\{w = ((1), \ldots, (n))\} = P_{\beta, r_0}\left((T_1, \ldots, T_n) \in \left\{t \in \mathbb{R}^n : t_{(1)} < \cdots < t_{(n)}\right\}\right)$$

$$= \prod_{i=1}^{n} \frac{e^{\beta \cdot \boldsymbol{x}_{(i)}}}{\sum_{j \in \mathcal{R}(t_{(i)})} e^{\beta \cdot \boldsymbol{x}_j}} = L_M(\beta; w, \boldsymbol{x}), \tag{4.20}$$

which does not depend on $r_0(t)$; therefore w is partially distribution constant with respect to $r_0(t)$. In (4.20), $\mathcal{R}(t_{(i)})$ denotes the set of the indices of the items at risk in the instant $t_{(i)}$, that is, $\mathcal{R}(t_{(i)}) = \{(i), (i+1), \ldots, (n)\}$.

The calculations for obtaining the marginal likelihood (4.20) are as follows.

$$P_{\beta, r_0}\left((T_1, \ldots, T_n) \in \left\{t \in \mathbb{R}^n : t_{(1)} < \cdots < t_{(n)}\right\}\right)$$

$$= \int_0^{+\infty} \int_{t_{(1)}}^{+\infty} \cdots \int_{t_{(n-1)}}^{+\infty} \prod_{i=1}^{n} p\left(t_{(i)}; \beta, r_0, \boldsymbol{x}_{(i)}\right) dt_{(n)} \cdots dt_{(1)}$$

$$= \int_0^{+\infty} \int_{t_{(1)}}^{+\infty} \cdots \int_{t_{(n-1)}}^{+\infty} \prod_{i=1}^{n} r_0(t_{(i)}) e^{\beta \cdot \boldsymbol{x}_{(i)}}$$

$$\times \exp\left\{-R_0(t_{(i)}) e^{\beta \cdot \boldsymbol{x}_{(i)}}\right\} dt_{(n)} \cdots dt_{(1)}.$$

Consider the innermost integral

$$\int_{t_{(n-1)}}^{+\infty} r_0(t_{(n)}) e^{\beta \cdot \boldsymbol{x}_{(n)}} \exp\left\{-R_0(t_{(n)}) e^{\beta \cdot \boldsymbol{x}_{(n)}}\right\} dt_{(n)}.$$

This is the survival function, $1 - F(\cdot)$, of the unit of index (n), evaluated at $t_{(n-1)}$,

$$1 - F(t_{(n-1)}; \beta, r_0(\cdot), \boldsymbol{x}_{(n)}) = \exp\left\{-R_0(t_{(n-1)})e^{\beta \cdot \boldsymbol{x}_{(n)}}\right\} .$$

The second step is calculating

$$\int_{t_{(n-2)}}^{+\infty} r_0(t_{(n-1)})e^{\beta \cdot \boldsymbol{x}_{(n-1)}} \exp\left\{-R_0(t_{(n-1)})\left(e^{\beta \cdot \boldsymbol{x}_{(n-1)}} + e^{\beta \cdot \boldsymbol{x}_{(n)}}\right)\right\} dt_{(n-1)}$$

$$= \frac{e^{\beta \cdot \boldsymbol{x}_{(n-1)}}}{e^{\beta \cdot \boldsymbol{x}_{(n-1)}} + e^{\beta \cdot \boldsymbol{x}_{(n)}}} \int_{t_{(n-2)}}^{+\infty} r_0(t_{(n-1)})\left(e^{\beta \cdot \boldsymbol{x}_{(n-1)}} + e^{\beta \cdot \boldsymbol{x}_{(n)}}\right)$$

$$\times \exp\left\{-R_0(t_{(n-1)})\left(e^{\beta \cdot \boldsymbol{x}_{(n-1)}} + e^{\beta \cdot \boldsymbol{x}_{(n)}}\right)\right\} dt_{(n-1)}$$

$$= \frac{e^{\beta \cdot \boldsymbol{x}_{(n-1)}}}{e^{\beta \cdot \boldsymbol{x}_{(n-1)}} + e^{\beta \cdot \boldsymbol{x}_{(n)}}} \exp\left\{-R_0(t_{(n-2)})\left(e^{\beta \cdot \boldsymbol{x}_{(n-1)}} + e^{\beta \cdot \boldsymbol{x}_{(n)}}\right)\right\} ,$$

where we have exploited the fact that the function to be integrated is proportional to the density corresponding to the failure rate $r_0(t)\left(e^{\beta \cdot \boldsymbol{x}_{(n-1)}} + e^{\beta \cdot \boldsymbol{x}_{(n)}}\right)$. The integral is the corresponding survival function evaluated at $t_{(n-2)}$. The same argument gives, for the next step,

$$\int_{t_{(n-3)}}^{+\infty} r_0(t_{(n-2)})e^{\beta \cdot \boldsymbol{x}_{(n-2)}} e^{-R_0(t_{(n-2)})\left\{e^{\beta \cdot \boldsymbol{x}_{(n-2)}} + e^{\beta \cdot \boldsymbol{x}_{(n-1)}} + e^{\beta \cdot \boldsymbol{x}_{(n)}}\right\}} dt_{(n-2)}$$

$$= \frac{e^{\beta \cdot \boldsymbol{x}_{(n-2)}}}{e^{\beta \cdot \boldsymbol{x}_{(n-2)}} + e^{\beta \cdot \boldsymbol{x}_{(n-1)}} + e^{\beta \cdot \boldsymbol{x}_{(n)}}}$$

$$\times \exp\left\{-R_0(t_{(n-3)})\left(e^{\beta \cdot \boldsymbol{x}_{(n-2)}} + e^{\beta \cdot \boldsymbol{x}_{(n-1)}} + e^{\beta \cdot \boldsymbol{x}_{(n)}}\right)\right\} .$$

The last step is to calculate

$$\int_0^{+\infty} r_0(t_{(1)})e^{\beta \cdot \boldsymbol{x}_{(1)}} \exp\left\{-R_0(t_{(1)})\left(e^{\beta \cdot \boldsymbol{x}_{(1)}} + \cdots + e^{\beta \cdot \boldsymbol{x}_{(n)}}\right)\right\} dt_{(1)}$$

$$= \frac{e^{\beta \cdot \boldsymbol{x}_{(1)}}}{e^{\beta \cdot \boldsymbol{x}_{(1)}} + \cdots + e^{\beta \cdot \boldsymbol{x}_{(n)}}} \int_0^{+\infty} r_0(t_{(1)})\left(e^{\beta \cdot \boldsymbol{x}_{(1)}} + \cdots + e^{\beta \cdot \boldsymbol{x}_{(n)}}\right)$$

$$\times \exp\left\{-R_0(t_{(1)})\left(e^{\beta \cdot \boldsymbol{x}_{(1)}} + \cdots + e^{\beta \cdot \boldsymbol{x}_{(n)}}\right)\right\} dt_{(1)} .$$

The integral on the right-hand side is equal to 1, because it is a survival function evaluated at 0. The factor obtained multiplies the analogous factors from the preceding steps. Thus,

$$P_{\beta,r_0}\{w = ((1),\ldots,(n))\} = \frac{e^{\beta \cdot \boldsymbol{x}_{(n-1)}}}{e^{\beta \cdot \boldsymbol{x}_{(n-1)}} + e^{\beta \cdot \boldsymbol{x}_{(n)}}}$$

$$\times \frac{e^{\beta \cdot \boldsymbol{x}_{(n-2)}}}{e^{\beta \cdot \boldsymbol{x}_{(n-2)}} + e^{\beta \cdot \boldsymbol{x}_{(n-1)}} + e^{\beta \cdot \boldsymbol{x}_{(n)}}} \times \cdots \times \frac{e^{\beta \cdot \boldsymbol{x}_{(1)}}}{e^{\beta \cdot \boldsymbol{x}_{(1)}} + \cdots + e^{\beta \cdot \boldsymbol{x}_{(n)}}},$$

which is the same as (4.20). The marginal likelihood (4.20), written as

$$L_M(\beta; w, \boldsymbol{x}) = \prod_{i=1}^{n} \frac{r_{T_i}(t_{(i)})}{\sum_{j \in \mathcal{R}(t_{(i)})} r_{T_i}(t_{(i)})},$$

has an immediate interpretation. The i-th factor is the density of the inverse rank (i) conditionally on $(1), \ldots, (i-1)$, that is, the probability that, among all the items at risk in the instant $t_{(i)}$, that with index (i) will fail first. Kalbfleisch and Prentice (1973) have obtained a marginal likelihood that extends (4.20) to deal with censored observations and with ties in the survival times. However, for these more complex cases partial likelihood, introduced in section 4.8, provides a flexible, and widely applicable, tool. $\triangle$

It will be clear from section 7.7 that the examples above have a unified interpretation. In composite group families there is a natural marginal likelihood which is based on the maximal invariant statistic.

4.5 Conditional Likelihood: Examples

The conditional likelihood function defined by (4.12) is a genuine likelihood in the reduced conditional model. Therefore, under regularity conditions, the score will have zero conditional expectation and the information identity will hold.

Example 4.10 *Independent Poisson observations with means in geometrical progression*
Let $y_0, y_1, \ldots, y_n$ be independent observations from Poisson distributions with mean $\zeta \tau^i$, $i = 0, 1, \ldots, n$. This model is useful in virology, where τ expresses the speed at which variations in the average number of observed events takes place. The joint density of the observations is

$$p_Y(y_0, y_1, \ldots, y_n; \zeta, \tau) = \prod_{i=0}^{n} \frac{(\zeta \tau^i)^{y_i} e^{-\zeta \tau^i}}{y_i!}.$$

The statistic $u = \sum_{i=0}^{n} y_i$ has a Poisson distribution with mean $\zeta \sum_{i=0}^{n} \tau^i = \zeta(1 - \tau^{n+1})/(1 - \tau)$. The conditional density of the observations given u is

$$p_{Y|U=u}(y_0, \ldots, y_n; u, \tau, \zeta) = \frac{\tau^{\Sigma i y_i} / \Pi y_i!}{(\sum \tau^i)^u / u!},$$

which does not depend on ζ. Furthermore, the marginal distribution of u does not contain any relevant information about τ when ζ is unknown, in the sense that $p_U(u; \hat{\zeta}_\tau, \tau)$ does not depend on τ. $\triangle$

Example 4.11 *Exponential distribution with unknown location*
Let $y_1, \ldots, y_n$ be independent observations from a distribution with density

$$p(y; \mu, \sigma) = \frac{1}{\sigma} e^{-(y-\mu)/\sigma} I_{[\mu, +\infty)}(y), \quad \mu \in \mathbb{R}, \ \sigma > 0.$$

Let σ be the parameter of interest. The likelihood equation for σ, based on the full likelihood, is

$$\left. \frac{\partial l(\mu, \sigma)}{\partial \sigma} \right|_{\mu = y_{(1)}} = 0,$$

that is

$$-\frac{n}{\sigma} + \frac{1}{\sigma^2} \sum_{i=1}^{n} (y_i - y_{(1)}) = 0.$$

The null expectation of the left-hand side is $-1/\sigma$. The maximum likelihood estimate of σ is $\hat{\sigma} = n^{-1} \sum_{i=1}^{n} (y_i - y_{(1)})$ and has bias equal to $-\sigma/n$. The sample minimum $y_{(1)}$ is a minimal sufficient statistic for μ, for every value of σ. It is easy to verify that the conditional log-likelihood function given $u = y_{(1)}$ is

$$l_c(\sigma) = -(n - 1) \log \sigma - \frac{1}{\sigma} \sum_{i=1}^{n} (y_i - y_{(1)}). \tag{4.21}$$

The estimating equation based on $l_c(\sigma)$ is unbiased. In this case, the estimator based on this equation is also unbiased; remember, however, that only unbiasedness of an estimating equation is parameterization invariant. $\triangle$

Example 4.12 *Rasch model (cont.)*
Let y_{ij}, $i = 1, \ldots, n$, $j = 1, 2$, be independent observations from Bernoulli trials with probabilities of success

$$\pi_{i1} = \frac{e^{\zeta_i}}{1 + e^{\zeta_i}}, \quad \pi_{i2} = \frac{e^{\zeta_i - \tau}}{1 + e^{\zeta_i - \tau}}.$$

Note that τ is the difference on the logit scale of the probabilities of success in the two trials, constant for each subject, that is

$$\tau = \log\left(\frac{\pi_{i1}}{1-\pi_{i1}}\right) - \log\left(\frac{\pi_{i2}}{1-\pi_{i2}}\right).$$

The quantity e^τ is the **odds ratio**. The log-likelihood for $\theta = (\tau, \zeta_1, \ldots, \zeta_n)$ is

$$l(\tau, \zeta_1, \ldots, \zeta_n) = \sum_{i=1}^{n} \zeta_i(y_{i1}+y_{i2}) - \tau\sum_{i=1}^{n} y_{i2} - \sum_{i=1}^{n} \log(1+e^{\zeta_i}) - \sum_{i=1}^{n} \log(1+e^{\zeta_i-\tau}).$$

$$(4.22)$$

From Example 4.2, this log-likelihood gives an estimator for τ which is not consistent. The statistic $u = (u_1, \ldots, u_n)$, with $u_i = y_{i1} + y_{i2}$, is minimal sufficient for $(\zeta_1, \ldots, \zeta_n)$ for each value of τ. Conditionally on $(u_1, \ldots, u_n)$, the y_{i2} are independent observations from Bernoulli trials with probability of success equal to zero if $u_i = 0$, equal to one if $u_i = 2$ and equal to $e^{-\tau}/(1+e^{-\tau})$ if $u_i = 1$. Suppose that, without loss of generality, $u_i = 1$ for the first n_u observations ($n_u < n$). The conditional log-likelihood function is

$$l_c(\tau) = -\tau\sum_{i=1}^{n_u} y_{i2} - n_u \log(1 + e^{-\tau}).$$

Here, as seems reasonable, the estimator of τ, based on $l_c(\tau)$, is consistent (Andersen, 1980, Chapter 6). $\qquad\qquad\triangle$

Example 4.13 *Neyman–Scott problem, exponential distribution*
Let y_{ij}, $i = 1, \ldots, n$, $j = 1, 2$, be independent observations from exponential distributions with expectation $(\tau+\zeta_i)^{-1}$ if $j = 1$ and ζ_i^{-1} if $j = 2$, $i = 1, \ldots, n$. Let τ be the parameter of interest; τ is the difference between the failure rates of y_{i1} and of y_{i2}. The maximum likelihood estimate of τ, based on the full likelihood, is the solution of the equation

$$\tau\sum_{i=1}^{n} (y_{i2} - y_{i1}) + 2n = \sum_{i=1}^{n} \sqrt{\tau^2(y_{i1} + y_{i2})^2 + 4}.$$

$$(4.23)$$

Since this estimating equation has bias of order $O(n)$, the maximum likelihood estimator of τ is not consistent. As in the previous example, a consistent estimator based on the conditional likelihood can be obtained. Indeed, the statistic $u = (u_1, \ldots, u_n)$, with $u_i = y_{i1} + y_{i2}$, is minimal sufficient for $(\zeta_1, \ldots, \zeta_n)$ for each value of τ. Since u_i, $i = 1, \ldots, n$, has density

$$p_{U_i}(u; \tau, \zeta_i) = \frac{\zeta_i(\tau + \zeta_i)}{\tau}(e^{-\zeta_i u} - e^{-(\zeta_i+\tau)u}), \qquad u > 0,$$

the conditional log-likelihood function is

$$l_c(\tau) = n \log \tau - \tau \sum_{i=1}^{n} y_{i1} - \sum_{i=1}^{n} \log(1 - e^{-\tau u_i}) . \qquad (4.24)$$

It is left to the reader to check the properties of $l_c(\tau)$ (see Problem 4.9). $\triangle$

The examples given in this section suggest that recourse to conditional likelihood is all the more advisable the larger the dimension of the nuisance parameter is. Eliminating nuisance parameters through conditioning is generally carried out when the reference statistical model is a multiparameter exponential family. For a more general discussion and further examples, see section 5.4.2.

4.6 Profile Likelihood

Constructing marginal and conditional likelihoods is only possible when partially distribution constant or partially sufficient statistics are available for the nuisance parameter. One quite general idea for defining a pseudo-likelihood for τ considers substituting the nuisance parameter with a consistent estimate in the original likelihood.

Definition 4.6 If, under a parametric statistical model with likelihood function $L(\theta)$, $\tau = \tau(\theta)$ is the parameter of interest, the profile likelihood for τ is the function

$$L_p(\tau) = \sup_{\theta|\tau} L(\theta) ,$$

where $\theta|\tau$ indicates all the values of θ such that $\tau(\theta) = \tau$. The profile log-likelihood function is denoted by $l_p(\tau)$.

Suppose, for simplicity, that $\theta = (\tau, \zeta)$. The profile likelihood becomes

$$L_p(\tau) = L(\tau, \hat{\zeta}_\tau) ,$$

where $\hat{\zeta}_\tau$ is the maximum likelihood estimate of ζ for fixed τ. Consider the partition of the score vector

$$l_* = (l_\tau, l_\zeta) ,$$

where $l_\tau = \partial l(\tau, \zeta)/\partial \tau$ is a k-dimensional vector and $l_\zeta = \partial l(\tau, \zeta)/\partial \zeta$ is a $(p - k)$-dimensional vector; under regularity conditions, $\hat{\zeta}_\tau$ is the solution in ζ of $l_\zeta(\tau, \zeta) = 0$.

The pseudo-likelihood $L_P(\tau)$ is not deduced from a density function, thus, it is not a genuine likelihood. However, the profile likelihood does have some interesting properties which make it resemble a true likelihood.

i) The maximum profile likelihood estimate is the same as the maximum likelihood estimate of τ based on $L(\tau, \zeta)$, $\hat{\tau}$, that is

$$\sup_{\tau} L_P(\tau) = L_P(\hat{\tau}) \ .$$

ii) The log-likelihood ratio combinant is the same as the combinant based on $L(\tau, \zeta)$ which is usually used for testing hypotheses on τ, with unknown ζ. In fact,

$$\begin{aligned} W_P = W_P(\tau) &= 2\left\{ l_P(\hat{\tau}) - l_P(\tau) \right\} \\ &= 2\left\{ l(\hat{\tau}, \hat{\zeta}) - l(\tau, \hat{\zeta}_\tau) \right\} \ . \end{aligned} \quad (4.25)$$

Under the usual regularity conditions, the null asymptotic distribution of W_P is χ_k^2. We give an outline of the proof below. Firstly, analogously with the partition of l_* into the components l_τ and l_ζ, consider the partitions of $i(\theta) = i(\tau, \zeta)$ and of its inverse

$$i = i(\theta) = \begin{pmatrix} i_{\tau\tau} & i_{\tau\zeta} \\ i_{\zeta\tau} & i_{\zeta\zeta} \end{pmatrix},$$

$$i^{-1} = i(\theta)^{-1} = \begin{pmatrix} i^{\tau\tau} & i^{\tau\zeta} \\ i^{\zeta\tau} & i^{\zeta\zeta} \end{pmatrix}.$$

Use the same partitions for the observed information matrix j and for j^{-1}. It is useful to remember the rule for inverting a block matrix, which gives

$$i^{\tau\tau} = \left(i_{\tau\tau} - i_{\tau\zeta} i_{\zeta\zeta}^{-1} i_{\zeta\tau} \right)^{-1}, \qquad i^{\tau\zeta} = -i^{\tau\tau} i_{\tau\zeta} i_{\zeta\zeta}^{-1} \ . \quad (4.26)$$

Using (3.42), we obtain for $W_P(\tau)$ the asymptotic representation

$$\begin{aligned} W_P(\tau) &= 2\left\{ l(\hat{\tau}, \hat{\zeta}) - l(\tau, \zeta) \right\} - 2\left\{ l(\tau, \hat{\zeta}_\tau) - l(\tau, \zeta) \right\} \\ &= l_*^T i^{-1} l_* - l_\zeta^T i_{\zeta\zeta}^{-1} l_\zeta + o_p(1) \ , \end{aligned} \quad (4.27)$$

where l_*, l_ζ, etc. are treated as column vectors. Let

$$W_{uP}(\tau, \zeta) = \left(l_\tau - i_{\tau\zeta} i_{\zeta\zeta}^{-1} l_\zeta \right)^T i^{\tau\tau} \left(l_\tau - i_{\tau\zeta} i_{\zeta\zeta}^{-1} l_\zeta \right) ; \quad (4.28)$$

then with straightforward algebra (following, e.g. Azzalini, 1996, section A.5.4) we have the identity

$$l_*^{\mathrm{T}} i^{-1} l_* - l_\zeta^{\mathrm{T}} i_{\zeta\zeta}^{-1} l_\zeta = W_{uP}(\tau, \zeta). \tag{4.29}$$

Since, under regularity conditions, $\hat{\theta} - \theta$ has a null asymptotic $N(0, i^{-1})$ distribution, the combinant

$$W_{eP}(\tau, \zeta) = (\hat{\tau} - \tau)^{\mathrm{T}} (i^{\tau\tau})^{-1} (\hat{\tau} - \tau) \tag{4.30}$$

has a null asymptotic χ_k^2 distribution. Using the expansion $\hat{\theta} - \theta = i(\theta)^{-1} l_* + o_p(n^{-1/2})$ (i.e. (3.39) extended to the multiparameter case), that is

$$\begin{pmatrix} \hat{\tau} - \tau \\ \hat{\zeta} - \zeta \end{pmatrix} = \begin{pmatrix} i^{\tau\tau} & i^{\tau\zeta} \\ i^{\zeta\tau} & i^{\zeta\zeta} \end{pmatrix} \begin{pmatrix} l_\tau \\ l_\zeta \end{pmatrix} + o_p(n^{-1/2}),$$

we can write

$$W_{eP}(\tau, \zeta) = (i^{\tau\tau} l_\tau + i^{\tau\zeta} l_\zeta)^{\mathrm{T}} (i^{\tau\tau})^{-1} (i^{\tau\tau} l_\tau + i^{\tau\zeta} l_\zeta) + o_p(1).$$

Therefore, using the second identity in (4.26) and relations (4.27) and (4.29), it is easy to see that $W_P = W_{eP}(\tau, \zeta) + o_p(1)$ and, hence, that W_P has null asymptotic χ_k^2 distribution, just as $W_{uP}(\tau, \zeta)$ does. Furthermore, the combinants $W_{uP}(\tau, \zeta)$ and $W_{eP}(\tau, \zeta)$ have a null asymptotic χ_k^2 distribution even if ζ is replaced by $\hat{\zeta}_\tau$. Thus, we obtain the combinants

$$W_{uP} = W_{uP}(\tau, \hat{\zeta}_\tau) = \left(l_\tau(\tau, \hat{\zeta}_\tau) \right)^{\mathrm{T}} i^{\tau\tau}(\tau, \hat{\zeta}_\tau) \, l_\tau(\tau, \hat{\zeta}_\tau) \tag{4.31}$$

and

$$W_{eP} = W_{eP}(\tau, \hat{\zeta}_\tau) = (\hat{\tau} - \tau)^{\mathrm{T}} \left(i^{\tau\tau}(\tau, \hat{\zeta}_\tau) \right)^{-1} (\hat{\tau} - \tau), \tag{4.32}$$

which are asymptotically equivalent versions of W_P, suitable for inference about τ with nuisance parameter ζ. In these statistics, the expected information i can be replaced by the observed information, j or $\hat{j}$, without changing the null asymptotic distribution. The acceptance regions of tests on τ based on W_P, W_{eP}, W_{uP} are therefore asymptotically similar, that is, they have an asymptotic level that does not depend on ζ. However, only inference procedures based on W_P and W_{uP} are invariant under interest-respecting reparameterizations. If τ is a scalar, the one-sided versions of W_P, W_{eP} and W_{uP} are

$$Z_P = \mathrm{sgn}(\hat{\tau} - \tau) \sqrt{2(l_P(\hat{\tau}) - l_P(\tau))}, \tag{4.33}$$

$$Z_{e_P} = (\hat{\tau} - \tau)/\sqrt{i^{\tau\tau}(\tau, \hat{\zeta}_\tau)}, \tag{4.34}$$

$$Z_{u_P} = l_\tau(\tau, \hat{\zeta}_\tau)\sqrt{i^{\tau\tau}(\tau, \hat{\zeta}_\tau)}. \tag{4.35}$$

iii) From the previous point, it follows that

$$P_{\tau,\zeta}\{2\left(l_P(\hat{\tau}) - l_P(\tau)\right) \leq \chi^2_{k;1-\alpha}\} \doteq 1 - \alpha\,,$$

with $\chi^2_{k;1-\alpha}$ equal to $(1-\alpha)$-th quantile of the χ^2_k distribution. Therefore, confidence regions based on the profile likelihood ratio test, or on an asymptotically equivalent version of it, have nominal asymptotic level $1 - \alpha$ for every value of ζ.

iv) The profile observed information is defined as

$$j_P(\tau) = -\frac{\partial^2}{\partial\tau\partial\tau^T} l_P(\tau) = -\frac{\partial^2}{\partial\tau\partial\tau^T} l(\tau, \hat{\zeta}_\tau)\,.$$

Note that

$$\frac{\partial}{\partial\tau}l_P(\tau) = l_\tau(\tau, \hat{\zeta}_\tau) + l_\zeta(\tau, \hat{\zeta}_\tau)\,\frac{\partial}{\partial\tau}\hat{\zeta}_\tau\,,$$

where the second summand on the right-hand side vanishes, because $l_\zeta(\tau, \hat{\zeta}_\tau) = 0$. Therefore, the profile score is

$$\frac{\partial}{\partial\tau}l_P(\tau) = l_\tau(\tau, \hat{\zeta}_\tau)\,. \tag{4.36}$$

The matrix of second partial derivatives is

$$\frac{\partial^2}{\partial\tau\partial\tau^T} l_P(\tau) = l_{\tau\tau}(\tau, \hat{\zeta}_\tau) + l_{\tau\zeta}(\tau, \hat{\zeta}_\tau)\frac{\partial}{\partial\tau}\hat{\zeta}_\tau\,.$$

The quantity $(\partial\hat{\zeta}_\tau)/\partial\tau$, can be obtained by differentiating the likelihood equation $l_\zeta(\tau, \hat{\zeta}_\tau) = 0$ with respect to τ. This gives

$$l_{\zeta\tau}(\tau, \hat{\zeta}_\tau) + l_{\zeta\zeta}(\tau, \hat{\zeta}_\tau)\frac{\partial}{\partial\tau}\hat{\zeta}_\tau = 0$$

from which

$$\frac{\partial}{\partial\tau}\hat{\zeta}_\tau = -(l_{\zeta\zeta}(\tau, \hat{\zeta}_\tau))^{-1}l_{\zeta\tau}(\tau, \hat{\zeta}_\tau)\,.$$

It follows that

$$j_P(\tau) = -(l_{\tau\tau} - l_{\tau\zeta}(l_{\zeta\zeta})^{-1}l_{\zeta\tau})\,, \tag{4.37}$$

where all the derivatives indicated are evaluated at $(\tau, \hat{\zeta}_\tau)$. Therefore,

$$(j_P(\tau))^{-1} = j^{\tau\tau}(\tau, \hat{\zeta}_\tau),$$

where $j^{\tau\tau}(\tau, \zeta)$ is the left upper block of the inverse of the overall observed information matrix.

Undoubtedly, properties *i)-iv)* make the profile likelihood an interesting tool for inference. However, $L_P(\tau)$ is not a genuine likelihood. The profile score function $\partial l_P(\tau)/\partial \tau$ does not have zero null expectation. In regular cases, the profile score has null expectation of order $O(1)$ (see section 9.4.6). When incidental parameters are present, as in Examples 4.12 and 4.13, and in Problems 4.1 and 4.9, the lack of consistency of the τ component of the overall maximum likelihood estimator $\hat{\theta}$, can be traced back to the fact that the profile score function has null expectation of order $O(n)$. Using $L_P(\tau)$ is the same as acting as if ζ were known and equal to $\hat{\zeta}_\tau$. This cannot be reasonable if the data do not contain a large amount of information about ζ, something that usually happens if the dimension of ζ is large.

To explicitly compensate for the lack of knowledge about ζ, various modified versions of the profile likelihood have been proposed in recent years. One example will be looked at in the next section, others are illustrated in section 11.6.

Example 4.14 *Normal distribution*
Let $y_1, \ldots, y_n$ be independent observations from a $N(\mu, \sigma^2)$ distribution. Let μ be the parameter of interest. The maximum likelihood estimate of σ^2 for fixed μ is $\hat{\sigma}_\mu^2 = \sum_{i=1}^n (y_i - \mu)^2/n = \hat{\sigma}^2 + (\mu - \hat{\mu})^2$, where $\hat{\sigma}^2 = (n-1)s_n^2/n$ is the maximum likelihood estimate of σ^2 with unknown μ. The profile log-likelihood is

$$l_P(\mu) = -\frac{n}{2}\log(\hat{\sigma}_\mu^2).$$

The combinant $W_P(\mu)$ is

$$W_P(\mu) = n\log\left(\frac{\hat{\sigma}_\mu^2}{\hat{\sigma}^2}\right).$$

A confidence interval with asymptotic level $1 - \alpha$ based on $W_P(\mu)$ is therefore,

$$\bar{y}_n \pm s_n \sqrt{\frac{n-1}{n}\left\{\exp\left(\frac{1}{n}\chi_{1;1-\alpha}^2\right) - 1\right\}}.$$

The confidence interval for μ based on Student's t distribution with $n - 1$ degrees of freedom, t_{n-1}, is $\bar{y}_n \pm n^{-1/2}s_n t_{n-1;1-\alpha/2}$. Both intervals have the asymptotic representation $\bar{y}_n \pm n^{-1/2}s_n z_{1-\alpha/2}\{1 + O(n^{-1})\}$, with different terms of order $O(n^{-1})$. More accurate approximations can be obtained using higher-order asymptotic methods (see Example 11.8). $\triangle$

4.7 Orthogonal Parameterization and Approximate Conditional Likelihood

If the likelihood function has separable parameters, calculating the profile likelihood and related quantities, such as $\hat{\zeta}_\tau, j_P(\tau)$, etc., is very much simplified, since

$$l(\theta) = l_1(\tau) + l_2(\zeta)$$

and, hence, $\hat{\zeta}_\tau = \hat{\zeta}$, $l_P(\tau) = l_1(\tau)$, $j_P(\tau) = -\partial^2 l_1(\tau)/\partial\tau\partial\tau^{\mathsf{T}}$. One way of trying to come closer to this situation is to use an **orthogonal parameterization**.

Definition 4.7 The components τ and ζ of θ are called **orthogonal parameters** if the components l_τ and l_ζ of the score vector are uncorrelated or, equivalently, if $i_{\zeta\tau} = 0$.

The main consequence of parameter orthogonality is that $\hat{\zeta}$ and $\hat{\tau}$ are asymptotically independent. Furthermore, the covariance matrix of $\hat{\tau}$ can be evaluated as if ζ were known. In fact,

$$i^{\tau\tau} = \left(i_{\tau\tau} - i_{\tau\zeta}(i_{\zeta\zeta})^{-1}i_{\zeta\tau}\right)^{-1} = (i_{\tau\tau})^{-1}, \quad \text{if } i_{\zeta\tau} = 0 .$$

In addition,

$$\hat{\zeta}_\tau - \hat{\zeta}' = O_p(n^{-1}) \tag{4.38}$$

while usually $\hat{\zeta}_\tau - \hat{\zeta} = O_p(n^{-1/2})$. Relation (4.38) can be justified as follows. Consider for simplicity $p = 2$ and $k = 1$. The log-likelihood function may be expanded around $(\hat{\tau}, \hat{\zeta})$ as

$$
\begin{aligned}
l(\tau, \zeta) &= l(\hat{\tau}, \hat{\zeta}) - \frac{1}{2}\left\{(\tau - \hat{\tau})^2\hat{\jmath}_{\tau\tau} + 2(\tau - \hat{\tau})(\zeta - \hat{\zeta})\hat{\jmath}_{\tau\zeta} + (\zeta - \hat{\zeta})^2\hat{\jmath}_{\zeta\zeta}\right\} \\
&\quad + O_p(n^{-1/2}) .
\end{aligned}
\tag{4.39}
$$

Differentiating both sides of (4.39) with respect to ζ gives

$$l_\zeta(\tau, \zeta) = -(\tau - \hat{\tau})\hat{\jmath}_{\tau\zeta} - (\zeta - \hat{\zeta})\hat{\jmath}_{\zeta\zeta} + O_p(1) .$$

This expansion gives a linearization of the likelihood equation for ζ with fixed τ. When equating the leading term on the right-hand side to zero it yields

$$
\begin{aligned}
\hat{\zeta}_\tau - \hat{\zeta} &= (\hat{\tau} - \tau)\hat{\jmath}_{\tau\zeta}(\hat{\jmath}_{\zeta\zeta})^{-1} + O_p(n^{-1}) \\
&= (\hat{\tau} - \tau)i_{\tau\zeta}(i_{\zeta\zeta})^{-1} + O_p(n^{-1}) .
\end{aligned}
$$

If $i_{\tau\zeta} = 0$ then (4.38) is obtained.

The approximation to a likelihood with separable parameters can also be easily seen in (4.39). If $i_{\tau\zeta} = 0$, $j_{\tau\zeta}$ and $\hat{\jmath}_{\tau\zeta}$ are of order $O_p(n^{1/2})$. Therefore,

$$
\begin{aligned}
l(\tau,\zeta) &= c - \frac{1}{2}(\tau - \hat{\tau})^2 \hat{\jmath}_{\tau\tau} - \frac{1}{2}(\zeta - \hat{\zeta})^2 \hat{\jmath}_{\zeta\zeta} + O_p(n^{-1/2}) \\
&= l_1(\tau) + l_2(\zeta) + O_p(n^{-1/2})
\end{aligned}
$$

locally, for θ in a neighbourhood of $\hat{\theta}$ with radius of order $O(n^{-1/2})$.

We briefly illustrate how to find an orthogonal parameterization. Again, for simplicity, take $p = 2$, $k = 1$ (the extension to the case $p > 2$, $k = 1$ is immediate). The aim is to find a new parameterization (τ, χ), with $\chi = \chi(\tau, \zeta)$, such that χ and τ are orthogonal. We denote by $l^\perp(\tau, \chi)$ the log-likelihood in the new parameterization,

$$
l^\perp(\tau, \chi) = l(\tau, \zeta(\tau, \chi)) .
$$

Differentiation gives

$$
\frac{\partial l^\perp}{\partial \tau} = \frac{\partial l}{\partial \tau} + \frac{\partial l}{\partial \zeta}\frac{\partial \zeta}{\partial \tau}
$$

$$
\frac{\partial^2 l^\perp}{\partial \tau \partial \chi} = \frac{\partial^2 l}{\partial \tau \partial \zeta}\frac{\partial \zeta}{\partial \chi} + \frac{\partial^2 l}{\partial \zeta^2}\frac{\partial \zeta}{\partial \tau}\frac{\partial \zeta}{\partial \chi} + \frac{\partial l}{\partial \zeta}\frac{\partial^2 \zeta}{\partial \tau \partial \chi} . \tag{4.40}
$$

The reparameterization (τ, χ) must satisfy $E\left(\partial^2 l^\perp / \partial \tau \partial \chi\right) = 0$. Taking the expectation of the right-hand side of (4.40), the contribution of the third summand vanishes. Therefore, the transformation $\chi(\tau, \zeta)$ must satisfy the equation

$$
\frac{\partial \zeta}{\partial \chi}\left(-i_{\zeta\tau} - i_{\zeta\zeta}\frac{\partial \zeta}{\partial \tau}\right) = 0
$$

that is

$$
i_{\zeta\tau} + i_{\zeta\zeta}\frac{\partial \zeta(\tau, \chi)}{\partial \tau} = 0 . \tag{4.41}
$$

When $p > 2$ and $k > 1$, it is not usually possible to obtain a *global* orthogonal parameterization, that is, one that is the same for each value of τ. It is possible, however, to find a *local* orthogonal parameterization, which satisfies orthogonality for a fixed $\tau = \tau_0$ (see Barndorff-Nielsen and Cox, 1994, p. 50).

Example 4.15 *Exponential distributions with proportional means*
Let Y_1, Y_2 be independent exponential random variables with expectation ζ and $\tau\zeta$, respectively. The aim is to find an orthogonal parameterization $(\tau, \chi(\tau, \zeta))$. Since $i_{\zeta\tau} = 1/(\zeta\tau)$, $i_{\zeta\zeta} = 2/\zeta^2$, condition (4.41) becomes

$$\frac{2}{\zeta^2} \frac{\partial \zeta(\tau, \chi)}{\partial \tau} + \frac{1}{\zeta\tau} = 0 ,$$

that is

$$\frac{\partial \zeta(\tau, \chi)}{\partial \tau} \sqrt{\tau} + \frac{1}{2} \frac{1}{\sqrt{\tau}} \zeta(\tau, \chi) = 0 .$$

Note that the left-hand side is the partial derivative with respect to τ of $g(\tau, \chi) = \zeta(\tau, \chi)\sqrt{\tau}$. This derivative is zero if $g(\tau, \chi) = g(\chi)$. Thus, we can choose $g(\chi) = \chi$ and put $\zeta\sqrt{\tau} = \chi$, from which $\zeta = \chi/\sqrt{\tau}$, $\zeta\tau = \chi\sqrt{\tau}$. $\triangle$

Cox and Reid (1987) defined a modified version of the profile likelihood by exploiting the consequences of orthogonality. If τ and ζ are orthogonal then, from (4.39),

$$l(\tau, \zeta) = l(\hat{\tau}, \hat{\zeta}) - \frac{1}{2}\left\{(\tau - \hat{\tau})^2 j_{\tau\tau}(\hat{\tau}, \hat{\zeta}) + (\zeta - \hat{\zeta})^{\mathrm{T}} i_{\zeta\zeta}(\tau, \zeta)(\zeta - \hat{\zeta})\right\} + O_p(n^{-1/2}) ,$$

hence $\hat{\zeta}$ is asymptotically partially sufficient for ζ. Thus, we could argue that for inference about τ the conditional model given $\hat{\zeta}$ may offer a more accurate pseudo-likelihood than the profile likelihood. The asymptotic distribution of $\hat{\zeta}$ can generally be approximated by $N(\zeta, j^{\zeta\zeta})$, where, because of orthogonality, $j^{\zeta\zeta} \doteq (j_{\zeta\zeta})^{-1}$. Therefore, the conditional log-likelihood given $\hat{\zeta}$ may be approximated as

$$\log p_{Y|\hat{\zeta}}(y; \hat{\zeta}, \tau, \zeta) \doteq c(y) + l(\tau, \zeta) - \frac{1}{2}\log |j_{\zeta\zeta}(\tau, \zeta)| + \frac{1}{2}(\hat{\zeta} - \zeta)^{\mathrm{T}}(j^{\zeta\zeta})^{-1}(\hat{\zeta} - \zeta) .$$

Clearly, this conditional log-likelihood is not independent of ζ, because $\hat{\zeta}$ is only asymptotically partially sufficient for ζ and, furthermore, an approximate distribution has been used. One natural solution is to replace ζ with $\hat{\zeta}_\tau$. By doing this, and by (4.38), the contribution of the quadratic form $(\hat{\zeta} - \hat{\zeta}_\tau)^{\mathrm{T}}(j^{\zeta\zeta})^{-1}(\hat{\zeta} - \hat{\zeta}_\tau)$ becomes of order $O_p(n^{-1})$ and can be neglected when compared to the contributions of the other two summands that depend on τ. Thus, we can define the approximate conditional log-likelihood

$$
\begin{aligned}
l_{AC}(\tau) &= l(\tau, \hat{\zeta}_\tau) - \frac{1}{2} \log |j_{\zeta\zeta}(\tau, \hat{\zeta}_\tau)| \\
&= l_P(\tau) - \frac{1}{2} \log |j_{\zeta\zeta}(\tau, \hat{\zeta}_\tau)| .
\end{aligned}
\tag{4.42}
$$

This is a penalized version of the profile log-likelihood, where the penalization takes into account the information about the nuisance parameter as τ varies. However, (4.42) does require an orthogonal parameterization, which certainly exists only for a scalar parameter of interest. A further serious drawback is that, although a one-to-one smooth function of ζ is still orthogonal to τ, the approximate conditional likelihood is not invariant under interest-respecting reparameterizations.

Example 4.16 *Normal linear regression*
Let $y = (y_1, \ldots, y_n)$ be a vector of independent observations each having a $N(\mu_i, \sigma^2)$ distribution, $i = 1, \ldots, n$, with $\mu_i = \beta \cdot x_i$, where β is a k-dimensional vector of regression coefficients and x_i is a vector of known constants. Let σ^2 be the parameter of interest. Because μ_i and σ^2 are orthogonal, β and σ^2 are also orthogonal. Let X be the $n \times k$ matrix of the explanatory variables, with x_i as the i-th row. The log-likelihood function is

$$
\begin{aligned}
l(\beta, \sigma^2) &= -\frac{n}{2} \log \sigma^2 - \frac{1}{2\sigma^2} (y - X\beta)^{\mathrm{T}} (y - X\beta) \\
&= -\frac{n}{2} \log \sigma^2 - \frac{1}{2\sigma^2} \left\{ SSE + (\hat{\beta} - \beta)^{\mathrm{T}} X^{\mathrm{T}} X (\hat{\beta} - \beta) \right\} ,
\end{aligned}
$$

where $\hat{\beta} = (X^{\mathrm{T}} X)^{-1} X^{\mathrm{T}} y$ and $SSE = (y - X\hat{\beta})^{\mathrm{T}} (y - X\hat{\beta})$ is the sum of the squares of the residuals. Since $\hat{\beta}_{\sigma^2} = \hat{\beta}$, the profile log-likelihood is

$$
l_P(\sigma^2) = -\frac{n}{2} \log \sigma^2 - \frac{1}{2\sigma^2} SSE .
$$

Furthermore, $j_{\beta\beta} = (\sigma^2)^{-1} (X^{\mathrm{T}} X)$, which does not depend on β. Therefore, because $|j_{\beta\beta}| = (\sigma^2)^{-k} |X^{\mathrm{T}} X|$, the approximate conditional log-likelihood (4.42) is

$$
l_{AC}(\sigma^2) = -\frac{1}{2} (n - k) \log \sigma^2 - \frac{1}{2\sigma^2} SSE .
$$

Unlike as for $l_P(\sigma^2)$, the score associated with $l_{AC}(\sigma^2)$ is an exactly unbiased estimating function. The corresponding estimator of σ^2 is the usual unbiased estimator, corrected for the degrees of freedom. Furthermore, $l_{AC}(\sigma^2)$ coincides with the marginal likelihood based on SSE.   △

Example 4.17 *Normal distribution (cont.)*
Under the same assumptions as in Example 4.14,

$$j_{\sigma^2\sigma^2}(\mu, \hat{\sigma}_\mu^2) = \frac{n}{2} \frac{1}{(\hat{\sigma}_\mu^2)^2} \ .$$

The approximate conditional log-likelihood is

$$l_{AC}(\mu) = -\frac{n-2}{2} \log(\hat{\sigma}_\mu^2) \ . \tag{4.43}$$

The combinant $2(l_{AC}(\hat{\mu}) - l_{AC}(\mu))$ has null asymptotic χ_1^2 distribution. A confidence interval with asymptotic level $1 - \alpha$ based on $l_{AC}(\mu)$ is therefore

$$\bar{y}_n \pm s_n \sqrt{\frac{n-1}{n} \left\{ \exp\left(\frac{\chi_{1;1-\alpha}^2}{n-2}\right) - 1 \right\}} \ .$$

This does not match the exact interval based on the t_{n-1} distribution. Furthermore, it has the asymptotic representation

$$\bar{y}_n \pm n^{-1/2} s_n z_{1-\alpha/2} \left\{ 1 + O(n^{-1}) \right\} \ ,$$

where the term of order $O(n^{-1})$ is different from that arising from the analogous representation of the t intervals. Hence, with respect to profile likelihood based confidence intervals considered in Example 4.14, there is no reduction in the asymptotic order of error in the approximation of the exact t intervals. See, however, Problem 4.22.   △

Example 4.18 *Exponential distribution with proportional means (cont.)*
Under random sampling of size n from the model in Example 4.15,

$$\begin{aligned} l_P(\tau) &= -2n \log \hat{\chi}_\tau \\ &= -n \log \tau - 2n \log \left(\bar{y}_1 + \frac{\bar{y}_2}{\tau} \right) , \end{aligned}$$

where $\bar{y}_1$ and $\bar{y}_2$ are the two sample means. In addition, $j_{\chi\chi}(\tau, \hat{\chi}_\tau) = 2n/\hat{\chi}_\tau^2$. Thus, the approximate conditional log-likelihood (4.42) is

$$\begin{aligned} l_{AC}(\tau) &= -2n \log \hat{\chi}_\tau + \log \hat{\chi}_\tau \\ &= -\left(n - \frac{1}{2}\right) \log \tau - (2n - 1) \log \left(\bar{y}_1 + \frac{\bar{y}_2}{\tau} \right) . \end{aligned}$$

Consider the alternative orthogonal parameterization, (τ, λ), with $\lambda = \log \chi$. The profile log-likelihood is obviously unchanged, while the adjustment term in (4.42) changes according to the rule in (1.19). In particular,

$$j_{\lambda\lambda}(\tau, \lambda) = j_{\chi\chi}(\tau, e^{\lambda})(e^{\lambda})^2 - l_{\chi}(\tau, e^{\lambda})e^{\lambda},$$

which, calculated at $\hat{\lambda}_{\tau} = \log \hat{\chi}_{\tau}$, reduces to $2n$ and can thus be neglected. Therefore, with this parameterization, the approximate conditional log-likelihood coincides with the profile log-likelihood. Furthermore, the profile log-likelihood is the same as the marginal log-likelihood based on $v = \bar{y}_2 / \bar{y}_1$. Indeed, v is distributed as $\tau F_{2n,2n}$, where F_{n_1,n_2} indicates the central F distribution with (n_1, n_2) degrees of freedom. The corresponding marginal likelihood is

$$L_M(\tau) = \left\{ \tau^n \left(1 + \frac{v}{\tau} \right)^{2n} \right\}^{-1}.$$

$\triangle$

4.8 Partial Likelihood

The term *partial likelihood* refers to a pseudo-likelihood which was introduced by Cox (1975) as a generalization of a key tool for inference about the regression coefficients in a proportional hazards model, in the presence of censored observations (Cox, 1972).

Certain assumptions about the distribution of the observations y have to be satisfied in order to construct a partial likelihood. In particular, assume that $y = (y_1, \ldots, y_n)$ is in a one-to-one correspondence with the sequence $(u_1, v_1, u_2, v_2, \ldots, u_m, v_m)$ and that the conditional distribution of V_i, given the preceding observations in the sequence, $(u_1, v_1, u_2, v_2, \ldots, u_i)$, only depends on the parameter of interest τ. Let us write $u^{(j)} = (u_1, \ldots, u_j)$ and $v^{(j)} = (v_1, \ldots, v_j)$, $j = 1, \ldots, m$. The likelihood can be factorized as

$$
\begin{aligned}
L(\theta; y) &= L(\theta; u^{(m)}, v^{(m)}) \\
&= \prod_{i=1}^{m} P_{U_i, V_i | U^{(i-1)}, V^{(i-1)}} (u_i, v_i; u^{(i-1)}, v^{(i-1)}, \tau, \zeta) \\
&= \prod_{i=1}^{m} P_{U_i | U^{(i-1)}, V^{(i-1)}} (u_i; u^{(i-1)}, v^{(i-1)}, \tau, \zeta) \\
&\qquad \times \prod_{i=1}^{m} P_{V_i | U^{(i)}, V^{(i-1)}} (v_i; u^{(i)}, v^{(i-1)}, \tau) \qquad (4.44)
\end{aligned}
$$

where, for $i = 1$, there is no conditioning on $(u^{(i-1)}, v^{(i-1)})$.

Definition 4.8 The factor

$$L_{PA}(\tau) = \prod_{i=1}^{m} p_{V_i | U^{(i)}, V^{(i-1)}} (v_i; u^{(i)}, v^{(i-1)}, \tau)$$

on the right-hand side of (4.44) is called the **partial likelihood** based on V. The corresponding **partial log-likelihood** is

$$l_{PA}(\tau) = \sum_{i=1}^{m} \log p_{V_i | U^{(i)}, V^{(i-1)}} (v_i; u^{(i)}, v^{(i-1)}, \tau). \qquad (4.45)$$

Partial likelihood is not a genuine likelihood. It is useful to underline where it differs from marginal and conditional likelihood. If $v^{(m)}$ is partially distribution constant for ζ, the marginal likelihood based on $v^{(m)}$ is

$$L_M(\tau; v^{(m)}) = \prod_{i=1}^{m} p_{V_i | V^{(i-1)}} (v_i; v^{(i-1)}, \tau),$$

which matches the partial likelihood only if $U^{(m)}$ and $V^{(m)}$ are independent. If $u^{(m)}$ is partially sufficient for ζ, the conditional likelihood given $u^{(m)}$ is

$$L_C(\tau; v^{(m)} | u^{(m)}) = \prod_{i=1}^{m} p_{V_i | V^{(i-1)}, U^{(m)}} (v_i; v^{(i-1)}, u^{(m)}, \tau),$$

which matches the partial likelihood only if V_i is independent of $(U_{i+1}, \ldots, U_m)$, for every $i = 1, \ldots, m$.

Example 4.19 *The proportional hazards model (cont.)*
Suppose that the data are n pairs (t_i, δ_i), $i = 1, \ldots, n$, where t_i denotes the observed lifetime for the i-th item and δ_i is an indicator whose value is 1 or 0 depending on whether t_i is either an uncensored lifetime or a censoring time. Assume that the censoring mechanism is uninformative (see Example 1.2). Suppose that the distribution of the lifetime T_i^* of the i-th item follows a proportional hazards model as in Example 4.9. The aim is to construct a likelihood function for β.

The likelihood function (1.22) is

$$L(\beta, r_0(\cdot)) = \prod_{i=1}^{n} \left\{ p_{T_i^*}(t_i; r_0(\cdot), \beta, \boldsymbol{x}_i) \right\}^{\delta_i} \left\{ 1 - F_{T_i^*}(t_i; r_0(\cdot), \beta, \boldsymbol{x}_i) \right\}^{1 - \delta_i},$$

where

$$p_{T_i^*}(t_i; r_0(\cdot), \beta, \boldsymbol{x}_i) = r_0(t_i)e^{\beta \cdot \boldsymbol{x}_i} \exp\left\{-R_0(t_i)e^{\beta \cdot \boldsymbol{x}_i}\right\}, \quad i = 1, \ldots, n.$$

This likelihood is useful for inference about β only so long as the failure rate is either known or has been modelled parametrically.

Let $m = \sum_{i=1}^{n} \delta_i$ be the number of uncensored observations and let $t_{(1)} < \cdots < t_{(m)}$ be the uncensored survival times arranged in increasing order. Indicate the corresponding vectors of explanatory variables by $\boldsymbol{x}_{(1)}, \ldots, \boldsymbol{x}_{(m)}$ and denote by $\mathcal{R}(t_{(j)})$ the set of indices of the observations (both censored and uncensored) which correspond to the items at risk at the time $t_{(j)}$. Then, let

- v_j be the indicator of which item has failed at $t_{(j)}$;

- u_j be the complete description of the behaviour of the system in the time interval $(t_{(j-1)}, t_{(j)}]$, apart from the information about which item failed at $t_{(j)}$; u_j describes, in particular, the censoring between the two successive failures.

Thus

$$p_{V_j | U^{(j)}, V^{(j-1)}}(v_j; u^{(j)}, v^{(j-1)}, \beta) = \frac{r(t_{(j)}; \boldsymbol{x}_{(j)}, \beta)}{\sum\limits_{i \in \mathcal{R}(t_{(j)})} r(t_{(j)}; \boldsymbol{x}_i, \beta)},$$

where $r(t_{(j)}; \boldsymbol{x}_{(j)}, \beta)$ denotes the failure rate of the item that has, effectively, failed. Therefore the partial likelihood is

$$L_{PA}(\beta) = \prod_{j=1}^{m} \frac{e^{\beta \cdot \boldsymbol{x}_{(j)}}}{\sum\limits_{i \in \mathcal{R}(t_{(j)})} e^{\beta \cdot \boldsymbol{x}_i}}. \tag{4.46}$$

If all observations are uncensored, (4.46) is the same as the marginal likelihood obtained in Example 4.9. $\triangle$

Example 4.20 *Disconnected segments of a Markov process*
This example is taken from Wong (1986, section 6.1). Suppose that the data are made up of m disjoint segments in a time series

$$(y_{n_1}, \ldots, y_{l_1}), (y_{n_2}, \ldots, y_{l_2}), \ldots, (y_{n_m}, \ldots, y_{l_m}),$$

with $n_1 < l_1 < n_2 < l_2 < \ldots < n_m < l_m$. Assume that, within each segment, the series is generated by a first-order autoregressive process, $y_t = \tau y_{t-1} + \varepsilon_t$

with $|\tau| < 1$, and that the ε_t are independent and normally distributed. The joint density of the observations is

$$
p_Y(y; \tau, \zeta) = \prod_{j=1}^{m} \left\{ p_{Y_{n_j}|Y_{l_{j-1}}} (y_{n_j}; y_{l_{j-1}}, \tau, \zeta) \prod_{t=n_j+1}^{l_j} p_{Y_t|Y_{t-1}} (y_t; y_{t-1}, \tau) \right\}
$$

$$
= \prod_{j=1}^{m} \left\{ p_{Y_{n_j}|Y_{l_{j-1}}} (y_{n_j}; y_{l_{j-1}}, \tau, \zeta) \left(\prod_{t=n_j+1}^{l_j} \frac{1}{\sqrt{2\pi}} e^{-\frac{1}{2}(y_t - \tau y_{t-1})^2} \right) \right\},
$$

where, for $j = 1$, there is no conditioning on Y_{l_0}. Note that no assumptions have been made about the stochastic behaviour of the series between successive segments, such behaviour might depend for instance on a nuisance parameter ζ. The above factorization identifies the partial log-likelihood

$$
l_{PA}(\tau) = -\frac{1}{2} \sum_{j=1}^{m} \sum_{t=n_j+1}^{l_j} (y_t - \tau y_{t-1})^2
$$

for inference about τ. See Wong (1986, section 6.1) for a discussion about the loss of efficiency in estimating τ which results from the reduction to partial log-likelihood. $\triangle$

In many respects partial likelihood does behave like a genuine likelihood. Suppose, for simplicity, that τ is a scalar. The score function obtained from (4.45) is

$$
l_{PA}(\tau)_* = \sum_{i=1}^{m} l_i(\tau; v_i, c^{(i)})_* \,,
$$

where

$$
l_i(\tau; v_i, c^{(i)})_* = l_{i*} = \frac{\partial}{\partial \tau} \log p_{V_i|C^{(i)}=c^{(i)}} (v_i; c^{(i)}, \tau)
$$

with $c^{(i)}$ as a short notation for the set $(u^{(i)}, v^{(i-1)})$ of the conditioning variables pertaining to v_i.

If the conditional density $p_{V_i|C^{(i)}=c^{(i)}} (v_i; c^{(i)}, \tau)$ satisfies the usual regularity conditions, then the identities

$$
E\left\{ l_i(\tau; v_i, c^{(i)})_* \mid c^{(i)} \right\} = 0
$$

and

$$
\mathrm{Var}\left\{ l_i(\tau; v_i, c^{(i)})_* \mid c^{(i)} \right\} = -E\left\{ \frac{\partial^2}{\partial \tau^2} \log p_{V_i|C^{(i)}=c^{(i)}} (v_i; c^{(i)}, \tau) \mid c^{(i)} \right\}
$$

hold. From the first identity we immediately conclude that, marginally, the null expectation of the partial score is zero. In addition, marginally, the summands l_{i*} of the partial score are uncorrelated. In fact, supposing without loss of generality $i < j$, we have

$$E(l_{i*}l_{j*}) = E\left(E(l_{i*}l_{j*} \mid c^{(j)})\right) = E\left(l_{i*}E(l_{j*} \mid c^{(j)})\right) = 0 \ .$$

In conclusion, marginally, the variance of the partial score is given by the sum of the variances of the single summands.

Under mild regularity conditions which are related to the dependence structure of the uncorrelated summands, a central limit theorem holds for $l_{P_A}(\tau)_*$, which justifies the normal approximation for the distribution of the score and of the partial maximum likelihood estimator $\hat\tau_{P_A}$. The asymptotic covariance matrix of $\hat\tau_{P_A}$ is consistently estimated by the inverse of

$$\left(-\frac{\partial^2}{\partial\tau^2}l_{P_A}(\tau)\right)\Bigg|_{\tau=\hat\tau_{P_A}} \ .$$

Under the same conditions $2(l_{P_A}(\hat\tau_{P_A}) - l_{P_A}(\tau))$ has the usual null asymptotic chi-squared distribution.

4.9 Quasi-Likelihood

For data y, consider a semiparametric or nonparametric statistical model $\mathcal{F}$, parameterized by θ. Let τ be a k-dimensional parameter of interest and let $q(y;\tau)$ be a k-dimensional combinant which is an unbiased estimating function, that is, such that

$$E_\theta(q(Y;\tau)) = 0 \qquad\qquad (4.47)$$

for every θ such that $\tau(\theta) = \tau$. Equation (4.47) makes the combinant $q(\cdot)$ analogous to a score vector. As was seen at the end of section 3.4.3, under regularity conditions the estimator of τ, based on the estimating equation $q(y;\tau) = 0$ is consistent and asymptotically normal, as is the maximum likelihood estimator within a parametric model. Thus the general question arises of whether it is possible to find a pseudo-log-likelihood $l_Q(\tau;y)$ which has $q(y;\tau)$ as its gradient. This question has an obvious answer if $k = 1$, given by

$$l_Q(\tau) = \int_c^\tau q(y;t)dt \ ,$$

with c an arbitrary constant. If τ is multi-dimensional and $q(\cdot)$ is continuously differentiable with respect to τ, then a necessary condition for $l_Q(\tau; y)$ to exist is that the matrix

$$-j_Q(\tau) = \frac{\partial}{\partial \tau^T} q(y; \tau)$$

is symmetric. One important example, where it is possible to obtain $l_Q(\tau)$ is given by generalized linear models and will be examined in section 6.4.2.

If $l_Q(\tau)$ exists and can be used to define tests and confidence regions in the usual manner, then $l_Q(\tau)$ is said to be a quasi-log-likelihood function based on the combinant $q(\cdot)$, which is the corresponding quasi-score.

Note that, unlike as happens with a genuine score, the identity, analogous to the information identity,

$$\text{Var}_\theta(q(Y; \tau)) = -E_\theta \left(\frac{\partial}{\partial \tau^T} q(Y; \tau) \right) \tag{4.48}$$

does not, in general, hold. However, it is possible to recover this identity through a nonsingular linear transformation of $q(\cdot)$ of the form $\tilde{q}(y; \tau) = Aq(y; \tau)$, where

$$A^T = - \left(\text{Var}_\theta(q(Y; \tau)) \right)^{-1} E_\theta \left(\frac{\partial}{\partial \tau^T} q(Y; \tau) \right) .$$

As is implicit in this equation, only the covariance matrix of $q(\cdot)$ and the expectation of the matrix of partial derivatives of $q(\cdot)$ with respect to τ are needed. Moreover, A must depend on τ only.

If a quasi-likelihood $L_Q(\tau)$ satisfies (4.48) many asymptotic considerations are simplified. For example, the matrix $j_Q(\tau)$ of quasi-observed information is connected in the usual way to the asymptotic covariance matrix of the estimator based on $q(\cdot)$.

If $k > 1$ a quasi-likelihood for τ, based on $q(\cdot)$, may not exist. Even in this case, it is possible to define a profile quasi-likelihood for a scalar component of τ, τ_1 say. Let τ_2 denote the remaining $k - 1$ components of τ. The corresponding partition of $q(\cdot)$ is $(q_1(\cdot), q_2(\cdot))$. Let $\hat{\tau}_2(\tau_1)$ denote the solution with respect to τ_2 of the partial estimating equation $q_2(y; \tau_1, \tau_2) = 0$. The profile quasi-log-likelihood for τ_1 is defined by

$$l_Q(\tau_1) = \int_c^{\tau_1} q_1(y; t, \hat{\tau}_2(t)) dt ,$$

up to an additive constant. Barndorff-Nielsen (1995c) shows that under regularity conditions, the log-likelihood ratio statistic and its signed version have the usual null asymptotic distribution.

4.10 Empirical Likelihood

Empirical likelihood, introduced by Owen (1988), is a nonparametric profile likelihood that can be used for inference about a finite number k of moments or smooth functions of moments. The most striking property of empirical likelihood is that the corresponding log-likelihood ratio statistic is asymptotically chi-squared, as is true for parametric likelihood. Confidence regions based on the empirical log-likelihood ratio statistic have an advantage over the usual confidence regions based on the asymptotic normality of sample moments, in that they are not necessarily symmetric and they do not include inadmissible parameter values.

Let $y = (y_1, \ldots, y_n)$ be independent observations from a univariate random variable with distribution function $F_0(y)$, expectation μ and finite variance. Suppose, for simplicity, that there are no ties. In section 3.6.2 it was seen that the empirical distribution function $\hat{F}_n(\cdot)$ is the generalized maximum likelihood estimate of $F_0(\cdot)$. In particular, $\hat{F}_n(\cdot)$ maximizes the **generalized likelihood function**

$$L(F) = \prod_{i=1}^{n} \left\{ F(y_i) - F(y_i^-) \right\} ,$$

among all the distribution functions F on $\mathbb{R}$. Note that $L(F) = 0$ if one of the values y_i is not a jump point of F. Furthermore, if F has large generalized likelihood, then it places its probability mass almost entirely on the observations $y_1, \ldots, y_n$. Indeed, if for $0 < c < 1$

$$\frac{L(F)}{L(\hat{F}_n)} \geq 1 - c > 0 ,$$

then, for the well-known inequality between the arithmetic and the geometric mean,

$$\frac{1}{n} \sum_{i=1}^{n} (F(y_i) - F(y_i^-)) \geq \left\{ \prod_{i=1}^{n} (F(y_i) - F(y_i^-)) \right\}^{1/n}$$

$$= \frac{1}{n} \left(\frac{L(F)}{L(\hat{F}_n)} \right)^{1/n} \geq \frac{1}{n} (1-c)^{1/n} ,$$

from which $\sum_{i=1}^{n} (F(y_i) - F(y_i^-)) \geq (1-c)^{1/n} \geq 1 + \log(1-c)/n = 1 - O(n^{-1})$. Therefore, at least asymptotically, $L(F)$ can be identified with a multinomial likelihood with weights $w = (w_1, \ldots, w_n)$ assigned to $(y_1, \ldots, y_n)$, with $\sum_{i=1}^{n} w_i = 1$, $w_i > 0$, so that

$$L(F) = \prod_{i=1}^{n} w_i \ . \tag{4.49}$$

Let μ be the parameter of interest, all the further aspects of distribution being treated as a nuisance parameter. It is natural to associate the profile likelihood deduced from (4.49) to μ as a pseudo-likelihood.

Definition 4.9 The function

$$L_E(\mu) = \sup_{w : \sum w_i y_i = \mu} \prod_{i=1}^{n} w_i \ , \qquad \mu \in (y_{(1)}, y_{(n)}) \ , \tag{4.50}$$

where the supremum is taken with respect to the vectors w of normalized positive weights, is said to be the **empirical likelihood** for μ.

Because of the assumptions on w, $L_E(\mu)$ is only defined for $\mu \in (y_{(1)}, y_{(n)})$. The subset of $\mathbb{R}^n$, where the supremum in (4.50) is sought, is closed and bounded and the function $\prod_{i=1}^{n} w_i$ is continuous. Therefore, the supremum is a maximum. This maximum is unique. Indeed, suppose that there are two different weight vectors $w^{(0)}$ and $w^{(1)}$, with $w_j^{(0)} \neq w_j^{(1)}$, which maximize $L_E(\mu)$ under the given constraints. For $\alpha \in (0, 1)$ the convex linear combination $w(\alpha) = \alpha w^{(0)} + (1 - \alpha) w^{(1)}$ also satisfies the constraints. Since $\alpha w_j^{(0)} + (1 - \alpha) w_j^{(1)} > (w_j^{(0)})^{\alpha} (w_j^{(1)})^{1-\alpha}$, then

$$\prod_{i=1}^{n} (\alpha w_i^{(0)} + (1 - \alpha) w_i^{(1)}) > \prod_{i=1}^{n} (w_i^{(0)})^{\alpha} (w_i^{(1)})^{1-\alpha} = L_E(\mu) \ ,$$

which contradicts our assumption about $w^{(0)}$ and $w^{(1)}$.

Since $\hat{F}_n(\cdot)$ maximizes $L(F)$, then

$$\hat{\mu}_E = \int y \, d\hat{F}_n(y) = \bar{y}_n \ .$$

This agrees with the properties of a profile likelihood.

In order to investigate the behaviour of the empirical likelihood when constructing confidence regions, it is convenient to translate the problem of constrained maximization (4.50) into one with an elementary solution. The constraints on the w_i imply that

$$\sum_{i=1}^{n} w_i (1 + \lambda(y_i - \mu)) = 1 \ , \tag{4.51}$$

where λ is a free variable, whose value has to be fixed so as to satisfy $w_i > 0$, $\sum w_i = 1$, $\sum w_i y_i = \mu$. Consider values of λ such that $1 + \lambda(y_i - \mu) > 0, i = 1, \ldots, n$. Since

$$\left(\prod_{i=1}^{n} w_i\right)^{1/n} = \left(\prod_{i=1}^{n} w_i(1 + \lambda(y_i - \mu))\frac{1}{1 + \lambda(y_i - \mu)}\right)^{1/n}$$

$$= \left(\prod_{i=1}^{n} w_i(1 + \lambda(y_i - \mu))\right)^{1/n} \left(\prod_{i=1}^{n} \frac{1}{1 + \lambda(y_i - \mu)}\right)^{1/n},$$

where only the first factor depends on w, the inequality between the arithmetic and the geometric mean shows that the maximum is obtained for

$$w_i(1 + \lambda(y_i - \mu)) = \text{constant},$$

that is, using (4.51), for

$$w_i = \frac{1}{n(1 + \lambda(y_i - \mu))}.$$

In this expression λ is to be chosen as a function of μ, so as to satisfy

$$\sum_{i=1}^{n} \frac{1}{n(1 + \lambda(y_i - \mu))} y_i = \mu,$$

that is

$$g(\lambda, \mu) = \sum_{i=1}^{n} \frac{y_i - \mu}{n(1 + \lambda(y_i - \mu))} = 0, \qquad (4.52)$$

which expresses λ as an implicit function of μ, $\lambda = \lambda(\mu)$. Consequently,

$$L_E(\mu) = \prod_{i=1}^{n} n^{-1} \{1 + \lambda(\mu)(y_i - \mu)\}^{-1}, \qquad \mu \in (y_{(1)}, y_{(n)}),$$

with $\lambda(\mu)$ solution of (4.52).

It is not difficult to check that, for fixed μ, $g(\lambda, \mu)$ is a decreasing function of λ. Furthermore, $g(0, \bar{y}_n) = 0$, so that $\lambda(\bar{y}_n) = 0$. A Taylor expansion of $g(\lambda, \mu)$ around $(0, \bar{y}_n)$ gives the local linearization

$$0 = g(\lambda(\mu), \mu) \doteq g(0, \bar{y}_n) + \lambda(\mu)g_\lambda(0, \bar{y}_n) + (\mu - \bar{y}_n)g_\mu(0, \bar{y}_n),$$

where $g_\lambda(\cdot)$ and $g_\mu(\cdot)$ denote the first-order partial derivatives of $g(\lambda,\mu)$. Therefore,

$$\lambda(\mu) \doteq (\bar{y}_n - \mu)\frac{g_\mu(0, \bar{y}_n)}{g_\lambda(0, \bar{y}_n)} \doteq \frac{\bar{y}_n - \mu}{s_n^2} \; , \tag{4.53}$$

so that $\lambda(\mu)$ is of order $O_p(n^{-1/2})$.

Using (4.53) it is simple to prove that the empirical score $l_E(\mu)_*$ and the empirical observed information $j_E(\mu)$ have the asymptotic representations

$$l_E(\mu)_* \doteq n \frac{\bar{y}_n - \mu}{s_n^2}$$

and

$$j_E(\mu) \doteq \frac{n}{s_n^2} \; .$$

These relations confirm the analogy between the empirical likelihood and a genuine likelihood. In fact, the leading term of the null expectation of the score is zero and the observed information asymptotically matches the null variance of the score.

The empirical log-likelihood ratio can be defined as

$$W_E(\mu) = -2\log\left\{L_E(\mu)/L_E(\hat{\mu}_E)\right\} \; . \tag{4.54}$$

Since $L_E(\hat{\mu}_E) = n^{-n}$,

$$W_E(\mu) = 2\sum_{i=1}^{n}\log\left\{1 + \lambda(\mu)(y_i - \mu)\right\} \; .$$

It may be proved that $\log L_E(\mu)$ is concave, thus confidence regions based on $W_E(\mu)$ are convex, that is intervals when μ is a scalar. Owen (1988) showed that the null asymptotic distribution of $W_E(\mu)$ is chi-squared with one degree of freedom. This can be checked using expansion (4.53). Taking into account that $\lambda(\mu)$ is of order $O_p(n^{-1/2})$, we can write

$$\begin{aligned}
W_E(\mu) &= 2\sum_{i=1}^{n}\log\left\{1 + \lambda(\mu)(y_i - \mu)\right\} \\
&= 2\sum_{i=1}^{n}\left\{\lambda(\mu)(y_i - \mu) - \frac{1}{2}\lambda(\mu)^2(y_i - \mu)^2 + O_p(n^{-3/2})\right\} \\
&= 2n\lambda(\mu)\left\{\bar{y}_n - \mu - \frac{1}{2}\lambda(\mu)(s_n^2 + (\bar{y}_n - \mu)^2)\right\} + O_p(n^{-1/2})
\end{aligned}$$

$$= 2n \frac{\bar{y}_n - \mu}{s_n^2} \left(\bar{y}_n - \mu - \frac{1}{2} \frac{\bar{y}_n - \mu}{s_n^2} s_n^2 \right) + O_p(n^{-1/2})$$

$$= \left(\frac{\bar{y}_n - \mu}{s_n/\sqrt{n}} \right)^2 + O_p(n^{-1/2}).$$

This shows that the signed version of $W_E(\mu)$ is equivalent to the first-order to the usual t statistic. Hence, the square of this latter statistic is asymptotically χ_1^2 and can be interpreted as a Wald version of $W_E(\mu)$. The further terms in the expansion of $W_E(\mu)$ are corrections for the asymmetry of the distribution of the sample mean.

4.11 Bibliographic Note

Nuisance parameters are so much a part of statistical models that even the informal notion of sufficiency, considered by Fisher (1920), took their presence into account. Otherwise it would not be possible to interpret the qualification *sufficient* for σ^2 which is given to the sample variance when sampling from a normal distribution. The need for a special theory has emerged slowly, starting from Bartlett (1936a, 1937).

For the extreme case of incidental parameters in Neyman–Scott problems, Lindsay (1980), Bickel (1982) and Kumon and Amari (1984) investigated quantifying the information about the parameter of interest carried by the data. For the Neyman–Scott problem in a semiparametric setting, see Small and Murdoch (1993). There are two main approaches to recovering consistency and the usual asymptotic properties in inference about the parameter of interest. The first exploits the notion of pseudo-likelihood and was introduced by Kalbfleisch and Sprott (1970), where likelihood methods are extended to marginal, conditional and profile likelihoods. Conditional likelihood is explored in depth in Andersen (1970, 1971, 1973). The second approach can be attributed to Kiefer and Wolfowitz (1956); the sequence of incidental parameters which correspond to the sequence of data is modelled as independent realizations of a probability distribution. For more recent developments in this direction see, Lindsay (1980, 1983a, 1983b, 1983c, 1995) and Pfanzagl (1993).

The Rasch model (Rasch, 1960) has played a special role in the development of methods for dealing with incidental parameters. See Andersen (1980, Chapter 6) for more details about inference based on conditional likelihood.

Partial sufficiency was introduced by Fraser (1956); see Barndorff-Nielsen and Blæsild (1975) for a detailed examination of this in the context of exponential families. The notion of partially distribution constant statistics can be attributed to Sandved (1965) and Sverdrup (1966). Barndorff-Nielsen (1978, Chapter 4) distinguishes between four types of reduction called, respectively, B-sufficiency (ancillarity), S-sufficiency (ancillarity), G-sufficiency (ancillarity), M-sufficiency (ancillarity). The term ancillarity is used in Barndorff-Nielsen (1978) with the same meaning as distribution constancy. The first concept corresponds to the usual definition of sufficiency (without nuisance parameters); the second corresponds to partial sufficiency; the third expresses invariance reduction with respect to nuisance parameters in the context of a group family (see Chapter 7) and the last, and most general, does not respect the principle of invariance under interest-respecting reparameterizations and is not discussed here. A further notion of L-sufficiency was introduced by Rémon (1984); lastly, Jørgensen (1993) defined I-sufficiency (ancillarity) and gave a definition of L-ancillarity, attributing it, as a personal communication, to Barndorff-Nielsen. The proposal made here, of partial non-information in the extended sense, section 4.2.3, is quite general. However, there are few indications that it may be useful (see Kalbfleisch and Sprott, 1974). For further contributions, see also, Dawid (1975), Sprott (1975), Basu (1977, 1978), Zhu and Reid (1994) and Severini (1995). Severini (1993) takes up the problem of constructing locally ancillary statistics with nuisance parameters. Further results are obtained by Severini (1994).

The construction of marginal and conditional likelihoods can be generalized to the case where the only statistics that are partially distribution constant or partially sufficient for the nuisance parameter depend on the parameter of interest τ. In this situation the Jacobian of the transformation does depend on τ and, therefore, cannot be neglected in the likelihood (see Kalbfleisch and Sprott, 1970, 1974, for further discussion and some examples). For an application of marginal likelihood to time series models, see Tunnicliffe-Wilson (1989).

Using the methods of Royall (1986) and Kent (1982), described in section 3.4.3, Stafford (1996) proposes a robust adjustment of the profile likelihood such that the adjusted likelihood satisfies the information identity (1.9), with error of order $O(n^{-1})$.

Huzurbazar (1950) did the first study on the consequences of parameter orthogonality. The derivation of the approximate conditional likelihood, as presented in section 4.7, is essentially heuristic. A more thorough treatment would require a full development of asymptotic methods: see Cox and Reid (1987); see also section 11.6. Cox and Reid (1993) introduced a version of the approximate conditional likelihood which does not require that the orthogonal parameterization should, explicitly, be found; it requires however integration of a function of null moments of log-likelihood derivatives. DiCiccio, Martin, Stern and Young (1996) propose a criterion for choosing between orthogonal parameterizations based on the requirement that the approximate conditional likelihood should satisfy the information identity (1.9) to a high asymptotic order. Liang (1987) and Ferguson, Reid and Cox (1991) studied the properties of estimating equations deduced from approximate conditional likelihood. For applications to time series models and comparisons with marginal likelihood, see Cruddas, Reid and Cox (1989) and Bellhouse (1990).

For an introduction to theoretical developments of partial likelihood, see Andersen, Borgan, Gill and Keiding (1993, section 2.7.3).

The notion quasi-likelihood was introduced by Wedderburn (1974) and developed by McCullagh (1983). It will be taken up again in section 6.4.2. For a survey, see Firth (1993).

The definition of empirical likelihood and the validity of the result on the asymptotic distribution can be extended to vector means and regular functions of vector means (Owen, 1990; Hall and La Scala, 1990). In addition, by suitable corrections, the accuracy of the asymptotic chi-squared approximation for the null distribution of $W_E(\mu)$ can be increased (DiCiccio, Hall and Romano, 1991). Qin and Lawless (1994) and Adimari (1996) extend the definition of empirical likelihood to nonlinear functionals. Murphy (1995b), Li (1995), Li, Hollander, McKeague and Yang (1996) and Adimari (1997) consider empirical likelihood for incomplete data.

Since the 1970's numerous other pseudo-likelihoods have been considered. Some of these are: the pseudo-likelihood of Besag (Besag, 1974, 1977, 1978, 1986; Ripley, 1988), which deals with models for spatial data, and that of Gong and Samaniego (1981) (see also Liang and Self, 1996); the bootstrap likelihood (Davison, Hinkley and Worton, 1992), which is in the spirit of empirical likelihood; dual likelihood (Mykland, 1995a), which associates a likelihood to a martingale estimating equation; composite likelihood (Lindsay, 1988) and projected likelihood (McLeish and Small, 1992), for semiparametric models; M-order likelihood (Azzalini, 1983), for stationary processes; penalized likelihood (Good and Gaskins, 1971; Green, 1987), for an infinite-dimensional parameter of interest such as a density or a regression function; the various instances of predictive likelihood (Butler, 1986, 1989; Bjørnstad, 1990).

4.12   Problems

4.1 With reference to the problem described in Example 4.1, show that the maximum likelihood estimator of σ^2, $\hat{\sigma}^2$, is not consistent. Furthermore, check that the statistic W has the form given in Example 4.1 and, using the expansion $\log(1 + x) = x - x^2/2 + \ldots$, show that the null expectation of W is of order $O(n)$.

[Section 4.1]

4.2 With reference to the Rasch model introduced in Example 4.2, show that, for $k = 2$, the maximum likelihood estimator of τ_1 converges in probability to $2\tau_1$.

[Section 4.1; Andersen, 1980, section 6.4]

4.3 Show that the pseudo-likelihoods $L_M(\tau)$ and $L_C(\tau)$, defined by (4.10) and (4.12), respectively, are invariant under interest-respecting reparameterizations.

[Section 4.3]

4.4 Do in detail the calculations that give equation (4.13) in Example 4.6. State whether the reduction to $|\bar{y}_n|$ is a reduction by sufficiency or not. Check that the score (4.14) has zero null expectation.

[Section 4.4]

4.5 Let $t_1, \ldots, t_n$ be independent observations from a proportional hazards model, with failure rate

$$r_{T_i}(t) = r_0(t)e^{\beta x_i}, \quad i = 1, \ldots, n,$$

with $\beta \in \mathbb{R}$, $x_i = 0$, for $i = 1, \ldots, n_1$, and $x_i = 1$ for $i = n_1 + 1, \ldots, n$ $(n_1 < n)$. Obtain the marginal likelihood for β and the marginal score statistic for testing the hypothesis of homogeneity of the two groups, $H_0 : \beta = 0$. Calculate the score statistic when $n = 8$, $n_1 = 4$, and it is known that in the vector of the order statistics, those which occupy the positions $(1, 3, 6, 7)$ belong to the group with $x_i = 0$.

[Section 4.4]

4.6 With reference to the problem in Example 4.7, show that $(u_1, \ldots, u_n)$ is sufficient for $(\mu_1, \ldots, \mu_n)$ for fixed σ^2 and that (4.15) is the conditional likelihood for σ^2 given $u = (u_1, \ldots, u_n)$.

[Section 4.5]

4.7 Referring to Example 4.10, show that the likelihood equation for τ based on the conditional likelihood is the same as that obtained by starting from the full likelihood $L(\tau, \zeta)$.

[Section 4.5]

4.8 Do the detailed calculations necessary to confirm the statements in Example 4.11.

[Section 4.5]

4.9 Do the detailed calculations which give the equation (4.23). Using Jensen's inequality with $g(u) = \sqrt{\tau^2 u^2 + 4}$, show that the estimating equation (4.23) has bias of order $O(n)$. Show that the estimating equation for τ based on the conditional likelihood (4.24) is unbiased.

[Section 4.5]

4.10 To get information about the relative difficulty of the exams in Probability and in Statistics, the marks obtained by a random sample of $n = 20$ students who passed the first time were recorded. For $i = 1, \ldots, 20$ and $j = 1, 2$, let $y_{ij} = 1$ if the individual mark was higher than $7/10$ and 0 otherwise; $j = 1$ refers to Probability. The data matrix $[y_{ij}]^{\mathrm{T}}$ is

$$\begin{pmatrix} 0 & 0 & 1 & 1 & 1 & 0 & 0 & 1 & 0 & 1 & 1 & 0 & 1 & 0 & 0 & 1 & 0 & 1 & 1 & 1 \\ 1 & 0 & 0 & 1 & 0 & 0 & 0 & 0 & 0 & 0 & 1 & 1 & 0 & 0 & 0 & 0 & 1 & 0 & 1 & 0 \end{pmatrix}.$$

Obtain a confidence interval for the relative difficulty of the two exams based on the Rasch model (see Examples 4.2 and 4.12).

[Section 4.5]

4.11 Let $y_{i1}, \ldots, y_{in_i}$, $i = 1, \ldots, m$, be independent observations of a negative binomial r.v. $NB(\zeta_i, \tau)$. Identify a partially sufficient statistic for $\zeta = (\zeta_1, \ldots, \zeta_n)$. Obtain the corresponding conditional likelihood for τ. For $m = 1$ and $n_1 = 12$, with data $(0, 1, 1, 1, 2, 2, 3, 3, 3, 3, 8, 12)$, compare, graphically, the conditional log-likelihood function with the summand neglected in the overall log-likelihood, evaluated at $\zeta_1 = \hat{\zeta}_1(\tau)$, the maximum likelihood estimate of ζ_1 for fixed τ.

[Section 4.5; Kalbfleisch and Sprott, 1974, Example 2.3]

4.12 Let $y_1, \ldots, y_m$ be independent observations from a binomial distribution with probability $p = \zeta$ and index $n = \tau$. Identify a partially sufficient statistic for ζ and obtain a conditional likelihood for τ. For $m = 2$, consider some values of (y_1, y_2) and study the corresponding graphs of the conditional log-likelihood function, comparing them with those of the profile likelihood.

[Sections 4.5 and 4.6]

4.13 Show that the profile log-likelihood is invariant under interest-respecting reparameterizations. Consider the combinants W_P, $W_{uP}(\tau, \hat{\zeta}_\tau)$ and $W_{eP}(\tau, \hat{\zeta}_\tau)$, defined, respectively, by (4.25), (4.31) and (4.32). Show that only the tests and the confidence regions based on the first two are invariant under interest-respecting reparameterizations.

[Section 4.6]

4.14 Let $y_1, \ldots, y_n$ be independent observations from a two-parameter Weibull distribution with p.d.f. (1.29). Obtain the three asymptotically equivalent forms of the likelihood ratio test on ν, when λ is a nuisance parameter.

[Section 4.6]

4.15 State which of the parameterizations considered in Problems 1.6 and 1.9 are orthogonal.

[Section 4.7]

4.16 Obtain the generalization of the orthogonality condition (4.41) to the case $p > 2$, $k = 1$.

[Section 4.7]

4.17 Show that, if (y_{i1}, y_{i2}), $i = 1, \ldots, n$, are n independent observations from the r.v. (Y_1, Y_2) as in Example 4.15, then

$$\hat{\chi} = \sqrt{\bar{y}_1 \bar{y}_2}, \quad \hat{\chi}_\tau = \frac{1}{2} \frac{1}{\sqrt{\tau}} (\tau \bar{y}_1 + \bar{y}_2),$$

with $\bar{y}_1 = \frac{1}{n} \sum_{i=1}^n y_{i1}$, $\bar{y}_2 = \frac{1}{n} \sum_{i=1}^n y_{i2}$. Show that the distribution of $\hat{\chi}_\tau$ only depends on χ. Show that $\hat{\chi}_\tau - \hat{\chi}$ has the same distribution as

$$\frac{\chi}{n} \left(\frac{X_{1n} + X_{2n}}{2} - \sqrt{X_{1n} X_{2n}} \right), \tag{4.55}$$

where X_{1n} and X_{2n} are independent gamma r.v.'s with shape parameter n and unit scale parameter. Using (4.55), show, directly, that $\hat{\chi}_\tau - \hat{\chi} = O_p(n^{-1})$.

[Section 4.7]

4.18 Show that if the parameters τ and χ are orthogonal, any one-to-one smooth function of τ is orthogonal to any one-to-one smooth function of χ.

[Section 4.7]

4.19 Suppose $p = 2, k = 1$ and assume that $\zeta > 0$ and that τ and ζ are orthogonal. Obtain the approximate conditional log-likelihood with reference to the parameterization (τ, ζ^2) and show that the correction term is different from that given in (4.42).

[Section 4.7]

4.20 Show that the first summand in (4.42), $l_p(\tau)$, is of order $O_p(n)$, while the second is of order $O_p(1)$.

[Section 4.7]

4.21 Show that, when $p = 2, k = 1$, the null asymptotic distribution of $2\{l_{AC}(\hat{\tau}) - l_{AC}(\tau)\}$ is χ_1^2.

[Section 4.7]

4.22 Under the same assumptions as in Example 4.14, examine, numerically, the coverage errors of the confidence intervals for μ with confidence level 0.95 based on the profile likelihood (Example 4.14) and on the approximate conditional likelihood (Example 4.17).

[Section 4.7]

4.23 Consider the Neyman–Scott problem illustrated in Example 4.7. Show that the approximate conditional log-likelihood for σ^2, with $(\mu_1, \ldots, \mu_n)$ a nuisance parameter, coincides with the marginal log-likelihood based on SS_w.

[Section 4.7]

4.24 Let $x_1, \ldots, x_n$ and $y_1, \ldots, y_n$ be independent observations from independent r.v.'s X and Y with expectations μ_x and μ_y and finite variances σ_x^2 and σ_y^2. Let $\tau = \mu_y / \mu_x$ be the parameter of interest. Starting form the unbiased estimating equation

$$q(\bar{x}_n, \bar{y}_n; \tau) = \bar{y}_n - \tau \bar{x}_n = 0 ,$$

obtain the expression of a quasi log-likelihood for τ.

[Section 4.9]

CHAPTER 5

EXPONENTIAL FAMILIES

5.1 Introduction

Among parametric statistical models, two broad classes, exponential families and group families, play prime roles, both because of their structure and because of their inferential properties. These classes were first highlighted by Fisher (1934) and are important for two main reasons. Above all, both contain a large number of statistical models which are useful in applications, either directly or as elements of a complex model. Furthermore, the assumption that the data generating probability model does belong to an exponential or a group family, makes it possible to define general inference procedures, which are simple and accurate.

The following points are especially important. Non-trivial sufficiency or conditionality reductions can be carried out, even in the presence of nuisance parameters. Likelihood quantities behave in a regular manner as regards existence and uniqueness of estimates, asymptotic distributions, etc.. In some cases optimal tests, confidence regions and estimators can be identified. The general methods for obtaining higher-order asymptotic approximations for the distribution of likelihood quantities are easy to apply (see e.g. section 10.10.2). And, lastly, the theory of exponential families is, in itself, an important tool for obtaining asymptotic expansions (see section 10.7).

This chapter presents some aspects of the theory of exponential families. Group families will be described in chapter 7. Chapter 6 deals with exponential dispersion families; these are an extension of exponential families which is important for the theory of generalized linear models and, also, for justifying quasi-likelihood methods.

5.2   Exponential Families of Order 1

Consider a non-degenerate random variable with support $\mathcal{Y} \subseteq \mathbb{R}$ and density $p_0(y)$ with respect to a measure μ^*. In the examples considered here μ^* will be either the Lebesgue measure or a counting measure. A parametric family of densities that includes $p_0(y)$ as a special case and whose elements have the same support $\mathcal{Y}$ may be built through **exponential tilting**. This term (introduced by Efron, 1981) is used because the densities in the family are, as a function of y, proportional to $e^{\theta y} p_0(y)$.

Consider, in particular, the mapping $\theta \rightarrow M_0(\theta)$, where

$$M_0(\theta) = \int_{\mathcal{Y}} e^{\theta y} p_0(y) d\mu^* , \qquad \theta \in \mathbb{R} ,$$

with values in $(0, +\infty]$. If $M_0(\theta)$ is finite for θ in a neighbourhood of the origin, it is the moment generating function of the distribution with density $p_0(y)$. Let

$$\tilde{\Theta} = \{\theta \in \mathbb{R} : \quad M_0(\theta) < +\infty\} . \tag{5.1}$$

Clearly, $\tilde{\Theta}$ is non-empty, and it is easy to show that $\tilde{\Theta}$ is convex (see Problem 5.1). We assume in the following that $\tilde{\Theta}$ is non-degenerate, i.e. that $\tilde{\Theta} \neq \{0\}$.

For every $\theta \in \tilde{\Theta}$,

$$p(y; \theta) = \frac{e^{\theta y} p_0(y)}{M_0(\theta)} = \exp\{\theta y - K(\theta)\} p_0(y) \tag{5.2}$$

is a probability density function, with $p(y; 0) = p_0(y)$. If $M_0(\theta)$ is finite in a neighbourhood of 0, the function $K(\theta) = K_0(\theta) = \log M_0(\theta)$ is the cumulant generating function of the distribution with density $p_0(y)$. The increased modelling flexibility with respect to $p_0(y)$, made possible by the parametric family with density (5.2), is easy to interpret. Indeed, under appropriate conditions (see Theorem 5.1 in section 5.3), as θ varies in $\tilde{\Theta}$, the expectation of the distributions with density (5.2) takes values in the whole open interval defined by the extremes of the support of $p_0(y)$. We shall also see, in section 5.5, that, under random sampling of size n from a distribution with density (5.2), there exists a one-dimensional complete sufficient statistic, and therefore inference is made easier.

The above construction leads to the following definition.

Definition 5.1 The parametric family

$$\mathcal{F}_{ne}^1 = \left\{ p(y; \theta) = \exp\{\theta y - K(\theta)\} p_0(y), \ y \in \mathcal{Y}, \ \theta \in \tilde{\Theta} \right\} , \tag{5.3}$$

where $K(\theta) = \log \int_{\mathcal{Y}} e^{\theta y} p_0(y) d\mu^*$ and $\tilde{\Theta} = \{\theta \in \mathbb{R} : K(\theta) < +\infty\}$, is said to be a **natural exponential family of order 1** generated by $p_0(y)$; $\tilde{\Theta}$ is called the **natural parameter space.**

All the elements of $\mathcal{F}_{ne}^1$ have the same support. In family (5.3) the parameter space, $\tilde{\Theta}$, is the widest possible: correspondingly, the family is said to be **full.** In some applications, it may prove convenient to assume, as a statistical model, a subfamily of (5.3), whose parameter space Θ is a proper subset of the natural parameter space, $\Theta \subset \tilde{\Theta}$. The family corresponding to Θ is said to be a **non-full** natural exponential family. Below, unless otherwise stated, the symbol $\mathcal{F}_{ne}^1$ will be used to indicate a full family, with $\tilde{\Theta}$ non-degenerate. The theory is more straightforward if the family is not only full, but also, regular.

Definition 5.2 The family $\mathcal{F}_{ne}^1$ is said to be **regular** if $\tilde{\Theta}$ is open.

Since $\tilde{\Theta}$ is non-degenerate, there is a one-to-one correspondence between the density $p_0(y)$ and the function $K(\theta)$, because the moment generating function uniquely identifies the distribution (see section 3.2.1). Therefore, if the likelihood $e^{\theta y - K(\theta)}$ is assigned for an $\mathcal{F}_{ne}^1$, the function $p_0(y)$ is uniquely determined.

Example 5.1 $\mathcal{F}_{ne}^1$ *generated by the uniform distribution on* $[0,1]$
Let $p_0(y)$ be the density of the uniform distribution on $[0,1]$. Since $M_0(\theta) = \int_0^1 e^{\theta y} dy = (e^\theta - 1)/\theta$, for $\theta \neq 0$, and $M_0(0) = \lim_{\theta \to 0}(e^\theta - 1)/\theta = 1$, we can write $M_0(\theta) = (e^\theta - 1)/\theta$, for $\theta \in \mathbb{R}$, meaning that, at zero, the function is defined by continuity. Then the $\mathcal{F}_{ne}^1$ generated by $p_0(y) = I_{[0,1]}(y)$ has density

$$p(y; \theta) = \exp\{\theta y - \log\left((e^\theta - 1)/\theta\right)\} I_{[0,1]}(y), \quad \theta \in \mathbb{R}.$$

Since, as always happens when $p_0(y)$ has bounded support, $\tilde{\Theta} = \mathbb{R}$, this $\mathcal{F}_{ne}^1$ is also regular. $\triangle$

Example 5.2 $\mathcal{F}_{ne}^1$ *generated by the Poisson distribution with unit mean*
Let $p_0(y) = e^{-1}(y!)^{-1}$, $y \in \mathbb{N}$. The corresponding moment generating function is

$$M_0(\theta) = \exp\{e^\theta - 1\}, \quad \tilde{\Theta} = \mathbb{R},$$

and exponential tilting gives

$$p(y; \theta) = e^{\theta y + (1 - e^\theta)} \frac{e^{-1}}{y!} = \frac{(e^\theta)^y e^{-e^\theta}}{y!}, \quad y \in \mathbb{N}, \theta \in \tilde{\Theta}.$$

This is the density of a Poisson distribution with mean e^θ. $\triangle$

Example 5.3 $\mathcal{F}_{ne}^1$ *generated by the standard normal distribution*
Let $p_0(y) = e^{-y^2/2}/\sqrt{2\pi}$, $y \in \mathbb{R}$. Then $M_0(\theta) = e^{\theta^2/2}$, $\tilde{\Theta} = \mathbb{R}$, and the $\mathcal{F}_{ne}^1$ generated by $p_0(y)$ has density

$$p(y; \theta) = e^{\theta y - \theta^2/2}\frac{1}{\sqrt{2\pi}}e^{-y^2/2} = \frac{1}{\sqrt{2\pi}}e^{-(y-\theta)^2/2}, \quad y \in \mathbb{R}, \ \theta \in \mathbb{R}.$$

Therefore, the family of $N(\theta, 1)$ distributions is obtained. This family is regular.   △

Example 5.4 $\mathcal{F}_{ne}^1$ *generated by a gamma distribution*
The generating element $p_0(y)$ can, in its turn, be an element of a one-parameter family. Exponential tilting usually then gives a two-parameter family. For example, let $p_0(y; \nu) = y^{\nu-1}e^{-y}/\Gamma(\nu)$, $y > 0$, $\nu > 0$; since $M_0(\theta) = (1 - \theta)^{-\nu}$ for $\theta < 1$, the natural parameter space is $\tilde{\Theta} = (-\infty, 1)$. The p.d.f. obtained by exponential tilting is

$$\begin{aligned} p(y; \theta, \nu) &= e^{\theta y + \nu \log(1-\theta)}y^{\nu-1}e^{-y}/\Gamma(\nu) \\ &= e^{-(1-\theta)y}y^{\nu-1}(1 - \theta)^\nu/\Gamma(\nu), \quad y > 0, \ \theta < 1. \end{aligned}$$

This is the p.d.f. of a gamma distribution with shape parameter ν and scale parameter $1/(1 - \theta)$. For every fixed ν, this gives a regular $\mathcal{F}_{ne}^1$.   △

Example 5.5 *Closure under exponential tilting*
The natural exponential family of order 1 generated by an element of a given $\mathcal{F}_{ne}^1$ coincides with the initial $\mathcal{F}_{ne}^1$. Therefore, a natural exponential family is closed under exponential tilting. Let

$$p_0(y; \phi) = \exp\{\phi y - K(\phi)\}p_0(y), \quad \phi \in \tilde{\Phi},$$

be an element of an $\mathcal{F}_{ne}^1$. The corresponding moment generating function is

$$\begin{aligned} M(\theta; \phi) &= \int_y e^{\theta y} \exp\{\phi y - K(\phi)\}p_0(y)\, d\mu^* \\ &= e^{-K(\phi)}\int_y e^{(\theta+\phi)y}p_0(y)\, d\mu^* \\ &= e^{-K(\phi)+K(\theta+\phi)}, \quad \theta + \phi \in \tilde{\Phi}. \end{aligned}$$

Therefore, $p_0(y; \phi)$ has cumulant generating function

$$K_Y(\theta; \phi) = K(\theta + \phi) - K(\phi)$$

and the natural exponential family generated by $p_0(y; \phi)$ has density

$$
\begin{aligned}
p(y; \theta, \phi) &= e^{\theta y - K(\theta + \phi) + K(\phi)} e^{\phi y - K(\phi)} p_0(y) \\
&= e^{(\theta + \phi)y - K(\theta + \phi)} p_0(y), \quad \theta + \phi \in \tilde{\Phi}.
\end{aligned}
$$

With the reparameterization $\theta' = \theta + \phi$ the initial $\mathcal{F}^1_{ne}$ is re-obtained. △

Because of closure under exponential tilting, an $\mathcal{F}^1_{ne}$ can be thought of as being generated by any one of its members. This means that, without any loss of generality, we can suppose that zero belongs to the natural parameter space. In fact, if $\theta_0 \in \tilde{\Theta}$, we could write

$$
p(y; \theta) = e^{\theta y - K_Y(\theta; \theta_0)} p(y; \theta_0)
$$

that, for $\theta = 0$, gives $p(y; \theta_0)$. Furthermore, the possibility of reproducing the whole family $\mathcal{F}^1_{ne}$, starting from any chosen element, offers an interesting analogy between natural exponential families and straight lines: once a direction is given, a straight line is determined by any one of its points.

Although it contains many well-known parametric families, the one-parameter class (5.3) is too restricted. The multiparameter extension will be considered in section 5.4. Here, we introduce extensions which are relevant to the one-parameter case.

a) Let Y be a non-degenerate random variable with density $p_0(y)$ and let $s(Y)$ be a statistic with cumulant generating function $K_S(\theta)$ for $\theta \in \tilde{\Theta}_S$ (the subscript S is omitted when this will not affect clarity). Then,

$$
p(y; \theta) = \exp\{\theta s(y) - K_S(\theta)\} p_0(y), \qquad y \in \mathcal{Y}, \, \theta \in \tilde{\Theta}_S, \tag{5.4}
$$

is a density for every $\theta \in \tilde{\Theta}_S$.

b) The parameter θ in (5.4) could be expressed as a function of a parameter $\phi \in \Phi$, with $\theta(\Phi) \subseteq \tilde{\Theta}_S$, giving

$$
\begin{aligned}
p(y; \phi) &= \exp\{\theta(\phi) s(y) - K_S(\theta(\phi))\} p_0(y) \\
&= \exp\{\theta(\phi) s(y) - G(\phi)\} h(y),
\end{aligned}
$$

where $G(\phi) = K_S(\theta(\phi))$ and $h(y) = p_0(y)$.

This motivates the following definition.

Definition 5.3 A full exponential family of order 1 is a class of densities of the form

$$\mathcal{F}_e^1 = \left\{ p(y; \phi) = \exp\{\theta(\phi)s(y) - G(\phi)\}h(y), \qquad y \in \mathcal{Y} \subseteq \mathbb{R}, \ \phi \in \Phi \right\}, \quad (5.5)$$

where $\mathcal{Y}$ does not depend on ϕ, $h(y) > 0$, $\theta(\cdot)$ is a function of ϕ, Φ is such that $\theta(\Phi) = \tilde{\Theta}$, $s(\cdot)$ is a scalar statistic, whose value is called the natural **observation**.

If Φ is such that $\theta(\Phi) \subset \tilde{\Theta}$, then the family is said to be non-full. For $\Phi \subseteq \mathbb{R}$ and $\theta(\phi)$ one-to-one, ϕ is simply a reparameterization of the model; if $\theta(\phi)$ is not one-to-one, ϕ is not an identifiable parameter.

An $\mathcal{F}_{ne}^1$ can be associated to every $\mathcal{F}_e^1$ by considering the family induced by the transformation $s(\cdot)$ and parameterized by the natural or canonical parameter $\theta = \theta(\phi)$. Suppose that $\theta(\phi)$ is one-to-one with inverse $\phi(\theta)$; if Y has density (5.5), then the density of $s(Y)$ is

$$p_s(s; \theta) = \exp\{\theta s - G(\phi(\theta))\}h_s(s) .$$

Table 5.1 shows the characteristics of the main one-parameter exponential families. The $IG(\phi, \lambda)$ distribution that appears in the table is the inverse Gaussian distribution. This distribution was first obtained in 1915 by the physicist Schrödinger as a description of the random time wherein the trajectory of a Brownian motion with positive drift (see Example 1.5), reaches an assigned positive level for the first time.

Note that $\tilde{\Theta}$ is an open set in all the families listed in table 5.1, except for $IG(\phi, \lambda)$ with fixed λ, where $\tilde{\Theta} = (-\infty, 0]$. Therefore, $IG(\theta, \lambda)$ is a full, but not a regular, family. The distribution which corresponds to $\theta = 0$ has density

$$p_0(y) = \frac{\sqrt{\lambda}}{\sqrt{2\pi}} y^{-3/2} e^{-\lambda/(2y)}, \qquad y > 0, \ \lambda > 0 .$$

This is a stable distribution with index $1/2$, see (3.21).

5.3 Mean Value Mapping and Variance Function

Consider a regular $\mathcal{F}_{ne}^1$ with density

$$p(y; \theta) = \exp\{\theta y - K(\theta)\}p_0(y), \qquad y \in \mathcal{Y}, \ \theta \in \tilde{\Theta} .$$

Let Y be a random variable with density $p(y; \theta)$ and let $M_Y(t; \theta)$ and $K_Y(t; \theta)$ be the corresponding moment, and cumulant, generating functions. As in

Table 5.1: *Characteristics of some well-known $\mathcal{F}_e^1$.*

Notation	$N(\phi,1)$	$P(\phi)$	$Bi(n,\phi)$	$Ga(\nu,\phi)$ $\nu > 0$ fixed	$IG(\phi,\lambda)$ $\lambda > 0$ fixed
$\mathcal{y}$	$\mathbb{R}$	$\mathbb{N}$	$\{0,1,\ldots,n\}$	$\mathbb{R}^+$	$\mathbb{R}^+$
$p(y;\phi)$	$\dfrac{e^{-(y-\phi)^2/2}}{\sqrt{2\pi}}$	$\dfrac{e^{-\phi}\phi^y}{y!}$	$\binom{n}{y}\phi^y(1-\phi)^{n-y}$	$\dfrac{y^{\nu-1}e^{-y/\phi}}{\phi^\nu\Gamma(\nu)}$	$\dfrac{\sqrt{\lambda}e^{\sqrt{\lambda\phi}}}{\sqrt{2\pi y^3}}e^{-(\lambda/y+\phi y)/2}$
$s(y)$	y	y	y	y	y
$\theta(\phi)$	ϕ	$\log(\phi)$	$\log\left(\dfrac{\phi}{1-\phi}\right)$	$-1/\phi$	$-\phi/2$
Φ	$\mathbb{R}$	$\mathbb{R}^+\setminus\{0\}$	$(0,1)$	$\mathbb{R}^+\setminus\{0\}$	$\mathbb{R}^+$
$\bar{\Theta}$	$\mathbb{R}$	$\mathbb{R}$	$\mathbb{R}$	$\mathbb{R}^-\setminus\{0\}$	$\mathbb{R}^-$
$G(\phi)$	$\phi^2/2$	ϕ	$-n\log(1-\phi)$	$\nu\log(\phi)$	$-\sqrt{\lambda\phi}$
$K_S(\theta)$	$\theta^2/2$	e^θ	$n\log\left(1+e^\theta\right)$	$-\nu\log(-\theta)$	$-\sqrt{-2\theta\lambda}$
$h(y)$	$\dfrac{e^{-y^2/2}}{\sqrt{2\pi}}$	$\dfrac{1}{y!}$	$\binom{n}{y}$	$y^{\nu-1}/\Gamma(\nu)$	$\dfrac{\sqrt{\lambda}}{\sqrt{2\pi y^3}}e^{-\lambda/(2y)}$

Example 5.5, it is easy to obtain

$$M_Y(t;\theta) = \exp\left\{K(\theta + t) - K(\theta)\right\}, \qquad (5.6)$$

$$K_Y(t;\theta) = K(\theta + t) - K(\theta). \qquad (5.7)$$

These generating functions are finite in a neighbourhood around the origin, since, in a regular family, θ is always an inner point of $\tilde{\Theta}$. If $\mathcal{F}_{ne}^1$ is not regular, relations (5.6) and (5.7) hold for $\theta \in \text{int } \tilde{\Theta}$, where int $\tilde{\Theta}$ denotes the set of inner points of $\tilde{\Theta}$. Therefore, moments and cumulants of every order exist for each $\theta \in \text{int } \tilde{\Theta}$, and the cumulants are given by

$$\kappa_r(Y) = \frac{d^r}{d\theta^r}K(\theta), \qquad (5.8)$$

so that, in particular,

$$E_\theta(Y) = \kappa_1(Y) = \frac{d}{d\theta}K(\theta), \qquad (5.9)$$

$$\text{Var}_\theta(Y) = \kappa_2(Y) = \frac{d^2}{d\theta^2}K(\theta). \qquad (5.10)$$

Where it is useful to shorten the notation, the derivatives of $K(\theta)$ are indicated by $K'(\theta)$, $K''(\theta)$, and in general by $K^{(r)}(\theta)$.

Let

$$\mu(\theta) = E_\theta(Y); \qquad (5.11)$$

then

$$\text{Var}_\theta(Y) = \frac{d}{d\theta}\mu(\theta). \qquad (5.12)$$

Since $\text{Var}_\theta(Y) > 0$ for every $\theta \in \tilde{\Theta}$, $\mu(\cdot)$ is a monotone increasing function and $K(\cdot)$ is a convex function. Furthermore, since $\mathcal{F}_{ne}^1$ is a family with a monotone likelihood ratio in y, its elements are stochastically ordered (see Lehmann, 1986, Lemma 3.2).

Now, consider an $\mathcal{F}_{ne}^1$ which is not, necessarily, regular. Relation (5.11) defines a function $\mu(\cdot)$, with domain int $\tilde{\Theta}$, which is called the **mean value mapping**. This is one-to-one and smooth, with range

$$M = \mu(\text{int } \tilde{\Theta}).$$

The set M is called the **mean space** or the **expectation space**.

In a vector space V the set

$$C(V) = \{v \in V : \quad v = \alpha v_1 + (1-\alpha)v_2, \quad \alpha \in [0,1], \; v_1, v_2 \in V\}$$

is called the closed convex hull of $V \subseteq V$. Let $C = C(\mathcal{Y})$ be the closed convex hull of the common support, $\mathcal{Y}$, of the elements of $\mathcal{F}^1_{ne}$. Because of well-known properties of expectation,

$$M \subseteq \text{int } C. \tag{5.13}$$

More can be said for a regular family.

Theorem 5.1 If $\mathcal{F}^1_{ne}$ is regular, then

$$M = \text{int } C. \tag{5.14}$$

Proof. Since

$$\frac{K(\theta)}{\theta} = \frac{1}{\theta} \log \int_{\mathcal{Y}} e^{\theta y} p_0(y) d\mu^* = \log \left\{ \int_{\mathcal{Y}} (e^y)^\theta \, p_0(y) d\mu^* \right\}^{1/\theta},$$

for $\theta \in \tilde{\Theta} \setminus \{0\}$, the function $K(\theta)/\theta$ is the logarithm of the power mean of order θ of e^Y, $\{E(e^{\theta Y})\}^{1/\theta}$, where Y has density $p_0(y)$. Therefore, by properties of power means (see Problems 5.6 and 5.7), the function $K(\theta)/\theta$ is strictly increasing with lower and upper bound equal, respectively, to the lower and upper bounds of the support $\mathcal{Y}$ of Y. More specifically, if $\mathcal{Y}$ is bounded, the extremes of $\mathcal{Y}$ are obtained as the limits of $K(\theta)/\theta$ as $\theta \to \pm\infty$. When $\tilde{\Theta}$ is bounded, $\mathcal{Y}$ must be unbounded and its extremes still correspond to the extreme values of $K(\theta)/\theta$. The same limiting results illustrated above for $K(\theta)/\theta$, hold for the strictly increasing function $\mu(\theta) = dK(\theta)/d\theta$. Indeed, on the one hand, by (5.13), $\mu(\theta)$ cannot exceed the extremes of $\mathcal{Y}$ and, on the other, since $K(\theta)/\theta$ is monotone, we have

$$\frac{d}{d\theta} \frac{K(\theta)}{\theta} = \frac{\theta\mu(\theta) - K(\theta)}{\theta^2} > 0,$$

that is,

$$\mu(\theta) > \frac{K(\theta)}{\theta}, \qquad \text{for } \theta > 0$$

and

$$\mu(\theta) < \frac{K(\theta)}{\theta}, \qquad \text{for } \theta < 0.$$

□

Theorem 5.1 gives a sufficient condition for the equality $M = \text{int } C$ to hold. For a necessary and sufficient condition a new definition is required.

Definition 5.4 The family $\mathcal{F}_{ne}^1$ is said to be **steep** if $|\mu(\theta)| \rightarrow +\infty$ for every sequence of points $\theta \in \mathrm{int}\,\tilde{\Theta}$ which converges to a boundary point of $\tilde{\Theta}$.

The implication

$$\mathcal{F}_{ne}^1 \text{ regular } \Rightarrow \mathcal{F}_{ne}^1 \text{ steep}$$

holds. In fact, the function $K(\theta)$ tends to infinity for any sequence of points $\theta \in \mathrm{int}\,\tilde{\Theta}$ which converges to a boundary point of $\tilde{\Theta}$, and, because $K(\theta)$ is both continuous and convex in $\mathrm{int}\,\tilde{\Theta}$, its derivative will tend to infinity too.

The reinforced version of Theorem 5.1 is, schematically,

$$\mathcal{F}_{ne}^1 \text{ steep } \iff M = \mathrm{int}\,C \ . \tag{5.15}$$

For a proof, see Barndorff-Nielsen (1978, Theorem 9.2, p. 142).

The condition $M = \mathrm{int}\,C$ establishes a remarkable duality between the sample space and the expectation space M (and between the sample space and $\mathrm{int}\,\tilde{\Theta}$ through the one-to-one correspondence between $\mathrm{int}\,\tilde{\Theta}$ and M). In particular, for every point in $\mathrm{int}\,C$, there is a unique $\theta \in \mathrm{int}\,\tilde{\Theta}$, and, hence, there is a unique distribution in $\mathcal{F}_{ne}^1$ which has that point as its mean value. This correspondence is crucial in maximum likelihood estimation and, more generally, in inference.

The following example, drawn from Letac (1990), shows that a non-regular family may be non-steep.

Example 5.6 *A family which is not steep*
Let $p_0(y) \propto e^{-|y|}/(1 + y^4)$, $y \in \mathbb{R}$. It is clear that $\tilde{\Theta} = [-1, 1]$. The function $K(\theta)$ cannot be written explicitly. However, for $\theta \in (-1, 1)$, we have

$$\mu(\theta) = \int_{-\infty}^{+\infty} y e^{-|y|+\theta y - K(\theta)}/(1 + y^4)\,dy$$

with

$$\lim_{\theta \to 1} \mu(\theta) = \int_{-\infty}^{+\infty} y e^{-|y|+y}/(1 + y^4)\,dy \bigg/ \int_{-\infty}^{+\infty} e^{-|y|+y}/(1 + y^4)\,dy = c < +\infty \ .$$

Therefore, $M = (-c, c)$, while $\mathrm{int}\,C = \mathbb{R}$.   △

The transformation $\mu(\cdot)$, which is one-to-one and smooth, provides a reparameterization of a regular $\mathcal{F}_{ne}^1$, called the **mean value parameterization**. If $\tilde{\Theta}$ is not open, the mean value parameterization can be defined by continuity at the boundary points of $\tilde{\Theta}$. Let us denote the inverse function of $\mu(\theta)$ by $\theta(\mu)$.

The cumulants $\kappa_r(Y; \theta)$ can be expressed as functions of μ. For instance, the variance can be written as

$$\text{Var}_{\theta(\mu)}(Y) = V(\mu) = \mu'(\theta)\big|_{\theta=\theta(\mu)} = \frac{d^2}{d\theta^2}K(\theta)\bigg|_{\theta=\theta(\mu)} \tag{5.16}$$

Definition 5.5 The function $V(\mu)$ defined on M by (5.16) is called the variance function and is indicated by $(M, V(\mu))$.

The importance of the variance function is related to the following characterization result (Morris, 1982).

Theorem 5.2 If Y has density that belongs to a $\mathcal{F}_{ne}^1$, then the variance function, $(M, V(\mu))$, will determine, uniquely, both $\tilde{\Theta}$ and the function $K(\theta)$, hence it will characterize a particular $\mathcal{F}_{ne}^1$ within the class of natural exponential families of order 1.

Proof. To prove the result, we obtain a representation of $K(\cdot)$ in terms of $V(\mu)$. Differentiation with respect to μ of the identity

$$\mu(\theta(\mu)) = \mu$$

gives

$$\mu'(\theta(\mu))\frac{d}{d\mu}\theta(\mu) = 1 ,$$

from which we have

$$\frac{d}{d\mu}\theta(\mu) = \frac{1}{\mu'(\theta(\mu))} = \frac{1}{V(\mu)} . \tag{5.17}$$

The derivative (5.17) is finite for every $\mu \in M$, because no element of $\mathcal{F}_{ne}^1$ is degenerate. Choose a value $\mu_0 \in M$. Then,

$$\int_{\mu_0}^{\mu} \frac{1}{V(m)}dm = \theta(\mu) - \theta(\mu_0) , \tag{5.18}$$

where we can suppose, without loss of generality, that $\theta(\mu_0) = 0$. Furthermore, consider the identity

$$K(\theta) = \int_0^{\theta} \mu(t)dt .$$

With the change of variable $m = \mu(t)$, and taking (5.17) into account, this identity becomes

$$K(\theta(\mu)) = \int_{\mu_0}^{\mu} m\frac{1}{V(m)}dm . \tag{5.19}$$

On the basis of (5.18) and (5.19) the function $K(\cdot)$ must satisfy the relation

$$K\left(\int_{\mu_0}^{\mu}\frac{1}{V(m)}dm\right) = \int_{\mu_0}^{\mu}\frac{m}{V(m)}dm\ . \qquad (5.20)$$

Given M and $V(\mu)$, then (5.20) explicitly expresses $K(\cdot)$ as a function of μ and, implicitly, defines $K(\theta)$. The natural parameter space $\tilde{\Theta}$ is obtained from (5.18), as μ varies in M. $\qquad\qquad\qquad\qquad\qquad\qquad\qquad\qquad\qquad\qquad\quad$ □

Example 5.7 *Parabolic variance function*
Let us find the $\mathcal{F}^1_{ne}$ with variance function

$$(M, V(\mu)) = ((0, +\infty), \mu^2)\ .$$

Consider $\mu_0 = 1$; from (5.20) we have

$$K\left(\int_1^{\mu}\frac{1}{m^2}dm\right) = \int_1^{\mu}\frac{1}{m}dm\ ,$$

that is

$$K\left(1 - \frac{1}{\mu}\right) = \log\mu\ .$$

With $\theta = 1 - 1/\mu$, we have $\tilde{\Theta} = (-\infty, 1)$ and $\mu(\theta) = (1 - \theta)^{-1}$, so that

$$K(\theta) = -\log(1 - \theta)$$

and

$$p(y; \theta) = \exp\{\theta y + \log(1 - \theta)\}p_0(y)\ .$$

In order to determine $p_0(y)$ we only need to identify the distribution which has $M_0(\theta) = (1 - \theta)^{-1}$ as its moment generating function. This is an exponential distribution with mean 1, i.e. with density $p_0(y) = e^{-y}I\{y \geq 0\}$. Therefore, if $M = (0, +\infty)$ and $V(\mu) = \mu^2$, then

$$p(y; \theta) = \exp\{\theta y + \log(1 - \theta)\}e^{-y}I\{y \geq 0\}\ ,$$

which is the density of an exponential distribution with mean $1/(1-\theta) > 0$. In order to check that the result does not depend on the choice $\mu_0 = 1$, consider $\mu_0 = 2$. Then (5.20) becomes

$$K\left(\int_2^{\mu}\frac{1}{m^2}dm\right) = \int_2^{\mu}\frac{1}{m}dm$$

Table 5.2: *Variance functions of some* $\mathcal{F}_{ne}^1$

$\mathcal{F}_e^1$	$N(\phi,1)$	$P(\phi)$	$Bi(n,\phi)$	$Ga(\nu,\phi)$	$IG(\phi,\lambda)$
M	$\mathbb{R}$	$\mathbb{R}^+ \setminus \{0\}$	$(0,n)$	$\mathbb{R}^+ \setminus \{0\}$	$\mathbb{R}^+ \setminus \{0\}$
$V(\mu)$	1	μ	$\mu\left(1 - \dfrac{\mu}{n}\right)$	$\dfrac{\mu^2}{\nu}$	$\dfrac{\mu^3}{\lambda}$

that is

$$K\left(\frac{1}{2} - \frac{1}{\mu}\right) = \log\mu - \log 2 .$$

With $\theta^* = (\mu - 2)/(2\mu)$ we obtain $\tilde{\Theta}^* = (-\infty, 1/2)$ and, since $\mu(\theta^*) = 2(1 - 2\theta^*)^{-1}$, we have

$$K(\theta^*) = -\log(1 - 2\theta^*) ,$$

that is $M_0(\theta^*) = (1 - 2\theta^*)^{-1}$, which is the moment generating function of the exponential distribution with density

$$p_0(y) = \frac{1}{2}e^{-y/2}I\{y \geq 0\} .$$

Therefore

$$\begin{aligned}
p(y; \theta^*) &= e^{\theta^* y + \log(1 - 2\theta^*)}\frac{1}{2}e^{-y/2}I\{y \geq 0\} \\
&= \left(\frac{1}{2} - \theta^*\right)e^{-\left(\frac{1}{2} - \theta^*\right)y}I\{y \geq 0\} , \qquad \frac{1}{2} - \theta^* > 0
\end{aligned}$$

i.e., with the reparameterization $\lambda = 1/2 - \theta^*$,

$$p(y; \lambda) = \lambda e^{-\lambda y}I\{y \geq 0\}, \qquad \lambda > 0 ,$$

the same family as obtained with $\mu_0 = 1$. △

Table 5.2 shows the variance functions for the five $\mathcal{F}_{ne}^1$ associated to the $\mathcal{F}_e^1$ considered in Table 5.1.

Some comment is required regarding Theorem 5.2.

i) In order to identify an $\mathcal{F}_{ne}^1$ both M and $V(\mu)$ must be known. In other words, the function $V(\mu)$ alone is not enough to identify $K(\cdot)$. For example, $V(\mu) = \mu^2$ characterizes two distinct exponential families, one with $M = (-\infty, 0)$, the other with $M = (0, +\infty)$. Since $V(\mu) = 0$ corresponds to a singularity in (5.20) this bifurcation is not surprising.

ii) The correspondence between $(M, V(\mu))$ and an $\mathcal{F}_{ne}^1$ only holds for natural exponential families and not for $\mathcal{F}_e^1$ families.

iii) Morris (1982) showed that only six natural exponential families exist which have a quadratic variance function, $V(\mu) = a + b\mu + c\mu^2$, where a, b, c are known constants. These families are: the normal, the Poisson, the gamma, the binomial, the negative binomial and a sixth family, the **generalized hyperbolic secant distribution**, with density

$$p(y; \theta) = \exp\{\theta y + \log(\cos(\theta))\} 2 \cosh(\pi y/2), \quad y \in \mathbb{R}, \ |\theta| < \pi/2 \,.$$

This is the density of $\pi^{-1} \log(X/(1 - X))$, where

$$X \sim Be\left(0.5 + \theta/\pi, \ 0.5 - \theta/\pi\right) \,,$$

with $Be(\alpha, \beta)$ denoting the beta distribution. The class of natural exponential families of order one with quadratic variance function is closed under linear transformations (see Problem 5.13) and convolutions (Morris, 1982, section 3).

5.4 Multiparameter Exponential Families

5.4.1 Definitions and Basic Results

The definitions given in section 5.2 for the one-parameter case can be quite easily extended to fit the multiparameter case.

Consider a non-degenerate distribution having density $p_0(y)$, $y \in \mathcal{Y} \subseteq \mathbb{R}^p$, $p > 1$, with respect to a dominating measure μ^*. Assume that $\mathcal{Y}$ is not contained within any subspace of $\mathbb{R}^p$ with dimension less than p. Consider the function with values in $(0, +\infty]$

$$M_0(\theta) = \int_{\mathcal{Y}} e^{\theta \cdot y} p_0(y) d\mu^* \,, \qquad \theta \in \mathbb{R}^p \,,$$

where $\theta \cdot y$ indicates the scalar product between θ and y. If $M_0(\theta)$ is finite in a neighbourhood of the origin, it is the moment generating function of the distribution with density $p_0(y)$. Let $\tilde{\Theta} = \{\theta \in \mathbb{R}^p : M_0(\theta) < +\infty\}$ be the **natural parameter space**. It is a non-empty convex set, as can be proved by retracing the same steps as in Problem 5.1. Therefore, if $\tilde{\Theta}$ is non-degenerate, it is genuinely p-dimensional, that is, it contains a p-dimensional rectangle.

Definition 5.6 The parametric family

$$\mathcal{F}_{ne}^{p} = \left\{ p(y; \theta) = \exp\{\theta \cdot y - K(\theta)\} p_0(y), \quad y \in \mathcal{Y} \subseteq \mathbb{R}^p, \quad \theta \in \tilde{\Theta} \right\}, \quad (5.21)$$

where $K(\theta) = \log \int_{\mathcal{Y}} e^{\theta \cdot y} p_0(y) d\mu^*$ and $\tilde{\Theta} = \{\theta \in \mathbb{R}^p : K(\theta) < +\infty\}$, is said to be the full natural exponential family of order p generated by $p_0(y)$.

The family $\mathcal{F}_{ne}^{p}$ is said to be regular if $\tilde{\Theta}$ is an open set of $\mathbb{R}^p$.

The following extensions allow us to define an exponential family of order p, such that the dimensions both of the sample space and of the parameter space do not necessarily coincide with p.

First, let Y be a d-dimensional random vector $d \geq 1$, with density $p_0(y)$; let $s_1(y), \ldots, s_p(y)$ be scalar statistics, with $(1, s_1(y), \ldots, s_p(y))$ linearly independent. Let $K_S(\theta)$ be the cumulant generating function of $(s_1(Y), \ldots, s_p(Y))$ and $\tilde{\Theta}_S = \{\theta \in \mathbb{R}^p : K_S(\theta) < +\infty\}$. Then the function

$$p(y; \theta) = \exp\{\theta \cdot s(y) - K_S(\theta)\} p_0(y), \quad y \in \mathcal{Y} \subseteq \mathbb{R}^d,$$

where $s(y) = (s_1(y), \ldots, s_p(y))$, is a density for every $\theta \in \tilde{\Theta}_S$.

In addition, we may allow θ to be a function of a parameter ϕ, i.e. $\theta = \theta(\phi)$, $\phi \in \Phi$, $\theta(\Phi) \subseteq \tilde{\Theta}_S$, and write

$$\begin{aligned} p(y; \phi) &= \exp\{\theta(\phi) \cdot s(y) - K_S(\theta(\phi))\} p_0(y) \\ &= \exp\{\theta(\phi) \cdot s(y) - G(\phi)\} h(y), \end{aligned}$$

where the representation is assumed to be minimal, i.e. such that there is no linear dependence between the components of $(1, \theta^1(\phi), \ldots, \theta^p(\phi))$ or, equivalently, between the components of $(1, s_1(y), \ldots, s_p(y))$.

The two steps of this extension motivate the following definition.

Definition 5.7 A class of densities of the form

$$\mathcal{F}_e^p = \left\{ p(y; \phi) = \exp\{\theta(\phi) \cdot s(y) - G(\phi)\} h(y), \quad y \in \mathcal{Y} \subseteq \mathbb{R}^d, \quad \phi \in \Phi \right\}, \tag{5.22}$$

where $h(y) > 0$, $\theta(\Phi) \subseteq \tilde{\Theta}_S$, is called an exponential family of order p. The notation $\mathcal{F}_e^p$ is understood to indicate a minimal representation.

Let $\Phi \subseteq \mathbb{R}^q$. There are various possible situations according to the dimensions q and p of the parameter spaces Φ and $\tilde{\Theta}_S$. If $q > p$, then the parameter ϕ is not identifiable: such a possibility should, therefore, be discarded. When $p = q$

and $\theta(\phi)$ is one-to-one, θ is a reparameterization; if $\theta(\phi)$ is not one-to-one ϕ is still non-identifiable. Lastly, when $q < p$, we have a (p,q) curved exponential family. This is a special case of a non-full family. The corresponding restriction of the natural parameter space is a curve in $\tilde{\Theta} \subseteq \mathbb{R}^p$, which shows that the parameter space is q-dimensional, with $q < p$. We use the symbol $\mathcal{F}_e^{p,q}$ to indicate a curved family. The symbol $\mathcal{F}_e^p$ is used only to indicate families whose parameter space is genuinely p-dimensional, that is, where int Φ is non-empty in $\mathbb{R}^p$. Curved exponential families are treated in more detail in section 5.10.

In relation to the family (5.22), θ is said to be the natural, or canonical parameter, and the statistics $s_1(y), \ldots, s_p(y)$ are called the natural observations. It is not difficult to check that when $q = p$ the statistic $s = (s_1(y), \ldots, s_p(y))$ has an $\mathcal{F}_{ne}^p$ distribution in the natural parameterization.

Consider the identity

$$p_S(s; \theta) = e^{(\theta - \theta_0) \cdot s - K_S(\theta) + K_S(\theta_0)} p_0(s) , \qquad \theta, \theta_0 \in \tilde{\Theta}_S , \qquad (5.23)$$

where $p_0(s)$ is the density of s for $\theta = \theta_0$. The representation (5.23) is convenient whenever it is simpler to study the distribution of s for $\theta = \theta_0$ than it is to study the more general case. This possibility is illustrated in the example below.

Example 5.8 *Wishart distribution*
Let $y_1, \ldots, y_n$ be independent realizations of a $N_d(0, \Sigma)$. The joint density of the $d \times n$ matrix of observations is

$$p(y; \Sigma) = \frac{1}{(2\pi)^{nd/2} |\Sigma|^{n/2}} \exp\left\{ -\frac{1}{2} \operatorname{tr}(\Sigma^{-1} D) \right\} ,$$

where $D = \sum_{i=1}^n y_i y_i^T$ (see, for example, Mardia, Kent and Bibby, 1979, section 4.1). This density belongs to an $\mathcal{F}_e^p$, with $p = d(d+1)/2$ and has, as natural observations, the p distinct elements of the symmetric matrix D. Exploiting (5.23) with θ_0 equal to the value of the natural parameter which corresponds to $\Sigma = I_d$, the density of D (or, more precisely, that of the distinct elements of D) can be written as

$$p_D(D; \Sigma) = |\Sigma|^{-n/2} e^{-\frac{1}{2} \operatorname{tr}(\Sigma^{-1} D)} e^{\frac{1}{2} \operatorname{tr}(D)} p_0(D) ,$$

where $p_0(D)$ is the density of the natural observations when the components of the d-variate normal distribution are independent with mean zero and unit variance. For $n \geq d$ this density is (see, for example, Rao, 1973, pp. 597-598)

$$p_0(D) = c_0 \exp\{-\frac{1}{2}\operatorname{tr}(D)\}|D|^{(n-d-1)/2}, \tag{5.24}$$

with

$$c_0 = \left\{2^{nd/2}\,\pi^{d(d-1)/2}\prod_{j=1}^{d}\Gamma\left(\frac{n-j+1}{2}\right)\right\}^{-1}.$$

Therefore,

$$p_D(D;\Sigma) = c_0\,|\Sigma|^{-n/2}\exp\{-\frac{1}{2}\operatorname{tr}(\Sigma^{-1}D)\}|D|^{(n-d-1)/2}. \tag{5.25}$$

A distribution with density (5.25) is called a Wishart distribution with n degrees of freedom and matrix Σ and is usually indicated by the symbol $W_d(n,\Sigma)$. $\triangle$

In the following, the most important properties of an $\mathcal{F}_{n_e}^p$ are studied. This is the same as concentrating on the distribution of the natural observations of an $\mathcal{F}_e^p$.

Let Y be a p-dimensional random variable with a density in (5.21). The moment generating function of Y is

$$M_Y(t;\theta) = \exp\{K(\theta+t) - K(\theta)\}$$

while the cumulant generating function is

$$K_Y(t;\theta) = K(\theta+t) - K(\theta).$$

The generic cumulant of order r is

$$\kappa^{i_1,\dots,i_r}(Y;\theta) = \frac{\partial^r K(\theta)}{\partial\theta^{i_1}\dots\partial\theta^{i_r}}.$$

In particular,

$$E_\theta(Y) = \mu(\theta) = \frac{\partial K(\theta)}{\partial\theta} = \left(\frac{\partial K(\theta)}{\partial\theta^1},\dots,\frac{\partial K(\theta)}{\partial\theta^p}\right)^{\mathrm{T}} \tag{5.26}$$

is the mean vector and

$$\operatorname{Var}_\theta(Y) = \frac{\partial^2 K(\theta)}{\partial\theta\,\partial\theta^{\mathrm{T}}} = \frac{\partial\mu(\theta)}{\partial\theta^{\mathrm{T}}} \tag{5.27}$$

is the covariance matrix of Y. Since $\text{Var}_\theta(Y)$ is positive definite, the transformation $\mu(\theta)$, with domain int $\tilde\Theta$ and values in $M = \mu(\text{int}\,\tilde\Theta)$, defined by (5.26), is one-to-one and smooth. Jørgensen (1987) extended Theorem 5.2 to the multiparameter case: a natural exponential family of order p is characterized, in the class of all the $\mathcal{F}^p_{ne}$, by its variance function.

In the multiparameter case too, we speak of a steep $\mathcal{F}^p_{ne}$ if definition 5.4 applies where $|\mu(\theta)|$ is understood to be the norm of the vector $\mu(\theta)$. One result of the theory of convex functions (Rockafellar, 1970, p. 258) allows the following generalization of the necessary and sufficient condition (5.15) to be proved.

Theorem 5.3 An $\mathcal{F}^p_{ne}$ is steep if, and only if, $M = \text{int}\,C$, where C is the closed convex hull of $\mathcal{Y}$.

Proof. See Barndorff-Nielsen (1978, Theorem 9.2). $\square$

Example 5.9 *Two-parameter gamma distribution*
The family $Ga(\nu,\lambda)$, with $\nu > 0$, $\lambda > 0$, is an $\mathcal{F}^2_e$ with $\theta^1(\nu,\lambda) = -1/\lambda$, $\theta^2(\nu,\lambda) = \nu$, $s_1(y) = y$, $s_2(y) = \log y$ and $G(\nu,\lambda) = \log\Gamma(\nu) + \nu\log\lambda$. Therefore, the natural parameter space is $\tilde\Theta = (\mathbb{R}^- \setminus \{0\}) \times (\mathbb{R}^+ \setminus \{0\})$ and the family $\mathcal{F}^2_{ne}$ induced by natural observations $(y,\log y)$ is regular. $\triangle$

Example 5.10 *Two-parameter inverse Gaussian distribution*
The family $IG(\nu,\lambda)$, with $\nu \geq 0$, $\lambda > 0$, is an $\mathcal{F}^2_e$ with $\theta^1(\nu,\lambda) = -\nu/2$, $\theta^2(\nu,\lambda) = -\lambda/2$, $s_1(y) = y$, $s_2(y) = 1/y$ and $G(\nu,\lambda) = -(\sqrt{\lambda\nu} + \log(\lambda)/2)$. The natural parameter space is $\tilde\Theta = \mathbb{R}^- \times (\mathbb{R}^- \setminus \{0\})$; since this is not an open set, the $\mathcal{F}^2_{ne}$ induced by $(y,1/y)$ is not regular. $\triangle$

5.4.2 Independence, Marginal and Conditional Distributions

The two results given below hold for the natural observations of an $\mathcal{F}^p_{ne}$.

Theorem 5.4 If the natural observations of an $\mathcal{F}^p_{ne}$ are independent for $\theta = \theta_0 \in \tilde\Theta$, then this is also true for every $\theta \in \tilde\Theta$.

Proof. A $\mathcal{F}^p_{ne}$ could be thought to have been generated by any one of its elements, and particularly by $p(y;\theta_0)$. From the assumption, $p(y;\theta_0) = p_{Y_1}(y_1;\theta_0)\cdots p_{Y_p}(y_p;\theta_0)$, so that

$$
\begin{aligned}
M_Y(\theta;\theta_0) &= M_{Y_1}(\theta^1;\theta_0)\cdots M_{Y_p}(\theta^p;\theta_0)\\
&= \exp\{K_{Y_1}(\theta^1;\theta_0) + \ldots + K_{Y_p}(\theta^p;\theta_0)\}.
\end{aligned}
$$

Therefore

$$
\begin{aligned}
p(y; \theta) &= e^{\theta \cdot y - K_Y(\theta; \theta_0)} p(y; \theta_0) \\
&= \exp\{\theta^1 y_1 + \ldots + \theta^p y_p - K_{Y_1}(\theta^1; \theta_0) - \ldots - K_{Y_p}(\theta^p; \theta_0)\} \\
&\quad \times p_{Y_1}(y_1; \theta_0) \cdots p_{Y_p}(y_p; \theta_0) \\
&= \exp\{\theta^1 y_1 - K_{Y_1}(\theta^1; \theta_0)\} p_{Y_1}(y_1; \theta_0) \cdots \\
&\quad \times \exp\{\theta^p y_p - K_{Y_p}(\theta^p; \theta_0)\} p_{Y_p}(y_p; \theta_0) .
\end{aligned}
$$

□

Theorem 5.5 If, for every $\theta \in \tilde{\Theta}$, the natural observations of a regular $\mathcal{F}^p_{ne}$ are uncorrelated, then they are also independent.

Proof. Suppose, for simplicity, that $p = 2$ and $0 \in \tilde{\Theta}$. The natural observations (y_1, y_2) are uncorrelated if and only if, for every $\theta = (\theta^1, \theta^2) \in \tilde{\Theta}$,

$$
\frac{\partial^2}{\partial \theta^1 \partial \theta^2} K(\theta^1, \theta^2) = 0 ,
$$

that is, for every $\theta \in \tilde{\Theta}$,

$$
\frac{\partial}{\partial \theta^2} \left(\frac{\partial}{\partial \theta^1} K(\theta^1, \theta^2) \right) = \frac{\partial}{\partial \theta^2} \mu_1(\theta) = 0 ,
$$

where $\mu_1(\theta) = \dfrac{\partial}{\partial \theta^1} K(\theta)$. Thus, $\mu_1(\theta)$ does not depend on θ^2, that is, for every θ,

$$
\mu_1(\theta) = \mu_1(\theta^1, \theta^2) = \mu_1(\theta_1, 0) .
$$

Symmetrically, an analogous result must also be true for $\mu_2(\theta)$. Therefore, $K(\theta^1, \theta^2)$ must be of the form

$$
K(\theta^1, \theta^2) = K_{Y_1}(\theta^1, 0) + K_{Y_2}(0, \theta^2) ,
$$

where $K_{Y_1}(\cdot, 0)$ and $K_{Y_2}(0, \cdot)$ are the cumulant generating functions of Y_1 and Y_2 for $\theta = 0$. It follows that the components of (Y_1, Y_2) are independent when $\theta = 0$, and, therefore, for every $\theta \in \tilde{\Theta}$ because of Theorem 5.4. □

Consider an $\mathcal{F}^p_e$ with canonical parameterization and natural observations $s(y)$, whose distributions form an $\mathcal{F}^p_{ne}$. Let $s(y) = (t(y), u(y))$ be a partition of the vector of the natural observations, where $t(\cdot)$ has k components and $u(\cdot)$ is $(p - k)$-dimensional. Consider the corresponding partition of the natural parameter $\theta = (\tau, \zeta)$. The density of a generic element of the family can be written as

$$p(y; \tau, \zeta) = \exp\{\tau \cdot t(y) + \zeta \cdot u(y) - K(\tau, \zeta)\}h(y) , \qquad (5.28)$$

with $(\tau, \zeta) \in \tilde{\Theta}$.

The following theorem is very important for model reduction, allowing inference about selected components of the natural parameter.

Theorem 5.6 For an $\mathcal{F}_e^p$ with density (5.28)

i) the family of marginal distributions of $U = u(Y)$ is an $\mathcal{F}_{ne}^{p-k}$ for every fixed value of τ, that is

$$p_U(u; \tau, \zeta) = \exp\{\zeta \cdot u - K_\tau(\zeta)\}h_\tau(u) ; \qquad (5.29)$$

ii) the family of conditional distributions of $T = t(Y)$ given $u(Y) = u$ is an $\mathcal{F}_{ne}^k$ and the conditional densities are independent of ζ, that is

$$p_{T|U=u}(t; u, \tau) = \exp\{\tau \cdot t - K_u(\tau)\}h_u(t) . \qquad (5.30)$$

Proof. Suppose that, without loss of generality, $0 \in \tilde{\Theta}$, so that

$$p(y; \tau, \zeta) = \exp\{\tau \cdot t(y) + \zeta \cdot u(y) - K(\tau, \zeta)\} p_0(y) ,$$

where $p_0(y)$ is the density that corresponds to $(\tau, \zeta) = (0, 0)$. The family of distributions of the natural observations $(t(Y), u(Y))$ is an $\mathcal{F}_{ne}^p$, with density

$$p_{T,U}(t, u; \tau, \zeta) = \exp\{\tau \cdot t + \zeta \cdot u - K(\tau, \zeta)\}p_0(t, u)$$

with respect to a dominating measure, $\mu^* = \mu^*_{T,U}$. The symbol $p_0(t, u)$ denotes the density of the natural observations when $(\tau, \zeta) = (0, 0)$. In order to show *i)*, write $p_0(t, u) = p_0(u)p_0(t|u)$, where $p_0(u)$ is the marginal density of U and $p_0(t|u)$ is the conditional density of T given $U = u$. Suppose in addition, without loss of generality, that the measure $\mu^*_{T,U}$ is a product measure, that is, such that if $B = B_\tau \times B_U$, with B_τ an event in the support $\mathcal{T}$ of $t(Y)$ and B_U an event in the support $\mathcal{U}$ of $u(Y)$, then $\mu^*_{T,U}(B) = \mu^*_\tau(B_\tau)\mu^*_U(B_U)$, for some measures μ^*_U and μ^*_τ. Thus, the marginal density of U with respect to μ^*_U is

$$p_U(u; \tau, \zeta) = e^{\zeta \cdot u - K(\tau, \zeta)}\left\{\int_{\mathcal{T}_u} e^{\tau \cdot t} p_0(t|u)d\mu^*_\tau\right\} p_0(u) , \qquad (5.31)$$

where $\mathcal{T}_u$ indicates the support of $p_0(t|u)$. The quantity within braces in (5.31) is the conditional moment generating function of T given $U = u$ when (T, U) has density $p_0(t, u)$. Let $K_u(\tau)$ indicate the corresponding cumulant generating function. Then (5.31) can be written in the form (5.29) with

$$
\begin{aligned}
K_\tau(\zeta) &= K(\tau, \zeta) , \\
h_\tau(u) &= e^{K_u(\tau)} p_0(u) ,
\end{aligned}
$$

where

$$
K_u(\tau) = \log \left\{ \int_{\mathcal{T}_u} e^{\tau \cdot t} p_0(t|u) d\mu_\tau^* \right\} = \log \left\{ E_0 \left(e^{\tau T} | U = u \right) \right\} . \tag{5.32}
$$

The moment generating function of U is given by

$$
M_U(v) = \exp \left\{ K(\tau, \zeta + v) - K(\tau, \zeta) \right\}, \quad \zeta + v \in \tilde{Z}_\tau \subseteq \mathbb{R}^{p-k} ,
$$

where $\tilde{Z}_\tau = \{ \zeta \in \mathbb{R}^{p-k} : (\tau, \zeta) \in \tilde{\Theta} \}$.

Part $ii)$ of the theorem is proved by noting that the conditional density of T given $U = u$ is given by the ratio

$$
\begin{aligned}
p_{T|U=u}(t; u, \tau, \zeta) &= \frac{p_{T,U}(t, u; \tau, \zeta)}{p_U(u; \tau, \zeta)} \\
&= \frac{\exp\{\tau \cdot t + \zeta \cdot u - K(\tau, \zeta)\} p_0(t, u)}{\exp\{\zeta \cdot u - K(\tau, \zeta)\} e^{K_u(\tau)} p_0(u)} \\
&= \exp\{\tau \cdot t - K_u(\tau)\} p_0(t|u) .
\end{aligned}
$$

$\square$

Theorem 5.6 has important consequences for data and model reduction. On the basis of $ii)$, u is partially sufficient for ζ. Therefore, if τ is the parameter of interest, a conditional likelihood based on (5.30) is

$$
L_c(\tau) = \exp\{\tau \cdot t - K_u(\tau)\} . \tag{5.33}
$$

This result makes it simpler to work with a natural parameterization. However, Theorem 5.6 only expresses a structural result and the function $K_u(\tau)$ is not given explicitly. One exception is when the marginal distribution of u is known. In this case, $K_u(\tau)$ can be found through comparison with (5.31). The conditional likelihood $L_c(\tau)$ will be studied further in section 5.7; an asymptotic approximation will be given in section 10.10.2.

Example 5.11 *Two-sample Poisson problem*
Let T and V be independent random variables with Poisson distribution with means μ and ν respectively. Their joint density is

$$p_{T,V}(t,v;\mu,\nu) = \exp\{t\log\mu + v\log\nu - (\mu+\nu)\}\frac{1}{t!}\frac{1}{v!}, \quad t,v = 0,1,\ldots,$$

and can also be written as

$$p_{T,V}(t,v;\mu,\nu) = \exp\{t\log\left(\frac{\mu}{\nu}\right) + (t+v)\log\nu - (\mu+\nu)\}\frac{1}{t!}\frac{1}{v!}.$$

Let $\tau = \log(\mu/\nu)$ and $\zeta = \log\nu$. The natural observations $(t,u) = (t,t+v)$ have joint density

$$p_{T,U}(t,u;\tau,\zeta) = \exp\{\tau t + \zeta u - e^\zeta(e^\tau+1)\}\frac{1}{t!}\frac{1}{(u-t)!}$$

for $u = 0,1,\ldots$ and $t = 0,1,\ldots,u$. Since U has a Poisson distribution with mean $\mu+\nu = e^\zeta(e^\tau+1)$, then

$$
\begin{aligned}
p_{T|U=u}(t;u,\tau,\zeta) &= \frac{e^{\tau t+\zeta u-e^{\tau+\zeta}-e^\zeta}u!}{e^{\zeta u-e^{\tau+\zeta}-e^\zeta}(e^\tau+1)^u t!(u-t)!}, \quad t = 0,1,\ldots,u, \\[2mm]
&= \binom{u}{t}e^{\tau t}\left(\frac{1}{e^\tau+1}\right)^u \\[2mm]
&= \binom{u}{t}\left(\frac{e^\tau}{e^\tau+1}\right)^t\left(\frac{1}{e^\tau+1}\right)^{u-t}.
\end{aligned}
$$

Thus, we find the well-known result

$$T|U = u \sim Bi(u, e^\tau/(e^\tau+1)).$$

This means that the conditional distribution of T given $U = u$ has density of the form (5.30) with $p = 2$, $k = 1$, $K_u(\tau) = u\log(e^\tau+1)$ and $h_u(t) = \binom{u}{t}$. The inference problem of comparing μ and ν can be reduced to an inference problem about τ with ζ treated as a nuisance parameter. Note that $\tau = 0$ if and only if $\mu - \nu = 0$ and that the the sign of τ coincides with the sign of $\mu - \nu$. The conditional likelihood (5.33) becomes

$$L_c(\tau) = \exp\{\tau t - u\log(e^\tau+1)\}.$$

It is easy to see that there is no relevant information about τ in the marginal distribution of u. More specifically, the corresponding log-likelihood summand is partially uninformative about τ in the extended sense (see section 4.2.3). If the parameter of interest were $\mu - \nu$, which cannot be expressed as a component in a natural parameterization, model reduction would be more difficult. $\triangle$

Example 5.12 *Log-odds ratio*
Let T and V be independent random variables with distributions $Bi(n, \phi_1)$ and $Bi(m, \phi_2)$, respectively. The joint density is

$$p_{T,V}(t, v; \theta^1, \theta^2) = \binom{n}{t}\binom{m}{v} e^{\theta^1 t + \theta^2 v - n \log(1+e^{\theta^1}) - m \log(1+e^{\theta^2})} ,$$

where $\theta^1 = \log\left(\frac{\phi_1}{1-\phi_1}\right)$ and $\theta^2 = \log\left(\frac{\phi_2}{1-\phi_2}\right)$. This p.d.f. can be written as

$$p_{T,V}(t, v; \theta^1, \theta^2) = \binom{n}{t}\binom{m}{v} e^{(\theta^1-\theta^2)t + \theta^2(t+v) - \{n \log(1+e^{\theta^1}) + m \log(1+e^{\theta^2})\}} ,$$

which is an $\mathcal{F}_{ne}^2$, with natural observations $(t, u) = (t, t + v)$ and natural parameter $(\tau, \zeta) = (\theta^1 - \theta^2, \theta^2)$. The parameter

$$\tau = \log\left(\frac{\phi_1}{1-\phi_1} \bigg/ \frac{\phi_2}{1-\phi_2}\right)$$

is the difference between the probabilities of success expressed on the logit scale. The p.d.f. of (T, U) is

$$p_{T,U}(t, u; \tau, \zeta) = \binom{n}{t}\binom{m}{u-t} e^{\tau t + \zeta u - \{n \log(1+e^{\tau+\zeta}) + m \log(1+e^{\zeta})\}} , \qquad (5.34)$$

for $u = 0, \ldots, m + n$ and $\max(0, u - m) \leq t < \min(u, n)$. Therefore, by Theorem 5.6, the conditional p.d.f. of T given $U = u$ belongs to an $\mathcal{F}_{ne}^1$ with generic element

$$p_{T|U=u}(t; u, \tau) = \exp\{\tau t - K_u(\tau)\} h_u(t) .$$

Even in this case it is not difficult to find the explicit expressions of $K_u(\tau)$ and $h_u(t)$. Since

$$p_U(u; \tau, \zeta) = \exp\left\{\zeta u - (n \log(1 + e^{\tau+\zeta}) + m \log(1 + e^{\zeta}))\right\} C_u(\tau) ,$$

where

$$C_u(\tau) = \sum_{t'=\max(0,u-m)}^{\min(u,n)} \binom{n}{t'}\binom{m}{u-t'} e^{\tau t'} ,$$

we have

$$p_{T|U=u}(t; u, \tau) = e^{\tau t} \binom{n}{t} \binom{m}{u-t} \Big/ C_u(\tau) . \qquad (5.35)$$

Therefore,

$$K_u(\tau) = \log C_u(\tau), \quad h_u(t) = \binom{n}{t} \binom{m}{u-t} .$$

For $\tau = 0$, (5.35) is a hypergeometric distribution; for $\tau \neq 0$, (5.35) is a non-central hypergeometric distribution. As in the preceding example, the p.d.f. (5.35) provides a conditional likelihood for inference about τ. $\triangle$

Example 5.13 *Conditional inference about the shape parameter of the gamma distribution*

Let $y_1, \ldots, y_n$ be independent observations from a $Ga(\nu, \lambda)$ distribution. Taking Example 5.9 into account, the joint density is

$$p_Y(y; \tau, \zeta) = \exp\left\{ \tau \sum_{i=1}^n \log y_i + \zeta \sum_{i=1}^n y_i - n \log \Gamma(\tau) + n\tau \log(-\zeta) \right\} \prod_{i=1}^n y_i^{-1} ,$$

where $\tau = \nu > 0$ and $\zeta = -1/\lambda < 0$. This density belongs to an $\mathcal{F}_e^2$. However, it is not so easy here as it was in the two preceding examples, to explicitly express the joint density of the natural observations $(t, u) = \left(\sum_{i=1}^n \log y_i, \sum_{i=1}^n y_i \right)$.
Using Theorem 5.6, we can say that the p.d.f. of $T = \sum \log y_i$ conditionally on $U = \sum y_i = u$ is of the form (5.30), but the explicit expressions of $K_u(\tau)$ and $h_u(t)$ cannot be obtained directly. It is known, however, that the marginal distribution of U is $Ga(n\nu, \lambda)$. Therefore, the conditional log-likelihood for ν is

$$l_c(\nu) = \nu t - n\nu \log u - n \log \Gamma(\nu) + \log \Gamma(n\nu) . \qquad (5.36)$$

Comparing this expression with (5.33), we obtain

$$K_u(\nu) = n \left\{ \nu \log u + \log \Gamma(\nu) - \log \Gamma(n\nu)/n \right\} . \qquad (5.37)$$

$\triangle$

5.5 Sufficiency and Completeness

One immediate application of the Neyman–Fisher factorization theorem shows that the natural observations of an $\mathcal{F}_e^p$ are a p-dimensional sufficient statistic.

This result is particularly important because the class of exponential families of order p is closed under random sampling of size n. Let $Y_1, \ldots, Y_n$ be independent copies of a random variable with density $p(y; \phi) \in \mathcal{F}_e^p$. The density of $Y = (Y_1, \ldots, Y_n)$ is

$$p_Y(y; \phi) = e^{\theta(\phi) \cdot \sum_{i=1}^n s(y_i) - nG(\phi)} \prod_{i=1}^n h(y_i), \quad y \in \mathcal{Y}^n, \ \phi \in \Phi.$$

This is again an $\mathcal{F}_e^p$. The sufficient statistic $s = (s_1(y), \ldots, s_p(y))$, with $s_r(y) = \sum_{i=1}^n s_r(y_i)$, $r = 1, \ldots, p$, has fixed dimension p whatever the sample size is. Since the minimal representation of $\mathcal{F}_e^p$ is adopted, $s(y)$ is a minimal sufficient statistic if $n \geq p$.

If $\theta(\phi)$ is one-to-one and smooth (therefore, the family is not curved), then, with the natural parameterization θ, the statistical model induced by s is an $\mathcal{F}_{ne}^p$. Furthermore, the reduction to s is exhaustive, because in the reduced $\mathcal{F}_{ne}^p$ model there are no non-trivial distribution constant statistics. This is one immediate consequence of the following theorem.

Theorem 5.7 An $\mathcal{F}_{ne}^p$ is complete, provided that int $\tilde{\Theta} \neq \emptyset$.

Proof. Suppose that there exists a scalar statistic $w(y)$ such that the set $\{y : w(y) \neq 0\}$ has positive measure μ^* and such that

$$\int_{\mathcal{Y}} w(y) e^{\theta \cdot y - K(\theta)} p_0(y) d\mu^* = 0 , \quad \text{for all } \theta \in \tilde{\Theta} . \tag{5.38}$$

Then we can define the densities

$$p_0^+(y) = \frac{p_0(y) w^+(y)}{\int_{\mathcal{Y}} p_0(y) w^+(y) d\mu^*}$$

and

$$p_0^-(y) = \frac{p_0(y) w^-(y)}{\int_{\mathcal{Y}} p_0(y) w^-(y) d\mu^*} ,$$

where $w^+(y) = \max\{w(y), 0\}$ and $w^-(y) = \max\{-w(y), 0\}$. These two densities have the same normalizing constant, by (5.38) with $\theta = 0$. Hence, from (5.38) it follows that

$$\int_{\mathcal{Y}} e^{\theta \cdot y} p_0^+(y) d\mu^* = \int_{\mathcal{Y}} e^{\theta \cdot y} p_0^-(y) d\mu^* ,$$

for all $\theta \in \tilde{\Theta}$. Consequently, the measure μ^* of the set $\{y : w(y) \neq 0\}$ must be zero, because of the characterization property of the moment generating function. $\qquad \square$

For simplicity, the above theorem is given with reference to an $\mathcal{F}_{ne}^p$, that is, to a full family. The same result holds for all the subfamilies with parameter space $\Theta \subset \tilde{\Theta}$, with a non-empty interior. This excludes the subfamilies, such as curved exponential families, where every point of Θ is a boundary point.

5.6 Likelihood and Exponential Families

Suppose that, perhaps after sufficiency reduction or reparameterization, the density of the observations is of an $\mathcal{F}_{ne}^p$ form

$$p_S(s; \theta) = e^{\theta \cdot s - K(\theta)} p_0(s), \qquad \theta \in \tilde{\Theta}, 0 \in \tilde{\Theta}.$$

Then, the log-likelihood is

$$l(\theta; s) = \log p_S(s; \theta) = \theta \cdot s - K(\theta). \tag{5.39}$$

The score vector is

$$l_*(\theta) = s - \frac{\partial K(\theta)}{\partial \theta} = s - E_\theta(S), \tag{5.40}$$

the deviation of s from its expectation. The observed information is

$$j(\theta) = -\frac{\partial}{\partial \theta^{\mathrm{T}}} l_*(\theta) = \frac{\partial^2 K(\theta)}{\partial \theta \partial \theta^{\mathrm{T}}} = \mathrm{Var}_\theta(S) \tag{5.41}$$

and, since it does not depend on s, it coincides with the expected information $i(\theta)$.

For an $\mathcal{F}_{ne}^p$, the maximum likelihood estimate of θ, $\hat{\theta}$, if it exists, is the solution of the likelihood equation

$$s - E_\theta(S) = 0. \tag{5.42}$$

This solution, if it exists, is unique in that the matrix

$$-\frac{\partial^2}{\partial \theta \partial \theta^{\mathrm{T}}} l(\theta; s) = \mathrm{Var}_\theta(S)$$

is positive definite. Existence of the maximum likelihood estimate is ensured if there exists a value of θ in $\tilde{\Theta}$ for which the expected value of the distribution of S is the observed value s.

Theorem 5.8 If an $\mathcal{F}^p_{ne}$ is steep, and therefore, *a fortiori*, if it is regular, the maximum likelihood estimate $\hat{\theta}$ exists and is the unique solution to the likelihood equation (5.42) if, and only if, $s \in \operatorname{int} C$, where C is the closed convex hull of the support of S.

Proof. The correspondence between the observed value of s and a value $\theta \in \tilde{\Theta}$ which makes s the expected value of the distribution of S is assured by Theorem 5.3. For the proof of non-existence of the maximum likelihood estimate when s belongs to the boundary of C, see Barndorff-Nielsen (1978, section 9.3). $\square$

Under the assumptions of Theorem 5.8, the maximum likelihood estimate exists with probability one, and is the solution to the equation (5.42) if, and only if, the boundary of C has probability zero.

When $\hat{\theta}$ exists with probability one,

$$E_{\hat{\theta}}(S) = \mu(\hat{\theta}) = s \ . \tag{5.43}$$

This is a fundamental identity for inference in an $\mathcal{F}^p_{ne}$. Because the transformation $\mu(\cdot)$ is one-to-one and smooth it follows, from (5.43), that the correspondence between $\hat{\theta}$ and s is one-to-one and smooth too. The maximum likelihood estimate $\hat{\theta}$ is, therefore, a minimal sufficient statistic. The Jacobian of the transformation $s \to \hat{\theta}$ is

$$1 \Big/ \left| \frac{\partial \mu(\hat{\theta})}{\partial \hat{\theta}^{\mathrm{T}}} \right|_{\hat{\theta}=\hat{\theta}(s)} = 1 \big/ |i(\hat{\theta})| \ . \tag{5.44}$$

Lastly, the log-likelihood can be written in the form $l(\theta; \theta)$. In fact,

$$l(\theta; s) = \theta \cdot s - K(\theta) = \theta \cdot \mu(\hat{\theta}) - K(\theta) = l(\theta; \hat{\theta}) \ . \tag{5.45}$$

5.7 Profile Likelihood and Mixed Parameterization

Suppose that after reduction by sufficiency or reparameterization the density of the observations is in an $\mathcal{F}^p_{ne}$ with density

$$p_{T,U}(t, u; \tau, \zeta) = e^{\tau \cdot t + \zeta \cdot u - K(\tau, \zeta)} p_0(t, u) \ , \qquad \theta = (\tau, \zeta) \in \tilde{\Theta}, \quad 0 \in \tilde{\Theta} \ .$$

Let τ be the parameter of interest, while ζ is a nuisance parameter.

The conditional log-likelihood function $l_c(\tau)$, based on (5.33), is a particularly appealing pseudo-likelihood for inference about τ because it is a genuine log-likelihood. Basing inference about τ upon $l_c(\tau)$ is the same as neglecting the log-likelihood summand deduced from the marginal density of U (see (5.31)), that is

$$l_U(\tau, \zeta) = \zeta \cdot u - K(\tau, \zeta) + K_u(\tau) .$$

On the basis of point $i)$ of Theorem 5.6, the statistical model for U is an $\mathcal{F}_{ne}^{p-k}$ for every fixed τ. If the family is regular, the maximum likelihood estimate of ζ, $\hat{\zeta}_\tau$, when it exists, is a one-to-one function of u. Therefore, the corresponding statistical model is saturated and u is I-distribution constant for τ (cf. section 4.2.3).

In addition, since $\hat{\zeta}_\tau$ satisfies the equation

$$\frac{\partial}{\partial \zeta} K(\tau, \zeta) \bigg|_{\zeta = \hat{\zeta}_\tau} = u , \tag{5.46}$$

we have

$$l_U(\tau, \hat{\zeta}_\tau) = \hat{\zeta}_\tau \cdot K_\zeta(\tau, \hat{\zeta}_\tau) - K(\tau, \hat{\zeta}_\tau) + K_u(\tau) , \tag{5.47}$$

where $K_\zeta(\tau, \zeta) = K_\zeta = (\partial/\partial\zeta) K(\tau, \zeta)$. Furthermore, let

$$\begin{aligned}
K_\tau(\tau, \zeta) &= K_\tau = (\partial/\partial\tau) K(\tau, \zeta) , \\
K_{\tau\tau}(\tau, \zeta) &= K_{\tau\tau} = (\partial^2/\partial\tau\partial\tau^T) K(\tau, \zeta) , \\
K_{\tau\zeta}(\tau, \zeta) &= K_{\tau\zeta} = (\partial^2/\partial\tau\partial\zeta^T) K(\tau, \zeta) , \\
K_{\zeta\tau}(\tau, \zeta) &= K_{\zeta\tau} = (K_{\tau\zeta}(\tau, \zeta))^T , \\
K_{\zeta\zeta}(\tau, \zeta) &= K_{\zeta\zeta} = (\partial^2/\partial\zeta\partial\zeta^T) K(\tau, \zeta) .
\end{aligned}$$

The maximum likelihood estimate of τ based on (5.47) is the solution in τ to the equation

$$\hat{\zeta}_\tau \left(K_{\zeta\tau}(\tau, \hat{\zeta}_\tau) + K_{\zeta\zeta}(\tau, \hat{\zeta}_\tau) \frac{\partial}{\partial\tau} \hat{\zeta}_\tau \right) - K_\tau(\tau, \hat{\zeta}_\tau) + \frac{\partial}{\partial\tau} K_u(\tau) = 0 . \tag{5.48}$$

By differentiating both sides of (5.46) with respect to τ we obtain

$$\frac{\partial}{\partial\tau} \hat{\zeta}_\tau = - \left(K_{\zeta\zeta}(\tau, \hat{\zeta}_\tau) \right)^{-1} K_{\zeta\tau}(\tau, \hat{\zeta}_\tau) , \tag{5.49}$$

which, when substituted into (5.48), gives

$$- K_\tau(\tau, \hat{\zeta}_\tau) + \frac{\partial}{\partial\tau} K_u(\tau) = 0 . \tag{5.50}$$

As will be seen in section 10.10.2, if $K(\tau, \zeta) = O(n)$, then for every τ,

$$\frac{\partial}{\partial \tau} K_u(\tau) = K_\tau(\tau, \hat{\zeta}_\tau) + O_p(1)$$

and, consequently, the estimating function on the left-hand side of (5.50) is always $O_p(1)$ no matter what τ is. Therefore, the corresponding estimating equation cannot provide a consistent estimator for τ. To conclude, the information about τ which can be deduced from the log-likelihood summand omitted in reduction by conditioning is negligible: asymptotically it is partially uninformative about τ in the extended sense (see section 4.2.3).

As mentioned in section 5.4, it may prove difficult to obtain an exact conditional likelihood. One possibility is to use the profile likelihood (see section 4.6). The profile log-likelihood is

$$l_p(\tau) = \tau \cdot t + \hat{\zeta}_\tau \cdot u - K(\tau, \hat{\zeta}_\tau) \, . \tag{5.51}$$

The profile score (see formula (4.36)) is

$$l_p(\tau)_* = t - \tilde{K}_\tau \, , \tag{5.52}$$

where the symbol $\tilde{f}$ indicates that the function $f(\tau, \zeta)$ is evaluated at $(\tau, \hat{\zeta}_\tau)$. Thus the profile score is the deviation of t from its expectation evaluated at $(\tau, \hat{\zeta}_\tau)$. It follows that, in general, the null expectation of the profile score is not zero even for exponential families. The observed profile information is (see formula (4.37))

$$j_p(\tau) = \tilde{K}_{\tau\tau} - \tilde{K}_{\tau\zeta} \left(\tilde{K}_{\zeta\zeta} \right)^{-1} \tilde{K}_{\zeta\tau} \, . \tag{5.53}$$

Because

$$i(\theta) = i(\tau, \zeta) = \begin{pmatrix} K_{\tau\tau} & K_{\tau\zeta} \\ K_{\zeta\tau} & K_{\zeta\zeta} \end{pmatrix},$$

$j_p(\tau)$ coincides with the upper left-hand $k \times k$ submatrix of $i(\theta)^{-1}$, evaluated at $(\tau, \hat{\zeta}_\tau)$. Furthermore, since $i(\tau, \zeta) = \text{Var}_{\tau, \zeta}(T, U)$, $j_p(\tau)$ is also the covariance matrix of the residuals of the linear regression of T on U, evaluated at $(\tau, \hat{\zeta}_\tau)$.

An orthogonal parameterization always exists for an $\mathcal{F}^p_{n_e}$, so that the approximate conditional likelihood (4.42) for τ can be obtained. The result is very peculiar, in that the parameter of interest may be multidimensional. So far, an $\mathcal{F}^p_{n_e}$ has been studied with the natural parameterization $\theta = (\tau, \zeta)$ and mention has been made of the possibility of parameterizing the model with the mean $\mu(\theta) = (\mu_t, \mu_u)$, where

$$\mu_t = \mu_t(\tau, \zeta) = K_\tau \, , \qquad \mu_u = \mu_u(\tau, \zeta) = K_\zeta \, .$$

Definition 5.8 The one-to-one smooth function

$$\psi = \psi(\tau, \zeta) = (\tau, \mu_u) \tag{5.54}$$

is called the **mixed parameterization** of an $\mathcal{F}_{ne}^p$.

The following results hold for the mixed parameterization.

Theorem 5.9 If an $\mathcal{F}_{ne}^p$ is regular or, at least, steep, the parameters τ and μ_u are variation independent, that is

$$(\tau, \mu_u)(\tilde{\Theta}) = \text{int } \tilde{T} \times \text{int } C_u \,,$$

where C_u denotes the closed convex hull of the support of the marginal component u of (t, u) and $\tilde{T}$ is the projection of the natural parameter space $\tilde{\Theta}$, $\tilde{T} = \{\tau : (\tau, \zeta) \in \tilde{\Theta} \text{ for some } \zeta\}$.

Proof. The subfamily of the $\mathcal{F}_{ne}^p$ obtained by fixing $\tau = \tau_0$ is an $\mathcal{F}_{ne}^{p-k}$. This subfamily has natural observations u and natural parameter ζ. The support of (t, u) does not depend on $\theta = (\tau, \zeta)$; hence, neither does C_u depend on θ. Since $\mathcal{F}_{ne}^p$ is regular (or steep), so is the subfamily $\mathcal{F}_{ne}^{p-k}$. Therefore, $\text{int } C_u = \mu_u(\text{int } \tilde{Z}_{\tau_0})$, where $\tilde{Z}_{\tau_0} = \{\zeta \in \mathbb{R}^{p-k} : (\tau_0, \zeta) \in \tilde{\Theta}\}$. But $\mu_u(\text{int } \tilde{Z}_{\tau_0})$ is the set of values that μ_u can take when $\tau = \tau_0$; and, since this coincides with $\text{int } C_u$, it does not vary as τ_0 varies in $\tilde{T}$. □

Theorem 5.10 The mixed parameterization $\psi = (\tau, \mu_u)$ is orthogonal. Specifically,

$$i^{\Psi}(\tau, \mu_u) = \begin{pmatrix} K_{\tau\tau} - K_{\tau\zeta}(K_{\zeta\zeta})^{-1}K_{\zeta\tau} & 0 \\ 0 & (K_{\zeta\zeta})^{-1} \end{pmatrix} , \tag{5.55}$$

where the quantities indicated are evaluated at (τ, ζ), with $\zeta = \zeta(\tau, \mu_u)$.

Proof. By applying (1.21) and partitioning the parameter and the information matrix into blocks, since $\mu_u(\tau, \zeta) = K_\zeta(\tau, \zeta)$, we have

$$\begin{aligned} [\theta_a^r] &= \left[\psi_r^a \big|_{\theta=\theta(\psi)}\right]^{-1} = \begin{pmatrix} I_k & 0 \\ K_{\zeta\tau} & K_{\zeta\zeta} \end{pmatrix}^{-1} \\ &= \begin{pmatrix} I_k & 0 \\ -(K_{\zeta\zeta})^{-1}K_{\zeta\tau} & (K_{\zeta\zeta})^{-1} \end{pmatrix} . \end{aligned}$$

Thus,

$$
\begin{aligned}
i^{\Psi}(\tau, \mu_u) &= \begin{pmatrix} I_k & -K_{\tau\zeta}(K_{\zeta\zeta})^{-1} \\ 0 & (K_{\zeta\zeta})^{-1} \end{pmatrix} \begin{pmatrix} K_{\tau\tau} & K_{\tau\zeta} \\ K_{\zeta\tau} & K_{\zeta\zeta} \end{pmatrix} \begin{pmatrix} I_k & 0 \\ -(K_{\zeta\zeta})^{-1}K_{\zeta\tau} & (K_{\zeta\zeta})^{-1} \end{pmatrix} \\
&= \begin{pmatrix} I_k & -K_{\tau\zeta}(K_{\zeta\zeta})^{-1} \\ 0 & (K_{\zeta\zeta})^{-1} \end{pmatrix} \begin{pmatrix} -K_{\tau\zeta}(K_{\zeta\zeta})^{-1}K_{\zeta\tau} + K_{\tau\tau} & K_{\tau\zeta}(K_{\zeta\zeta})^{-1} \\ 0 & I_{p-k} \end{pmatrix} \\
&= \begin{pmatrix} K_{\tau\tau} - K_{\tau\zeta}(K_{\zeta\zeta})^{-1}K_{\zeta\tau} & 0 \\ 0 & (K_{\zeta\zeta})^{-1} \end{pmatrix} .
\end{aligned}
$$

$\square$

In a natural exponential family, the expression of the approximate conditional log-likelihood for τ (4.42) is an immediate consequence of Theorem 5.10. We only have to observe that, by (1.19), and since the maximum likelihood estimate of μ_u for a fixed value of τ, $\hat{\mu}_{u\tau}$, coincides with $\hat{\mu}_u$,

$$
j^{\Psi}_{\mu_u \mu_u}(\tau, \hat{\mu}_u) = (K_{\zeta\zeta}(\tau, \zeta(\tau, \hat{\mu}_u)))^{-1} .
$$

Hence,

$$
l_{AC}(\tau) = l_P(\tau) + \frac{1}{2} \log |\tilde{K}_{\zeta\zeta}| . \tag{5.56}
$$

In section 10.10.2 we will show that if $K(\tau, \zeta) = O(n)$, then the approximate conditional log-likelihood (5.56) approximates the exact conditional log-likelihood based on (5.33) with error of order $O(n^{-1})$, while the profile log-likelihood $l_P(\tau)$ (5.51) has error of order $O(1)$.

5.8 Procedures with Finite-Sample Optimality Properties

Exponential families are in a privileged position as far as the existence of optimum inference procedures is concerned. However, it should be stressed that, even here, optimal solutions only exist for some special inference problems.

As regards the optimum theory of point estimation, (see section 3.5.2), in a family where the sufficient statistic s is complete, this statistic is an unbiased estimator with uniformly minimum variance of its own expectation. In other words, in the mean value parameterization, the optimal unbiased estimator is immediately available. For other parameterizations, if there is an unbiased estimator, the Rao–Blackwell–Lehmann–Scheffé theorem guarantees that there is an efficient estimator and characterizes its structure. However, often its explicit expression is not immediate: for an asymptotic approximation, see section 10.10.2, Example 11.13. Lastly, one interesting result is that the Cramér–Rao lower bound is only reached within an exponential family with the mean parameterization (see, for example, Lehmann, 1983, section 2.6).

Below we will give some details of optimality theory of testing hypotheses under exponential families. The basic notations and results reviewed in section 3.5.3 will be used. Optimal procedures are systematically available only when the parameter of interest is a scalar. This has two aspects: *i*) there exist uniformly most powerful tests for exponential families of order 1; *ii*) for exponential families of order $p > 1$ there exist uniformly most powerful similar tests if the parameter of interest is a scalar linear combination of the components of the natural parameter.

5.8.1   Testing Hypotheses: One-Parameter Case

Let $Y_1, \ldots, Y_n$ be independent copies of a random variable with density in an $\mathcal{F}_e^1$,

$$p(y; \theta) = e^{\theta s(y) - K(\theta)} h(y), \quad y \in \mathcal{Y}, \ \theta \in \tilde{\Theta} \subseteq \mathbb{R}.$$

Then $Y = (Y_1, \ldots, Y_n)$ has density

$$p_Y(y; \theta) = e^{\theta \Sigma s(y_i) - n K(\theta)} \Pi h(y_i),$$

which has a monotone likelihood ratio in $s = \Sigma s(y_i)$. Therefore, there exists a uniformly most powerful level α test for

$$H_0 : \ \theta \leq \theta_0 \quad \text{versus} \quad H_1 : \ \theta > \theta_0. \tag{5.57}$$

This will reject H_0 with probability

$$\pi(y) = \begin{cases} 1 & \text{if } s > s_\alpha \\ \gamma & \text{if } s = s_\alpha \\ 0 & \text{if } s < s_\alpha, \end{cases} \tag{5.58}$$

where the values s_α and γ are chosen in such a way as to give

$$E_{\theta_0}\{\pi(Y)\} = \alpha \tag{5.59}$$

(Lehmann, 1986, Theorem 3.2).

There exists also an optimal test for

$$H_0 : \ \theta = \theta_0 \quad \text{versus} \quad H_1 : \ \theta \neq \theta_0 \tag{5.60}$$

so long as attention is restricted to the class of unbiased tests. A level α test $\pi(y)$ is unbiased if

$$E_\theta\{\pi(Y)\} \geq \alpha \qquad \text{for each } \theta \in \tilde{\Theta} \tag{5.61}$$

and

$$E_{\theta_0}\{\pi(Y)\} = \alpha.$$

Relation (5.61) imposes a constraint on the power function $\beta(\theta) = E_\theta\{\pi(Y)\}$ to have a minimum at θ_0. The uniformly most powerful unbiased test for (5.60) is defined by

$$\pi(y) = \begin{cases} 1 & \text{if } s < c_1 \text{ o } s > c_2 \\ \gamma_i & \text{if } s = c_i, \ i = 1, 2 \\ 0 & \text{if } c_1 < s < c_2, \end{cases} \tag{5.62}$$

where the constants $c_1, c_2, \gamma_1, \gamma_2$ are to be chosen so as to satisfy

$$E_{\theta_0}\{\pi(Y)\} = \alpha, \quad E_{\theta_0}\{s(Y)\pi(Y)\} = \alpha E_{\theta_0}\{s(Y)\} \tag{5.63}$$

(Lehmann, 1986, section 4.2).

Because finding such a test is not straightforward, exact unbiasedness is usually neglected and the acceptance region is taken as being the intersection of the acceptance regions of the two one-sided tests of H_0, versus $\theta > \theta_0$ and versus $\theta < \theta_0$, each with level $\alpha/2$. If the null distribution of $s(Y)$ is symmetrical, this procedure coincides with the optimum procedure.

The optimal tests (5.58) and (5.62) have nested acceptance regions, thus they define uniformly most accurate confidence intervals through (3.67).

5.8.2 Testing Hypotheses: Multiparameter Case

Let $Y_1, \ldots, Y_n$ be independent copies of a random variable with density in an $\mathcal{F}_e^p$

$$p(y; \tau, \zeta) = e^{\tau \cdot t(y) + \zeta \cdot u(y) - K(\tau, \zeta)} h(y), \quad y \in \mathcal{Y}, \ (\tau, \zeta) \in \tilde{\Theta}, \tag{5.64}$$

where τ is k-dimensional and ζ is $(p - k)$-dimensional. Usually, there is no optimal level α test for hypotheses on the global parameter (τ, ζ). The exception being some hypotheses on τ when $k = 1$, that is, when the parameter of interest is a scalar, while the dimension of ζ can be greater than one. The main reason for this exception is that, on the basis of Theorem 5.6, the conditional distributions of T given $U = u$ form an $\mathcal{F}_{ne}^1$, hence the tests described in subsection 5.8.1 can be used: rejection regions or P-values are determined on the basis of the conditional distribution of T given $U = u$ under H_0. The

main results can be summarized as follows. For a more detailed discussion, see
Lehmann (1986, sections 4.3 and 4.4).

For

$$H_0 : \tau \leq \tau_0 \quad \text{versus} \quad H_1 : \tau > \tau_0$$

there exists a level α uniformly most powerful test among tests which are similar on the boundary, hence, *a fortiori* among unbiased tests. In this case
the boundary Υ is the set $\{(\tau, \zeta) \in \tilde{\Theta} : \tau = \tau_0\}$. Because the corresponding subfamily is complete, every similar test has Neyman structure thus, the
global optimum problem can be reduced to the corresponding optimum problem within the conditional subfamily with density (5.30). Hence, the uniformly
most powerful level α similar test has critical function

$$\pi(y) = \begin{cases} 1 & \text{if } t > t_\alpha \\ \gamma & \text{if } t = t_\alpha \\ 0 & \text{if } t < t_\alpha , \end{cases} \tag{5.65}$$

with t_α and γ determined so that

$$E_{\tau_0}\{\pi(Y)|U = u\} = \alpha . \tag{5.66}$$

Analogously, there exists a level α uniformly most powerful unbiased test
for

$$H_0 : \tau = \tau_0 \quad \text{versus} \quad H_1 : \tau \neq \tau_0 .$$

The critical function which corresponds to (5.62) in the conditional model is

$$\pi(y) = \begin{cases} 1 & \text{if } t < c_1 \text{ o } t > c_2 \\ \gamma_i & \text{if } t = c_i, \ i = 1, 2 \\ 0 & \text{if } c_1 < t < c_2 , \end{cases} \tag{5.67}$$

with c_i and $\gamma_i, i = 1, 2$, such as to satisfy the conditions

$$E_{\tau_0}\{\pi(Y)|U = u\} = \alpha$$

$$\tag{5.68}$$

$$E_{\tau_0}\{t(Y)\pi(Y)|U = u\} = \alpha E_{\tau_0}\{t(Y)|U = u\} .$$

Thus, the Neyman–Pearson theory gives optimal tests that are conditional
ones. The aim of conditioning here is very different from that in the Fisherian
context. Here, it is not a question of ensuring that inference is relevant, rather,
it is merely a question of ensuring that the formal constraint of unbiasedness

is satisfied. In fact, the conditioning statistic is partially distribution constant for τ only in the extended sense (cf. section 5.7). From the Fisherian point of view, optimal conditional tests are well founded to the same extent as is the reduction to conditional likelihood.

Example 5.14 *Optimal similar test on the shape parameter of the gamma distribution*

Let $Y_1, \ldots, Y_n$ be independent observations from a $Ga(\tau, \lambda)$ distribution. We wish to obtain a level α test for $H_0 : \tau \leq \tau_0$ versus $H_1 : \tau > \tau_0$. When $\tau_0 = 1$ we are testing for exponentiality against gamma alternatives. Since the joint density is

$$p_Y(y; \tau, \zeta) = \exp\{(\tau - 1)\Sigma \log y_i + \zeta\Sigma y_i - n(\log \Gamma(\tau) - \tau \log(-\zeta))\} \, ,$$

where $\zeta = -1/\lambda$, the uniformly most powerful level α similar test rejects H_0 if

$$t = \Sigma \log y_i > t_{\alpha,u} \, ,$$

where $u = \sum y_i$ and

$$P_{\tau_0}\{\Sigma \log Y_i > t_{\alpha,u} \,|\, \Sigma Y_i = u\} = \alpha \, .$$

Thus, this test is the same as that which rejects H_0 for large values of the statistic

$$\log(R) = \frac{\Sigma \log y_i}{n} - \log\left(\frac{u}{n}\right) \tag{5.69}$$

in the conditional distribution given u. Since R is the ratio between the geometric mean and the arithmetic mean of the sample observations, its distribution does not depend on the scale parameter λ. Because of Basu's theorem (Theorem 2.2), R and U are independent and conditioning is carried out automatically when referring to the marginal distribution of R. See Keating, Glaser, Ketchum (1990) for more about this distribution. $\qquad \triangle$

Example 5.15 *Fisher's exact test*

Let $y_i = (x_i, z_i)$, $i = 1, \ldots, n$, be independent observations from a two-dimensional variable (X, Z) with binary components. Usually, the data are represented in a 2×2 table as below,

	Z=0	Z=1	
X=0	n_{00}	n_{01}	$n_{0\bullet}$
X=1	n_{10}	n_{11}	$n_{1\bullet}$
	$n_{\bullet 0}$	$n_{\bullet 1}$	n

with $n_{00} = \sum_{i=1}^{n}(1 - x_i)(1 - z_i)$, $n_{01} = \sum_{i=1}^{n}(1 - x_i)z_i$, $n_{10} = \sum_{i=1}^{n}x_i(1 - z_i)$, $n_{11} = \sum_{i=1}^{n} x_i z_i$; the usual notation, $n_{0\bullet}$, $n_{1\bullet}$, $n_{\bullet 0}$ and $n_{\bullet 1}$ is used for the marginal totals.

Let $\phi_{jk} = P(X_i = j, Z_i = k)$, $j, k = 0, 1$, be the cell probabilities. Y_i has a multinomial distribution with index 1 and parameter $\phi = (\phi_{00}, \phi_{01}, \phi_{10}, \phi_{11})$, where $\phi_{jk} > 0$ and $\phi_{00} + \phi_{01} + \phi_{10} + \phi_{11} = 1$. The summary of the data in a 2×2 table corresponds to the sufficiency reduction to $s = (n_{00}, n_{01}, n_{10}, n_{11})$; the distribution of s is multinomial with index n and parameter ϕ, i.e.

$$
\begin{aligned}
p_S(s; \phi) &= \frac{n!}{n_{00}! n_{01}! n_{10}! n_{11}!} \phi_{00}^{n_{00}} \phi_{01}^{n_{01}} \phi_{10}^{n_{10}} \phi_{11}^{n_{11}} \\
&= \frac{n!}{n_{00}! n_{01}! n_{10}! n_{11}!} \phi_{11}^{n} \exp\left\{ n_{00} \log(\frac{\phi_{00}}{\phi_{11}}) + n_{01} \log(\frac{\phi_{01}}{\phi_{11}}) + n_{10} \log(\frac{\phi_{10}}{\phi_{11}}) \right\} \\
&= \frac{n!}{n_{00}! n_{01}! n_{10}! n_{11}!} \phi_{11}^{n} \exp\left\{ n_{00} \left(\log(\frac{\phi_{00}}{\phi_{11}}) - \log(\frac{\phi_{01}}{\phi_{11}}) - \log(\frac{\phi_{10}}{\phi_{11}}) \right) \right. \\
&\qquad \left. + (n_{01} + n_{00}) \log(\frac{\phi_{01}}{\phi_{11}}) + (n_{10} + n_{00}) \log(\frac{\phi_{10}}{\phi_{11}}) \right\} \\
&= \frac{n!}{n_{00}! n_{01}! n_{10}! n_{11}!} \phi_{11}^{n} \exp\left\{ n_{00} \log \Delta + n_{0\bullet} \log(\frac{\phi_{01}}{\phi_{11}}) + n_{\bullet 0} \log(\frac{\phi_{10}}{\phi_{11}}) \right\},
\end{aligned}
$$

where

$$
\Delta = \frac{\phi_{00}\phi_{11}}{\phi_{01}\phi_{10}}
$$

is the odds ratio. Writing $\tau = \log \Delta$, $\zeta_1 = \log(\phi_{01}/\phi_{11})$, $\zeta_2 = \log(\phi_{10}/\phi_{11})$, the model is seen to be an $\mathcal{F}_{ne}^{3}$ with density

$$
\begin{aligned}
p(n_{00}, n_{0\bullet}, n_{\bullet 0}; \tau, \zeta_1, \zeta_2) &= \frac{n!}{n_{00}!(n_{0\bullet} - n_{00})!(n_{\bullet 0} - n_{00})!(n - n_{0\bullet} - n_{\bullet 0} + n_{00})!} \\
&\quad \times \exp\left\{ n_{00}\tau + n_{0\bullet}\zeta_1 + n_{\bullet 0}\zeta_2 - n\left(\tau + \zeta_1 + \zeta_2 \right. \right. \\
&\qquad \left. \left. + \log\left(1 + e^{-\tau - \zeta_1 - \zeta_2}(1 + e^{\zeta_1} + e^{\zeta_2})\right)\right)\right\},
\end{aligned}
$$

whose parameters are variation independent.

Let $\phi_{0\bullet} = \phi_{00} + \phi_{01}$ and $\phi_{\bullet 0} = \phi_{00} + \phi_{10}$ be the marginal probabilities. For fixed values of $\phi_{0\bullet}$ and $\phi_{\bullet 0}$, Δ is an increasing function of ϕ_{00}. Furthermore, $\Delta = 1$ if and only if $\phi_{00} = \phi_{0\bullet}\phi_{\bullet 0}$, that is, if X and Z are independent. Values of Δ greater than 1 correspond to positive association, while values smaller than 1 correspond to negative association. Since Δ is a monotone function of a natural parameter in the density of the sufficient statistic, there exists a uniformly most powerful similar test for independence against positive or negative association, i.e. for

$$H_0 : \Delta = 1 \quad \text{versus} \quad H_1 : \Delta > 1 .$$

This test rejects the null hypothesis if

$$n_{00} > c_{\alpha;u} ,$$

where $u = (n_{0\bullet}, n_{\bullet 0})$. The conditional distribution of n_{00} given $U = u$ can be obtained in two steps.

i) First, find the conditional distribution of (n_{00}, n_{10}) given $n_{0\bullet}$. Since $n_{0\bullet}$ has a marginal $Bi(n, \phi_{0\bullet})$ distribution, we have

$$p(n_{00}, n_{10} | n_{0\bullet}) = \binom{n_{0\bullet}}{n_{00}} \binom{n_{1\bullet}}{n_{10}} \left(\frac{\phi_{00}}{\phi_{0\bullet}}\right)^{n_{00}} \left(\frac{\phi_{01}}{\phi_{0\bullet}}\right)^{n_{01}} \left(\frac{\phi_{10}}{\phi_{1\bullet}}\right)^{n_{10}} \left(\frac{\phi_{11}}{\phi_{1\bullet}}\right)^{n_{11}} .$$

Therefore, conditionally on $n_{0\bullet}$, the variables n_{00} and n_{10} are independent and distributed according to $Bi(n_{0\bullet}, \phi_{00}/\phi_{0\bullet})$ and $Bi(n_{1\bullet}, \phi_{10}/\phi_{1\bullet})$, respectively.

ii) Now, consider, again conditionally on $n_{0\bullet}$, the distribution of n_{00} given $n_{00} + n_{10}$. This can be obtained immediately from Example 5.12, and is a non-central hypergeometric distribution with density

$$p(n_{00} | n_{0\bullet}, n_{\bullet 0}; \Delta) = \binom{n_{0\bullet}}{n_{00}} \binom{n - n_{0\bullet}}{n_{\bullet 0} - n_{00}} e^{n_{00} \log \Delta} \Big/ C_u(\Delta) ,$$

where

$$C_u(\Delta) = \sum_{m=a}^{b} \binom{n_{0\bullet}}{m} \binom{n - n_{0\bullet}}{n_{\bullet 0} - m} e^{m \log \Delta}$$

with $a = \max(0, n_{\bullet 0} - n_{1\bullet})$ and $b = \min(n_{0\bullet}, n_{\bullet 0})$. The null distribution, corresponding to independence is, therefore, hypergeometric and has support $\{a, a+1, \ldots, b\}$ and density

$$p(n_{00} | n_{0\bullet}, n_{\bullet 0}; 1) = \frac{\binom{n_{0\bullet}}{n_{00}} \binom{n - n_{0\bullet}}{n_{\bullet 0} - n_{00}}}{\binom{n}{n_{\bullet 0}}} .$$

Note that, because it is based on a statistic with discrete distribution, the optimal test is usually randomized. In order to avoid randomization, one will usually choose the critical value $c_{\alpha;u}$ which corresponds to the largest level

$\alpha' < \alpha$ which can be reached exactly. This solution is known as **Fisher's exact test** (Fisher, 1958). The actual overall level α can, thus, be quite different from the nominal level. Some authors emphasize this as a drawback and have sought suitable modifications. However, no such problem arises if the conclusions of the test are expressed as a P-value. $\triangle$

To conclude, we recall that the situations where an optimal test can be found are very particular, also when an exponential family is involved. For example, usually an optimal test does not exist when the parameter of interest cannot be expressed as a linear combination of natural parameters. Furthermore, even when there exists an optimal test, determining its rejection region or its P-value requires that the marginal or the conditional distributions of the sufficient statistics should be known. As regards this, Theorem 5.6 only offers information on the general structure of the distributions. Algorithms and corresponding software packages have been developed for some discrete distributions which arise in models for contingency tables or for regression problems with binary response. See Agresti (1992) for a survey of exact conditional inference for contingency tables. To overcome the difficulties mentioned above, the likelihood ratio tests and their asymptotically equivalent versions are often used in applications. Higher-order asymptotic methods often offer very accurate solutions: see sections 10.6, 10.10.2 and Chapter 11.

5.9   First-Order Asymptotic Theory

Let $Y_1, \ldots, Y_n$ be independent copies of a random variable with density in an $\mathcal{F}_e^p$

$$p(y; \tau, \zeta) = e^{\tau \cdot t(y) + \zeta \cdot u(y) - K(\tau, \zeta)} h(y) , \tag{5.70}$$

with $(\tau, \zeta) \in \tilde{\Theta} \subseteq \mathbb{R}^p$ and τ, as usual, k-dimensional. Since $(t(Y_i), u(Y_i))$ has finite moments of any order for every $(\tau, \zeta) \in \mathrm{int}\, \tilde{\Theta}$, then the multivariate central limit theorem gives

$$\frac{1}{\sqrt{n}} \left(\begin{array}{c} \sum_{i=1}^n t(Y_i) - nK_\tau(\tau, \zeta) \\ \sum_{i=1}^n u(Y_i) - nK_\zeta(\tau, \zeta) \end{array} \right) \xrightarrow{d} N_p(0, i_1) , \tag{5.71}$$

where

$$i_1 = i_1(\tau, \zeta) = \left(\begin{array}{cc} K_{\tau\tau}(\tau, \zeta) & K_{\tau\zeta}(\tau, \zeta) \\ K_{\zeta\tau}(\tau, \zeta) & K_{\zeta\zeta}(\tau, \zeta) \end{array} \right) .$$

The asymptotic distribution of the maximum likelihood estimator can be easily obtained directly from (5.71). In fact, $(\hat{\tau}, \hat{\zeta})$ is a one-to-one function of (t, u)

defined by the equations $nK_\tau(\hat{\tau}, \hat{\zeta}) = t$ and $nK_\zeta(\hat{\tau}, \hat{\zeta}) = u$. The Jacobian of the transformation $(\hat{\tau}, \hat{\zeta}) = (\hat{\tau}(t, u), \hat{\zeta}(t, u))$ is $\hat{i}^{-1} = (n i_1(\hat{\tau}, \hat{\zeta}))^{-1}$. Using the multivariate delta method, (Appendix A, Theorem A.15), since the asymptotic covariance matrix is $i_1^{-1} i_1 i_1^{-1} = i_1^{-1}$, we have

$$\sqrt{n} \begin{pmatrix} \hat{\tau} - \tau \\ \hat{\zeta} - \zeta \end{pmatrix} \xrightarrow{d} N_p(0, i_1^{-1}) \tag{5.72}$$

and, in particular,

$$\sqrt{n}(\hat{\tau} - \tau) \xrightarrow{d} N_k (0, i_1^{\tau\tau}) ,$$

where $i_1^{\tau\tau} = \left\{ K_{\tau\tau}(\tau, \zeta) - K_{\tau\zeta}(\tau, \zeta)(K_{\zeta\zeta}(\tau, \zeta))^{-1} K_{\zeta\tau}(\tau, \zeta) \right\}^{-1}$.

These results allow the asymptotic distribution of the likelihood ratio test, and of its asymptotically equivalent versions, to be found directly. The likelihood ratio test for

$$H_0 : \theta = \theta_0 \qquad \text{versus} \qquad H_1 : \theta \neq \theta_0 ,$$

with $\theta = (\tau, \zeta)$, is

$$\begin{aligned} W &= 2 \left\{ l(\hat{\theta}) - l(\theta_0) \right\} \\ &= 2 \left\{ (\hat{\tau} - \tau_0) \cdot t + (\hat{\zeta} - \zeta_0) \cdot u - n \left(K(\hat{\tau}, \hat{\zeta}) - K(\tau_0, \zeta_0) \right) \right\} , \end{aligned}$$

where $t = \sum t(y_i)$ and $u = \sum u(y_i)$. The usual asymptotic expansions $W = W_e + o_p(1)$ and $W = W_u + o_p(1)$ also hold, with

$$W_e = \left((\hat{\tau} - \tau_0)^{\mathrm{T}}, (\hat{\zeta} - \zeta_0)^{\mathrm{T}} \right) i(\hat{\tau}, \hat{\zeta}) \begin{pmatrix} \hat{\tau} - \tau_0 \\ \hat{\zeta} - \zeta_0 \end{pmatrix}$$

and

$$W_u = \left((t - nK_\tau(\tau_0, \zeta_0))^{\mathrm{T}}, (u - nK_\zeta(\tau_0, \zeta_0))^{\mathrm{T}} \right) (i(\tau_0, \zeta_0))^{-1} \begin{pmatrix} t - nK_\tau(\tau_0, \zeta_0) \\ u - nK_\zeta(\tau_0, \zeta_0) \end{pmatrix} .$$

The relations between W, W_e and W_u and the expressions of W_e and W_u provide here a direct justification of the well-known asymptotic null distribution results

$$W, W_e, W_u \xrightarrow{d} \chi_p^2 .$$

When τ is the parameter of interest, to test

$$H_0 : \tau = \tau_0 \qquad \text{versus} \quad H_1 : \tau \neq \tau_0 ,$$

the likelihood ratio test can be based on one of the three, asymptotically equivalent, combinants (4.25), (4.31) and (4.32), evaluated at $\tau = \tau_0$. These take now the form

$$W_P(\tau) = 2 \left\{ (\hat{\tau} - \tau) \cdot t + (\hat{\zeta} - \hat{\zeta}_\tau) \cdot u + nK(\tau, \hat{\zeta}_\tau) - nK(\hat{\tau}, \hat{\zeta}) \right\}, \quad (5.73)$$

$$W_{uP}(\tau, \hat{\zeta}_\tau) = n^{-1} \left(t - nK_\tau(\tau, \hat{\zeta}_\tau) \right)^T \tilde{\imath}_1^{\tau\tau} \left(t - nK_\tau(\tau, \hat{\zeta}_\tau) \right), \quad (5.74)$$

$$W_{eP}(\tau, \hat{\zeta}_\tau) = n(\hat{\tau} - \tau)^T (\tilde{\imath}_1^{\tau\tau})^{-1} (\hat{\tau} - \tau), \quad (5.75)$$

where

$$W_P(\tau) = W_{uP}(\tau, \hat{\zeta}_\tau) + o_p(1) = W_{eP}(\tau, \hat{\zeta}_\tau) + o_p(1) .$$

The expansion $W_P(\tau) = W_{uP}(\tau, \hat{\zeta}_\tau) + o_p(1)$ provides the basic argument to show the asymptotic null distribution results

$$W_P(\tau), \, W_{uP}(\tau, \hat{\zeta}_\tau), \, W_{eP}(\tau, \hat{\zeta}_\tau) \xrightarrow{d} \chi_k^2 .$$

The following theorem offers an approximation for the conditional null distribution of T given $U = u$ (see the general comment at the end of section 5.8.2). This is a conditional central limit theorem, by Holst (1981).

Theorem 5.11 Consider random sampling of size n from one element of a regular exponential family with density (5.70), and let $T = \sum t(Y_i)$ and $U = \sum u(Y_i)$, then

$$\left(\frac{T}{\sqrt{n}} - E_\tau \left(\frac{T}{\sqrt{n}} \middle| U = u \right) \right) \xrightarrow{d} N_k \left(0, \mathrm{Var}_\tau \left(\frac{T}{\sqrt{n}} \middle| U = u \right) \right) .$$

Proof. See Holst (1981).   □

The approximations

$$E_\tau \left(\frac{T}{\sqrt{n}} \middle| U = u \right) = \sqrt{n} \tilde{K}_\tau + O(n^{-1/2}) ,$$

$$\mathrm{Var}_\tau \left(\frac{T}{\sqrt{n}} \middle| U = u \right) = (\tilde{\imath}_1^{\tau\tau})^{-1} + O(n^{-1}) ,$$

which will be obtained in section 10.10.2 (formulae (10.59) and (10.60)), might be useful, and, together with Theorem 5.11, give the asymptotic approximation

$$\frac{T}{\sqrt{n}} \middle| U = u \sim N_k \left(\sqrt{n} \tilde{K}_\tau, (\tilde{\imath}_1^{\tau\tau})^{-1} \right) .$$

5.10 Curved Exponential Families

Most of the discussion in this chapter has concentrated upon exponential families where both the vector of the natural observations, and the minimal sufficient statistic, have the same dimension as the parameter. In section 5.4.1, as a
comment on Definition 5.7, curved exponential families, $\mathcal{F}_e^{p,q}$, were mentioned.
In a curved exponential family the dimension p of the vector of natural observations is greater than the dimension q of the parameter space Φ. This class is
particularly interesting for data and model reduction. It has already been seen
that in a curved exponential family the maximum likelihood estimate cannot
be a sufficient statistic. Furthermore, since the vector of natural observations
is not necessarily complete, then there is the possibility of a non-trivial reduction by conditioning on an ancillary statistic, so as to recover the information
lost in the reduction to the maximum likelihood estimate (see section 2.8).

Firstly, remember that the elements of a curved exponential family $\mathcal{F}_e^{p,q}$,
have density

$$p(y;\phi) = \exp\{\theta(\phi) \cdot s(y) - G(\phi)\}h(y) , \qquad (5.76)$$

where $y \in \mathcal{Y} \subseteq \mathbb{R}^d$, $\phi \in \Phi \subseteq \mathbb{R}^q$, $s(y)$ and $\theta(\phi)$ have dimension $p > q$. Assume
that the representation (5.76) is minimal. An $\mathcal{F}_e^{p,q}$ can always be thought of
as a subfamily of a full exponential family of order p, $\mathcal{F}_e^p$, with density

$$p(y;\theta) = \exp\{\theta \cdot s(y) - K(\theta)\}h(y) , \qquad (5.77)$$

with $\theta \in \tilde{\Theta}$, where $\tilde{\Theta}$ is the natural parameter space

$$\tilde{\Theta} = \left\{ \theta \in \mathbb{R}^p : \int_{\mathcal{Y}} e^{\theta \cdot s(y)}h(y)d\mu^* < +\infty \right\} .$$

The normalizing constant $G(\phi)$ in (5.76) is, thus, $G(\phi) = K(\theta(\phi))$. Assume
in the following, that the function $\theta(\phi)$ defines a one-to-one and smooth correspondence between $\Phi \in \mathbb{R}^q$ and $\theta(\Phi) \subset \tilde{\Theta}$; furthermore, assume that the
matrix of first-order partial derivatives of $\theta(\phi)$ with respect to ϕ has full rank
q, for every $\phi \in \Phi$. The $\mathcal{F}_e^{p,q}$ with density (5.76) is said to be **embedded**
in the $\mathcal{F}_e^p$ with density (5.77). This latter is also called the **ambient family**.
The requirement that the representation (5.76) should be minimal makes it
impossible for $\theta(\Phi)$ to be a convex subset of $\tilde{\Theta}$.

From a general point of view, the study of curved exponential families
is interesting in that a sufficiently smooth parametric family can be locally
approximated, to a suitable order, by a curved exponential family (Efron,

1975). For simplicity, consider a generic one-parameter statistical model $\mathcal{F}$ specified as $(\mathcal{Y}, p(y; \phi), \Phi)$, with $\Phi \subseteq \mathbb{R}$, and let $\phi_0 \in \Phi$. Indicating by $l_j(\phi_0)$ the derivative of order j of the log-likelihood $l(\phi) = \log(p(y; \phi))$, evaluated at ϕ_0, consider the $\mathcal{F}_e^p$ with density

$$p^l(y; \theta) = \exp\{l(\phi_0) + \theta_1 l_1(\phi_0) + \ldots + \theta_p l_p(\phi_0) - K(\theta)\}, \qquad (5.78)$$

where $K(\theta)$ is a normalizing constant. With the constraint $\theta_j = (\phi - \phi_0)^j / j!$, $j = 1, \ldots, p$, the densities (5.78), for $\phi \in \Phi$, form a $(p, 1)$ curved exponential family, such that $p^l(y; \theta(\phi_0)) = p(y; \phi_0)$, whose log-likelihood has derivative $l_j(\phi_0)$ at $\theta(\phi_0)$. Thus, we could say that the derivatives of the log-likelihood evaluated at ϕ_0 are a locally sufficient statistic.

One can establish a link between a generic parametric family and a curved exponential family from another point of view too. It is possible, by means of a finite discretization of the support, to represent a parametric family with a q-dimensional parameter by a multinomial distribution, whose cell probabilities are a function of the parameter of the model. Therefore, if the number $p+1$ of the multinomial cells is greater than $q + 1$, a (p, q) curved exponential family is obtained.

The $\mathcal{F}_e^{p,q}$ class is closed under random sampling. If $Y_1, \ldots, Y_n$ are independent copies of a random variable with density (5.76), it is immediate to check that the joint density of $(Y_1, \ldots, Y_n)$ still belongs to an $\mathcal{F}_e^{p,q}$. Furthermore, $\sum_{i=1}^n s(y_i)$ is a p-dimensional sufficient statistic, which is also minimal if $n \geq p$.

The natural observation $s(y)$ has density

$$p_S(s; \phi) = \exp\{\theta(\phi) \cdot s - K(\theta(\phi))\} h_s(s) \qquad (5.79)$$

and the corresponding cumulant generating function is

$$\begin{aligned} K_S(t; \phi) &= \log\left(\int_S e^{(\theta(\phi)+t)\cdot s - K(\theta(\phi))} h_s(s) d\mu^*\right) \\ &= K(\theta(\phi) + t) - K(\theta(\phi)). \end{aligned} \qquad (5.80)$$

In the following, indices $r, s, \ldots$, with $r, s, \ldots = 1, \ldots, q$, are used to indicate the components of ϕ; in addition, $a, b, \ldots$, with $a, b, \ldots = 1, \ldots, p$, are used as indices for θ and $s(y)$. For every ϕ such that $\theta(\phi) \in \text{int } \tilde{\Theta}$, the generic cumulant of order m of $s(Y)$ is

$$\kappa^{a_1, \ldots, a_m} = \left. \frac{\partial^m}{\partial \theta^{a_1} \ldots \partial \theta^{a_m}} K(\theta) \right|_{\theta = \theta(\phi)}. \qquad (5.81)$$

In particular, $s(Y)$ has expectation

$$E_\phi\{s(Y)\} = [\kappa^a(\theta(\phi))] = \left.\frac{\partial}{\partial\theta}K(\theta)\right|_{\theta=\theta(\phi)} = \mu(\theta(\phi)), \qquad (5.82)$$

where $\mu(\theta)$ denotes the mean value parameter of the embedding $\mathcal{F}_e^p$. Restricting the natural parameter space to $\theta(\Phi)$ entails that the expectation space is restricted to the subset $M_\Phi = \mu(\theta(\Phi))$ of $M = \mu(\text{int}\,\tilde{\Theta})$.

Example 5.16 *Normal parabola*
Let $y = (y_1, \ldots, y_n)$ be n independent observations from a $N(\mu, \sigma^2)$ distribution. With the natural parameterization, the joint density is

$$p_Y(y; \theta) = \exp\left\{\theta_1 \sum y_i + \theta_2 \sum y_i^2 - n\left(-\frac{\theta_1^2}{2\theta_2} - \log(-2\theta_2)\right)/2\right\},$$

where $(\theta_1, \theta_2) \in \mathbb{R} \times (\mathbb{R}^- \setminus \{0\})$, $\theta_1 = \mu/\sigma^2$, $\theta_2 = -1/(2\sigma^2)$. The natural observations $(\sum y_i, \sum y_i^2)$ have mean

$$\mu(\theta) = \left(-n\frac{\theta_1}{2\theta_2}, -\frac{n}{2}\left(\frac{1}{\theta_2} - \frac{\theta_1^2}{2\theta_2^2}\right)\right).$$

The normal parabola is the curved subfamily obtained by imposing the constraint $\mu = \sigma > 0$. Let $\phi = 1/\sigma$; the restriction imposed on the natural parameter space $\tilde{\Theta} = \mathbb{R} \times (\mathbb{R}^- \setminus \{0\})$ is given by the equation $(\theta_1, \theta_2) = (\phi, -\phi^2/2)$: this is one branch of a parabola. The corresponding restriction in the expectation space is $M_\Phi = \{(\mu_1, \mu_2) : \mu_2 = 2\mu_1^2/n, \mu_1 > 0\}$. $\qquad\triangle$

In the statistical model with density (5.79), the log-likelihood function is

$$l(\phi) = \theta(\phi) \cdot s - K(\theta(\phi)). \qquad (5.83)$$

The corresponding score vector has elements

$$l_r(\phi) = \sum_{a=1}^{p} \{s^a - \kappa^a(\theta(\phi))\}\,\theta_r^a(\phi), \qquad (5.84)$$

where s^a indicates the generic component of s and $\theta_r^a(\phi) = (\partial/\partial\phi^r)\theta^a(\phi)$. Differentiating (5.84) with respect to ϕ^s, gives the generic element of the observed information matrix

$$j_{rs} = -l_{rs}(\phi) = \sum_{a,b=1}^{p} \kappa^{a,b}(\theta(\phi))\theta_r^a(\phi)\theta_s^b(\phi) - \sum_{a=1}^{p}\{s^a - \kappa^a(\theta(\phi))\}\,\theta_{rs}^a(\phi),$$

$$(5.85)$$

where $\theta_{rs}^a(\phi) = (\partial^2/\partial\phi^r\partial\phi^s)\theta^a(\phi)$. Therefore, (see (5.82)) the expected information matrix has the generic element

$$i_{rs} = \sum_{a,b=1}^p \kappa^{a,b}(\theta(\phi))\theta_r^a(\phi)\theta_s^b(\phi) . \qquad (5.86)$$

Note the formal analogy between these formulae and those obtained in section 1.4.3 with reference to reparameterization.

Let C indicate the closed convex hull of the support of the natural observations $s(y)$; this set is the same both for the curved family $\mathcal{F}_e^{p,q}$ and for the corresponding ambient family. However, because of the non-linear constraints in the natural parameter space, there is no longer a one-to-one correspondence between int C and the expectation space. Therefore, an observed value $s(y)$ will not necessarily coincide with the mean of an element of the curved family. This has immediate consequences for the interpretation of the maximum likelihood estimates in an $\mathcal{F}_e^{p,q}$. In particular, we can guess that the maximum likelihood estimate should be identified as the point in M_Φ which has minimum distance, in a suitable metric, from the observed value of s.

A sufficient condition for the existence of the maximum likelihood estimate of ϕ is that there exists the maximum likelihood estimate of θ in the ambient family, for the observed value of s. This is the same as requiring that the $\mathcal{F}_e^p$ ambient family is steep and that $s \in$ int C. For details, see Brown (1986, Theorem 5.7).

From (5.84) the likelihood equation is

$$\sum_{a=1}^p \left\{s^a - \kappa^a(\theta(\phi))\right\} \theta_r^a(\phi) = 0 , \qquad r = 1,\ldots,q . \qquad (5.87)$$

The vectors $[\theta_1^a],\ldots,[\theta_q^a]$ are a basis of the q-dimensional tangent space to the curve $\theta(\phi)$, for every ϕ. Thus, equation (5.87) expresses an orthogonality condition, at $\hat\phi$, between this tangent space and the vector $[s^a - \kappa^a(\theta(\phi))]$. In other words, for a given s, a value ϕ on the curve $\theta(\Phi)$ could be a maximum likelihood estimate only if the difference vector of s and the mean $\mu(\theta(\phi))$ is orthogonal to the curve at that point. The example below will illustrate this interpretation.

Example 5.17 *Normal parabola (cont.)*
With $(s_1, s_2) = (\sum y_i, \sum y_i^2)$, since $((\partial/\partial\phi)\theta_1, (\partial/\partial\phi)\theta_2) = (1, -\phi)$, the likeli-

hood equation for ϕ is

$$\left(s_1 - \frac{n}{\phi}\right) + \left(s_2 - \frac{2n}{\phi^2}\right)(-\phi) = 0 . \qquad (5.88)$$

Since the vectors $(1, -\phi)$ and $(1, 1/\phi)$ are orthogonal, (5.88) is satisfied if and only if the vectors $(s_1 - n/\phi, s_2 - 2n/\phi^2)$ and $(1, 1/\phi)$ are parallel. This means that the values (s_1, s_2) which give the same m.l.e. $\hat{\phi}$ can be represented geometrically as points on the straight line applied to $\mu(\theta(\hat{\phi})) = (n/\hat{\phi}, 2n/\hat{\phi}^2)$ with direction $(1, 1/\hat{\phi})$; this has equation $s_2 = n/\hat{\phi}^2 + s_1/\hat{\phi}$. $\qquad \triangle$

For a fixed value of s, the likelihood equation (5.87) can, usually, have more than one solution. Vice versa, a fixed value of ϕ is the maximum likelihood estimate for all the values of s which belong to the $(p - q)$-dimensional linear subspace defined by the q linear constraints (5.87).

The orthogonality condition (5.87), when referred to the expectation space, may be written as

$$(s - \mu(\phi))^{\mathrm{T}} \frac{\partial \theta}{\partial \mu}\bigg|_{\mu=\mu(\phi)} \frac{\partial \mu}{\partial \phi} = 0 . \qquad (5.89)$$

Remember that

$$\frac{\partial \theta}{\partial \mu}\bigg|_{\mu=\mu(\phi)} = [\kappa^{a,b}(\theta(\phi))]^{-1} = (\mathrm{Var}_\phi(S))^{-1} .$$

Therefore, by (5.89) the vector $s - \mu(\phi)$ and the curve M_Φ should be orthogonal in the metric defined by the inverse of the covariance matrix of S, $\mathrm{Var}_\phi(S)$. In this metric, $\hat{\phi}$ corresponds to the point in M_Φ which has minimum distance from the observed value of s. Thus, $\mu(\hat{\phi})$ is the orthogonal projection of s on M_Φ if the matrix of the scalar product is $(\mathrm{Var}_\phi(S))^{-1}$. Observe that the usual concepts of vector space and scalar product are being used here in a local sense: the tangent space to the curve M_Φ and the metric vary as ϕ varies in Φ.

Because $\hat{\phi}$ is not a sufficient statistic, it would be reasonable to ask whether there exist general criteria, under an $\mathcal{F}_e^{p,q}$, for finding ancillary statistics. From an abstract point of view, the construction of an ancillary is clear with reference to the space of natural observations and to its subspace M_Φ. A curved coordinate system in the natural observation space should be found, with M_Φ as a component; the remaining component identifies an auxiliary statistic, which is ancillary if it is distribution constant. Specific instances have been illustrated in Examples 2.12, 2.13 and 2.14. However, there are no general results, at least, not unless we abandon the requirement of exact ancillarity and seek approximate solutions (cf. section 2.8). One of the best-known proposals is the Efron–Hinkley ancillary (Efron and Hinkley, 1978). Here we will only give the main result; further details can be found in the bibliographic note, section 5.11.

Consider for simplicity a one-parameter model with parameter ϕ, and consider the approximating curved exponential family (5.78), with $\phi_0 = \hat{\phi}$. Then $(\hat{\phi}, \hat{\jmath})$ is a locally sufficient statistic. The Efron–Hinkley ancillary statistic is, essentially, an asymptotic standardization of the ratio between observed and expected information, evaluated at $\hat{\phi}$. This gives the function of $(\hat{\phi}, \hat{\jmath})$

$$a_{EH} = \left(\frac{\hat{\jmath}}{i} - 1\right) \bigg/ \sqrt{\operatorname{Var}_{\hat{\phi}}\left(\frac{\hat{\jmath}}{i}\right)} \doteq \left(\frac{\hat{\jmath}}{i} - 1\right) \big/ \hat{\gamma} \,, \tag{5.90}$$

where

$$\gamma = \gamma(\phi) = \frac{1}{i^{3/2}} \left\{ i(\nu_{2,2} - i^2) - \nu_{1,2}^2 \right\} \,, \tag{5.91}$$

with $\nu_{j,k} = E_\phi(l_j(\phi)l_k(\phi))$, $\nu_j = E_\phi(l_j(\phi))$. It is not difficult to show (Efron and Hinkley, 1978, section 8) that under random sampling of size n, the statistic a_{EH} has standard normal asymptotic distribution with error of order $O(n^{-1/2})$, for every value of ϕ. Therefore, it is asymptotically distribution constant.

If the statistical model is an $\mathcal{F}_e^{p,1}$, the quantity γ has an interesting geometrical interpretation. In fact it represents the curvature in $\mathbb{R}^p$ of $\theta(\phi)$, using the covariance matrix of the natural observations of the ambient $\mathcal{F}_e^p$ as the matrix of the scalar product, that is, the matrix $K_{\theta\theta} = [\kappa^{a,b}(\theta(\phi))]$. Let $\theta'(\phi)$ and $\theta''(\phi)$ denote, respectively, the column vectors of first and second-order partial derivatives of $\theta(\phi)$ with respect to ϕ. Consider the symmetric 2×2 matrix $D = [d_{ij}]$, with $d_{11} = \theta'(\phi)^{\mathrm{T}} K_{\theta\theta} \theta'(\phi)$, $d_{12} = d_{21} = \theta'(\phi)^{\mathrm{T}} K_{\theta\theta} \theta''(\phi)$, $d_{22} = \theta''(\phi)^{\mathrm{T}} K_{\theta\theta} \theta''(\phi)$. The curvature of $\theta(\phi)$ at ϕ is given by the rate of variation of the direction of $\theta(\phi)$ with respect to the length of the cord that joins the two points. It is not difficult to show (see, for example, Efron, 1975, section 2) that, in the metric induced by $K_{\theta\theta}$, the curvature is equal to

$$\gamma = \gamma(\phi) = \sqrt{|D|/(d_{11})^3} \tag{5.92}$$

and that the quantities defined by (5.91) and (5.92) are the same. The curvature is zero in a full exponential family, that is, for $p = 1$. This provides a further reason for justifying the analogy between exponential families and straight lines, which was mentioned as a comment to Example 5.5. For the definitions of curvature and of their statistical interpretation in the multiparameter case, see Amari (1988) and Kass (1989). Research on conditioning on approximate ancillary statistics is ongoing, but important stimulus has come from the development of asymptotic techniques for the approximation of distributions (see Chapters 10 and 11). Later on, attention will mainly be paid to the concept of stable statistical procedures, which are of special interest for applications (cf. section 2.8).

5.11 Bibliographic Note

Fisher (1934) observed that, under regularity conditions, non-trivial reduction by sufficiency is only possible in exponential families. This statement was later taken up by other authors and shown, fully, by Dynkin (1951). The main references for the material dealt with in this chapter are Barndorff-Nielsen (1978) and Brown (1986). For the theory of testing hypotheses, the classic source is Lehmann (1986, Chapters 4 and 5). For an alternative method for introducing an $\mathcal{F}_e^1$, see Example 2.4.

Tweedie (1957) carried out the first detailed study of the probabilistic and statistical properties of the inverse Gaussian distribution. Two recent monographs on this distribution are Chhikara and Folks (1989) and Seshadri (1993); one example of the use of the inverse Gaussian in applications is given by Desmond and Chapman (1993).

Letac and Mora (1990) have identified all the natural exponential families with cubic variance function, thus extending the original classification by Morris (1982). From the point of view of applications, the variance function highlights an element of rigidity associated with the use of an $\mathcal{F}_{ne}^1$ as a statistical model: the variance is determined by the mean. Exponential dispersion families, which will be discussed in the next chapter, are a more flexible class of parametric models. Efron (1986b), Gelfand and Dalal (1990) proposed further parametric families again in the attempt to improve flexibility. See also the bibliographic note to Chapter 6.

The appropriateness of conditioning, that is the basis for the Fisher's exact test in 2×2 tables, has been the subject of a long debate: see Yates (1984) and Upton (1992) for some general comments.

Curved exponential families were first introduced and studied by Efron (1975, 1978). For a formalization using differential geometry, see Amari (1985, part II). Methods for constructing approximate ancillaries in curved exponential families are described e.g. in Barndorff-Nielsen and Cox (1994, sections 7.2 and 7.3) and in McCullagh (1987, section 8.3); see also Amari (1985, Chapter 7). The idea that curved exponential families could be adopted as approximating models for any parametric model was suggested by Efron (1975). For formulations that exploit differential geometry, see Amari (1987), and Barndorff-Nielsen and Jupp (1989).

5.12   Problems

5.1 Show that the set $\tilde{\Theta}$ defined by (5.1) is convex (in this case an interval); in other words, if $M_0(\theta_0) < +\infty$ and $M_0(\theta_1) < +\infty$, with $\theta_0 < \theta_1$, then $M_0(\theta) < +\infty$ for every $\theta \in [\theta_0, \theta_1]$ too. (*Hint*: write $\theta = \alpha\theta_0 + (1 - \alpha)\theta_1, \alpha \in [0, 1]$, and use the inequality between the geometric and the arithmetic mean.)

[Section 5.2]

5.2 Given an $\mathcal{F}_{ne}^1$ with density $p(y; \theta) = e^{\theta y - \theta^2/2} p_0(y)$, identify $p_0(y)$ and the support $\mathcal{Y}$.

[Section 5.2]

5.3 Obtain the $\mathcal{F}_{ne}^1$ generated by the binomial distribution $Bi(n, 1/2)$.

[Section 5.2]

5.4 Say whether it is possible to construct a non-degenerate $\mathcal{F}_{ne}^1$ taking the log-normal distribution which corresponds to the $N(0, 1)$ as the generating element.

[Section 5.2]

5.5 Calculate the cumulants, up to the fourth order, for the $\mathcal{F}_e^1$ given in Table 5.1. Examine what happens in correspondence with the boundary points of $\tilde{\Theta}$ whenever the family is not regular.

[Section 5.3]

5.6 Let Y be a positive non-degenerate r.v.. Let $U = Y^s$, $s > 0$. Using Jensen's inequality, show that, if $\alpha > 1$ or $\alpha < -1$, $E(U^\alpha) > \{E(U)\}^\alpha$. Put $\alpha = t/s$, with $t > s$, or with $t < 0$ and $|t| > s$; deduce that $\{E(Y^r)\}^{1/r}$ is a strictly increasing function of r.

[Section 5.3]

5.7 Under the same assumptions as in Problem 5.6, show that, if Y has finite moments of any order, the limit of $\{E(Y^r)\}^{1/r}$ as $r \to +\infty$ is equal to the upper bound of the support of Y. (*Hint*: Use Markov inequality for $E(Y^r)$.)

[Section 5.3]

5.8 Show that the family $IG(\phi, \lambda)$, with fixed $\lambda > 0$, (see table 5.1) which is non-regular, is, nonetheless, steep.

[Section 5.3]

5.9 Let Z be a r.v. with density $p_z(z) = 2\exp(-2z)$, for $z > 0$. Let $Y = e^Z$. Consider the $\mathcal{F}^1_{ne}$ generated by $p_Y(y)$. Show that this family is not steep.

[Section 5.3]

5.10 Obtain the variance functions given in Table 5.2.

[Section 5.3]

5.11 Check that the variance function associated with the negative binomial distribution is quadratic.

[Section 5.3]

5.12 Determine the natural exponential family with variance function

$$(M, V(\mu)) = ((-\infty, 0), \mu^2) .$$

[Section 5.3]

5.13 Show that, if Y has distribution belonging to an $\mathcal{F}^1_{ne}$ with a quadratic variance function, the random variable $Y^* = (Y - b)/c$, $c \neq 0$, also has distribution belonging to an $\mathcal{F}^1_{ne}$ with a quadratic variance function.

[Section 5.3]

5.14 Let Y be a one-dimensional r.v. with density belonging to an $\mathcal{F}^p_{ne}$. Let $c \in \text{int } \mathcal{Y}$ be a given constant. Show that the statistical model for $Z = \min(Y, c)$ is still an $\mathcal{F}^p_{ne}$. Determine the c.g.f. of Z.

[Section 5.4.1]

5.15 Let (Y, δ) be a mixed type r.v. like those described in Example 1.2. In particular, let $Y = \min(T, c)$, where T is a r.v. with density $p_T(t) = e^{-t}I_{[0,+\infty)}(t)$ and c is a known positive constant; remember that $\delta = I_{(0,c)}(T)$. Let

$$(s_1(y, \delta), s_2(y, \delta)) = (\delta y + (1 - \delta)c, \delta)$$

and indicate the joint density of $(s_1(Y, \delta), s_2(Y, \delta))$ by $p_0(s_1, s_2)$. Obtain the $\mathcal{F}^2_{ne}$ generated by $p_0(s_1, s_2)$.

[Section 5.4.1]

5.16 Obtain the mean vectors and the covariance matrices for the natural observations of the $\mathcal{F}_e^2$ considered in the Examples 5.9 and 5.10.

[Section 5.4.1]

5.17 Show that a natural exponential family of order p is closed under exponential tilting, that is, that the natural exponential family of order p generated by any element of a given $\mathcal{F}_{n_e}^p$ coincides with the initial $\mathcal{F}_{n_e}^p$.

[Section 5.4.1]

5.18 State in which examples in section 4.5 the conditional likelihood obtained is a special case of (5.33). Find the corresponding $\mathcal{F}_{n_e}^p$ and the function $K_u(\tau)$ for each case.

[Section 5.4.2]

5.19 Consider the model induced by random sampling of size n from each of the exponential families in Table 5.1. Find the minimal sufficient statistic and see if it is complete.

[Section 5.5]

5.20 Consider the von Mises distribution introduced in Example 2.14. Show that, for a fixed λ, this is a $(2,1)$ curved exponential family and that this structure is maintained under random sampling of size n. Furthermore, show that the family, induced by the natural observations, is not complete.

[Section 5.5]

5.21 Consider the n-dimensional distribution in Example 5.16. Show that the minimal sufficient statistic is not complete.

[Section 5.5]

5.22 Study the problem of the existence and the uniqueness of the m.l.e. for the exponential families described in Table 5.1 considering separately the two cases:
i) a single observation;
ii) random sampling of size $n > 1$.

[Section 5.6]

5.23 Let Y_1, Y_2, Y_3 be independent random variables, with $Y_i \sim Bi(1, \phi_i)$, where $\phi_i = e^{\theta^1 + \theta^2 x_i}/(1 + e^{\theta^1 + \theta^2 x_i})$ and $x_1 = -1$, $x_2 = 0$, $x_3 = 1$. Find the minimal sufficient statistic s and say for which values of s the m.l.e. of $\theta = (\theta^1, \theta^2)$ does exist.

[Section 5.6]

5.24 Obtain the representation (5.45) of the log-likelihood under random sampling from each of the exponential families described in Table 5.1.

[Section 5.6]

5.25 Check that $j_p(\tau)$ as defined by (5.53) coincides with the covariance matrix of the residuals of the linear regression of t on u, evaluated at $(\tau, \hat{\zeta}_\tau)$.

[Section 5.7]

5.26 With reference to Example 5.11, find the expression of the profile likelihood for τ and compare it with the conditional likelihood.

5.27 Check that (5.54) defines a reparameterization.

[Section 5.7]

5.28 Under the same assumptions as in Example 5.13, obtain the expression of the approximate conditional log-likelihood (5.56) if the parameter of interest is ν. Using Stirling's approximation (see Example 9.8), check that, with respect to the conditional log-likelihood (5.36), the error is of order $O(n^{-1})$.

[Section 5.7]

5.29 Obtain the uniformly most powerful unbiased test for problems (5.57) and (5.60) under random sampling of size n for each of the exponential families in Table 5.1. In particular, for $Ga(\nu, \phi)$ obtain both the test on ν with ϕ known and that on ϕ with ν known. Proceed in the same way with $IG(\phi, \lambda)$. For each distribution, provide some indication on finding the exact null distribution (for the inverse Gaussian, see Seshadri, 1993).

[Section 5.8.1]

5.30 Obtain the likelihood ratio tests for the same testing problems as in Problem 5.29. Compare the tests obtained with the optimal tests.

[Section 5.8.1]

5.31 Using the results in Example 5.13, find the cumulant generating function of $-\log(R)$, where R is defined by (5.69).

[Section 5.8.2]

5.32 Under the same assumptions and notation as in Example 5.11, obtain the uniformly most powerful level α similar test for equality of two Poisson distributions, that is, for

$$H_0 : \mu = \nu \quad \text{versus} \quad H_1 : \mu \neq \nu .$$

[Section 5.8.2]

5.33 Under the same assumptions and notation as in Example 5.12, obtain the uniformly most powerful level α similar test for equality of two binomial distributions with the same index, that is, for

$$H_0 : \phi_1 = \phi_2 \quad \text{versus} \quad H_1 : \phi_1 \neq \phi_2 .$$

[Section 5.8.2]

5.34 Study the relationship between the optimal tests given by (5.65) and (5.67) and the tests based on the profile likelihood.

[Section 5.8.2]

5.35 Let $y_1, \ldots, y_n$ be independent realizations of a r.v. with support $\mathcal{Y} \subseteq \mathbb{R}$ and with density $p_Y(y)$ with respect to a dominating measure μ^*. Consider the problem of testing $H_0 : p_Y(y) \in \mathcal{F}_e^p$ against $H_1 : p_Y(y) \in \mathcal{F}_e^q$, where $\mathcal{F}_e^p$ and $\mathcal{F}_e^q$ are two exponential families of order p and q, respectively. Suppose that $\mathcal{F}_e^p$ and $\mathcal{F}_e^q$ are **separate families**, that is, are such that no element of $\mathcal{F}_e^p$ can be obtained as the limit in distribution of elements of $\mathcal{F}_e^q$. Let

$$p(y; \zeta) = e^{\zeta \cdot s(y) - K(\zeta)} h(y) , \qquad \zeta \in \tilde{Z} \subseteq \mathbb{R}^p ,$$

and

$$p(y; \phi) = e^{\phi \cdot t(y) - \Lambda(\phi)} k(y) , \qquad \phi \in \tilde{\Phi} \subseteq \mathbb{R}^q ,$$

be the densities of a generic element of $\mathcal{F}_e^p$ or of $\mathcal{F}_e^q$, respectively. Let $\mathcal{F}$ be the union family $\mathcal{F}_e^p \bigcup \mathcal{F}_e^q$, with density

$$p_Y(y) = c(\zeta, \phi, \tau) p(y; \zeta)^{1-\tau} p(y; \phi)^\tau , \qquad \tau = 0, 1 , \tag{5.93}$$

where $c(\zeta, \phi, \tau)$ is a normalizing constant. Find a minimal sufficient statistic for $\mathcal{F}$. Show that (5.93) is a probability density for every $\tau \in [0, 1]$ and indicate with $\bar{\mathcal{F}}$ the corresponding extended family. Show that $\bar{\mathcal{F}}$ has the same reduction by sufficiency as $\mathcal{F}$. Determine the conditions for the existence of a level α uniformly most powerful similar test for $H_0 : \tau = 0$ versus $H_1 : \tau > 0$ and give the expression of the test statistic. Show that the test obtained is level α uniformly most powerful similar for the original problem.

[Section 5.8.2; Pace and Salvan, 1990]

5.36 Let $Y_1, \ldots, Y_n$ be i.i.d. random variables with continuous distribution on $\mathbb{R}$ and density in an $\mathcal{F}_e^2$, of the form

$$p(y; \tau, \zeta) = \exp \left\{ \tau t(y) + \zeta u(y) - K(\tau, \zeta) \right\} h(y) ,$$

where $\theta = (\tau, \zeta) \in \tilde{\Theta}$ and $\tilde{\Theta}$ is the natural parameter space. Let (τ, μ_u) be the mixed parameterization. Assume that ζ may be represented as $\zeta = -\tau g'(\mu_u)$, with $g(\cdot)$ a suitable function. Show that, if $u(\cdot)$ is one-to-one, then the r.v.'s $U_n = \sum u(Y_i)$ and $\bar{T}_n = \sum t(Y_i) - ng(\sum u(Y_i)/n)$ are independent. Furthermore, the distribution of $\bar{T}_n$ does not depend on ζ. Discuss the value of this result to define uniformly most powerful similar tests based on $\bar{T}_n$ for hypotheses on τ.

[Section 5.8.2; Bar-Lev and Reiser, 1982]

5.37 Obtain an approximation for the null distribution of the optimal test for $H_0 : \tau \leq \tau_0$ versus $H_1 : \tau > \tau_0$ under the same assumptions as in Example 5.14. Compare the resulting approximate test with the one-sided test Z_{u_P} (4.35) based on the profile likelihood.

[Section 5.9]

5.38 Obtain an approximation for the null distribution of the optimal test for $H_0 : \lambda \leq \lambda_0$ versus $H_1 : \lambda > \lambda_0$ under random sampling of size n from a $IG(\phi, \lambda)$ distribution. Compare the resulting approximate test with the one-sided test Z_{u_P} (4.35) obtained from the profile likelihood.

[Section 5.9]

5.39 Let $Y = (Y_1, Y_2, Y_3)$, with $Y_1 + Y_2 + Y_3 = n$, be distributed according to a trinomial distribution with index n and parameter (π_1, π_2). This is an element of the $\mathcal{F}_e^2$ with density

$$p(y; \pi_1, \pi_2) = \frac{n!}{y_1! y_2! y_3!} \pi_1^{y_1} \pi_2^{y_2} (1 - \pi_1 - \pi_2)^{y_3} ,$$

where $\pi_1, \pi_2 > 0$ and $\pi_1 + \pi_2 < 1$. Identify the natural observations and the natural parameters. The Hardy–Weinberg model is defined as a subfamily of such an $\mathcal{F}_e^2$ obtained with the constraints $\pi_1 = \phi^2$, $\pi_2 = 2\phi(1 - \phi)$, with $\phi \in (0, 1)$. Show that the constraint introduced into the natural parameter space is linear and consequently the corresponding subfamily, written in the minimal form, is an $\mathcal{F}_e^1$.

[Section 5.10]

5.40 Consider Example 1.4 and, in particular, the homogeneous Poisson process that gives the likelihood (1.27); assume that t_0 is a realization of a positive random variable with distribution independent of λ. Show that the likelihood (1.27) is that of an $\mathcal{F}_e^{2,1}$. Find a minimal sufficient statistic in the form $(\hat{\lambda}, a)$, with a ancillary.

[Section 5.10]

5.41 Consider Example 1.5 and the likelihood function (1.28); assume that t_0 is a realization of a positive random variable with distribution independent of μ. Show that the likelihood (1.28) is that of an $\mathcal{F}_e^{2,1}$. Find the minimal sufficient statistic in the form $(\hat{\mu}, a)$, with a ancillary.

[Section 5.10]

5.42 With reference to the $\mathcal{F}_{ne}^2$ obtained in Problem 5.15, consider the curved subfamily defined by $(\theta_1, \theta_2) = (-\phi, -\log(\phi))$, with $\phi > 0$. Show that the class of densities thus obtained correspond to the r.v. (S_1, S_2) if T has density $p_T(t; \phi) = \phi e^{-\phi t} I_{[0,+\infty)}(t)$. Write the likelihood equation for ϕ in the form (5.87) and discuss its geometrical interpretation.

[Section 5.10; Brown, 1986, section 5.14]

5.43 Show that the quantity γ defined by (5.91) is parameterization invariant.

[Section 5.10]

CHAPTER 6

EXPONENTIAL DISPERSION FAMILIES AND GENERALIZED LINEAR MODELS

6.1 Introduction

The class of dispersion families, first introduced by Jørgensen (1983), is a broad extension of exponential families. A dispersion family has densities of the form

$$p(y; \theta, \lambda) = a(\lambda, y) \exp\{\lambda q(y; \theta)\} , \qquad (6.1)$$

where λ is a scalar parameter and θ is p-dimensional. Exponential dispersion families (Jørgensen, 1987), where $q(y; \theta) = \theta \cdot y - K(\theta)$, are a subclass of dispersion families, that still represents an extension with respect to exponential families. Exponential dispersion families provide a large class of probability distributions for errors in generalized linear models and make it possible for the latter to be treated in a unified manner.

Generalized linear models are a class of parametric regression models introduced by Nelder and Wedderburn (1972), with the aim of extending the usual normal linear regression model. With respect to this latter, generalization concerns not only the distribution of errors, but also the possibility of a non-linear relation between the mean response and a linear combination of the covariates. Because of the close link with the theory of exponential families, these models represent one of the most important examples of parametric statistical models which combine flexibility for applications with simplicity of inference.

As with exponential families, abstract generation of exponential dispersion families takes place uniquely starting from a given moment generating function. Extension of an $\mathcal{F}_{ne}^p$ is obtained through the introduction of an additional parameter, with independent variation with respect to natural parameters. Although the extended family has the same mean value function as the original

$\mathcal{F}^p_{ne}$, its covariance matrix is modified by a scale factor. As was mentioned in section 5.11, this is one way of overcoming one aspect of the inherent rigidity of exponential family models, which arises because of the mean-(co)variance relation.

6.2 Exponential Dispersion Families

Let X be a non-degenerate random variable with density $p_0(x)$, $x \in \mathsf{X} \subseteq \mathbb{R}^p$. Let $K(\theta) = \log(E_0(e^{\theta X}))$ be the cumulant generating function of X, which is assumed to exist. The $\mathcal{F}^p_{ne}$ generated by $p_0(x)$ has density

$$p(x; \theta) = e^{\theta \cdot x - K(\theta)} p_0(x), \quad \theta \in \tilde{\Theta}, \tag{6.2}$$

where $\tilde{\Theta}$ is the natural parameter space.

The construction of exponential dispersion families can be introduced as follows. The sample mean $\bar{X}_n$ of n independent copies of a random variable distributed as in (6.2) has density

$$p(x; \theta) = e^{n(\theta \cdot x - K(\theta))} h_n(x) . \tag{6.3}$$

$\bar{X}_n$ still has expectation $\mu(\theta) = (\partial/\partial\theta) K(\theta)$ and covariance matrix $\mathrm{Var}(\bar{X}_n) = n^{-1}(\partial/\partial\theta^{\mathrm{T}})\mu(\theta) = n^{-1}(\partial^2/\partial\theta\partial\theta^{\mathrm{T}})K(\theta)$, that is, $\bar{X}_n$ has the same expectation as X and a covariance matrix which is modified by a scale factor n^{-1}. Thus we could think of defining a parametric family with a covariance matrix not uniquely defined by the mean, by replacing n in (6.3) with a parameter $\lambda > 0$, which can be interpreted as a calibrating factor for dispersion, and is analogous to sample size. However, it should be made clear that, while a distribution with cumulant generating function $nK(\theta)$ certainly exists for every $n \in \mathbb{N} \setminus \{0\}$, there is no guarantee that $\lambda K(\theta)$ will be a cumulant generating function of a random variable, for every $\lambda \in \mathbb{R}^+ \setminus \{0\}$. This is especially true for $\lambda \in (0, 1)$.

Firstly, consider the set Λ of the values $\lambda > 0$ such that the function $\lambda K(\theta)$ is the cumulant generating function of a distribution with density $p_0(x; \lambda)$, $x \in \mathsf{X} \subseteq \mathbb{R}^p$. The set Λ is closed with respect to the sum: if λ_1 and λ_2 are elements of Λ, so too is $\lambda_1 + \lambda_2$ an element of Λ. This property is a consequence of the closure under convolution of the family of densities $\{p_0(x; \lambda), \lambda \in \Lambda\}$, which can be seen as being generated by convolution starting from $p_0(x) = p_0(x; 1)$. Next, consider the $\mathcal{F}^p_{ne}$ generated, through exponential tilting, by $p_0(x; \lambda)$ which has density

$$p(x; \theta, \lambda) = \exp\{\theta \cdot x - \lambda K(\theta)\} p_0(x; \lambda), \quad x \in \mathsf{X} \subseteq \mathbb{R}^p . \tag{6.4}$$

Lastly, the density of $Y = X/\lambda$, where X is distributed according to (6.4), is

$$p_Y(y; \theta, \lambda) = \exp\{\lambda\{\theta \cdot y - K(\theta)\}\}\tilde{p}_0(y; \lambda), \qquad y \in \mathcal{Y}_\lambda \subseteq \mathbb{R}^p, \qquad (6.5)$$

where $\tilde{p}_0(y; \lambda)$ represents the density of Y when X has density $p_0(x; \lambda)$. The support $\mathcal{Y}_\lambda$ of $Y = X/\lambda$ usually depends on λ. If, however, the support of X is $\mathbb{R}^p$ or is a cone in $\mathbb{R}^p$ (X is such that if $x \in X$ also $\lambda x \in X$, for every $\lambda > 0$), then $\mathcal{Y}_\lambda = \mathcal{Y} = X$. This is the very interesting case for statistical modelling, though from the outset it excludes distributions for integer-valued data, unless the precision parameter λ is considered as fixed (see Examples 6.5 and 6.6). On the other hand, the definition of quasi-likelihood in section 6.4.2 is motivated by the class (6.5) and gives a tool for inference suitable for discrete data too.

By construction, the parameters θ and λ are variation independent, that is, $(\theta, \lambda) \in \tilde{\Theta} \times \Lambda$. Without essential loss of generality with respect to (6.5), we give the following definition, wherein it is assumed that the support does not depend on λ.

Definition 6.1 A class of densities

$$\mathcal{F}^p_{ed} = \left\{ p(y; \theta, \lambda) = a(\lambda, y)e^{\lambda\{\theta \cdot y - K(\theta)\}}, \; y \in \mathcal{Y} \subseteq \mathbb{R}^p, \; \theta \in \tilde{\Theta}, \; \lambda \in \Lambda \right\}, \quad (6.6)$$

with $a(\lambda, y)$ a known and positive function, $\tilde{\Theta} \subseteq \mathbb{R}^p$, $\Lambda \subseteq \mathbb{R}^+ \setminus \{0\}$, is called an **exponential dispersion family of order p**.

In this chapter, unlike in the rest of the book, the symbol θ does not refer to the global parameter of the statistical model. This not only conforms to the usual notation on exponential dispersion families, but also allows analogies and generalizations with respect to exponential families to be made more clearly. Family (6.6) gives an $\mathcal{F}^p_e$ for every fixed value of λ. But it does not generally define an exponential family as λ varies in Λ.

The following terminology is used with reference to an $\mathcal{F}^p_{ed}$ in the form (6.6). The parameter θ is called the **natural parameter** (or the canonical parameter) and takes values in the natural parameter space $\tilde{\Theta}$. The function $K(\cdot)$ is called the **cumulant generator**; $K(\cdot)$ is not the cumulant generating function for (6.6). Lastly, the parameter λ is called the **precision parameter**.

It is interesting to note that $\Lambda = \mathbb{R}^+ \setminus \{0\}$ if, and only if, $(\exp(K(\theta)))^\lambda$ is a moment generating function for every $\lambda > 0$. This implies that the distribution with density $p_0(x; \lambda)$ is infinitely divisible (see section 3.2.6); indeed, its characteristic function is $C(t) = (\exp\{K(it)\})^\lambda$.

Example 6.1 $\mathcal{F}_{ed}^1$ *generated by the* $IG(1,1)$ *distribution*
Let $X \sim IG(1,1)$ (see Table 5.1), with density

$$p_0(x) = \frac{1}{\sqrt{2\pi x^3}}\, e^{-(x^{-1}+x-2)/2}\, , \quad x > 0\, .$$

From Table 5.1 we can deduce that X has c.g.f.

$$K_X(\theta) = K(\theta) = 1 - \sqrt{1 - 2\theta}\, .$$

It is not difficult to check that the function

$$\lambda K(\theta) = \lambda(1 - \sqrt{1 - 2\theta})$$

is the c.g.f. of a $IG(1,\lambda^2)$ distribution; this is a c.g.f. for every $\lambda > 0$, so that $\Lambda = \mathbb{R}^+ \setminus \{0\}$. The $IG(1,\lambda^2)$ density is

$$p_0(x;\lambda) = \frac{\lambda e^\lambda}{\sqrt{2\pi x^3}} e^{-\frac{1}{2}\left(\frac{\lambda^2}{x}+x\right)}\, , \quad x > 0\, ,$$

and the $\mathcal{F}_{ne}^1$ generated by $p_0(x;\lambda)$ through exponential tilting has density

$$p(x;\theta,\lambda) = e^{\theta x - \lambda(1-\sqrt{1-2\theta})} p_0(x;\lambda)\, .$$

Thus the density of $Y = X/\lambda$, where $X \sim p_0(x;\theta,\lambda)$, is

$$
\begin{aligned}
p_Y(y;\theta,\lambda) &= \lambda e^{\lambda(\theta y - (1-\sqrt{1-2\theta}))}\frac{\lambda e^\lambda}{\sqrt{2\pi(\lambda y)^3}} e^{-\frac{1}{2}\left(\frac{\lambda}{y}+\lambda y\right)} \\
&= \frac{\sqrt{\lambda}}{\sqrt{2\pi y^3}} e^{\lambda\sqrt{1-2\theta}} e^{-\frac{1}{2}\left(\frac{\lambda}{y}+\lambda(1-2\theta)y\right)}\, ,
\end{aligned}
$$

for $y > 0$. With the reparameterization $\phi = \lambda(1 - 2\theta)$, we obtain the $IG(\phi,\lambda)$ density. $\triangle$

If Y is a p-dimensional random variable with density belonging to family (6.6), its moment generating function is

$$
\begin{aligned}
M_Y(t;\theta,\lambda) &= \int_y e^{t\cdot y + \lambda\{\theta\cdot y - K(\theta)\}} a(\lambda,y) d\mu^* \\
&= e^{\lambda\{K(\theta+t/\lambda)-K(\theta)\}} \int_y e^{\lambda\{(\theta+t/\lambda)\cdot y - K(\theta+t/\lambda)\}} a(\lambda,y) d\mu^* \\
&= \exp\{\lambda(K(\theta + t/\lambda) - K(\theta))\}\, .
\end{aligned}
$$

The last integral above is equal to one, because it is the integral, on the whole support, of a density which belongs to an $\mathcal{F}^p_{ed}$ with natural parameter $\theta + t/\lambda$. Thus, the cumulant generating function of Y is

$$K_Y(t; \theta, \lambda) = \lambda \left\{ K\left(\theta + \frac{t}{\lambda}\right) - K(\theta) \right\} , \tag{6.7}$$

and a generic joint cumulant of Y of order r is

$$\kappa^{i_1, \ldots, i_r} = \frac{\partial^r K(\theta)}{\partial \theta^{i_1} \cdots \partial \theta^{i_r}} \lambda^{1-r}, \quad \theta \in \text{int } \tilde{\Theta}, \quad r = 1, 2, \ldots . \tag{6.8}$$

From (6.8) Y has mean value

$$\mu(\theta) = \frac{\partial K(\theta)}{\partial \theta} = \left(\frac{\partial K(\cdot)}{\partial \theta^1}, \ldots, \frac{\partial K(\cdot)}{\partial \theta^p} \right) . \tag{6.9}$$

As anticipated, $\mu(\theta)$ does not depend on λ and coincides with the mean value mapping for the $\mathcal{F}^p_{ne}$ defined by (6.2). The function $\mu(\cdot)$ is one-to-one and smooth because of the results about natural exponential families in section 5.3. Thus an $\mathcal{F}^p_{ed}$ with $\tilde{\Theta}$ open can be reparameterized using the mean $\mu = \mu(\theta)$ instead of the natural parameter θ. Furthermore, within the class of all the $\mathcal{F}^p_{ne}$, a specific $\mathcal{F}^p_{ne}$ is characterized by its variance function

$$V(\mu) = \left. \frac{\partial^2 K(\theta)}{\partial \theta \partial \theta^{\mathrm{T}}} \right|_{\theta = \theta(\mu)}$$

(cf. section 5.4.1). Since the $\mathcal{F}^p_{ne}$ with density (6.2) determines, uniquely, the $\mathcal{F}^p_{ed}$ with cumulant generating function (6.7), the function $V(\mu)$ will also characterize a specific $\mathcal{F}^p_{ed}$ within the class of all the $\mathcal{F}^p_{ed}$ (Jørgensen, 1987, Theorem 1). From (6.8),

$$\text{Var}(Y) = \frac{1}{\lambda} V(\mu) . \tag{6.10}$$

In an $\mathcal{F}^p_{ed}$, the parameters θ and λ are orthogonal. In fact, if y is an observation from an element of (6.6), the log-likelihood function is

$$l(\theta, \lambda) = \log a(\lambda, y) + \lambda \{ \theta \cdot y - K(\theta) \} .$$

Thus,

$$\frac{\partial l(\theta, \lambda)}{\partial \theta} = \lambda \{ y - \mu(\theta) \}$$

and

$$\frac{\partial^2 l(\theta, \lambda)}{\partial \theta \partial \lambda} = y - \mu(\theta) \tag{6.11}$$

(both equalities concern $p \times 1$ vectors). Consider the partition

$$i = i(\theta, \lambda) = \begin{pmatrix} i_{\theta\theta} & i_{\theta\lambda} \\ i_{\lambda\theta} & i_{\lambda\lambda} \end{pmatrix}$$

of the expected information matrix; θ and λ are orthogonal if $i_{\theta\lambda} = (i_{\lambda\theta})^{\mathrm{T}} = 0$. This follows from (6.11), since

$$i_{\theta\lambda} = E_{\theta,\lambda}(Y - \mu(\theta)) = 0 \ . \tag{6.12}$$

6.3   Parameterization (μ, σ^2) and Convolution Properties

Sometimes it may prove convenient to refer to the reparameterization of an $\mathcal{F}_{ed}^p$ with (μ, σ^2), where $\mu = \mu(\theta)$ and $\sigma^2 = 1/\lambda$. The parameter σ^2 is called the **dispersion parameter**. Note that orthogonality of μ and σ^2 follows from orthogonality of θ and λ. When referring to this new parameterization the notation

$$Y \sim ED_p(\mu, \sigma^2 V(\mu))$$

is used to indicate that Y has density $p(y; \theta, \lambda)$ which belongs to an $\mathcal{F}_{ed}^p$ with $\theta = \theta(\mu)$, $\lambda = 1/\sigma^2$ and variance function $V(\mu)$; remember that $\theta(\mu)$ indicates the inverse function of $\mu(\theta)$. Observe the explicit analogy with the notation $N_p(\mu, \Sigma)$ used to indicate normal distributions. With the parameterization (μ, σ^2), the cumulant generating function of Y (6.7) is written as

$$K_Y(t; \mu, \sigma^2) = \frac{1}{\sigma^2} \{ K(\theta(\mu) + \sigma^2 t) - K(\theta(\mu)) \} \ . \tag{6.13}$$

The (μ, σ^2) parameterization is particularly important in that it provides the starting point for defining generalized linear models (see section 6.4). For these, in close analogy with normal regression, the usual assumption is that there are n independent observations y_i, $i = 1, \ldots, n$, with y_i realization of an $ED_1(\mu_i, \sigma^2 V(\mu_i))$, and with μ_i a function of a linear combination of k explanatory variables. To this end it is interesting to observe that each of the exponential families described in Table 5.1 is also an $ED_1(\mu, \sigma^2 V(\mu))$ distribution. This follows immediately from Example 6.1 also, for the inverse Gaussian distribution.

Example 6.2 *Normal distribution*

Let $Y \sim N(\mu, \sigma^2)$. The density of Y may be written as

$$p(y; \mu, \sigma^2) = \frac{1}{\sqrt{2\pi}\sigma} e^{-y^2/2\sigma^2} \exp\left\{-\frac{\mu^2}{2\sigma^2} + \frac{\mu}{\sigma^2}y\right\}$$

$$= \frac{1}{\sqrt{2\pi}\sigma} e^{-y^2/2\sigma^2} \exp\left\{\frac{1}{\sigma^2}(\mu y - \mu^2/2)\right\}, \qquad (6.14)$$

$y \in \mathbb{R}$. Density (6.14) is of the same form as in (6.6) with $\theta = \theta(\mu) = \mu$, $\lambda = 1/\sigma^2$ and $K(\theta) = \theta^2/2$. $\triangle$

Example 6.3 *Gamma distribution*

Let $Y \sim Ga(\nu, \phi)$ with density

$$p(y; \nu, \phi) = \frac{y^{\nu-1} e^{-y/\phi}}{\phi^\nu \Gamma(\nu)}$$

$$= \exp\{-y/\phi - \nu \log \phi\} \frac{y^{\nu-1}}{\Gamma(\nu)}, \quad y > 0, \ \nu, \phi > 0.$$

Since $E(Y) = \mu = \nu\phi$, we can write the density with the parameterization (μ, ν), as

$$p(y; \nu, \mu) = \exp\left\{-\frac{\nu}{\mu}y + \nu \log \nu - \nu \log \mu\right\} \frac{y^{\nu-1}}{\Gamma(\nu)}$$

$$= \exp\left\{\nu\left(-\frac{1}{\mu}y - \log \mu\right)\right\} \frac{y^{\nu-1}\nu^\nu}{\Gamma(\nu)}$$

which is the density in an $\mathcal{F}^1_{ed}$ with $\theta = -1/\mu = -1/(\nu\phi)$, $\lambda = \nu$ and $K(\theta) = -\log(-\theta)$. Alternatively, we can say that $Y \sim ED_1(\mu, \sigma^2 V(\mu))$, with $\mu = \nu\phi$, $\sigma^2 = 1/\nu$, $V(\mu) = \mu^2$, $\theta(\mu) = -1/\mu$. In other words, the density of $Y \sim Ga(\nu, \phi)$ can be written in the form

$$p(y; \mu, \sigma^2) = \exp\left\{\frac{1}{\sigma^2}\left(-\frac{1}{\mu}y - \log \mu\right)\right\} \frac{y^{1/\sigma^2-1}}{\Gamma(\frac{1}{\sigma^2})(\sigma^2)^{1/\sigma^2}}. \qquad (6.15)$$

$\triangle$

Example 6.4 *Inverse Gaussian distribution*

Let $Y \sim IG(\phi, \lambda)$ with density

$$p(y; \phi, \lambda) = \frac{\sqrt{\lambda}}{\sqrt{2\pi}} y^{-3/2} e^{\sqrt{\lambda\phi}} \exp\left\{-\frac{1}{2}\left(\frac{\lambda}{y} + \phi y\right)\right\}, \quad y > 0, \ \lambda > 0, \ \phi \geq 0.$$

The cumulant generating function of Y is (see Table 5.1)

$$K_Y(t; \phi, \lambda) = -\sqrt{\lambda(\phi - 2t)} + \sqrt{\lambda\phi} \,,$$

therefore,

$$\mu = K_Y'(0; \phi, \lambda) = \sqrt{\frac{\lambda}{\phi}} \,,$$

furthermore,

$$K_Y''(0; \phi, \lambda) = \lambda^2(\lambda\phi)^{-3/2} = \frac{1}{\lambda}\left(\frac{\lambda}{\phi}\right)^{3/2} = \frac{1}{\lambda}\mu^3.$$

With the reparameterization $(\mu, \sigma^2) = (\sqrt{\lambda/\phi}, 1/\lambda)$, the density of Y can be written as

$$\begin{aligned}
p(y; \mu, \sigma^2) &= \frac{1}{\sqrt{2\pi}\sigma} y^{-3/2} e^{1/(\mu\sigma^2)} \exp\left\{-\frac{1}{2}\left(\frac{1}{\sigma^2 y} + \frac{1}{\sigma^2\mu^2}y\right)\right\} \\
&= \exp\left\{\frac{1}{\sigma^2}\left(\frac{1}{\mu^2}y + \frac{1}{\mu}\right)\right\} \frac{1}{\sqrt{2\pi}\sigma} y^{-3/2} e^{-1/(2\sigma^2 y)} \,, \quad (6.16)
\end{aligned}$$

so that $Y \sim ED_1(\mu, \sigma^2 V(\mu))$, with $V(\mu) = \mu^3$. The natural parameterization (θ, λ) is $\theta = \theta(\mu) = 1/\mu^2$ and $\lambda = 1/\sigma^2$. $\triangle$

Example 6.5 *Poisson distribution*
Let $Y \sim P(\mu)$ with density

$$p(y; \mu) = \exp\{y \log \mu - \mu\}\frac{1}{y!}, \quad y \in \mathbb{N}, \, \mu > 0. \quad (6.17)$$

The density (6.17) is an $ED_1(\mu, \sigma^2 V(\mu))$, with $\sigma^2 = 1$, $V(\mu) = \mu$. The natural parameterization is $\theta = \log\mu$. $\triangle$

Example 6.6 *Binomial distribution*
Let $X \sim Bi(n, \phi)$. The random variable $Y = X/n$ has density

$$p(y; n, \phi) = \binom{n}{ny} \exp\left\{ny \log\left(\frac{\phi}{1-\phi}\right) + n \log(1 - \phi)\right\}.$$

Since $E(Y) = \mu = \phi$ and $\text{Var}(Y) = \phi(1 - \phi)/n = \mu(1 - \mu)/n$, we let $\mu = \phi$, $\sigma^2 = 1/n$ and write the density of Y in the form

$$p(y; \mu, \sigma^2) = \binom{n}{ny} \exp\left\{\frac{1}{\sigma^2}\left(\log\left(\frac{\mu}{1-\mu}\right)y + \log(1 - \mu)\right)\right\}, \quad (6.18)$$

which is an $ED_1(\mu, \sigma^2 V(\mu))$ with $V(\mu) = \mu(1-\mu)$. The natural parameterization is obtained with $\theta = \theta(\mu) = \log(\mu/(1-\mu))$. $\triangle$

Note that both the examples of discrete distributions considered above are exponential dispersion families with known dispersion parameter.

One important aspect of the analogy between an $ED_p(\mu, \sigma^2 V(\mu))$ family and the $N_p(\mu, \Sigma)$ distribution is expressed by the following theorem which states a closure property under specific linear combinations of independent variables with the same variance function.

Theorem 6.1 Let $Y_1, \ldots, Y_n$ be independent random variables with

$$Y_i \sim ED_p\left(\mu(\theta), \frac{\sigma^2}{w_i} V(\mu(\theta))\right), \quad i = 1, \ldots, n,$$

where $w_1, \ldots, w_n \in \Lambda$ are known constants. Let $w_\bullet = \sum w_i$, then

$$\frac{1}{w_\bullet} \sum_{i=1}^n w_i Y_i \sim ED_p\left(\mu(\theta), \frac{\sigma^2}{w_\bullet} V(\mu(\theta))\right). \tag{6.19}$$

Proof. If $Y_i \sim ED_p\left(\mu(\theta), \frac{\sigma^2}{w_i} V(\mu(\theta))\right)$, then, from (6.13),

$$K_{Y_i}\left(t; \theta, \frac{\sigma^2}{w_i}\right) = \frac{w_i}{\sigma^2}\left\{K\left(\theta + \frac{\sigma^2}{w_i}t\right) - K(\theta)\right\}.$$

Therefore, $w_i Y_i$ has cumulant generating function

$$K_{w_i Y_i}\left(t; \theta, \frac{\sigma^2}{w_i}\right) = K_{Y_i}\left(tw_i; \theta, \frac{\sigma^2}{w_i}\right) = \frac{w_i}{\sigma^2}\{K(\theta + \sigma^2 t) - K(\theta)\},$$

and, consequently, $\sum_{i=1}^n w_i Y_i$ has cumulant generating function

$$K_{\Sigma w_i Y_i}(t; \theta, \sigma^2) = \frac{w_\bullet}{\sigma^2}\left\{K(\theta + \sigma^2 t) - K(\theta)\right\}.$$

Thus, $\sum w_i Y_i / w_\bullet$ has cumulant generating function

$$K_{\Sigma w_i Y_i / w_\bullet}(t; \theta, \sigma^2) = K_{\Sigma w_i Y_i}\left(\frac{t}{w_\bullet}; \theta, \sigma^2\right) = \frac{w_\bullet}{\sigma^2}\left\{K\left(\theta + \frac{\sigma^2}{w_\bullet}t\right) - K(\theta)\right\}.$$

$\square$

One immediate consequence of Theorem 6.1 is that if $Y_1, \ldots, Y_n$ are independent and identically distributed according to $ED_p(\mu, \sigma^2 V(\mu))$, then

$$\bar{Y}_n = \frac{1}{n} \sum Y_i \sim ED_p\left(\mu, \frac{\sigma^2}{n} V(\mu)\right) . \tag{6.20}$$

Closure under convolution for independent normal, gamma, Poisson and binomial distributions are special instances of the result (6.20). Furthermore, if $Y_1, \ldots, Y_n$ are independent and identically distributed $IG(\phi, \lambda)$, then $\bar{Y}_n \sim IG(n\phi, n\lambda)$. In fact, the parameters (ϕ', λ') of $\bar{Y}_n$ must satisfy the identities $\lambda'/\phi' = \lambda/\phi$ and $(\lambda')^{-1}(\lambda'/\phi')^{3/2} = (n\lambda)^{-1}(\lambda/\phi)^{3/2}$.

From the central limit theorem,

$$\sqrt{\frac{n}{\sigma^2}}(\bar{Y}_n - \mu) \xrightarrow{d} N_p(0, V(\mu))$$

as $n \to +\infty$ or, equivalently, as $\sigma^2 \to 0$. Therefore, on the basis of (6.20), for $Y \sim ED_p(\mu, \sigma^2 V(\mu))$ the approximation

$$\frac{Y - \mu}{\sqrt{\sigma^2}} \dot{\sim} N_p(0, V(\mu)) \tag{6.21}$$

holds for $\sigma^2 \to 0$. With reference to asymptotic results obtained as $\sigma^2 \to 0$, Jørgensen (1987) introduced the term small dispersion asymptotics.

6.4 Generalized Linear Models

The classical normal linear regression model is specified by the assumptions

$$Y_1, \ldots, Y_n \quad \text{univariate independent r.v.'s,} \tag{6.22}$$

$$E(Y_i) = \mu_i = \beta^{\mathrm{T}} x_i = \beta_1 x_{i1} + \cdots + \beta_k x_{ik}, \tag{6.23}$$

$$Y_i \sim N(\mu_i, \sigma^2) , \tag{6.24}$$

with $x_i = (x_{i1}, \ldots, x_{ik})^{\mathrm{T}}$ a known vector of fixed explanatory variables for the i-th observation, $i = 1, \ldots, n$, and $\beta = (\beta_1, \ldots, \beta_k)^{\mathrm{T}}$ a vector of unknown regression coefficients.

In a generalized linear model (g.l.m.) the assumption of independence (6.22) is retained and assumptions (6.23) and (6.24) are generalized as follows:

$$g(E(Y_i)) = g(\mu_i) = \beta^{\mathrm{T}} x_i , \tag{6.25}$$

with $g(\cdot)$ a one-to-one smooth function called the link function;

$$Y_i \sim ED_1(\mu_i, \sigma^2 V(\mu_i)) . \tag{6.26}$$

Assumption (6.25) is particularly useful whenever the expectation space $M = \mu(\text{int } \tilde{\Theta})$ does not coincide with $\mathbb{R}$, as happens, for example for the binomial, Poisson, gamma and inverse Gaussian distributions. If a function $g(\cdot)$ is chosen which maps M on the whole real line, then no restriction needs to be imposed on β, whose parameter space can be $\mathbb{R}^k$. For example, in the case of a binary response, one has $M = (0,1)$ and a possible link function with range $\mathbb{R}$ is $g(\mu) = \log(\mu/(1-\mu))$. When compared with the traditional approach, which suggests transforming the response onto a scale which ranges on the whole real line and where the distribution is approximately normal, generalized linear models offer increased flexibility in modelling.

Because the variance of Y_i is

$$\text{Var}(Y_i) = \sigma^2 V(\mu_i) = \sigma^2 V(g^{-1}(\beta^T x_i)), \tag{6.27}$$

models with unequal variances are possible under generalized linear models. However, while there is a degree of flexibility when modelling the mean, this is not so for the variance of Y_i. According to (6.27), the variance must depend on the same combination of explanatory variables as are used to model the mean. On the other hand, the assumption of independence of the observations is not relaxed, at least in this basic formulation (see, however, the bibliographic note, section 6.6).

For each specification of $ED_1(\mu_i, \sigma^2 V(\mu_i))$ in (6.26), one privileged link function is

$$g(\mu) = \theta(\mu) , \tag{6.28}$$

which assumes that the natural parameter θ of an $\mathcal{F}^1_{ed}$ is a linear combination of explanatory variables with coefficients β, $\theta_i = \beta^T x_i$, $i = 1, \ldots, n$. The function (6.28) is called the **canonical** link. In generalized linear models with normal, gamma, inverse Gaussian, Poisson and binomial distribution of errors, the canonical link can be immediately deduced from Examples 6.2-6.6. For ease of reference, these functions are shown in Table 6.1.

Some differentiation rules will be useful in the following. If $g(\mu_i) = \beta^T x_i$, then $\mu_i = g^{-1}(\beta^T x_i)$; therefore the i-th component of θ can be expressed as $\theta_i = \theta(\mu_i) = \theta(g^{-1}(\beta^T x_i))$. From (5.17), $\theta'(\mu) = 1/V(\mu)$. The quantity $\eta_i = \beta^T x_i$ is usually called the **linear predictor**. Thus, we obtain

$$\frac{\partial \theta_i}{\partial \beta_r} = \frac{\partial \theta_i}{\partial \mu_i} \frac{\partial \mu_i}{\partial \eta_i} \frac{\partial \eta_i}{\partial \beta_r} = \frac{1}{V(\mu_i)} \frac{1}{g'(\mu_i)} x_{ir}, \quad r = 1, \ldots, k, \tag{6.29}$$

Table 6.1: *Canonical link functions for some generalized linear models.*

Distribution	$N(\mu, \sigma^2)$	$Ga(\nu, \phi)$	$IG(\phi, \lambda)$	$P(\phi)$	$Bi(n, \phi)/n$
μ	μ	$\nu\phi$	$\sqrt{\lambda/\phi}$	ϕ	ϕ
σ^2	σ^2	ν^{-1}	λ^{-1}	1	n^{-1}
$V(\mu)$	1	μ^2	μ^3	μ	$\mu(1-\mu)$
$\theta(\mu)$	μ	$-\mu^{-1}$	μ^{-2}	$\log\mu$	$\log(\mu/(1-\mu))$

where the quantities indicated are evaluated at $\mu_i = g^{-1}(\beta^T x_i)$. If $g(\cdot)$ is the canonical link, then $g(\mu) = \theta(\mu)$ hence $g'(\mu) = \dfrac{1}{V(\mu)}$, from which

$$\frac{\partial \theta_i}{\partial \beta_r} = x_{ir}, \quad r = 1, \ldots, k. \tag{6.30}$$

This is obvious since if $g(\cdot)$ is the canonical link, $\theta_i = \beta^T x_i$.

6.4.1 Likelihood and Sufficiency

Let $Y_1, \ldots, Y_n$ be random variables whose joint distribution follows assumptions (6.22), (6.25) and (6.26). Then $Y = (Y_1, \ldots, Y_n)$ has density

$$p_Y(y; \beta, \sigma^2) = \exp\left\{ \frac{1}{\sigma^2} \sum_{i=1}^n \{\theta(\mu_i)y_i - K(\theta(\mu_i))\} \right\} \prod_{i=1}^n a\left(\frac{1}{\sigma^2}, y_i\right), \tag{6.31}$$

where $\theta(\mu_i) = \theta(g^{-1}(\beta^T x_i))$. Thus, generally, there is no minimal sufficient statistic with dimension smaller than n, even when, σ^2 is known. However, if $g(\mu) = \theta(\mu)$, that is, if $g(\cdot)$ is the canonical link, then, for every fixed value of σ^2, Y has density in an $\mathcal{F}_e^k$ and there is a k-dimensional minimal sufficient statistic for β. In particular, (6.31) simplifies as

$$p_Y(y; \beta, \sigma^2) = \exp\left\{ \frac{1}{\sigma^2} \left(\beta^T \sum_{i=1}^n x_i y_i - \sum_{i=1}^n K(\beta^T x_i) \right) \right\} \prod_{i=1}^n a\left(\frac{1}{\sigma^2}, y_i\right)$$

$$\tag{6.32}$$

and the k-dimensional statistic

$$\sum_{i=1}^n x_i y_i = \left(\sum_{i=1}^n x_{i1} y_i, \ldots, \sum_{i=1}^n x_{ik} y_i \right)^T$$

is minimal sufficient for β for every fixed value of σ^2. Some generalized linear models with the canonical link can be written as an exponential family even when σ^2 is unknown (see Problems 6.9 and 6.10).

From (6.31) the log-likelihood function is

$$l(\beta, \sigma^2) = \frac{1}{\sigma^2} \left\{ \sum_{i=1}^{n} (y_i \theta_i - K(\theta_i)) \right\} + \sum_{i=1}^{n} \log a \left(\frac{1}{\sigma^2}, y_i \right), \qquad (6.33)$$

where $\theta_i = \theta(\mu_i) = \theta(g^{-1}(\beta^T x_i))$.

The score vector has components

$$l_r = \frac{\partial l(\beta, \sigma^2)}{\partial \beta_r} = \frac{1}{\sigma^2} \sum_{i=1}^{n} \left(y_i \frac{\partial \theta_i}{\partial \beta_r} - \frac{\partial K(\theta_i)}{\partial \beta_r} \right), \qquad r = 1, \dots, k \quad (6.34)$$

$$l_{\sigma^2} = \frac{\partial l(\beta, \sigma^2)}{\partial \sigma^2} = -\frac{1}{\sigma^4} \sum_{i=1}^{n} (y_i \theta_i - K(\theta_i)) - \frac{1}{\sigma^4} \sum_{i=1}^{n} \frac{a'\left(\frac{1}{\sigma^2}, y_i\right)}{a\left(\frac{1}{\sigma^2}, y_i\right)} \quad (6.35)$$

where

$$a'\left(\frac{1}{\sigma^2}, y_i\right) = \frac{\partial a(\lambda, y_i)}{\partial \lambda}\bigg|_{\lambda = 1/\sigma^2}.$$

Since

$$\frac{\partial K(\theta_i)}{\partial \beta_r} = K'(\theta_i) \frac{\partial \theta_i}{\partial \beta_r} = \mu_i \frac{\partial \theta_i}{\partial \beta_r},$$

l_r may be written in the form

$$l_r = \frac{1}{\sigma^2} \sum_{i=1}^{n} (y_i - \mu_i) \frac{\partial \theta_i}{\partial \beta_r}, \qquad r = 1, \dots, k, \qquad (6.36)$$

so that the maximum likelihood estimate of β for a fixed value of σ^2, $\hat{\beta}_{\sigma^2}$, does not depend on σ^2 and coincides with the unconstrained maximum likelihood estimate $\hat{\beta}$. Using (6.29), the likelihood equations (6.36) for β are

$$\sum_{i=1}^{n} \frac{y_i - \mu_i}{V(\mu_i) g'(\mu_i)} x_{ir} = 0, \qquad r = 1, \dots, k. \qquad (6.37)$$

With the canonical link, $g'(\mu_i) = 1/V(\mu_i)$ and equations (6.37) take the simple form

$$\sum_{i=1}^{n} y_i x_{ir} = \sum_{i=1}^{n} \mu_i x_{ir}. \qquad (6.38)$$

These equations agree with the general structure of likelihood equations in exponential families (see (5.42)): the observed value of the minimal sufficient statistic is equated to its expectation. Equations (6.37) admit an interesting geometrical interpretation, which is fairly similar to that of condition (5.89) under a curved exponential family. From (6.29),

$$\frac{\partial \theta_i}{\partial \beta_r} = \frac{1}{V(\mu_i)} \frac{\partial \mu_i}{\partial \beta_r} ,$$

so that (6.37) can be written in matrix notation as

$$(y - \mu)^T V^{-1} D = 0 , \tag{6.39}$$

where $(y - \mu)^T = (y_1 - \mu_1, \ldots, y_n - \mu_n)$,

$$V = \text{diag}(V(\mu_i)) , \quad i = 1, \ldots, n , \tag{6.40}$$

and

$$D = [d_{ir}] = \left[\frac{\partial \mu_i}{\partial \beta_r} \right] .$$

The notation $[a_{ij}]$ is used, both here and below, to denote the matrix with elements a_{ij}; analogously, $[a_i]$ will be used to denote the vector with elements a_i.

Let $\mathcal{M}$ indicate the mean space

$$\mathcal{M} = \left\{ \mu \in \mathbb{R}^n : \mu_i = g^{-1}(\beta^T x_i), \beta \in \mathbb{R}^k \right\} . \tag{6.41}$$

The n-dimensional vectors $[\partial \mu_i / \partial \beta_1], \ldots, [\partial \mu_i / \partial \beta_k]$, which form the columns in D, are a basis of the k-dimensional tangent space to $\mathcal{M}$ for a fixed value of β. According to (6.39), at the point $\hat{\mu} = [g^{-1}(\hat{\beta}^T x_i)]$ the vector $y - \mu$ must be orthogonal to the tangent space at $\mathcal{M}$ in the metric defined by the matrix V^{-1}; in this metric, $\hat{\mu}$ is the point in $\mathcal{M}$ with the minimum distance from y.

As regards the existence and uniqueness of $\hat{\beta}$, if the link is the canonical one, the theory of exponential families applies. Sufficient conditions in the general case, with normal, Poisson, binomial and gamma distributions are given in Wedderburn (1976).

Now, we wish to obtain the expressions of the observed and expected information for β. Since β and σ^2 are orthogonal, we can proceed as if σ^2 were known. By differentiating both sides of (6.36) with respect to β_s, we obtain

$$j_{rs} = -l_{rs} = \frac{1}{\sigma^2} \sum_{i=1}^{n} \left\{ \frac{\partial \mu_i}{\partial \beta_s} \frac{\partial \theta_i}{\partial \beta_r} - (y_i - \mu_i) \frac{\partial^2 \theta_i}{\partial \beta_r \partial \beta_s} \right\} . \tag{6.42}$$

With the canonical link, $\partial^2\theta_i/(\partial\beta_r\partial\beta_s) = 0$ and the expression for j_{rs} is simplified into

$$j_{rs} = \frac{1}{\sigma^2}\sum_{i=1}^{n}\frac{\partial\mu_i}{\partial\beta_s}\frac{\partial\theta_i}{\partial\beta_r},$$

which is non-random and coincides with its expectation. In general, since $E(Y_i) = \mu_i$, from (6.42), we obtain

$$i_{rs} = E(j_{rs}) = \frac{1}{\sigma^2}\sum_{i=1}^{n}\frac{\partial\mu_i}{\partial\beta_s}\frac{\partial\theta_i}{\partial\beta_r}$$

$$= \frac{1}{\sigma^2}\sum_{i=1}^{n}\frac{1}{g'(\mu_i)}x_{is}\frac{1}{V(\mu_i)}\frac{1}{g'(\mu_i)}x_{ir} \tag{6.43}$$

that is

$$i_{rs} = \frac{1}{\sigma^2}\sum_{i=1}^{n}\frac{x_{ir}x_{is}}{(g'(\mu_i))^2 V(\mu_i)}. \tag{6.44}$$

Relation (6.44) is often written using matrix notation as

$$i_{\beta\beta} = \frac{1}{\sigma^2}X^{T}WX, \tag{6.45}$$

where

$$W = \text{diag}(w_i), \quad \text{with} \quad w_i = \frac{1}{(g'(\mu_i))^2 V(\mu_i)} \tag{6.46}$$

and X is the matrix of the explanatory variables, which has x_i as its i-th row. If, in particular, the link is canonical,

$$i_{rs} = \frac{1}{\sigma^2}\sum_{i=1}^{n}x_{ir}x_{is}V(\mu_i)$$

and, in matrix form,

$$i_{\beta\beta} = \frac{1}{\sigma^2}X^{T}VX,$$

with V defined by (6.40) .

Asymptotic normality of the maximum likelihood estimator gives

$$\hat{\beta} \sim N_k(\beta, \sigma^2(X^{T}WX)^{-1}), \tag{6.47}$$

for large n. Therefore, a consistent estimate of the covariance matrix of β, when σ^2 is known, is $\sigma^2(X^{T}\hat{W}X)^{-1}$, where $\hat{W}$ indicates the matrix W calculated at

$\beta = \hat{\beta}$. If σ^2 is unknown, it should be replaced by a consistent estimator, such as the maximum likelihood estimator, or that given by formula (6.50) below.

The likelihood equations (6.37) do not usually have explicit solutions. They should be solved by iterative methods such as the Newton–Raphson method. With $l_* = [l_r]$ and $j = [-l_{rs}]$, the $(m+1)$-th iteration is

$$\hat{\beta}^{(m+1)} = \hat{\beta}^{(m)} + \left\{ j(\hat{\beta}^{(m)}) \right\}^{-1} l_*(\hat{\beta}^{(m)}) .$$

In the expression above, the observed information can be replaced by the expected information (Fisher scoring method). This maintains the convergence of the algorithm and simplifies the expressions. Thus,

$$\hat{\beta}^{(m+1)} = \hat{\beta}^{(m)} + \left\{ i_{\beta\beta}(\hat{\beta}^{(m)}) \right\}^{-1} l_*(\hat{\beta}^{(m)})$$

or, also,

$$i_{\beta\beta}(\hat{\beta}^{(m)})\hat{\beta}^{(m+1)} = i_{\beta\beta}(\hat{\beta}^{(m)})\hat{\beta}^{(m)} + l_*(\hat{\beta}^{(m)}) . \qquad (6.48)$$

Using (6.29) and (6.46) we can write (6.36) as

$$l_r = \frac{1}{\sigma^2} \sum_{i=1}^{n} x_{ir}(y_i - \mu_i)w_i g'(\mu_i) , \quad r = 1, \ldots, k ,$$

or, in matrix notation,

$$l_* = \frac{1}{\sigma^2} X^{\mathrm{T}} W u ,$$

with $u = ((y_1 - \mu_1)g'(\mu_1), \ldots, (y_n - \mu_n)g'(\mu_n))^{\mathrm{T}}$. Therefore, using (6.45), equation (6.48) becomes

$$X^{\mathrm{T}} W X \hat{\beta}^{(m+1)} = X^{\mathrm{T}} W z , \qquad (6.49)$$

where $z = X\hat{\beta}^{(m)} + u = \left[\hat{\beta}^{(m)T} x_i + (y_i - \mu_i)g'(\mu_i) \right]$ is usually called the **adjusted dependent variable**; the quantities W and z are evaluated at $\hat{\beta}^{(m)}$. Observe that (6.49) formally coincides with normal equations for weighted least squares estimators; this means that calculations are greatly simplified, something which is very useful for numerical calculations. A natural, and generally useful, initial value would be either $\hat{\mu}_i^{(0)} = y_i$ or $\hat{\mu}_i^{(0)} = \max(\varepsilon, y_i)$, $\varepsilon > 0$, if, for example, $g(\mu) = \log\mu$.

So far, nothing has been said about the estimation of σ^2. Obviously, the maximum likelihood estimate, $\hat{\sigma}^2$, based on (6.35) with β replaced by $\hat{\beta}$, could be used. However, usually alternative estimators are suggested (see for example McCullagh and Nelder, 1989, p. 295) because of the possible numerical instability of $\hat{\sigma}^2$ and its lack of robustness under even slight model misspecification.

Estimators based on the method of moments are often used for σ^2. Since $\mathrm{Var}(Y_i) = \sigma^2 V(\mu_i)$, if β is known

$$\frac{1}{n} \sum_{i=1}^{n} \frac{(y_i - \mu_i)^2}{V(\mu_i)}$$

is an unbiased estimator of σ^2. If the expected values μ_i are replaced with their estimates based on $\hat{\beta}$, then the following adjusted estimator is obtained

$$\tilde{\sigma}^2 = \frac{1}{n-k} \sum_{i=1}^{n} \frac{(y_i - \hat{\mu}_i)^2}{V(\hat{\mu}_i)} . \tag{6.50}$$

Generally, $\tilde{\sigma}^2$ is consistent. If $g(\mu) = \mu$ and $V(\mu) = 1$ then $\tilde{\sigma}^2$ is unbiased (the assumptions become those of the classical normal linear regression model).

Another drawback of maximum likelihood estimation of σ^2 is that the profile likelihood estimating equation is biased, which is particularly important if k is large. This suggests the use of a suitable pseudo-likelihood for σ^2. With the canonical link, for every fixed value of σ^2 the model with densities (6.32) is an $\mathcal{F}_e^k$, with natural parameter β and natural observations $\sum y_i x_i$, which are, therefore, a partially sufficient statistic for β if σ^2 is unknown. In this case, we could define a conditional likelihood for σ^2, based on the density of the observations Y_i conditionally on the observed value of $U = \sum Y_i x_i$. The conditional log-likelihood has the general expression

$$l_c(\sigma^2) = \sum_{i=1}^{n} \log \left(a\left(\frac{1}{\sigma^2}, y_i\right) \right) - \log \left\{ \int_{\mathcal{Y}_u} \prod_{i=1}^{n} a\left(\frac{1}{\sigma^2}, y_i\right) dy_i \right\} , \tag{6.51}$$

where $\mathcal{Y}_u$ represents the set of values y_i, $i = 1, \ldots, n$, which give the value u of U. In general, (6.51) cannot be written in a closed form (see Problem 6.4 for a special case) and it is more convenient to use the approximate conditional likelihood (4.42). This is particularly easy to calculate because the parameterization (β, σ^2) is orthogonal; furthermore, the link may well be different from the canonical one. From (6.42),

$$|j_{\beta\beta}| = |[j_{rs}]| = (\sigma^2)^{-k} c(\beta, y, X) ,$$

where the function $c(\beta, y, X)$ does not depend on σ^2. In addition, the maximum likelihood estimate of β for fixed σ^2 does not depend on σ^2, that is $\hat{\beta}_{\sigma^2} = \hat{\beta}$. Therefore, the approximate conditional log-likelihood (4.42) takes the form

$$l_{AC}(\sigma^2) = l(\hat{\beta}, \sigma^2) + \frac{k}{2}\log(\sigma^2) , \qquad (6.52)$$

with $l(\beta, \sigma^2)$ given by (6.33).

6.4.2 Quasi-Likelihood

It is easy to check that the likelihood equation (6.37) for β is an unbiased estimating equation provided that $E(Y_i) = \mu_i = g^{-1}(\beta^T x_i)$. In other words, the parametric assumption $Y_i \sim ED_1(\mu_i, \sigma^2 V(\mu_i))$ could not even be satisfied, only the assumption about expectations is essential. Furthermore, the only further distributional feature that must be known in order to calculate the left-hand side of (6.37) is the variance function $V(\mu)$. Therefore it would be interesting to study the properties of the estimator $\hat{\beta}$, the solution to (6.37), under the weaker second-order assumptions

$$E(Y_i) \quad = \quad \mu_i(\beta) = g^{-1}(\beta^T x_i) ,$$

$$\mathrm{Var}(Y_i) \quad = \quad \sigma^2 V(\mu_i) , \qquad (6.53)$$

$$\mathrm{Cov}(Y_i, Y_j) \quad = \quad 0 \quad \text{if} \quad i \neq j .$$

The semiparametric statistical model specified by (6.53) is called a quasi-likelihood model; indeed, the assumptions (6.53) enable a quasi-likelihood function to be defined for β (see section 4.9) based on the combinant

$$q(y; \beta) = [l_r] = \left[\frac{1}{\sigma^2}\sum_{i=1}^{n}\frac{(y_i - \mu_i)}{V(\mu_i)}\frac{\partial\mu_i}{\partial\beta_r}\right] .$$

Since the matrix $(\partial q(y; \beta))/(\partial\beta^T) = [l_{rs}]$ is symmetric, there exists a function $l_Q(\beta)$ which has $q(y; \beta)$ as its gradient. For example, we could consider the function

$$l_Q(\beta) = \sum_{i=1}^{n}\int_{y_i}^{\mu_i}\frac{y_i - t}{\sigma^2 V(t)}dt . \qquad (6.54)$$

If $V(\mu_i) = 1$ and $g(\mu) = \mu$ assumptions (6.53) match the usual second-order assumptions of a classical linear model. On the other hand, if $V(\mu) = \mu^2$ we obtain a multiplicative model, $Y_i = \mu_i \varepsilon_i$, with $E(\varepsilon_i) = 1$ and $\text{Var}(\varepsilon_i) = \sigma^2$.

The model specified by (6.53) is equally suitable for dealing with both continuous and discrete data. In particular, for counts or proportions and for continuous waiting times, the second-order assumptions (6.53) offer an increase in flexibility with respect to the usual parametric specifications based, respectively, on the Poisson, the binomial and the exponential distributions. These parametric specifications consider models such as $Y_i \sim ED_1(\mu_i, \sigma^2 V(\mu_i))$, with $\sigma^2 = 1$ and $V(\mu_i) = \mu_i$ in the Poisson case, $\sigma^2 = n^{-1}$ and $V(\mu_i) = \mu_i(1 - \mu_i)$ if $nY_i \sim Bi(n, \mu_i)$, $\sigma^2 = 1$ and $V(\mu_i) = \mu_i^2$ in the exponential case. However, in applications it is often more plausible to assume $\text{Var}(Y_i) = \sigma^2 V(\mu_i)$, with $\sigma^2 > 1$, that is, that the data show **overdispersion** with respect to the corresponding parametric model where the variance of the observations is entirely determined by the expected value. The case of **underdispersion**, $\sigma^2 < 1$, is less important in applications, but can be dealt with under the model (6.53) as well.

It is not difficult to check that, even under the weaker assumptions (6.53), the information identity

$$E(l_r l_s) = -E(l_{rs}) \tag{6.55}$$

still holds. This identity follows by observing that

$$
E(l_r l_s) = \frac{1}{\sigma^4} E \left\{ \sum_{i=1}^n \frac{(Y_i - \mu_i)}{V(\mu_i)} \frac{\partial \mu_i}{\partial \beta_r} \sum_{j=1}^n \frac{(Y_j - \mu_j)}{V(\mu_j)} \frac{\partial \mu_j}{\partial \beta_s} \right\}
$$

$$
= \frac{1}{\sigma^4} \sum_{i=1}^n \sum_{j=1}^n \frac{\partial \mu_i}{\partial \beta_r} \frac{\partial \mu_j}{\partial \beta_s} \frac{1}{V(\mu_i)} \frac{1}{V(\mu_j)} E\left\{ (Y_i - \mu_i)(Y_j - \mu_j) \right\} \ .
$$

Since $E\{(Y_i - \mu_i)(Y_j - \mu_j)\}$ is zero for $i \neq j$, and is $\sigma^2 V(\mu_i)$ for $i = j$, we obtain

$$
E(l_r l_s) = \frac{1}{\sigma^2} \sum_{i=1}^n \frac{1}{V(\mu_i)} \frac{\partial \mu_i}{\partial \beta_r} \frac{\partial \mu_i}{\partial \beta_s} \ .
$$

On the other hand,

$$
\frac{\partial}{\partial \beta_s} l_r = l_{rs} = \frac{1}{\sigma^2} \sum_{i=1}^n \left\{ (y_i - \mu_i) \frac{\partial}{\partial \beta_s} \left(\frac{1}{V(\mu_i)} \frac{\partial \mu_i}{\partial \beta_r} \right) - \frac{1}{V(\mu_i)} \frac{\partial \mu_i}{\partial \beta_r} \frac{\partial \mu_i}{\partial \beta_s} \right\} ,
$$

$$\tag{6.56}$$

thus, the result is obtained by comparison, because the expected value of the first summand within braces in (6.56) is zero.

The identities $E(l_r) = 0$ and $E(l_r l_s) = -E(l_{rs})$ offer the key arguments for proving the asymptotic properties of the maximum likelihood estimator, especially as regards asymptotic normality. Therefore, even under second-order assumptions, both weak consistency of $\hat{\beta}$ and validity of the approximation $\hat{\beta} \sim N_k(\beta, \sigma^2(X^T W X)^{-1})$ for large n, are maintained. In this sense, the estimator of β based on the estimating equation (6.37) is robust. If σ^2 is unknown, as happens for example if possible overdispersion is taken into account, then an estimate of σ^2 must be available in order to be able to evaluate the asymptotic variance of $\hat{\beta}$. The estimator (6.50) based on the method of moments is still suitable, as it remains consistent under the quasi-likelihood assumptions (6.53).

6.4.3 Deviance Tests

Let $y_1, \ldots, y_n$ be observations from a generalized linear model, and assume, for simplicity, that σ^2 is known. Consider the partition of β, $\beta^T = (\beta_A^T, \beta_B^T)$, with

$$\beta_A^T = (\beta_1, \ldots, \beta_h) \qquad \text{and} \qquad \beta_B^T = (\beta_{h+1}, \ldots, \beta_k),$$

and suppose that we wish to test $H_0 : \beta_B = 0$ against $H_1 : \beta_B \neq 0$.

In the normal linear regression model, $Y \sim N_n(X\beta, \sigma^2 I_n)$, when σ^2 is known, the likelihood ratio test rejects the null hypothesis for large values of the statistic $(SSE_{H_0} - SSE)/\sigma^2$, where SSE_{H_0} and SSE are the residual sums of squares in, respectively, the constrained and the full model. Under H_0, this statistic has a χ^2_{k-h} distribution.

Under the assumptions of a generalized linear model, the statistic

$$W = 2\{l(\hat{\beta}_A, \hat{\beta}_B, \sigma^2) - l(\tilde{\beta}_A, 0, \sigma^2)\}$$

has a null asymptotic χ^2_{k-h} distribution; the symbol $\tilde{\beta}_A$ is used above to denote the maximum likelihood estimate of β_A when $\beta_B = 0$. The formal analogy with the normal linear model can be highlighted by writing (6.33) as

$$\begin{aligned}
l(\beta, \sigma^2) &= l^M(\mu, \sigma^2) \\
&= \frac{1}{\sigma^2} \sum_{i=1}^n \{y_i \theta(\mu_i) - K(\theta(\mu_i))\} + \sum \log a\left(\frac{1}{\sigma^2}, y_i\right),
\end{aligned}$$

with $\theta(\mu_i) = \theta(g^{-1}(\beta^T x_i))$, and defining

$$D(y; \hat{\mu}) = 2\sigma^2\left\{l^M(y, \sigma^2) - l^M(\hat{\mu}, \sigma^2)\right\}$$

$$= 2\sum_{i=1}^{n}\{y_i\,(\theta(y_i) - \theta(\mu_i)) - (K(\theta(y_i)) - K(\theta(\mu_i)))\}\;.\;(6.57)$$

The quantity $D(y;\hat\mu)$ is called the deviance. The symbol $l^M(\mu,\sigma^2)$ is used to denote the log-likelihood with the parameterization (μ,σ^2) when μ is not constrained to belong to the mean space (6.41). In particular, $l^M(y,\sigma^2)$ is obtained with $\mu_i = y_i$, $i = 1,\ldots,n$, i.e. by fitting the saturated regression model with $k = n$; in this case, μ is a one-to-one function of β and, by (6.37), $\hat\mu = y$. Thus, the difference $l^M(y,\sigma^2) - l^M(\hat\mu,\sigma^2)$ represents a measure of the reduction in goodness-of-fit from the saturated model to that with $k < n$ explanatory variables. Therefore, $D(y;\hat\mu)$ admits an interpretation which is analogous to that of the residual deviance SSE in a classical linear model. The log-likelihood ratio statistic can be written as

$$W = \frac{D(y,\hat\mu_0) - D(y,\hat\mu)}{\sigma^2},\qquad(6.58)$$

where $\hat\mu_0 = \mu(\bar\beta_A, 0)$. If σ^2 is unknown, it must be substituted in (6.58) by a consistent estimate (see section 6.4.1); the null asymptotic distribution remains χ^2_{k-h}.

Jørgensen (1987) obtained some interesting small dispersion asymptotic results which are fairly similar to the usual results of exact null distributions for likelihood-based tests in a normal regression model. For example, as $\sigma^2 \to 0$, the statistic

$$F = \frac{(D(y,\hat\mu_0) - D(y,\hat\mu))/(k-h)}{D(y,\hat\mu)/(n-k)}$$

has null asymptotic $F_{k-h,n-k}$ distribution.

To test $H_0 : \beta_B = 0$ under second-order assumptions, we could use the statistic analogous to (6.58) based on the quasi-likelihood function (6.54),

$$W_Q = 2\left\{l_Q(\hat\beta_A,\hat\beta_B) - l_Q(\bar\beta_A, 0)\right\}.$$

The statistic W_Q has null asymptotic χ^2_{k-h} distribution, both when σ^2 is known and when σ^2 is replaced by the estimator (6.50). The asymptotically equivalent forms based on combinants (4.32) and (4.31) are the same as those of the parametric case.

6.5 Generalized Linear Models for Binary Data

Let $Y_1,\ldots,Y_n$ be independent $Bi(1,\mu_i)$ random variables. Suppose that we wish to specify a model which expresses the probability of *success*, $\mu_i = P\{Y_i = $

1}, as a function of k explanatory variables with values $(x_{i1}, \ldots, x_{ik})$, $i = 1, \ldots, n$, with $k < n$. Since $Y_i \sim ED_1(\mu_i, V(\mu_i))$, with $V(\mu_i) = \mu_i(1 - \mu_i)$ (see Example 6.6), the statistical model is a generalized linear model if

$$g(\mu_i) = \beta^T x_i, \quad x_i = (x_{i1}, \ldots, x_{ik})^T,$$

that is, if the expectation of Y_i is assumed to be a function of the explanatory variables through the linear predictor $\eta_i = \beta^T x_i$. The link $g(\cdot)$ is usually chosen as an increasing function $g : [0, 1] \to \mathbb{R}$. The following link functions are often used in applications:

- logistic (canonical link)

$$g(\mu_i) = \log\left(\frac{\mu_i}{1 - \mu_i}\right) = \beta^T x_i, \tag{6.59}$$

   with inverse

$$\mu_i = \frac{e^{\beta^T x_i}}{1 + e^{\beta^T x_i}}; \tag{6.60}$$

   this generalized linear model is called the **logistic regression model** or, sometimes, the **logit** model;

- probit

$$g(\mu_i) = \Phi^{-1}(\mu_i) = \beta^T x_i, \tag{6.61}$$

   with inverse

$$\mu_i = \Phi(\beta^T x_i), \tag{6.62}$$

   where $\Phi(\cdot)$ is the standard normal distribution function; the corresponding generalized linear model is called the **probit** model;

- complementary log-log

$$g(\mu_i) = \log\{-\log(1 - \mu_i)\} = \beta^T x_i, \tag{6.63}$$

   with inverse

$$\mu_i = 1 - \exp\{-\exp(\beta^T x_i)\}; \tag{6.64}$$

- log-log

$$g(\mu_i) = -\log(-\log(\mu_i)),$$

   with inverse

$$\mu_i = \exp\{-\exp(-\beta^T x_i)\}.$$

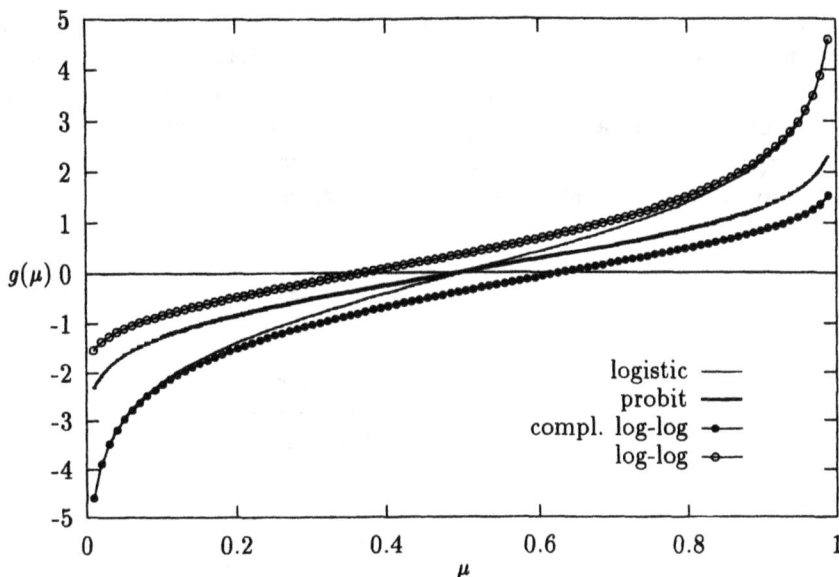

Figure 6.1: *Link functions for binary data.*

All these link functions are increasing; their behaviour is shown in Figure 6.1.

The relationship between logit and probit is almost linear for $0.1 \leq \mu_i \leq 0.9$; the complementary log-log function tends to $+\infty$ as $\mu_i \to 1$ more slowly than the others; the log-log function is rarely used because of its behaviour for $\mu_i < 0.5$.

The logistic link offers some advantages. It can be interpreted in terms of log-odds. Moreover, inference is simplified because the model is an $\mathcal{F}_{ne}^k$. The logistic model may be used for analyzing retrospective data as well, which is important for epidemiological studies (Breslow and Day, 1980). To illustrate this last point, consider a generic subject, and consider the fact that he or she catches a specific disease as a *success*. With E indicate the event *the subject catches the disease* and with $\bar{E}$ indicate its complementary. Let x be a vector of covariates. The logistic regression model (with intercept) assumes

$$P(E|x) = \frac{e^{\alpha + \beta^{\mathsf{T}} x}}{1 + e^{\alpha + \beta^{\mathsf{T}} x}} \, . \tag{6.65}$$

In a retrospective study one fixes the number of cases (subjects who have caught the disease) and the number of controls (subjects who have not caught the disease), to be randomly extracted from a population. Let Z be the binary indicator of whether a subject participates in the study and let $\pi_0 = P(Z = 1|E)$, $\pi_1 = P(Z = 1|\bar{E})$ be the known sampling fractions for cases and controls, respectively. Assume that π_0 and π_1 do not depend on x. Thus, according to Bayes' theorem,

$$
\begin{aligned}
P(E|Z = 1, x) &= \frac{P(E|x)P(Z = 1|E)}{P(Z = 1|E)P(E|x) + P(Z = 1|\bar{E})P(\bar{E}|x)} \\[2mm]
&= \frac{\pi_0 e^{\alpha + \beta^T x} \Big/ \left(1 + e^{\alpha + \beta^T x}\right)}{\left(\pi_0 e^{\alpha + \beta^T x} + \pi_1\right) \Big/ \left(1 + e^{\alpha + \beta^T x}\right)} \\[2mm]
&= \frac{(\pi_0/\pi_1)e^{\alpha + \beta^T x}}{1 + (\pi_0/\pi_1)e^{\alpha + \beta^T x}} = \frac{e^{\alpha^* + \beta^T x}}{1 + e^{\alpha^* + \beta^T x}},
\end{aligned}
$$

with $\alpha^* = \log(\pi_0/\pi_1) + \alpha$. Hence, provided that the model has an intercept term, the estimate of β is not influenced by the fact that the data are collected retrospectively. Note that the usual regression models are formulated with a prospective study in mind, that is a study where the probability of being included in the sample does not depend on the value of the response.

Example 6.7 *Logistic regression model and contingency tables*
As in all linear models, it is possible to deal with categorical explanatory variables by inserting dummy variables. Consider the simplest situation with a single binary explanatory variable. Thus, observations are subdivided into two groups, for example, treated and not treated subjects. Let $x_i = 1$ if the i-th subject is treated and $x_i = 0$ if the i-th subject is not treated. The data can be displayed as in the 2×2 table below

	treated	not treated	
success	n_{11}	n_{12}	$n_{1\bullet}$
unsuccess	n_{21}	n_{22}	$n_{2\bullet}$
	$n_{\bullet 1}$	$n_{\bullet 2}$	n

with $n_{\bullet 1} = \sum_{i=1}^{n} x_i$ and $n_{\bullet 2} = \sum_{i=1}^{n}(1 - x_i)$. If the vector $(y_1, \ldots, y_n)$ represents the observations on the response variable, $n_{1\bullet} = \sum_{i=1}^{n} y_i$ gives the total number of successes and $n_{2\bullet} = \sum_{i=1}^{n}(1 - y_i) = n - n_{1\bullet}$ the total number of unsuccesses.

The quantity $n_{11} = \sum_{i=1}^{n} x_i y_i$ represents the number of successes among treated subjects. The logit model assumes

$$P(Y_i = 1| x_i) = \frac{e^{\alpha+\beta x_i}}{1 + e^{\alpha+\beta x_i}},$$

that is,

$$P(Y_i = 1| x_i = 1) = \frac{e^{\alpha+\beta}}{1 + e^{\alpha+\beta}}$$

and

$$P(Y_i = 1| x_i = 0) = \frac{e^{\alpha}}{1 + e^{\alpha}}.$$

The log-odds ratio is

$$\log\left(\frac{P(Y_i = 1|x_i = 1)/P(Y_i = 0|x_i = 1)}{P(Y_i = 1|x_i = 0)/P(Y_i = 0|x_i = 0)}\right) = \beta.$$

The model can easily be extended to deal with data which are classified according to the levels of a further categorical variable, Z, with J levels. In other words, we can assume that the observations y_{ij}, $i = 1, \ldots, n$, $j = 1, \ldots, J$, are independent realizations of

$$Y_{ij} \sim Bi\left(1, \frac{e^{\alpha_j+\beta_j x_{ij}}}{1 + e^{\alpha_j+\beta_j x_{ij}}}\right). \tag{6.66}$$

This assumption is the same as formulating a logistic regression model for each of the 2×2 tables which are obtained by fixing the level of Z. If $\beta_1 = \cdots = \beta_J = \beta$, then β represents the common log-odds ratio for all these tables. See Davison (1992) for the problem of testing the hypothesis of homogeneity $H_0 : \beta_1 = \cdots = \beta_J$. $\triangle$

The log-likelihood function for a general logistic regression model is

$$
\begin{aligned}
l(\beta) &= \log\left\{\prod_{i=1}^{n}\left(\frac{e^{\beta^T x_i}}{1 + e^{\beta^T x_i}}\right)^{y_i}\left(\frac{1}{1 + e^{\beta^T x_i}}\right)^{1-y_i}\right\} \\
&= \beta^T \sum_{i=1}^{n} x_i y_i - \sum_{i=1}^{n} \log\left(1 + e^{\beta^T x_i}\right) \\
&= \sum_{j=1}^{k} \beta_j \sum_{i=1}^{n} x_{ij} y_i - \sum_{i=1}^{n} \log\left(1 + e^{\sum_{j=1}^{k} \beta_j x_{ij}}\right).
\end{aligned}
\tag{6.67}
$$

The minimal sufficient statistic is $s = (s_1, \ldots, s_k)^{\mathrm{T}}$, with $s_j = \sum_{i=1}^{n} x_{ij} y_i$, and it has distribution in an $\mathcal{F}_{ne}^{k}$. The results for exponential families regarding marginal and conditional distributions, existence of the maximum likelihood estimate of β, etc., apply for the statistical model induced by s. In particular,

$$p_s(s; \beta) = \frac{C(s) e^{\beta^{\mathrm{T}} s}}{\prod_{i=1}^{n} \left(1 + e^{\beta^{\mathrm{T}} x_i}\right)} ,$$

where $C(s)$ represents the number of distinct subsets of $(x_1, \ldots, x_n)$ such that $\sum_{i=1}^{n} x_i y_i = s$. The statistical model for s is a regular $\mathcal{F}_{ne}^{k}$ because $\beta \in \mathbb{R}^k$, hence, from the results in section 5.6, the maximum likelihood estimate of β exists so long as $s \in \mathrm{int}\, C$, where C represents the closed convex hull of the support of S.

With the exception of some very simple cases (see, for example, Problem 5.23), it may not be easy to see whether s belongs to the boundary of C. If the model includes an intercept term, then a necessary and sufficient condition for s to belong to $\mathrm{int}\, C$ is that the sets

$$A = \left\{ x : x = \sum_{i=1}^{n} k_i y_i x_i,\ k_i > 0 \right\},$$

which represents the open cone generated by the vectors x_i which correspond to successes, and

$$B = \left\{ x : x = \sum_{i=1}^{n} k_i (1 - y_i) x_i,\ k_i > 0 \right\},$$

which represents the open cone generated by the vectors x_i which correspond to unsuccesses, do not have an empty intersection (Silvapulle, 1981). The same condition is still necessary and sufficient for the maximum likelihood estimate of β to exist also for the probit model with an intercept. The example below, drawn from Silvapulle (1981), illustrates such a condition for the existence of $\hat{\beta}$ in a logistic regression model.

Example 6.8 *Logistic regression: existence of the m.l.e.*
A binary variable has been observed on a sample of 35 units, classified on the basis of the values of a covariate X. The table below shows the number of successes and unsuccesses for each level of X.

X	0	1	2	4	5	7	10	
success	0	0	1	1	3	2	1	8
unsuccess	18	8	1	0	0	0	0	27
	18	8	2	1	3	2	1	35

We wish to know whether the maximum likelihood estimate $(\hat{\beta}_1, \hat{\beta}_2)$ of the parameters of the logistic regression model

$$P(Y_i = 1 | X_i = x_i) = \frac{e^{\beta_1 + \beta_2 x_i}}{1 + e^{\beta_1 + \beta_2 x_i}} , \qquad i = 1, \ldots, 35 ,$$

exists. The two-dimensional vectors of the explanatory variables for the observations which correspond to successes, which generate the cone A, are

$$\begin{pmatrix} 1 \\ 2 \end{pmatrix} \quad \begin{pmatrix} 1 \\ 4 \end{pmatrix} \quad \begin{pmatrix} 1 \\ 5 \end{pmatrix} \quad \begin{pmatrix} 1 \\ 5 \end{pmatrix} \quad \begin{pmatrix} 1 \\ 5 \end{pmatrix} \quad \begin{pmatrix} 1 \\ 7 \end{pmatrix} \quad \begin{pmatrix} 1 \\ 7 \end{pmatrix} \quad \begin{pmatrix} 1 \\ 10 \end{pmatrix} .$$

The vectors of the explanatory variables for the observations which correspond to unsuccesses, which generate cone B, are

$$\underbrace{\begin{pmatrix} 1 \\ 0 \end{pmatrix} \cdots \begin{pmatrix} 1 \\ 0 \end{pmatrix}}_{18 \text{ times}} \underbrace{\begin{pmatrix} 1 \\ 1 \end{pmatrix} \cdots \begin{pmatrix} 1 \\ 1 \end{pmatrix}}_{8 \text{ times}} \begin{pmatrix} 1 \\ 2 \end{pmatrix} .$$

We can immediately check (see, also, Silvapulle, 1981, for a graphic representation) that $A \cap B = \emptyset$; therefore there is no finite maximum likelihood estimate of (β_1, β_2). $\triangle$

As a final point, we mention how the presence of overdispersion can be justified in a regression problem with a binary response. In such a situation it would be wise to base analyses on the quasi-likelihood model (6.53). Take the case where, for some combinations of levels of the explanatory variables, several observations on the response are available. The data may be represented in the form of proportions $y_1/m_1, \ldots, y_n/m_n$, with $y_i = 0, 1, \ldots, m_i$, $i = 1, \ldots, n$, where m_i is the number of observations in the i-th class, identified by a combination of levels of the explanatory variables, and y_i is the number of successes in the same class. The logistic regression model assumes

$$Y_i \sim Bi\left(m_i, \frac{e^{\beta^T x_i}}{1 + e^{\beta^T x_i}} \right) , \qquad i = 1, \ldots, n .$$

One can introduce various models which allow $\text{Var}(Y_i) > m_i \mu_i (1 - \mu_i)$, with $\mu_i = e^{\beta^T x_i} / \left(1 + e^{\beta^T x_i} \right)$. A typical situation occurs when the population is naturally subdivided into subpopulations which are not accounted for by the explanatory variables. The units which belong to the same class i, defined by the level of the explanatory variables, may be heterogeneous in that they belong to different subgroups. For simplicity, suppose that $m_1 = \cdots = m_n = m$ and each sample of size m is composed of $J = m/h$ subsamples (one for each subgroup of the population) of size h. Furthermore, suppose that in each subsample, the number of successes is

$$Z_{ij} \sim Bi(h, \pi_{ij}), \quad j = 1, \ldots, J, \quad i = 1, \ldots, n.$$

One simple model to take into account lack of homogeneity among subpopulations could be based on the assumption that, in the i-th class, the variability of probability π_{ij} is described by a probability distribution such that

$$
\begin{aligned}
E(\pi_{ij}) &= \pi_i \\
\mathrm{Var}(\pi_{ij}) &= \tau^2 \pi_i(1 - \pi_i),
\end{aligned}
$$

where $\tau = 0$ if there is homogeneity. The total number of successes in the i-th class is

$$Y_i = Z_{i1} + \cdots + Z_{iJ}$$

and we have

$$E(Y_i) = m\pi_i, \tag{6.68}$$

$$\mathrm{Var}(Y_i) = m\pi_i(1 - \pi_i)\{1 + (h - 1)\tau^2\} = m\pi_i(1 - \pi_i)\sigma^2, \tag{6.69}$$

with $\sigma^2 = 1 + (h - 1)\tau^2$ the dispersion parameter. Observe that $\sigma^2 = 1$ only if $\tau = 0$ or if $h = 1$.

One further source of overdispersion may be dependence among the binary variables of which Y_i is the sum (see, for example, Cox and Snell, 1989, section 3.2.2).

6.6 Bibliographic Note

The introduction of exponential dispersion families in section 6.2 is based on Jørgensen (1986, 1987); see Jørgensen (1992) for a survey of recent developments and also for a technique for constructing stochastic processes with stationary and independent increments which exploits Theorem 6.1.

The main reference for generalized linear models is the book by McCullagh and Nelder (1989) which treats in detail some important examples of generalized linear models not included here: models for politomous data and log-linear models (*ibidem*, Chapters 5 and 6). An elementary introduction to generalized linear models is Dobson (1990); for a brief survey, see Firth (1991). For the evaluation of models in terms of analysis of residuals and of tests for goodness-of-fit, see Pierce and Schafer (1986), McCullagh and Nelder (1989, sections 2.3 and 2.4, Chapter 12) and Davison and Tsai (1992). In particular, a test of goodness-of-fit can be based on the deviance (6.57) or on the **Pearson** statistic (6.50): see the contributions by McCullagh (1985, 1986a), Jørgensen (1987) and Farrington (1996) for the asymptotic distribution of this statistic. Greater flexibility in the choice of the link can be obtained by considering parametric families of link functions (see Problem 6.25; Pregibon, 1980; McCullagh and Nelder, 1989, section 11.3); furthermore, this makes evaluation of the goodness-of-fit of a fixed link function which belongs to a parametric class more direct. Cheng and Wu (1994) propose a test of goodness-of-fit of a parametric family of link functions. Mallick and Gelfand (1994) consider inference for generalized linear models with a semiparametric specification, which assumes that the link function could be any strictly increasing function.

Quasi-likelihood was introduced by Wedderburn (1974); McCullagh (1983) studied the asymptotic properties of statistics based on quasi-likelihood. An introduction, together with extensions to dependent observations can be found in McCullagh and Nelder (1989, Chapter 9); for a brief summary, see McCullagh (1986b). Recent surveys are McCullagh (1991) and Firth (1993). Various extensions of the quasi-likelihood model have been proposed in recent years with the aim of allowing more flexible modelling of the response variance: for example, the so-called moments method, described by Moore (1986), and the proposals examined in Nelder and Lee (1992) and in Yanez and Wilson (1995).

Most of section 6.5 is based on McCullagh and Nelder (1989, Chapter 4). An exhaustive monograph, dedicated to the analysis of categorical variables, is Agresti (1990). Albert and Anderson (1984) and Santner and Duffy (1986) extend Silvapulle's (1981) condition for the existence of the maximum likelihood estimate to multinomial logistic regression models.

6.7 Problems

6.1 Show that in a dispersion family with density (6.1) the combinant $q(y; \theta)$ is an unbiased estimating function for θ for every λ. Furthermore, show that the parameters θ and λ are orthogonal.

[Section 6.1]

6.2 Determine the $\mathcal{F}_{ed}^p$ generated by $X \sim N_p(0, I_p)$ by convolution, exponential tilting (6.4) and rescaling (6.5).

[Section 6.2]

6.3 Let $p_0(x) = p(x; \theta_0, \lambda_0)$ be one element of the family (6.6). Consider the family with density $p_0(x; \lambda)$ generated by $p_0(x)$ by convolution. Then consider the $\mathcal{F}_{ne}^p$ generated by $p_0(x; \lambda)$ through exponential tilting. Let X be a r.v. with density $p(x; \theta, \lambda)$ which belongs to this exponential family; find the density of $Y = X/\lambda$. Deduce that an $\mathcal{F}_{ed}^p$ could be thought to be generated by any one of its elements by means of the above extension steps.

[Section 6.2]

6.4 Let $y_1, \ldots, y_n$ be independent and identically distributed observations from a random variable with density in an $\mathcal{F}_{ed}^1$. Exploiting Theorem 6.1, obtain a conditional likelihood for λ and discuss the absence of information about λ in the neglected likelihood factor.

[Section 6.3; Jørgensen, 1993]

6.5 Show that closure under convolution of the $Ga(\nu, 1)$ family is a corollary of Theorem 6.1.

[Section 6.3]

6.6 Show that in a g.l.m. the dispersion parameter σ^2 is orthogonal to the regression parameters $\beta = (\beta_1, \ldots, \beta_k)$.

[Section 6.4]

6.7 Write the assumptions for a g.l.m. with gamma errors, that is with $Y_i \sim ED_1(\mu_i, \sigma^2\mu_i^2)$. Discuss the choice of the canonical link from the point of view of applications.

[Section 6.4]

6.8 Let Y be a scalar r.v. with density $p_Y(y; \theta)$, with $\theta \in \Theta \subseteq \mathbb{R}$. Since this is a one-parameter model, problems of overdispersion may arise in applications. A general approximate model may be specified to take possible overdispersion into account, by assuming that θ is a random variable with distribution specified only in terms of expectation and variance. More specifically,

$$E(\theta) = \xi, \qquad \text{Var}(\theta) = \frac{\tau}{\sqrt{n}} ,$$

with $\tau > 0$. The density $p_Y(y;\theta)$ should be understood as being a conditional density given θ. With $\mu(\theta)$ and $\sigma^2(\theta)$ indicate the mean and the variance of Y given θ. Show that the expansion

$$\text{Var}(Y) = \sigma^2(\xi) + \frac{\tau}{\sqrt{n}} \left\{ \frac{1}{2} \frac{\partial^2 \sigma^2(\xi)}{\partial \xi^2} + \left(\frac{\partial \mu(\xi)}{\partial \xi} \right)^2 \right\} + o(n^{-1/2})$$

holds for the marginal variance Y and interpret τ as the parameter which governs overdispersion. Obtain an analogous asymptotic expansion for the marginal density of Y up to, and including, terms of order $O(n^{-1/2})$. Given n i.i.d. observations of a r.v. with density in the approximated statistical model, obtain the locally most powerful test for $H_0 : \tau = 0$ versus $H_1 : \tau > 0$. Interpret the test statistic obtained.

[Section 6.4; Cox, 1983]

6.9 Identify a minimal sufficient statistic for (β, σ^2) in the g.l.m. with the canonical link where the n independent variables $Y_i \sim ED_1(\mu_i, \sigma^2 V(\mu_i))$ have a gamma distribution.

[Section 6.4.1]

6.10 Identify a minimal sufficient statistic for (β, σ^2) in the g.l.m. with the canonical link where the n independent variables $Y_i \sim ED_1(\mu_i, \sigma^2 V(\mu_i))$ have an inverse Gaussian distribution.

[Section 6.4.1]

6.11 Determine the estimating equation for σ^2 which is based on the approximate conditional log-likelihood (6.52) for a g l m. with gamma errors and any link function.

[Section 6.4.1]

6.12 Obtain the expression of $\tilde{\sigma}^2$ given by (6.50) for the generalized linear model based on the assumptions

$$Y_i \sim ED_1(\mu_i, \sigma^2 \mu_i^2) \quad \text{and} \quad Y_i \sim ED_1(\mu_i, \sigma^2 \mu_i^3),$$

respectively, in both cases using the canonical link.

[Section 6.4.1]

6.13 Let Y_{ij}, $i = 1, \ldots, n$, $j = 1, \ldots, p$, be independent random variables with Poisson distribution with mean μ_{ij}. Write the g.l.m. for Y_{ij} using the canonical link and the linear predictor $\eta_{ij} = \alpha_i + \beta^T x_{ij}$, where x_{ij} is a k-dimensional vector of explanatory variables for y_{ij}. Identify a conditional likelihood for β and discuss whether it is useful to condition in this situation.

[Section 6.4.1; McCullagh and Nelder, 1989, section 6.4.2]

6.14 Determine the quasi-likelihood function (6.54) for the following specifications of the variance function: $V(\mu) = 1$, $V(\mu) = \mu$, $V(\mu) = \mu^2$, $V(\mu) = \mu^3$, $V(\mu) = \mu^\gamma, \gamma > 0$, $V(\mu) = \mu(1 - \mu)$, $V(\mu) = \mu^2(1 - \mu^2)$. State whether, and in which cases, one reobtains the log-likelihood functions which correspond to known parametric models.

[Section 6.4.2]

6.15 Show that the estimating equations for β based on the assumptions (6.53) are unbiased even when the function $V(\mu)$ is not correctly specified. Using Royall's method, illustrated in section 3.4.3, obtain an estimator for the asymptotic covariance matrix of $\hat{\beta}$, which is consistent even when $V(\mu)$ is not correctly specified.

[Section 6.4.2]

6.16 Obtain the expression of the statistics (4.31) and (4.32) for testing H_0: $\beta_B = 0$ versus H_1: $\beta_B \neq 0$ as in section 6.4.3, both with known σ^2 and with unknown σ^2.

[Section 6.4.3]

6.17 Write the contribution of a single observation to the deviance (6.57) for the generalized linear models with the canonical links summarized in Table 6.1.

[Section 6.4.3]

6.18 With the same notation as in section 6.4.3, consider the problem of testing H_0: $\beta_B = 0$ versus H_1: $\beta_B \neq 0$ under the second-order assumptions (6.53). Show that the statistics based on the combinants (4.31) and (4.32), deduced from the quasi-log-likelihood (6.54), match the statistics obtained in Problem 6.16 for the parametric case.

[Section 6.4.3]

6.19 Explain why the probit model, unlike the logistic regression model, cannot be used for analyzing retrospective data.

[Section 6.5]

6.20 Write the log-likelihood function (6.67) for the special case of a logistic regression model with intercept and with one binary explanatory variable (see Example 6.7).

[Section 6.5]

6.21 Let $y_1, \ldots, y_n$ be independent realizations of

$$Y_i \sim Bi\left(m_i, \frac{e^{\beta^{\mathrm{T}} x_i}}{1 + e^{\beta^{\mathrm{T}} x_i}}\right), \quad i = 1, \ldots, n.$$

Write the likelihood function for β.

[Section 6.5]

6.22 Carry out the detailed calculations which give (6.68) and (6.69).

[Section 6.5]

6.23 Obtain the expression of the estimator $\tilde{\sigma}^2$, given by (6.50), of the dispersion parameter σ^2 for the model specified by (6.68) and (6.69).

[Section 6.5]

6.24 Let Y_{ij}, $i = 1, \ldots, n$, $j = 1, \ldots, J$, be independent variables distributed according to (6.66) with $\beta_1 = \cdots = \beta_J = \beta$; identify a suitable pseudo-likelihood for β.

[Section 6.5]

6.25 Consider a g.l.m. for a binary response with a link function specified as

$$g(\mu_i) = \log\left\{\frac{1}{\delta}\left(\left(\frac{1}{1 - \mu_i}\right)^{\delta} - 1\right)\right\},$$

where $\delta > 0$ is an unknown parameter. Define a test statistic for the null hypothesis that the link function is the canonical link versus alternatives in the direction of the complementary log-log link.

[Section 6.5; Pregibon, 1980; McCullagh and Nelder, 1989, section 11.3]

CHAPTER 7

GROUP FAMILIES

7.1 Introduction

As mentioned in section 5.1, group families are a second important class of statistical models. With regard to identifying the information contained in the data, group families represent the natural setting where both the conditionality principle, and marginal likelihood to eliminate nuisance parameters can be applied. Hence, the theory of group families is complementary to that of exponential families where, from the point of view of data and model reduction, the key aspects are sufficiency, and conditioning to eliminate nuisance parameters.

Let Y be a non-degenerate random variable with support $\mathcal{Y} \subseteq \mathbb{R}$, density $p_0(y)$ and distribution function $F_0(y)$. Consider the class of probability distributions induced by real measurable transformations $g(Y)$, with $g(\cdot)$ one-to-one. This class is made up of probability distributions on $\mathbb{R}$ whose support has the same cardinality as $\mathcal{Y}$. By restricting the transformations $g(\cdot)$ to suitable subclasses, it is possible to find a simple and easily interpretable method for constructing families of probability distributions which can be used as statistical models. If the identity transformation falls within the chosen subclass, the family will contain $p_0(y)$ and we can say that it is generated by $p_0(y)$ under the given subclass of transformations of the sample space.

Example 7.1 *Location family generated by $p_0(\cdot)$*
Let Y be a non-degenerate r.v. with density $p_0(y)$ and d.f. $F_0(y)$, $y \in \mathcal{Y} \subseteq \mathbb{R}$. The random variable $\mu + Y$, $\mu \in \mathbb{R}$, has d.f. $F(y; \mu) = F_0(y - \mu)$. The family

$$\mathcal{F}_L = \{p(y; \mu), \quad \mu \in \mathbb{R}, \quad y \in \mathcal{Y}_\mu\}, \tag{7.1}$$

where

$$p(y; \mu) = p_0(y - \mu)$$

and $\mathcal{Y}_\mu = \mu + \mathcal{Y} = \{y \in \mathbb{R} : y - \mu \in \mathcal{Y}\}$ is the support of $\mu + Y$, is called the location family generated by $p_0(\cdot)$. Observe that $p_0(y) = p(y; 0)$. $\triangle$

Example 7.2 *Scale family generated by $p_0(\cdot)$*
Consider the family of densities induced by the transformations σY, with $\sigma > 0$. The r.v. σY has d.f. $F(y; \sigma) = F_0(y/\sigma)$. If $p_0(\cdot)$ is a density with respect to the Lebesgue measure, we obtain the family of densities

$$\mathcal{F}_S = \{p(y; \sigma), \quad \sigma > 0, \quad y \in \mathcal{Y}_\sigma\} , \tag{7.2}$$

where

$$p(y; \sigma) = \frac{1}{\sigma} p_0 \left(\frac{y}{\sigma}\right)$$

and $\mathcal{Y}_\sigma = \sigma \mathcal{Y} = \{y \in \mathbb{R} : y/\sigma \in \mathcal{Y}\}$. The class (7.2) is said to be the **scale** family generated by $p_0(\cdot)$. $\triangle$

Example 7.3 *Scale and location family generated by $p_0(\cdot)$*
By combining the two classes of transformations illustrated in the above examples, we can define the **scale** and **location** family generated by $p_0(\cdot)$ as the family of the densities of the random variables $\mu + \sigma Y$, with $\mu \in \mathbb{R}$ and $\sigma > 0$, where Y has density $p_0(y)$. If $p_0(\cdot)$ is a density with respect to the Lebesgue measure, we obtain the family

$$\mathcal{F}_{SL} = \{p(y; \mu, \sigma), \quad \mu \in \mathbb{R}, \ \sigma > 0, \quad y \in \mathcal{Y}_{\mu, \sigma}\} , \tag{7.3}$$

with

$$p(y; \mu, \sigma) = \frac{1}{\sigma} p_0 \left(\frac{y - \mu}{\sigma}\right)$$

and $\mathcal{Y}_{\mu, \sigma} = \mu + \sigma \mathcal{Y} = \{y \in \mathbb{R} : (y - \mu)/\sigma \in \mathcal{Y}\}$. $\triangle$

Note that a closure property holds for each of the parametric classes considered in the preceding examples: each family can be rebuilt starting from any one of its elements by applying the same class of transformations. Note, too, that the support of a random variable which has a density that belongs to any one of the three classes (7.1)–(7.3) usually depends on the parameter. In particular, both in a location family and in a scale and location family $\mathcal{Y}_\mu = \mathcal{Y}$ and $\mathcal{Y}_{\mu,\sigma} = \mathcal{Y}$ only if $\mathcal{Y} = \mathbb{R}$; in a scale family $\mathcal{Y}_\sigma = \mathcal{Y}$ only if $\mathcal{Y} = \mathbb{R}$ or $\mathcal{Y} = \mathbb{R}^+$, $\mathcal{Y} = \mathbb{R}^-$. Except in the special case where the generating element Y has zero mean or unit variance, the parameters μ and σ, respectively, do not coincide with the mean and the standard deviation of a generic element of the family. Table 7.1 shows the density functions for some important scale and location families. The scale and location family $EV(\mu, \sigma)$ could be thought of as being generated by the density $p_0(y) = \exp\{y - e^y\}$, $y \in \mathbb{R}$, whose distribution function is $F_0(y) = 1 - e^{-e^y}$. Note that $F_0(y)$ coincides with the distribution function of $-Z$, where Z has extreme value distribution Λ_3, see (B.4). Furthermore, if $Y \sim EV(\mu, \sigma)$, the transformed variable $X = \exp(Y)$ has a Weibull distribution with density (1.29) and parameters $\nu = 1/\sigma$ and $\lambda = e^{-\mu}$. The term *extreme value distribution*, will, from now on, be used in a restricted sense, to indicate the $EV(\mu, \sigma)$ distribution.

As the following example will show, the three classes (7.1), (7.2) and (7.3) not only have similar constructions, but also share interesting properties for data and model reduction.

Example 7.4 *Model reduction in scale or location families*
Let $y_1, \ldots, y_n$ be independent realizations of a random variable with density belonging to one of the three classes (7.1), (7.2) or (7.3), and let $\bar{y}_n = n^{-1} \sum y_i$, $s_n^2 = \sum(y_i - \bar{y}_n)^2/(n - 1)$ be the sample mean and variance.

i) The statistic $c_1 = (y_1 - \bar{y}_n, \ldots, y_n - \bar{y}_n)$ is distribution constant in the family (7.1) hence, according to the conditionality principle, inference about μ should be carried out conditionally on c_1;

ii) the statistic $c_2 = \left(\dfrac{y_1}{\sqrt{s_n^2}}, \ldots, \dfrac{y_n}{\sqrt{s_n^2}} \right)$ is distribution constant in the family (7.2) hence, inference about σ should be conditional on c_2;

iii) the statistic $c_3 = \left(\dfrac{y_1 - \bar{y}_n}{\sqrt{s_n^2}}, \ldots, \dfrac{y_n - \bar{y}_n}{\sqrt{s_n^2}} \right)$ is distribution constant in the family (7.3) and inference about (μ, σ) should be carried out conditionally on c_3;

iv) the distribution of the statistic $c_4 = (y_1 - \bar{y}_n, \ldots, y_n - \bar{y}_n)$ depends only on σ in the family (7.3); hence the marginal model induced by c_4 could be useful for inference about σ when μ is a nuisance parameter. $\triangle$

This chapter will be concerned with finding an answer to the two questions raised by the preceding examples. In the first place, how to extend the techniques adopted for building parametric families in Examples 7.1, 7.2 and 7.3 by considering a generic class of transformations. This will be taken up in sections 7.2, 7.3 and 7.4. Data and model reduction and the definition of

Table 7.1: *Some scale and location families.*

Distribution	Probability density	Support $\mathcal{Y}_{\mu,\sigma}$
Normal: $N(\mu, \sigma^2)$	$\dfrac{1}{\sigma}\dfrac{1}{\sqrt{2\pi}}e^{-\frac{1}{2}\left(\frac{y-\mu}{\sigma}\right)^2}$	$\mathbb{R}$
Laplace: $La(\mu, \sigma)$	$\dfrac{1}{2\sigma}e^{-|y-\mu|/\sigma}$	$\mathbb{R}$
Cauchy: $Cau(\mu, \sigma)$	$\dfrac{1}{\sigma\pi}\dfrac{1}{1+\left(\frac{y-\mu}{\sigma}\right)^2}$	$\mathbb{R}$
Logistic: $Lo(\mu, \sigma)$	$\dfrac{1}{\sigma}\dfrac{e^{-(y-\mu)/\sigma}}{\left\{1+e^{-(y-\mu)/\sigma}\right\}^2}$	$\mathbb{R}$
Exponential: $Exp(\mu, \sigma)$	$\dfrac{1}{\sigma}e^{-(y-\mu)/\sigma}$	$[\mu, +\infty)$
Uniform: $U(\mu, \mu+\sigma)$	$\dfrac{1}{\sigma}$	$[\mu, \mu+\sigma]$
Extreme Value: $EV(\mu, \sigma)$	$\dfrac{1}{\sigma}\exp\left\{\dfrac{y-\mu}{\sigma}-e^{(y-\mu)/\sigma}\right\}$	$\mathbb{R}$

appropriate inference procedures will be dealt with in sections 7.5 and 7.6, both in the case of global interest and in that of partial interest.

7.2 Groups of Transformations

In each of the three Examples 7.1, 7.2 and 7.3, the family of densities is generated by $p_0(y)$ under a class $\mathcal{G}$ of one-to-one transformations which acts on $\mathbb{R}$ and has two properties.

1. **Closure under composition.** Application of the transformation $g_1 \in \mathcal{G}$ followed by application of the transformation $g_2 \in \mathcal{G}$ is called the composition of g_1 and g_2 and is indicated by $g_2 \circ g_1$; more explicitly, we have $g_2 \circ g_1(y) = g_2(g_1(y))$. The class $\mathcal{G}$ is said to be closed under composition if

$$g_1, g_2 \in \mathcal{G} \quad \Rightarrow \quad g_2 \circ g_1 \in \mathcal{G} . \tag{7.4}$$

In Example 7.1, we have

$$\mathcal{G} = \{g(y) = \mu + y, \ y \in \mathcal{Y} \subseteq \mathbb{R}, \ \mu \in \mathbb{R}\} \tag{7.5}$$

and, denoting by $g_1(y) = \mu_1 + y$ and $g_2(y) = \mu_2 + y$ two elements of $\mathcal{G}$,

$$g_2 \circ g_1(y) = \mu_2 + (\mu_1 + y) = (\mu_1 + \mu_2) + y = \nu + y,$$

where $\nu = \mu_1 + \mu_2 \in \mathbb{R}$.

For the scale transformations which belong to the class

$$\mathcal{G} = \{g(y) = \sigma y, \ y \in \mathcal{Y} \subseteq \mathbb{R}, \ \sigma > 0\} \tag{7.6}$$

considered in Example 7.2, if $g_1(y) = \sigma_1 y$ and $g_2(y) = \sigma_2 y$ are two elements of $\mathcal{G}$, then

$$g_2 \circ g_1(y) = \sigma_2(\sigma_1 y) = \sigma_2 \sigma_1 y = \xi y,$$

where $\xi = \sigma_2 \sigma_1 > 0$.

Lastly, consider the scale and location transformations in Example 7.3, that is, the class

$$\mathcal{G} = \{g(y) = \mu + \sigma y, \ y \in \mathcal{Y} \subseteq \mathbb{R}, \ \mu \in \mathbb{R}, \ \sigma > 0\} . \tag{7.7}$$

If $g_1(y) = \mu_1 + \sigma_1 y$ and $g_2(y) = \mu_2 + \sigma_2 y$ are two elements of $\mathcal{G}$, then

$$g_2 \circ g_1(y) = \mu_2 + \sigma_2(\mu_1 + \sigma_1 y) = \sigma_2 \mu_1 + \mu_2 + \sigma_2 \sigma_1 y = \nu + \xi y,$$

where $\nu = \sigma_2 \mu_1 + \mu_2 \in \mathbb{R}$ and $\xi = \sigma_2 \sigma_1 > 0$.

2. **Closure under inversion.** Let $e = e(y) = y$ be the identity transformation. The inverse function of g, indicated by g^{-1}, is such that
$g^{-1} \circ g(y) = e(y)$. The class $\mathcal{G}$ is said to be **closed under inversion** if

$$g \in \mathcal{G} \quad \Rightarrow \quad g^{-1} \in \mathcal{G} . \tag{7.8}$$

With the class (7.5), if $g(y) = y + \mu$, we have $g^{-1}(y) = y - \mu$. For
the scale transformations which belong to (7.6), if $g(y) = \sigma y$ we have
$g^{-1}(y) = y/\sigma$. For the scale and location transformations (7.7), the
inverse of $g(y) = \mu + \sigma y$ is $g^{-1}(y) = (y - \mu)/\sigma$. Thus, in all three cases
the inverse function is still one element of the corresponding class.

Definition 7.1 A non-empty class of one-to-one transformations $\mathcal{G}$, which
acts on the space $\mathcal{Y}$, is called a **transformation group** (or a group of transformations) if it is closed both under composition and under inversion.

Observe that if a class $\mathcal{G}$ of transformations is closed under composition or
inversion, then this property is not usually maintained in a subclass of $\mathcal{G}$. For
example, the class of location transformations restricted to $|\mu| < 1$ would not
satisfy (7.4). Indeed, for $\mu_1 = 1/2$ and $\mu_2 = 2/3$ we have $g_2 \circ g_1(y) = 7/6 + y$
and, obviously, $|7/6| > 1$.

Definition 7.2 Let $\mathcal{G}$ be a transformation group on $\mathcal{Y}$. A subset $\mathcal{G}_0$ of $\mathcal{G}$
is called a **subgroup of** $\mathcal{G}$ if it is closed both under composition and under
inversion.

E.g., the group (7.6) is a subgroup of (7.7).

The two lemmas below give some immediate consequences of the group
structure of $\mathcal{G}$: first, the identity transformation belongs to $\mathcal{G}$ and, second, the
composition of an element of $\mathcal{G}$ with its inverse is commutative. Because of
the latter property, although composition is not usually commutative, we still
do not need to distinguish between the *left inverse* and the *right inverse*.

Lemma 7.1 If $\mathcal{G}$ is a group of transformations, the identity transformation
$e = e(y) = y$ belongs to $\mathcal{G}$ and e is the unique element of $\mathcal{G}$ such that $e \circ g =
g \circ e = g$ for every $g \in \mathcal{G}$.

Proof. A group $\mathcal{G}$ is a non-empty class. Let g be one of its elements. Because of closure under inversion, there will be an element g^{-1} in $\mathcal{G}$ such that
$g^{-1} \circ g = e$. Then, e is an element of $\mathcal{G}$ because of closure under composition. If

e' were another identity transformation, then $e' \circ e = e' = e$, hence uniqueness is shown. □

Lemma 7.2 If $\mathcal{G}$ is a group of transformations, $g \circ g^{-1} = g^{-1} \circ g$.

Proof. If we apply the transformation g to both sides of $g^{-1} \circ g = e$, we get $g \circ g^{-1} \circ g = g \circ e = g$; thus, the lemma is proved if we remember that the identity transformation e is the only element of $\mathcal{G}$ such that $e \circ g = g \circ e = g$, for every $g \in \mathcal{G}$. □

Definition 7.3 A group $\mathcal{G}$ such that $g_2 \circ g_1 = g_1 \circ g_2$ for every $g_1, g_2 \in \mathcal{G}$ is said to be commutative.

Note that both (7.5) and (7.6) are commutative groups, while the group of scale and location transformations (7.7) is not.

The following property will be used repeatedly in subsequent sections (the proof is left as an exercise: Problem 7.3). If $\mathcal{G}$ is a transformation group on $\mathcal{Y}$ and if $g_0 \in \mathcal{G}$, then, as g varies in $\mathcal{G}$, so too does $g \circ g_0$ vary throughout $\mathcal{G}$. Hence, for every $g^* \in \mathcal{G}$ there is a transformation $g \in \mathcal{G}$ such that $g \circ g_0 = g^*$.

The simplest transformation group consists only of the identity transformation. The following examples illustrate some of the groups of transformations that, in addition to those considered in Examples 7.1-7.3, are among the most interesting in statistics.

Example 7.5 *The group of permutations on* $\mathbb{R}^n$
Let $y \in \mathcal{Y} = \mathbb{R}^n$ and let $g(y)$ be a permutation of the coordinates of y; in other words, $g(y_1, \ldots, y_n) = (y_{i_1}, \ldots, y_{i_n})$, where $(i_1, \ldots, i_n)$ is a permutation of the indices $(1, \ldots, n)$. The composition of two permutations is, itself, a permutation. Furthermore, the inverse function of a permutation is, itself, a permutation. Thus, the class of permutations on $\mathcal{Y}$ is a group of transformations. This group has finite cardinality $n!$. △

Example 7.6 *The group of monotone transformations on* $\mathbb{R}$
Let $\mathcal{Y} = \mathbb{R}$ and let $\mathcal{G}$ be the class of all the real monotone and differentiable functions of a real variable, so that if $g \in \mathcal{G}$ then g' has a constant sign. If f and g belong to $\mathcal{G}$, then $(f \circ g)' = (f' \circ g)g'$ will also have a constant sign. Furthermore, the derivative of the inverse function $(g^{-1})' = 1/g'$ has also a constant sign. Hence, the class of monotone differentiable transformations $\mathcal{G}$ on $\mathbb{R}$ is a group. △

Example 7.7 *The group of orthogonal transformations on* $\mathbb{R}^n$

Let $y \in \mathcal{Y} = \mathbb{R}^n$; consider the transformations $\mathbb{R}^n \to \mathbb{R}^n$ of the form Ay with A an orthogonal $n \times n$ matrix, that is, such that $A^{-1} = A^{\mathrm{T}}$. If A and B are orthogonal $n \times n$ matrices then the product matrix AB is also orthogonal, because $(AB)^{-1} = B^{-1}A^{-1} = B^{\mathrm{T}}A^{\mathrm{T}} = (AB)^{\mathrm{T}}$. Furthermore, if A is orthogonal then so too is its inverse, A^{-1}, because $(A^{-1})^{-1} = (A^{\mathrm{T}})^{-1} = (A^{-1})^{\mathrm{T}}$. Thus, the class under consideration is a group of transformations. $\triangle$

A group may be generated through a composition of subgroups. Let $\mathcal{G}_1$ and $\mathcal{G}_2$ be two groups of transformations on $\mathcal{Y}$. The class of transformations

$$\mathcal{G} = \mathcal{G}_1 \circ \mathcal{G}_2 \;=\; \left\{ g_1^1 \circ g_2^1 \circ \ldots \circ g_1^m \circ g_2^m, \; g_1^i \in \mathcal{G}_1, g_2^i \in \mathcal{G}_2, \right.$$
$$\left. i = 1, \ldots, m \,, m = 1, 2, \ldots \right\}$$

is a group (the proof is left to the reader: Problem 7.6) called the **group generated by the subgroups** $\mathcal{G}_1$ and $\mathcal{G}_2$.

7.3   Orbits and Maximal Invariants

Let $\mathcal{G}$ be a transformation group on $\mathcal{Y}$. The action of the group $\mathcal{G}$ on $\mathcal{Y}$ identifies some particular subsets of the space $\mathcal{Y}$.

Definition 7.4 Let $y \in \mathcal{Y}$. The set

$$G(y) = \{ y' \in \mathcal{Y} : y' = g(y), \quad g \in \mathcal{G} \}$$

is said to be the **orbit generated by** y.

All the points of $\mathcal{Y}$ that can be reached starting from y through a transformation $g \in \mathcal{G}$, belong to the set $G(y)$. Since $\mathcal{G}$ is a group, the orbits represent equivalence classes. Indeed, let us say that y is equivalent to x, in symbols $y \sim x$, if there exists $g \in \mathcal{G}$ such that $x = g(y)$. This relation is

– *reflexive*: $y \sim y$; the identity transformation belongs to $\mathcal{G}$;

– *symmetric*: if $y \sim x$ also $x \sim y$; in fact, if $y \sim x$ there exists $g \in \mathcal{G}$ such that $x = g(y)$, but because $\mathcal{G}$ is a group, $y = g^{-1}(x)$, with $g^{-1} \in \mathcal{G}$;

– *transitive*: if $y \sim x$ and $x \sim z$ also $y \sim z$; if, in fact, $x = g_1(y)$ and $z = g_2(x)$, then $z = g_2 \circ g_1(y)$; but $g_2 \circ g_1 \in \mathcal{G}$ because of closure under composition.

Since two equivalence classes, if they do not coincide, have empty intersection, the orbits form a partition of $\mathcal{Y}$, called the **orbital decomposition**. One can choose a point v in each orbit, called the orbit representative. Assume for simplicity that the group $\mathcal{G}$ is free, that is such that $g \neq g_1$ implies $g(y) \neq g_1(y)$. Then, each point $y \in \mathcal{Y}$ can be identified by two elements: the representative v of the orbit that y belongs to, and the element $g \in \mathcal{G}$, such that $g(v) = y$. If $\mathcal{G}$ were not free, a unique representation of y would have to be recovered by restricting attention to a suitable subgroup (the so-called free subgroup).

Definition 7.5 The group $\mathcal{G}$ is said to be transitive if $\mathcal{Y}$ has only one orbit, that is, if for each pair of elements y, y' of $\mathcal{Y}$ there exists $g \in \mathcal{G}$ such that $y' = g(y)$.

If the group $\mathcal{G}$ is transitive, the space $\mathcal{Y}$ can be completely reconstructed by the action of $\mathcal{G}$ on any of its points. As an example, the transformation group on $\mathbb{R}^n \setminus N$, where $N = \{y \in \mathbb{R}^n : y_1 = y_2 = \cdots = y_n\}$ with a generic element

$$g(y_1, \ldots, y_n) = (\mu + \sigma y_1, \ldots, \mu + \sigma y_n), \qquad \mu \in \mathbb{R}, \quad \sigma > 0,$$

is transitive only for $n = 1$ and is free only for $n > 1$ (Problem 7.9).

Those mappings defined on $\mathcal{Y}$ whose contours respect the orbits induced in $\mathcal{Y}$ by a group $\mathcal{G}$ are particularly important.

Definition 7.6 A function $u(\cdot)$ defined on $\mathcal{Y}$ with range $\mathcal{U}$ is said to be invariant under $\mathcal{G}$ if $u(g(y)) = u(y)$ for every $g \in \mathcal{G}$.

In other words, $u(\cdot)$ is said to be invariant if it assumes a constant value on the orbits of $\mathcal{G}$. If $\mathcal{G}$ is transitive, the only invariant functions are constant functions.

Definition 7.7 A transformation $v(\cdot)$ defined on $\mathcal{Y}$ with range $\mathcal{V}$ is said to be maximal invariant under $\mathcal{G}$ if it is invariant and if the implication

$$v(y) = v(y') \Rightarrow \text{there exists } g \in \mathcal{G} \text{ such that } y' = g(y)$$

holds.

For brevity, a maximal invariant transformation is usually called a maximal invariant. Thus, a maximal invariant is a function that has a constant value on each orbit and different values on distinct orbits; it contains all, and only, the information about the orbit y belongs to. If v is maximal invariant under $\mathcal{G}$ and the function $w(v)$ from $\mathcal{V}$ into a range space $\mathcal{W}$ is one-to-one, then w will also be maximal invariant. All maximal invariants induce the same orbital decomposition of $\mathcal{Y}$ and, in this sense, are equivalent: one specific maximal invariant is said to be a representation of the maximal invariant. For example, the function that associates the orbit representative $v(y)$ to each $y \in \mathcal{Y}$ is a representation of the maximal invariant.

Theorem 7.1 Let $v(\cdot)$ be a maximal invariant under $\mathcal{G}$. A necessary and sufficient condition for a function $u(y)$ to be invariant is that $u(y)$ only depends on y through $v(\cdot)$, that is, $u(\cdot)$ is invariant if, and only if, there exists a function $h(\cdot)$, such that $u(y) = h(v(y))$.

Proof.

i) *(Sufficiency.)* If $u(\cdot)$ is a function of y only through $v(y)$, it must necessarily be invariant because of the invariance of $v(\cdot)$.

ii) *(Necessity.)* Let y and y' be two points of $\mathcal{Y}$ such that $v(y) = v(y')$. Since $v(\cdot)$ is maximal invariant there must be a transformation $g \in \mathcal{G}$ such that $y' = g(y)$; furthermore, because $u(\cdot)$ is invariant, $u(y) = u(y')$. Therefore, whenever $v(y) = v(y')$ then $u(y) = u(y')$ too, which is the same as stating that $u(\cdot)$ must be a function of y only through $v(\cdot)$. $\qquad\square$

The following examples illustrate how to find maximal invariants under some groups of transformations.

Example 7.8 *Maximal invariant with respect to the group of location transformations on $\mathbb{R}^n$*
Let $\mathcal{Y} = \mathbb{R}^n$, $n > 1$, and consider the group of location transformations with generic element

$$g(y) = g(y_1, \ldots, y_n) = (\mu + y_1, \ldots, \mu + y_n), \quad \mu \in \mathbb{R} . \qquad (7.9)$$

The set of the differences

$$u(y) = (y_1 - \bar{y}_n, \ldots, y_n - \bar{y}_n) \qquad (7.10)$$

is invariant. The transformation (7.10) is also maximal invariant. Indeed, if y and y' are such that

$$y_i - \bar{y}_n = y'_i - \bar{y}'_n, \quad i = 1, \ldots, n ,$$

then

$$y'_i = (\bar{y}'_n - \bar{y}_n) + y_i = \nu + y_i , \quad i = 1, \ldots, n ,$$

with $\nu = \bar{y}'_n - \bar{y}_n$, and hence y' can be obtained starting from y through a transformation as (7.9); therefore, y and y' belong to the same orbit. It should be noted that the representation (7.10) of the maximal invariant is not unique. For example, other representations are

$$u_1(y) = (y_1 - y_2, y_2 - y_3, \ldots, y_{n-1} - y_n), \tag{7.11}$$

$$u_2(y) = (y_1 - y_n, \ldots, y_{n-1} - y_n). \tag{7.12}$$

It is easy to check that the transformations (7.10), (7.11) and (7.12) induce the same partition of $\mathcal{Y}$. If in fact y and y' belong to the same orbit, i.e. if $y' = \mu 1_n + y$, where 1_n indicates the vector of $\mathbb{R}^n$ with all the components equal to one, then

$$u(y) = u(y'), \quad u_1(y) = u_1(y'), \quad u_2(y) = u_2(y').$$

Vice versa, if two points give the same value of $u(\cdot)$ they will also produce the same value of $u_1(\cdot)$ and of $u_2(\cdot)$. If $n = 1$ the group $\mathcal{G}$ is transitive, and the only invariant functions are constant functions. $\triangle$

Example 7.9 *Maximal invariant with respect to the group of scale transformations on $\mathbb{R}^n$ $(n > 1)$*
Let $\mathcal{Y} = \mathbb{R}^n \setminus N$, where $N = \{y \in \mathbb{R}^n : y_1 = y_2 = \cdots = y_n\}$. Consider the group $\mathcal{G}$ of transformations with generic element

$$g(y) = (\sigma y_1, \ldots, \sigma y_n), \quad \sigma > 0. \tag{7.13}$$

The following are equivalent representations of the maximal invariant

$$v(y) = \left(\frac{y_1}{\sqrt{\sum y_i^2}}, \ldots, \frac{y_n}{\sqrt{\sum y_i^2}} \right), \tag{7.14}$$

$$v_1(y) = \left(\frac{y_1}{y_{(n)} - y_{(1)}}, \ldots, \frac{y_n}{y_{(n)} - y_{(1)}} \right). \tag{7.15}$$

Observe that in (7.14) the sum of the squares of the components is equal to one, while in (7.15) $\max_i \{y_i/(y_{(n)} - y_{(1)})\} - \min_i \{y_i/(y_{(n)} - y_{(1)})\} = 1$. $\triangle$

Example 7.10 *Maximal invariant with respect to the group of scale and location transformations on $\mathbb{R}^n$*
Let $\mathcal{Y} = \mathbb{R}^n \setminus N$, where $N = \{y \in \mathbb{R}^n : y_1 = y_2 = \cdots = y_n\}$. Consider the group $\mathcal{G}$ with generic element

$$g(y) = (\mu + \sigma y_1, \ldots, \mu + \sigma y_n), \quad \mu \in \mathbb{R}, \sigma > 0. \tag{7.16}$$

If $n \geq 2$ this group is not transitive (Problem 7.9), therefore, there is a non-trivial maximal invariant. This can be obtained in two steps (see also Problem 7.14).

1. Let $x = (x_1, \ldots, x_n) = u(y) = (y_1 - \bar{y}_n, \ldots, y_n - \bar{y}_n)$. The subgroup of scale transformations on $\mathcal{Y}$ also induces a group of scale transformations in the space of the values of x; indeed

$$
\begin{aligned}
u(\sigma y) &= (\sigma y_1 - \sigma \bar{y}_n, \ldots, \sigma y_n - \sigma \bar{y}_n) \\
&= (\sigma(y_1 - \bar{y}_n), \ldots, \sigma(y_n - \bar{y}_n)) = (\sigma x_1, \ldots, \sigma x_n).
\end{aligned}
$$

2. Thus, returning to Example 7.9, the maximal invariant is

$$
\begin{aligned}
z = v(x) &= \left(\frac{x_1}{\sqrt{\sum x_i^2}}, \ldots, \frac{x_n}{\sqrt{\sum x_i^2}} \right) \\
&= \left(\frac{y_1 - \bar{y}_n}{\sqrt{(n-1)s_n^2}}, \ldots, \frac{y_n - \bar{y}_n}{\sqrt{(n-1)s_n^2}} \right). \tag{7.17}
\end{aligned}
$$

Note that z has $n - 2$ functionally independent components.

It should be pointed out here, that if the transformations $u(\cdot)$ and $v(\cdot)$ are applied in a reverse order, then they do not give the maximal invariant. Indeed, if we let

$$
v(y) = \left(\frac{y_1}{\sqrt{\sum y_i^2}}, \ldots, \frac{y_n}{\sqrt{\sum y_i^2}} \right),
$$

we have

$$
v(y + \mu \mathbf{1}_n) = (v_1, \ldots, v_n) = \left(\frac{y_1 + \mu}{\sqrt{\sum(y_i + \mu)^2}}, \ldots, \frac{y_n + \mu}{\sqrt{\sum(y_i + \mu)^2}} \right)
$$

and $(v_1 - \bar{v}_n, \ldots, v_n - \bar{v}_n)$ is not invariant. A further important representation of the maximal invariant is defined in section 7.6.2 (formula (7.32)). $\triangle$

Example 7.11 *Maximal invariant with respect to the group of permutations on* $\mathbb{R}^n$

Let $\mathcal{Y} = \mathbb{R}^n$ and, for simplicity, only consider the points $y = (y_1, \ldots, y_n) \in \mathcal{Y}$ whose components are distinct. The group $\mathcal{G}$ of the permutations of the coordinates of y contains $n!$ elements. One maximal invariant is the order statistics $y_{\text{ord}} = (y_{(1)}, \ldots, y_{(n)})$. Indeed, y_{ord} is invariant under permutations of the components of y and, if two points of $\mathbb{R}^n$ have the same value of y_{ord}, they can be obtained from each other through a permutation of the coordinates. $\triangle$

Example 7.12 *Maximal invariant with respect to the group of increasing, continuous transformations*
Let $\mathcal{Y} = \mathbb{R}^n$ and consider the group $\mathcal{G}$ with generic element

$$g(y) = (f(y_1), \ldots, f(y_n)), \tag{7.18}$$

with $f(\cdot)$ a real function defined on $\mathbb{R}$ which is continuous and strictly increasing. Focus attention on the points of $\mathbb{R}^n$ with distinct coordinates and imagine the coordinates of y as being represented as n points on $\mathbb{R}$. Since every increasing transformation retains the order of $(y_1, \ldots, y_n)$, the set of ranks (or, equivalently, of inverse ranks, see Example 4.9) is invariant. Vice versa, if y and y' have the same set of ranks, then there is a monotone function $f(\cdot) : \mathbb{R} \to \mathbb{R}$, such that $y_i' = f(y_i)$. For example, we can take the function made up of segments of straight lines passing through $(y_1, y_1'), (y_2, y_2'), \ldots, (y_n, y_n')$ as $f(\cdot)$. Thus, the set of ranks is a maximal invariant. $\triangle$

Example 7.13 *Maximal invariant with respect to the group of orthogonal transformations on $\mathbb{R}^n$*
Let $\mathcal{Y} = \mathbb{R}^n$ and let $\mathcal{G}$ be the group of transformations of $\mathcal{Y}$ with generic element

$$g(y) = Ay, \tag{7.19}$$

with A an orthogonal matrix. Two points $y, y' \in \mathbb{R}^n$ can be obtained, one from the other, through an orthogonal transformation if, and only if, they are the same distance from the origin. More precisely, if $y' = Ay$ we have

$$\|y'\|^2 = y'^{\mathrm{T}} y' = y^{\mathrm{T}} A^{\mathrm{T}} A y = y^{\mathrm{T}} y = \|y\|^2 .$$

Vice versa, any two vectors of $\mathbb{R}^n$, y and y', can always be obtained from each other through a linear transformation; they have the same norm if the matrix that defines the transformation is orthogonal. Thus $r^2 = \sum y_i^2$ is a maximal invariant. $\triangle$

7.4   Simple Group Families

Let Y be a random variable with support $\mathcal{Y}$ and density $p_0(y)$ and let $\mathcal{G}$ be a group of measurable transformations of $\mathcal{Y}$ onto itself. Let gp_0 indicate the density of $g(Y)$ (obviously, gp_0 is not $g(p_0)$).

Definition 7.8 The class of densities

$$\mathcal{F}_\mathcal{G} = \{gp_0, \, g \in \mathcal{G}\} \tag{7.20}$$

is called a **simple group family** generated by $p_0(\cdot)$ under the action of the group $\mathcal{G}$.

So long as it is not ambiguous, the abbreviated term **group family** will be used instead of *simple group family*. In the following, we suppose that if g_1 and g_2 are two elements of $\mathcal{G}$ such that $g_1 \neq g_2$, then also $g_1 p_0 \neq g_2 p_0$, so that the model $\mathcal{F}_\mathcal{G}$ is identifiable. If the transformations in $\mathcal{G}$ are of the form $g(y) = g^0(y; \theta)$, with fixed $g^0(\cdot; \cdot)$ and $\theta \in \Theta \subseteq \mathbb{R}^p$, then $\mathcal{F}_\mathcal{G}$ is a parametric family. The example below will show how, as for exponential and exponential dispersion families, the closure property is also satisfied for group families: therefore a group family can be thought to be generated by any one of its elements.

Example 7.14 *Closure property*
Consider family (7.20) and let $p_1(y)$ be a density belonging to $\mathcal{F}_\mathcal{G}$. The family $\mathcal{F}_\mathcal{G}^1 = \{gp_1, \, g \in \mathcal{G}\}$ coincides with $\mathcal{F}_\mathcal{G}$. In fact, if $p_1(y) \in \mathcal{F}_\mathcal{G}$ there will be a transformation $g_1 \in \mathcal{G}$ such that $p_1 = g_1 p_0$. Thus, we could represent $\mathcal{F}_\mathcal{G}^1$ in the form $\mathcal{F}_\mathcal{G}^1 = \{g(g_1 p_0), \, g \in \mathcal{G}\}$. Since $g(g_1 p_0) = (g \circ g_1)p_0$ and $\{g \circ g_1, \, g \in \mathcal{G}\} = \mathcal{G}$ (see section 7.2 and Problem 7.3), then $\mathcal{F}_\mathcal{G}^1 = \mathcal{F}_\mathcal{G}$. $\triangle$

It might be useful to know how to check whether a given family of densities can be generated by one of its elements under the action of a group $\mathcal{G}$ of transformations of the sample space.

Definition 7.9 A statistical model $\mathcal{F} = \{p(y; \theta), \, y \in \mathcal{Y}, \, \theta \in \Theta\}$ is called an **invariant family** under a group $\mathcal{G}$ of transformations if, for every density that belongs to $\mathcal{F}$ and for every $g \in \mathcal{G}$, there exists a mapping $\bar{g} : \Theta \to \Theta$ such that

$$gp(y; \theta) = p(y; \bar{g}\theta), \quad \text{with } \bar{g}\theta \in \Theta \text{ and } \bar{g}\Theta = \Theta \,.$$

According to this definition, the family $\mathcal{F}$ is invariant under $\mathcal{G}$ if $gp(y; \theta)$ is still an element of $\mathcal{F}$ with parameter value $\bar{g}\theta$ and if, as θ varies in Θ, so too does

$\bar{g}\theta$ vary in all Θ. In other words, the class of transformations must not change the parameter space.

If $\mathcal{F}$ is an invariant family under $\mathcal{G}$, then $\mathcal{G}$ induces a class $\bar{\mathcal{G}}$ of transformations $\bar{g}$ on Θ. If, in addition, $\mathcal{F}$ satisfies the identifiability condition, then the function $\bar{g} : \Theta \to \Theta$, defined above, is one-to-one. Indeed, if $\bar{g}\theta_1 = \bar{g}\theta_2$ we will have $P_{\bar{g}\theta_1}(gA) = P_{\bar{g}\theta_2}(gA)$ for every event $A \subseteq \mathcal{Y}$, where $gA = \{g(y) : y \in A\}$. Furthermore, because $P_{\bar{g}\theta}(gA) = P_\theta(A)$ according to the definition of $\bar{g}$, if $\bar{g}\theta_1 = \bar{g}\theta_2$ then $\theta_1 = \theta_2$. It is not difficult to check that the class $\bar{\mathcal{G}}$ of one-to-one transformations $\bar{g} : \Theta \to \Theta$ is a group (Problem 7.12).

A connection does exist between group families and invariant families, that is, between Definitions 7.8 and 7.9. Consider a family $\mathcal{F}$ which is invariant under a group $\mathcal{G}$ such that the group $\bar{\mathcal{G}}$ induced in the parameter space is transitive. Let $p_0(y) = p(y; \theta_0)$, with $\theta_0 \in \Theta$ fixed arbitrarily. Then, the simple group family generated by $p_0(y)$ under the action of $\mathcal{G}$ is

$$\mathcal{F}_{\mathcal{G}} = \{gp_0(y), \ g \in \mathcal{G}\} = \{p(y; \bar{g}\theta_0), \ \bar{g} \in \bar{\mathcal{G}}\} = \mathcal{F},$$

hence, $\mathcal{F}$ is a group family. On the contrary, if $\mathcal{F}_{\mathcal{G}}$ is a group family, it is invariant under $\mathcal{G}$; the parameter θ is put in one-to-one correspondence with the elements of $\mathcal{G}$ and, because of the closure property (see Example 7.14), the action of $\mathcal{G}$ on $\mathcal{F}$ is transitive. Thus, a family of densities $\mathcal{F}$ is a group family if, and only if, there is a group $\mathcal{G}$ that acts transitively on $\mathcal{F}$, that is, if $\bar{\mathcal{G}}$ acts transitively on Θ.

Example 7.15 *Scale and location families*
The class (7.3) is the group family generated by $p_0(\cdot)$ under the action of the group $\mathcal{G}$ of scale and location transformations on $\mathbb{R}$. Since q is identified in $\mathcal{G}$ by the pair $(\mu, \sigma) \in \mathbb{R} \times (\mathbb{R}^+ \setminus \{0\})$, families obtained in this way are two-parameter families. Furthermore, if Y has density $p(y; \mu, \sigma)$ which belongs to the family (7.3), $g(Y) = \mu' + \sigma'Y$ will have density $p(y; \mu' + \sigma'\mu, \sigma'\sigma)$ and therefore $\bar{g}\theta = \bar{g}(\mu, \sigma) = (\mu' + \sigma'\mu, \sigma'\sigma)$. The group $\bar{\mathcal{G}}$ of transformations $\bar{g}(\mu, \sigma)$, induced on the parameter space $\Theta = \{(\mu, \sigma) \in \mathbb{R} \times (\mathbb{R}^+ \setminus \{0\})\}$, is transitive. $\triangle$

Example 7.16 *Multivariate normal distribution*
Let $\mathcal{Y} = \mathbb{R}^d$ and let $Y = (Y_1, \ldots, Y_d)^{\mathrm{T}}$ be a random vector having i.i.d. components with a $N(0, 1)$ distribution; let $p_0(y)$ indicate the density of Y. Consider the group $\mathcal{G}$ of affine transformations of $\mathbb{R}^d$ onto itself

$$\mu + BY, \tag{7.21}$$

with $\boldsymbol{\mu} = (\mu_1, \ldots, \mu_d)^T$ and B a $d \times d$ non-singular matrix. The transformed variable $\boldsymbol{\mu} + BY$ has density

$$p(y; \boldsymbol{\mu}, \Sigma) = \frac{1}{(2\pi)^{d/2}|\Sigma|^{1/2}} \exp\left\{-\frac{1}{2}(y - \boldsymbol{\mu})^T \Sigma^{-1}(y - \boldsymbol{\mu})\right\},$$

where $\Sigma = BB^T = \text{Var}(\boldsymbol{\mu} + BY)$. The group family $\mathcal{F} = \{gp_0, g \in \mathcal{G}\}$ coincides with the family of d-variate normal distributions. Note that, unlike as happens with the scale and location transformations considered in Example 7.15, the same d-variate normal distribution can be obtained by (7.21) with two different matrices B_1 and B_2, provided that $B_1 B_1^T = B_2 B_2^T$. Thus, it is possible to obtain the family of d-variate normal distributions by restricting attention to a subgroup of the transformations (7.21); for example, that of the transformations $\boldsymbol{\mu} + By$ with B a lower triangular matrix. Without any restriction, it would not be possible to establish a one-to-one correspondence between the parameter $(\boldsymbol{\mu}, \Sigma)$ and the element of $\mathcal{G}$ which gives a fixed d-variate normal distribution. $\triangle$

Example 7.17 *Linear model*
Let $U = (U_1, \ldots, U_n)$ be a random vector with fixed density $p_0(\cdot)$ and consider the transformations

$$Y_i = \mu_i + \sigma U_i, \quad i = 1, \ldots, n, \tag{7.22}$$

where $\sigma > 0$ and the location vector $\boldsymbol{\mu} = (\mu_1, \ldots, \mu_n)^T$ is constrained to belong to a k-dimensional subspace $(k < n)$ of $\mathbb{R}^n$, that is, $\boldsymbol{\mu} = \beta_1 x_1 + \cdots + \beta_k x_k$ with $\beta_1, \ldots, \beta_k$ arbitrary scalars and $x_1, \ldots, x_k$ linearly independent fixed vectors of $\mathbb{R}^n$. If $(U_1, \ldots, U_n) \sim N_n(0, I_n)$ we obtain the classical linear model with normal errors. $\triangle$

Example 7.18 *Sampling from a scale and location family*
Let $U = (U_1, \ldots, U_n)$, where the components U_i are i.i.d. with density $p_0(u)$, $u \in \mathcal{U} \subseteq \mathbb{R}$. Consider the group family generated by p_0 under the action of the group $\mathcal{G}$ with generic element

$$g(u_1, \ldots, u_n) = (\mu + \sigma u_1, \ldots, \mu + \sigma u_n), \quad \mu \in \mathbb{R}, \ \sigma > 0. \tag{7.23}$$

The family obtained in this way coincides with the family of densities of a random sample of size n from an element of the scale and location family (7.3). $\triangle$

Example 7.19 *The nonparametric class of continuous distributions with support $\mathbb{R}$*

Let Y be a r.v. with density $p_0(y)$, with respect to the Lebesgue measure, and support $\mathbb{R}$; for example, $Y \sim N(0, 1)$. Let $\mathcal{G}$ be the set of the differentiable functions $g : \mathbb{R} \to \mathbb{R}$, with positive derivative and such that $\lim_{y \to -\infty} g(y) = -\infty$ and $\lim_{y \to +\infty} g(y) = +\infty$. It is not difficult to check that $\mathcal{G}$ is a group. The group family $\mathcal{F} = \{gp_0, \ g \in \mathcal{G}\}$ is the nonparametric class of all the absolutely continuous distributions with support $\mathbb{R}$. $\triangle$

Example 7.20 *The von Mises distribution*

Consider the von Mises distribution introduced in Example 2.14 with density

$$p(y; \mu, \lambda) = \frac{1}{2\pi I_0(\lambda)} \exp\{\lambda \cos(y - \mu)\} \, ,$$

where $y \in (0, 2\pi)$, $\mu \in (0, 2\pi)$ and $\lambda > 0$. Let us consider the transformation to Cartesian coordinates, that is, let $X_1 = \cos Y$, $X_2 = \sin Y$. The Jacobian determinant is 1, and the density is transformed into

$$p_X(x; \xi, \lambda) = \frac{1}{2\pi I_0(\lambda)} \exp\{\lambda(x_1\xi_1 + x_2\xi_2)\} = \frac{1}{2\pi I_0(\lambda)} \exp\{\lambda x \cdot \xi\} \, ,$$

where $x = (x_1, x_2)^T$, $\xi = (\xi_1, \xi_2)^T$, $\xi_1 = \cos\mu$, $\xi_2 = \sin\mu$ and $x_1^2 + x_2^2 = \xi_1^2 + \xi_2^2 = 1$. It is easy to see that the von Mises family is invariant under the group $\mathcal{G}$ of orthogonal transformations on $\mathbb{R}^2$. In fact, $Z = AX$, with A an orthogonal 2×2 matrix, has density

$$\begin{aligned} p_z(z) &= \frac{1}{2\pi I_0(\lambda)} \exp\{\lambda(A^T z) \cdot \xi\} \\ &= \frac{1}{2\pi I_0(\lambda)} \exp\{\lambda z \cdot A\xi\} = p_X(z; A\xi, \lambda) \, . \end{aligned}$$

As A varies in the class of the orthogonal 2×2 matrices, we obtain all the vectors ξ on the unit circle, that is, the group $\bar{\mathcal{G}}$ of the transformations induced in the parameter space has, as its generic element, $\bar{g}(\xi, \lambda) = (A\xi, \lambda)$ and is transitive for fixed λ. Hence, with fixed precision parameter λ, the von Mises family is a simple group family. $\triangle$

While in Examples 7.15-7.19 a group family was *built* starting from a density p_0 and a group $\mathcal{G}$ of transformations, in the example above we *checked* that a given parametric family is a group family, through a check on the invariance of $\mathcal{F}$ and on the transitivity of $\bar{\mathcal{G}}$.

Some of the families considered in this chapter are both exponential and group families. In particular, with reference to the scale and location families illustrated in Table 7.1, this is true both for the normal distribution and for the exponential distribution (with fixed μ). Ferguson (1962) has shown that the only univariate distributions which have this property are the normal distribution and the gamma distribution with fixed location and shape parameters. One important example, in the multivariate case, is that of the Wishart distribution with density (5.25) (see Example 5.8), which may be obtained as a group family generated by the density (5.24), under the action of the group of linear transformations on the space of symmetric and positive definite $d \times d$ matrices (see also Barndorff-Nielsen, 1988, section 2.5).

7.5 Composite Group Families

As shown in Example 5.4 with regard to exponential families, so too, with families generated by groups of transformations, one can generate a family starting from a parametric class of densities $\mathcal{F}_0$. Example 7.14 showed that if $\mathcal{F}_0$ is, in its turn, a group family, and if the transformation group involved in the two constructions is the same, then one will obtain the initial family $\mathcal{F}_0$ (closure property) hence, there is no extension of the model.

Suppose that a parametric family

$$\mathcal{F}_0 = \{p_0(y; \tau),\ \tau \in \mathrm{T} \subseteq \mathbb{R}^k,\ y \in \mathcal{Y} \subseteq \mathbb{R}^d\} \tag{7.24}$$

has been assigned and that $\mathcal{G}$ is a transformation group $g : \mathcal{Y} \to \mathcal{Y}$ such that $\mathcal{F}_0$ is not invariant under $\mathcal{G}$. Let $gp_0(y; \tau)$ indicate the density of $g(Y)$ when Y has density $p_0(y; \tau)$.

Definition 7.10 The class of densities

$$\mathcal{F} = \{gp_0(y; \tau),\ \tau \in \mathrm{T},\ g \in \mathcal{G},\ y \in \mathcal{Y}\} \tag{7.25}$$

is said to be the **composite group family** generated by $\mathcal{F}_0$ under the action of $\mathcal{G}$.

If the distributions of $g_1(Y)$ and $g_2(Y)$ only coincide for $g_1 = g_2$ then the family (7.25) can be parameterized by $\theta = (\tau, g)$, with $\theta \in \Theta = \mathrm{T} \times \mathcal{G}$, and we can write

$$gp_0(y; \tau) = p(y; \tau, g) = p(y; \theta)\ .$$

Note that $\mathcal{F}$ is usually a semiparametric model; it is a parametric model whenever the elements of $\mathcal{G}$ can be identified through a finite number of real parameters.

By construction, each subfamily of $\mathcal{F}$ obtained by fixing the parameter τ is a simple group family and, hence, is invariant under $\mathcal{G}$. This means that $\mathcal{F}$ is invariant under $\mathcal{G}$. Furthermore, if g and g_0 are two elements of $\mathcal{G}$, we have

$$gp(y; \tau, g_0) = g \circ g_0 p_0(y; \tau) = p(y; \tau, g \circ g_0) \,,$$

so that the group $\mathcal{G}$ induces a class $\bar{\mathcal{G}}$ of transformations of the parameter space Θ with generic element

$$\bar{g}\theta = \bar{g}(\tau, g_0) = (\tau, g \circ g_0) \,. \tag{7.26}$$

The component τ of the parameter θ is called the **shape parameter** or **index parameter** and it is a maximal invariant under the group $\bar{\mathcal{G}}$ which acts on Θ as defined by (7.26).

As g varies in $\mathcal{G}$, $g \circ g_0$ varies throughout $\mathcal{G}$ (Problem 7.3). However, the component τ of θ remains unchanged under the action of $\bar{\mathcal{G}}$. Hence, $\bar{\mathcal{G}}$ does not act transitively on Θ. Thus a family $\mathcal{F}$ is a composite group family if, and only if, there is a group $\mathcal{G}$ with respect to which $\mathcal{F}$ is invariant. If such a group also acts transitively on $\mathcal{F}$, or rather, if $\bar{\mathcal{G}}$ acts transitively on Θ, then $\mathcal{F}$ is a simple group family. We can write

$$\mathcal{F} = \bigcup_{\tau \in T} \mathcal{F}_\tau \,,$$

where $\mathcal{F}_\tau = \{gp_0(y; \tau), g \in \mathcal{G}\}$. Hence, a composite group family is a union of simple group families. If Y_1 and Y_2 have densities that belong to $\mathcal{F}$, then a transformation $g \in \mathcal{G}$ exists, such that $Y_2 = g(Y_1)$ only if the densities of Y_1 and Y_2 belong to the same subfamily $\mathcal{F}_\tau$. One example of a composite group family is the von Mises distribution of Example 7.20, when λ is not fixed.

Example 7.21 *Gamma distribution*
Let

$$\mathcal{F}_0 = \left\{ p_0(y; \tau) = \frac{y^{\tau-1}e^{-y}}{\Gamma(\tau)}, \quad y > 0 \,, \tau \in \mathbb{R}^+ \setminus \{0\} \right\}.$$

Let $\mathcal{G}$ be the group of scale transformations on $\mathbb{R}^+ \setminus \{0\}$ as defined by (7.6). The composite group family generated by $\mathcal{F}_0$ through the action of $\mathcal{G}$ is

$$\mathcal{F} = \left\{ p(y; \tau, \sigma) = \frac{1}{\sigma} p_0 \left(\frac{y}{\sigma}; \tau \right) = \frac{y^{\tau-1}e^{-y/\sigma}}{\sigma^\tau \Gamma(\tau)}, \quad y > 0 \,, \tau > 0 \,, \sigma > 0 \right\},$$

the family of $Ga\,(\tau,\sigma)$ distributions. Furthermore, if $g(y) = \xi y$, $\xi > 0$, then $\bar{g}(\tau,\sigma) = (\tau,\xi\sigma)$. $\qquad\qquad\qquad\qquad\qquad\qquad\qquad\qquad\qquad\qquad\qquad\qquad\qquad$ $\triangle$

Example 7.22 *Proportional hazards model (cont.)*
The proportional hazards model is a composite group family. Indeed, the semiparametric model specified by the failure rate (1.2) is invariant under the group $\mathcal{G}$ of differentiable and strictly increasing transformations (Kalbfleisch and Prentice, 1973). In order to illustrate this, first observe that if T is a non-negative r.v. with density $p_T(t)$ and if $g(t)$ is a strictly increasing, differentiable function, such that $\lim_{t\to 0} g(t) = 0$ and $\lim_{t\to +\infty} g(t) = +\infty$, then, when T has failure rate $r_T(t)$, the transformed variable $g(T)$ has failure rate

$$r_{g(T)}(t) = \frac{p_T(g^{-1}(t))}{1 - F_T(g^{-1}(t))} \left(\frac{dg^{-1}(t)}{dt}\right) = r_T(g^{-1}(t)) \left(\frac{dg^{-1}(t)}{dt}\right).$$

Thus, if T has failure rate $r_0(t)e^{\beta^T x}$, $g(T)$ has failure rate

$$r_0(g^{-1}(t)) \left(\frac{dg^{-1}(t)}{dt}\right) e^{\beta^T x}.$$

In conclusion, if T is distributed according to a proportional hazards model then so too will the distribution of $g(T)$ belong to such a model. Furthermore,

$$\bar{g}\theta = \bar{g}(\beta_1,\ldots,\beta_k,r_0(t)) = \left(\beta_1,\ldots,\beta_k,r_0(g^{-1}(t))\left(\frac{dg^{-1}(t)}{dt}\right)\right),$$

and $\bar{g}$ leaves β unchanged, hence, does not act transitively on the (infinite-dimensional) parameter space. The proportional hazards model is, therefore, a composite group family generated by the family $\mathcal{F}_0$ of the r.v.'s with failure rate $e^{\beta^T x}$, $\beta \in \mathbb{R}^k$, that is, generated by the family of exponential distributions with mean $e^{-\beta^T x}$, $\beta \in \mathbb{R}^k$, and with $\mathcal{G}$ defined as above. Indeed, any smooth function $r_0(t)$ could be seen as the derivative of the inverse of a transformation belonging to $\mathcal{G}$. $\qquad\qquad\qquad\qquad\qquad\qquad\qquad\qquad\qquad\qquad\qquad\qquad\qquad\qquad$ $\triangle$

7.6 Inference in Simple Group Families

7.6.1 *Data and Model Reduction*

Let $\mathcal{F}_{\mathcal{G}} = \{p(y;\theta), \theta \in \Theta, y \in \mathcal{Y}\}$ be a simple group family with respect to the group $\mathcal{G}$. Random sampling of size n from an element of $\mathcal{F}_{\mathcal{G}}$ induces the family

$$\mathcal{F}_{\mathcal{G}}^n = \left\{\prod_{i=1}^n p(y_i;\theta),\ \theta \in \Theta,\ y \in \mathcal{Y}^n\right\}, \qquad (7.27)$$

where $\mathcal{Y}^n$ is the Cartesian product of n copies of $\mathcal{Y}$. The class of densities $\mathcal{F}_{\mathcal{G}}^n$ is a simple group family under the action of the induced group $\mathcal{G}^n$, with generic element $g^n(y_1, \ldots, y_n) = (g(y_1), \ldots, g(y_n))$, $g \in \mathcal{G}$ (see Example 7.18 and Problem 7.22). The group $\mathcal{G}^n$ induces a group $\bar{\mathcal{G}}^n$ which acts transitively on Θ. However, $\mathcal{G}^n$ does not usually act transitively on $\mathcal{Y}^n$. This implies that there is a non-trivial orbital decomposition of $\mathcal{Y}^n$ and that, consequently, it is possible to apply a reduction by conditioning according to Theorem 7.2 below. For simplicity, a statistic defined by a function which is invariant under $\mathcal{G}$ will be termed an invariant statistic under $\mathcal{G}$.

Theorem 7.2 Let $\mathcal{F}_{\mathcal{G}} = \{p(y; \theta), y \in \mathcal{Y}, \theta \in \Theta\}$ be a simple group family with respect to the group $\mathcal{G}$. If $v(y)$ is an invariant statistic under $\mathcal{G}$, then $v(Y)$ is distribution constant within $\mathcal{F}_{\mathcal{G}}$.

Proof. Let Y_0 be a random variable with density $p_0(y) \in \mathcal{F}_{\mathcal{G}}$; more specifically, let $p_0(y) = p(y; \theta_0)$, $\theta_0 \in \Theta$. Hence, $v(g(Y_0))$ has the same distribution as $v(Y)$, where Y has density $p(y; \bar{g}\theta_0)$. Since the group $\bar{\mathcal{G}}$ is transitive there will be a transformation $\bar{g} \in \bar{\mathcal{G}}$ for every $\theta \in \Theta$ such that $\theta = \bar{g}\theta_0$. Thus, as g varies in $\mathcal{G}$, $\bar{g}\theta_0$ will take all the values in Θ. On the other hand, since $v(y)$ is invariant, $v(g(y)) = v(y)$, for every $g \in \mathcal{G}$, hence, $v(g(Y_0)) \sim v(Y_0)$, for every $g \in \mathcal{G}$. Therefore, $v(Y)$ has the same distribution for every $\theta \in \Theta$. $\square$

The converse of Theorem 7.2 does not hold. Even though a statistic is distribution constant, or ancillary, this does not necessarily mean that it is invariant, as the counter-example below will show (Padmanabhan, 1977).

Example 7.23 *A non-invariant distribution constant statistic*
Let y_1 and y_2 be independent realizations of $N(\mu, 1)$ and let

$$c(y_1, y_2) = \begin{cases} y_1 - y_2 & \text{if } y_1 + y_2 \geq 1 \\ y_2 - y_1 & \text{if } y_1 + y_2 < 1 \ . \end{cases}$$

The statistic $c(y_1, y_2)$ is clearly distribution constant; however it is not invariant under location changes, that is, the identity $c(\nu + y_1, \nu + y_2) = c(y_1, y_2)$ for each $\nu \in \mathbb{R}$ is not satisfied. $\triangle$

On the basis of Theorem 7.2, in a simple group family, the partition of the sample space identified by the orbital decomposition makes it possible to isolate one portion of the data which is uninformative about θ. Knowing which orbit y^{obs} belongs to contributes no information about θ and the random experiment that produces the data y can be represented as a two-stage experiment. The first stage identifies the orbit y belongs to through the orbit representative v, the second stage gives $y = g(v)$, where $g \in \mathcal{G}$ is assumed to be in one-to-one correspondence with the elements of Θ. We then have the factorization

$$p_Y(y; \theta) = p_V(v(y)) p_{Y|V=v}(y; v, \theta) , \tag{7.28}$$

and, according to the conditionality principle with respect to distribution constant statistics (see section 2.6), inference about θ should be carried out conditionally on the observed value of v, that is, conditionally on the orbit $G(y)$ that the data belong to.

There are some analogies between model reduction based on a sufficient statistic and model reduction when an invariant statistic is available. In both cases, a partition of the sample space is identified. However, in the case of sufficiency, so long as $\mathcal{F}$ has been correctly specified, the orbit representative, i.e. the sufficient statistic, contains all the information about θ, while, conditioning on the element of the partition does not offer any information about θ. The opposite is true when the partition is induced by an invariant statistic.

It might happen that, for a given $\mathcal{F}$, both sufficiency and invariance reductions are possible. If the reduction based on a sufficient statistic $s(y)$ is carried out first, then a further reduction by invariance is possible whenever the model induced by $s(y)$ is still a group family. This property is guaranteed by a suitable representation of the sufficient statistic, hence, the definition below.

Definition 7.11 Let $\mathcal{G}$ be a transformation group on $\mathcal{Y}$. A function $s : \mathcal{Y} \to \mathcal{S}$ is said to be **equivariant** under $\mathcal{G}$ if the equality $s(y) = s(y')$ for some $y, y' \in \mathcal{Y}$ implies the equality $s(g(y)) = s(g(y'))$, for every $g \in \mathcal{G}$.

In other words, $s(\cdot)$ is equivariant if, when two points in the sample space have the same value of $s(\cdot)$, then equality of $s(\cdot)$ is maintained throughout the orbits generated by the two points. Note that invariance is a trivial case of equivariance.

If the function $s : \mathcal{Y} \to \mathcal{S}$ is equivariant under $\mathcal{G}$, it is possible to define a group $\tilde{\mathcal{G}}$ on $\mathcal{S}$, induced by $\mathcal{G}$, in a natural manner. Indeed, we can define the generic element of $\tilde{\mathcal{G}}$, which corresponds to $g \in \mathcal{G}$, as

$$\tilde{g}(s) = s(g(y_0)) , \tag{7.29}$$

where y_0 is such that $s(y_0) = s$. If $s(\cdot)$ is equivariant, then (7.29) is a good definition, in the sense that it does not depend on the choice of y_0 in the inverse image of s. In fact, if $y_1 \neq y_0$ is such that $s(y_1) = s$, we have, by the equivariance of $s(\cdot)$, $s(g(y_0)) = s(g(y_1))$, for each $g \in \mathcal{G}$. The group structure of $\tilde{\mathcal{G}}$ is an immediate consequence of definition (7.29) and of the group structure of $\mathcal{G}$.

Example 7.24 *Equivariant transformations under the group of scale and location transformations on* $\mathbb{R}^n$
Let $\mathcal{G}$ be the group which acts on $\mathbb{R}^n$ with generic element (7.23). The function $s(y) = s(y_1, \ldots, y_n) = (\sum y_i, \sum y_i^2) = (s_1, s_2)$ is equivariant. Indeed, if $y, y' \in \mathbb{R}^n$ are such that $s(y) = s(y')$, since

$$s(g(y)) = \left(n\mu + \sigma \sum y_i, n\mu^2 + 2\mu\sigma \sum y_i + \sigma^2 \sum y_i^2 \right),$$

then necessarily $s(g(y)) = s(g(y'))$ too. The group $\tilde{\mathcal{G}}$ induced by $\mathcal{G}$ on $\mathbb{R} \times \mathbb{R}^+$ has generic element $\tilde{g}(s_1, s_2) = (n\mu + \sigma s_1, n\mu^2 + 2\mu\sigma s_1 + \sigma^2 s_2)$. Other equivariant statistics can be found immediately. One example is $s' = (s_1', s_2') = (\bar{y}_n, s_n)$, whose induced group $\tilde{\mathcal{G}}$ has $\tilde{g}(s') = (\mu + \sigma s_1', \sigma s_2')$ as its generic element. △

If $\mathcal{F}_{\mathcal{G}}$ is a simple group family, the model induced by an equivariant statistic s with support $\mathcal{S}$ is a simple group family with respect to the group $\tilde{\mathcal{G}}$ on $\mathcal{S}$. Let $\mathcal{F}_{\tilde{\mathcal{G}}} = \{p_s(s; \theta), \theta \in \Theta\}$ be the model induced by s. If $s(Y)$ has density $p_s(s; \theta)$, the random variable $\tilde{g}(s(Y)) = s(g(Y))$ will have density $p_s(s; \tilde{g}\theta)$, with $\tilde{g}\theta \in \Theta$ because of the invariance of $\mathcal{F}_{\mathcal{G}}$. Furthermore, the transformation group induced by $\tilde{\mathcal{G}}$ on Θ coincides with $\tilde{\mathcal{G}}$, which is transitive. Hence, if the statistical model permits reduction both by sufficiency and by invariance, then it is possible to carry out reduction by invariance after reduction by sufficiency so long as we can choose an equivariant representation of the sufficient statistic.

Example 7.25 *Uniform distribution on* $[\theta, \theta + 1]$
Let $y_1, \ldots, y_n$ be independent realizations of a r.v. with $U(\theta, \theta+1)$ distribution, $\theta \in \mathbb{R}$. The statistic $s = (s_1, s_2) = (y_{(1)}, y_{(n)})$ is minimal sufficient. The family of $U(\theta, \theta + 1)$ distributions is a simple group family under the group $\mathcal{G}$ of location transformations on $\mathbb{R}$. The group of transformations of $\mathbb{R}^n$ with generic element $g^n(y) = (y_1 + \mu, \ldots, y_n + \mu)$, $\mu \in \mathbb{R}$, is associated with random sampling of size n. Since $s(g^n(y)) = (s_1 + \mu, s_2 + \mu)$, the statistic s is equivariant. Its distribution belongs to a simple group family, generated by the distribution of the minimum and the maximum of a random sample of size n from $U(0, 1)$ under the action of the transformation group

$$\tilde{\mathcal{G}} = \{(\mu + s_1, \mu + s_2), \quad \mu \in \mathbb{R}\}$$

on the sample space $\mathcal{S} = \{s_1, s_2 \in \mathbb{R}^2 : \quad 0 \le s_1 \le s_2 \le 1\}$. The sample range $v = s_2 - s_1 = y_{(n)} - y_{(1)}$ is a maximal invariant under $\tilde{\mathcal{G}}$ (see Example 2.10).

$\triangle$

If the statistic s is an estimator of θ, that is, if the range space of s is Θ, the term *equivariant* usually has a more limited meaning than in definition 7.11. We say that $\tilde{\theta}(y)$ is an **equivariant estimator under the group** $\mathcal{G}$ if (7.29) holds for $\tilde{\mathcal{G}} = \bar{\mathcal{G}}$, that is, if

$$\tilde{\theta}(g(y)) = \bar{g}(\tilde{\theta}(y)), \tag{7.30}$$

for each $g \in \mathcal{G}$. The property (7.30) coincides with equivariance under the reparameterization $\bar{g}(\theta)$. Since the maximum likelihood estimate $\hat{\theta}$ of θ is equivariant under reparameterizations and invariant under one-to-one transformations of y, then it will, in particular, satisfy (7.30).

7.6.2 Likelihood and Scale and Location Families

Theorem 7.3 below (Burridge, 1981) gives a sufficient condition for the maximum likelihood estimate of the parameter (μ, σ) to be unique, if it exists.

Theorem 7.3 Let $y_1, \ldots, y_n$ be independent realizations of a random variable with density in a scale and location family,

$$p(y; \mu, \sigma) = \frac{1}{\sigma} p_0 \left(\frac{y - \mu}{\sigma} \right),$$

with $(\mu, \sigma) \in \mathbb{R} \times (\mathbb{R}^+ \setminus \{0\})$ and $y \in \mathcal{Y} = \mathbb{R}$. Suppose that the function

$$q_0(y) = -\log(p_0(y))$$

has continuous first and second derivatives and that the second derivative is strictly positive for each $y \in \mathbb{R}$. Then, if the likelihood equation has a solution this is unique and is the maximum likelihood estimate.

Proof. The log-likelihood function is

$$l(\mu, \sigma) = -n \log \sigma - \sum_{i=1}^{n} q_0 \left(\frac{y_i - \mu}{\sigma} \right).$$

With the reparameterization $(\lambda, \varphi) = (1/\sigma, \mu/\sigma)$, the log-likelihood becomes

$$l(\lambda, \varphi) = n \log \lambda - \sum_{i=1}^{n} q_0(\lambda y_i - \varphi).$$

The aim is to prove that the function $f(\lambda, \varphi) = -l(\lambda, \varphi)$ is strictly convex (that is, that the log-likelihood is strictly concave). We have

$$
\begin{aligned}
f_\lambda &= \frac{\partial}{\partial \lambda} f(\lambda, \varphi) = -\frac{n}{\lambda} + \sum y_i q_0'(\lambda y_i - \varphi) \,, \\
f_\varphi &= \frac{\partial}{\partial \varphi} f(\lambda, \varphi) = -\sum q_0'(\lambda y_i - \varphi) \,, \\
f_{\lambda\lambda} &= \frac{\partial^2}{\partial \lambda^2} f(\lambda, \varphi) = \frac{n}{\lambda^2} + \sum y_i^2 q_0''(\lambda y_i - \varphi) \,, \\
f_{\lambda\varphi} &= \frac{\partial^2}{\partial \lambda \partial \varphi} f(\lambda, \varphi) = -\sum y_i q_0''(\lambda y_i - \varphi) \,, \\
f_{\varphi\varphi} &= \frac{\partial^2}{\partial \varphi^2} f(\lambda, \varphi) = \sum q_0''(\lambda y_i - \varphi) \,.
\end{aligned}
$$

Note that because $q_0''(\cdot) > 0$, the derivatives $f_{\lambda\lambda}$ and $f_{\varphi\varphi}$ are strictly positive. The determinant of the Hessian matrix of $f(\lambda, \varphi)$ is

$$
\begin{aligned}
\Delta &= f_{\lambda\lambda} f_{\varphi\varphi} - (f_{\lambda\varphi})^2 \\
&= (n/\lambda^2) \sum q_0''(\lambda y_i - \varphi) \\
&\quad + \left(\sum y_i^2 q_0''(\lambda y_i - \varphi) \right) \left(\sum q_0''(\lambda y_i - \varphi) \right) - \left(\sum y_i q_0''(\lambda y_i - \varphi) \right)^2 \,,
\end{aligned}
$$

and is strictly positive. Indeed, $(n/\lambda^2) \sum q_0''(\lambda y_i - \varphi)$ is always positive and

$$
\left(\sum y_i q_0''(\lambda y_i - \varphi) \right)^2 \leq \left(\sum y_i^2 q_0''(\lambda y_i - \varphi) \right) \left(\sum q_0''(\lambda y_i - \varphi) \right)
$$

follows immediately from the Cauchy–Schwarz inequality. $\square$

Let $(\hat{\mu}, \hat{\sigma})$ be the maximum likelihood estimate of (μ, σ), which we assume is finite and unique with probability one. Hence there is a one-to-one correspondence

$$
(y_1, \ldots, y_n) \longleftrightarrow (\hat{\mu}, \hat{\sigma}, a) \,, \tag{7.31}
$$

where

$$
a = \left(\frac{y_1 - \hat{\mu}}{\hat{\sigma}}, \ldots, \frac{y_n - \hat{\mu}}{\hat{\sigma}} \right) \tag{7.32}
$$

is the sample configuration (Fisher, 1934). In fact, $y_i = \hat{\mu} + \hat{\sigma} a_i$, $i = 1, \ldots, n$, where $a_i = (y_i - \hat{\mu})/\hat{\sigma}$. Let $q_0(y) = -\log(p_0(y))$. The log-likelihood $l(\mu, \sigma; y)$

can be written in the form

$$l(\mu, \sigma; \hat{\mu}, \hat{\sigma}, a) \;=\; -n \log \sigma - \sum_{i=1}^{n} q_0 \left(\frac{y_i - \mu}{\sigma} \right)$$

$$\qquad\qquad =\; -n \log \sigma - \sum_{i=1}^{n} q_0 \left(\frac{\hat{\sigma}}{\sigma} a_i + \frac{\hat{\mu} - \mu}{\sigma} \right). \qquad (7.33)$$

Note that of all the quantities a_i only $n - 2$ are functionally independent since the likelihood equations

$$l_\mu = (\partial/\partial\mu)l(\mu, \sigma) = 0, \quad l_\sigma = (\partial/\partial\sigma)l(\mu, \sigma) = 0$$

give the two constraints

$$\sum_{i=1}^{n} q_0'(a_i) \;=\; 0, \qquad\qquad\qquad (7.34)$$

$$\sum_{i=1}^{n} a_i q_0'(a_i) \;=\; n. \qquad\qquad\qquad (7.35)$$

Representation (7.33) of log-likelihood is very important in the light of the following theorem which summarizes the main results for inference in scale and location families. The theorem is given here with reference to the maximum likelihood estimator $(\hat{\mu}, \hat{\sigma})$, but the same results hold for any equivariant estimator of (μ, σ) (see Problem 7.28).

Theorem 7.4 Let $y_1, \ldots, y_n$ be independent realizations of a continuous random variable Y with a density belonging to a scale and location family, that is, $p(y; \mu, \sigma) = \sigma^{-1} p_0 \left((y - \mu)/\sigma \right)$, $(y - \mu)/\sigma \in \mathcal{Y}$, $\mu \in \mathbb{R}$, $\sigma > 0$. Assume that the maximum likelihood estimate $(\hat{\mu}, \hat{\sigma})$ of (μ, σ), based on $y = (y_1, \ldots, y_n)$, both exists and is finite. Then

i) The configuration statistic a, defined by (7.32), is maximal invariant with respect to the group $\mathcal{G}^n$ of scale and location transformations of the sample space. Thus, a is distribution constant, and is ancillary whenever $(y_{(1)}, \ldots, y_{(n)})$ is a minimal sufficient statistic.

ii) The joint density of $(\hat{\mu}, \hat{\sigma}, a_1, \ldots, a_{n-2})$ is

$$p(\hat{\mu}, \hat{\sigma}, a_1, \ldots, a_{n-2}; \mu, \sigma) = \frac{\hat{\sigma}^{n-2}}{\sigma^n} \left\{ \prod_{i=1}^{n} p_0 \left(\frac{\hat{\mu} - \mu}{\sigma} + \frac{\hat{\sigma}}{\sigma} a_i \right) \right\} h_n(a),$$

$$\qquad\qquad\qquad\qquad\qquad\qquad\qquad\qquad\qquad\qquad\qquad (7.36)$$

where a_{n-1} and a_n are expressed as a function of $(a_1, \ldots, a_{n-2})$, e.g. through the relations (7.34) and (7.35); the function $h_n(a)$ only depends on n and on $(a_1, \ldots, a_{n-2})$.

iii) The combinant

$$(q_1, q_2) = \left(\frac{\hat{\mu} - \mu}{\hat{\sigma}}, \frac{\hat{\sigma}}{\sigma}\right)$$

is a pivotal quantity whose conditional density, given a, is

$$p_{Q_1,Q_2|A=a}(q_1, q_2; a) = c(a) \, q_2^{n-1} \prod_{i=1}^{n} p_0(q_1 q_2 + q_2 a_i) \,, \tag{7.37}$$

where $c(a)$ is a normalizing constant.

Proof. The transformation $\bar{g}$ induced in the parameter space by $g^n(y) = (\nu + \xi y_1, \ldots, \nu + \xi y_n)$, with $\nu \in \mathbb{R}$, $\xi > 0$, is $\bar{g}(\mu, \sigma) = (\nu + \xi\mu, \xi\sigma)$. The estimate $(\hat{\mu}, \hat{\sigma})$ based on $g^n(y)$ is $(\nu + \xi\hat{\mu}, \xi\hat{\sigma})$ because of the equivariance of the maximum likelihood estimate. From this it immediately follows that the statistic a is invariant, that is, that $a(\nu + \xi y_1, \ldots, \nu + \xi y_n) = a(y_1, \ldots, y_n)$. We have

$$
\begin{aligned}
a(\nu + \xi y_1, \ldots, \nu + \xi y_n) &= \left(\frac{\nu + \xi y_1 - \nu - \xi\hat{\mu}}{\xi\hat{\sigma}}, \ldots, \frac{\nu + \xi y_n - \nu - \xi\hat{\mu}}{\xi\hat{\sigma}}\right) \\
&= \left(\frac{y_1 - \hat{\mu}}{\hat{\sigma}}, \ldots, \frac{y_n - \hat{\mu}}{\hat{\sigma}}\right) = a(y_1, \ldots, y_n) \,.
\end{aligned}
$$

The statistic a is maximal invariant too, that is, if $a(y) = a(y')$ then there exist $\nu \in \mathbb{R}$ and $\xi > 0$, such that $y_i' = \nu + \xi y_i$, $i = 1, \ldots, n$. Indeed, if $a(y) = a(y')$, then

$$\frac{y_i' - \hat{\mu}(y')}{\hat{\sigma}(y')} = \frac{y_i - \hat{\mu}(y)}{\hat{\sigma}(y)} \,, \qquad i = 1, \ldots, n \,,$$

from which

$$y_i' = \hat{\mu}(y') - \frac{\hat{\sigma}(y')}{\hat{\sigma}(y)}\hat{\mu}(y) + \frac{\hat{\sigma}(y')}{\hat{\sigma}(y)}y_i = \nu + \xi y_i \,,$$

with $\nu = \hat{\mu}(y') - (\hat{\sigma}(y')/\hat{\sigma}(y))\hat{\mu}(y)$ and $\xi = \hat{\sigma}(y')/\hat{\sigma}(y)$. This concludes the proof of part *i)*.

Consider the one-to-one function

$$(y_1, \ldots, y_n) \longmapsto (\hat{\mu}, \hat{\sigma}, a_1, \ldots, a_{n-2}) \,.$$

Since

$$p_Y(y_1, \ldots, y_n; \mu, \sigma) = \frac{1}{\sigma^n} \prod_{i=1}^{n} p_0 \left(\frac{y_i - \mu}{\sigma} \right),$$

and because $y_i = \hat{\mu} + \hat{\sigma} a_i$, the joint density of $(\hat{\mu}, \hat{\sigma}, a_1, \ldots, a_{n-2})$ is

$$p(\hat{\mu}, \hat{\sigma}, a_1, \ldots, a_{n-2}; \mu, \sigma) = \frac{1}{\sigma^n} \prod_{i=1}^{n} p_0 \left(\frac{\hat{\mu} - \mu}{\sigma} + \frac{\hat{\sigma}}{\sigma} a_i \right) |J|,$$

where $|J|$ is the Jacobian determinant of $(y_1, \ldots, y_n)$ expressed as a function of $(\hat{\mu}, \hat{\sigma}, a_1, \ldots, a_{n-2})$. From the representation

$$\begin{cases} y_1 = \hat{\mu} + \hat{\sigma} a_1 \\ \vdots \\ y_{n-2} = \hat{\mu} + \hat{\sigma} a_{n-2} \\ y_{n-1} = \hat{\mu} + \hat{\sigma} a_{n-1}(a_1, \ldots, a_{n-2}) \\ y_n = \hat{\mu} + \hat{\sigma} a_n(a_1, \ldots, a_{n-2}) \end{cases}$$

we obtain

$$J = \begin{pmatrix} 1 & a_1 & \hat{\sigma} & \cdots & 0 \\ \vdots & \vdots & \vdots & \ddots & \vdots \\ 1 & a_{n-2} & 0 & \cdots & \hat{\sigma} \\ 1 & f_1(a) & \hat{\sigma} f_{1,1}(a) & \cdots & \hat{\sigma} f_{1,n-2}(a) \\ 1 & f_2(a) & \hat{\sigma} f_{2,1}(a) & \cdots & \hat{\sigma} f_{2,n-2}(a) \end{pmatrix}$$

$$= \begin{pmatrix} 1_{n-2} & a^* & \hat{\sigma} I_{n-2} \\ 1_2 & f(a) & \hat{\sigma} F(a) \end{pmatrix},$$

where

$$\begin{aligned} a^* &= (a_1, \ldots, a_{n-2})^{\mathrm{T}} \\ f(a) &= (f_1(a), f_2(a))^{\mathrm{T}}, \\ f_1(a) &= a_{n-1}(a_1, \ldots, a_{n-2}), \\ f_2(a) &= a_n(a_1, \ldots, a_{n-2}), \\ F(a) &= [f_{i,j}(a)], \\ f_{i,j}(a) &= (\partial f_i(a))/(\partial a_j), \quad i = 1, 2; j = 1, \ldots, n-2. \end{aligned}$$

Remember the formula for the determinant of partitioned matrices

$$\begin{vmatrix} A & B \\ C & D \end{vmatrix} = |B| |C - DB^{-1}A|,$$

which holds so long as B is square and non-singular. Hence, with $\hat{\sigma} I_{n-2}$ as B,

$$
\begin{aligned}
|J| &= \hat{\sigma}^{n-2} \left| (1_2 \ f(a)) - \hat{\sigma} F(a) \frac{1}{\hat{\sigma}} I_{n-2} (1_{n-2} \ a^*) \right| \\
&= \hat{\sigma}^{n-2} h_n(a),
\end{aligned}
$$

with $h_n(a) = |(1_2 \ f(a)) - F(a)(1_{n-2} \ a^*)|$, which completes the proof of part *ii)*.

Let us denote by $p_A(a)$ the marginal density of $(a_1, \ldots, a_{n-2})$. Then

$$
p_{\hat{\mu}, \hat{\sigma}|A=a}(\hat{\mu}, \hat{\sigma}; \mu, \sigma, a) = c(a) \frac{\hat{\sigma}^{n-2}}{\sigma^n} \prod_{i=1}^{n} p_0 \left(\frac{\hat{\mu} - \mu}{\sigma} + \frac{\hat{\sigma}}{\sigma} a_i \right), \tag{7.38}
$$

where $c(a) = h_n(a)/p_A(a)$ can be interpreted as a normalizing constant. Lastly, since the function $(q_1, q_2)(\hat{\mu}, \hat{\sigma})$, given by

$$
q_1 = \frac{\hat{\mu} - \mu}{\hat{\sigma}}, \qquad q_2 = \frac{\hat{\sigma}}{\sigma},
$$

has $(\hat{\mu}, \hat{\sigma}) = (\sigma q_1 q_2 + \mu, \sigma q_2)$ as its inverse function, with Jacobian

$$
J = \begin{pmatrix} \sigma q_2 & \sigma q_1 \\ 0 & \sigma \end{pmatrix}
$$

giving $|J| = \sigma^2 q_2$, then

$$
p_{Q_1, Q_2|A=a}(q_1, q_2; a) = c(a) \frac{q_2^{n-2}}{\sigma^2} \left\{ \prod_{i=1}^{n} p_0(q_1 q_2 + q_2 a_i) \right\} \sigma^2 q_2,
$$

and the proof of part *iii)* is complete. □

Theorem 7.4 means that the conditionality principle can be applied to the statistical model induced by random sampling from a scale and location family. The reference distribution is $p_{\hat{\theta}|A=a}(\hat{\theta}; \theta, a)$, given by (7.38). The one-to-one transformation (q_1, q_2) of $(\hat{\mu}, \hat{\sigma})$ is useful in that, starting from (7.37), we can immediately obtain a conditional confidence region for (μ, σ). The constant $c(a)$ in (7.37) and (7.38) has the general expression

$$
c(a)^{-1} = \int_0^{+\infty} \int_{-\infty}^{+\infty} q_2^{n-1} \prod_{i=1}^{n} p_0(q_1 q_2 + q_2 a_i) dq_1 \, dq_2
$$

and can be determined numerically. See Lawless (1972, 1973, 1978) for applications to Cauchy, logistic, Weibull and extreme value distributions, Kappenman (1975) for Laplace distribution, Fraser (1976, 1979) for linear models.

Example 7.26 *Conditional inference in an $EV(\mu, \sigma)$ distribution*
Let $y_1, \ldots, y_n$ be independent realizations of a random variable Y with density

$$p(y; \mu, \sigma) = \frac{1}{\sigma} \exp \left\{ \frac{(y - \mu)}{\sigma} - e^{(y - \mu)/\sigma} \right\}, \qquad y \in \mathbb{R}, \, \mu \in \mathbb{R}, \, \sigma > 0.$$

The log-likelihood function is

$$l(\mu, \sigma) = -n \log \sigma + \frac{1}{\sigma} \sum_{i=1}^{n} y_i - n \frac{\mu}{\sigma} - \sum_{i=1}^{n} e^{(y_i - \mu)/\sigma}.$$

The score vector has components

$$l_\mu = \frac{1}{\sigma} \left(-n + \sum_{i=1}^{n} \exp\{(y_i - \mu)/\sigma\} \right),$$

$$l_\sigma = -\frac{n}{\sigma} - \frac{1}{\sigma^2} \sum_{i=1}^{n} y_i + \frac{n\mu}{\sigma^2} + \frac{1}{\sigma^2} \sum_{i=1}^{n} (y_i - \mu) \exp\{(y_i - \mu)/\sigma\}.$$

Assumptions of Theorem 7.3 hold and the maximum likelihood estimate of (μ, σ) is $(\hat{\mu}, \hat{\sigma})$, with

$$\hat{\mu} = \hat{\sigma} \log \left(\frac{\sum_{i=1}^{n} e^{y_i/\hat{\sigma}}}{n} \right)$$

and $\hat{\sigma}$ solution of the equation

$$\frac{\sum_{i=1}^{n} y_i \exp\{y_i/\hat{\sigma}\}}{\sum_{i=1}^{n} \exp\{y_i/\hat{\sigma}\}} - \frac{\sum_{i=1}^{n} y_i}{n} - \hat{\sigma} = 0. \tag{7.39}$$

Note that $\hat{\sigma} = 1/\hat{\nu}$, where ν is the shape parameter of the Weibull distribution $X = \exp\{Y\}$ (see section 7.1). Let $q_1 = (\hat{\mu} - \mu)/\hat{\sigma}$, $q_2 = \hat{\sigma}/\sigma$; then, from (7.37),

$$
\begin{aligned}
p_{Q_1, Q_2 | A = a}(q_1, q_2; a) &= c(a) q_2^{n-1} \prod_{i=1}^{n} \exp \left\{ (q_1 q_2 + q_2 a_i) - e^{q_1 q_2 + q_2 a_i} \right\} \\
&= c(a) q_2^{n-1} \exp \left\{ n q_1 q_2 + q_2 \sum_{i=1}^{n} a_i - e^{q_1 q_2} \sum_{i=1}^{n} e^{q_2 a_i} \right\}.
\end{aligned}
$$

Conditional inference about σ, or equally about the shape parameter $\nu = 1/\sigma$ of the Weibull distribution, could be based on the marginal distribution of q_2, conditionally on a. By integrating the joint conditional density with respect to q_1 we obtain

$$P_{Q_2|A=a}(q_2; a) = c(a)q_2^{n-1} \int_{-\infty}^{+\infty} e^{nq_1q_2}e^{q_2\Sigma a_i}e^{-e^{q_1q_2}\Sigma e^{q_2a_i}}\,dq_1 \qquad (7.40)$$

for $q_2 \geq 0$. Consider the change of variable $w = e^{q_1q_2}\sum e^{q_2a_i}$; from

$$e^{q_1q_2} = \frac{w}{\sum_{i=1}^{n}e^{q_2a_i}},$$

we get

$$e^{q_1q_2}q_2\,dq_1 = \frac{dw}{\sum_{i=1}^{n}e^{q_2a_i}},$$

from which follows

$$dq_1 = \frac{1}{q_2}e^{-q_1q_2}\frac{1}{\sum_{i=1}^{n}e^{q_2a_i}}dw = \frac{1}{q_2w}dw.$$

Now (7.40) becomes

$$
\begin{aligned}
P_{Q_2|A=a}(q_2; a) &= \frac{c(a)q_2^{n-2}}{\sum_{i=1}^{n}e^{q_2a_i}}e^{q_2\Sigma a_i}\int_{0}^{+\infty}\frac{w^{n-1}}{(\sum e^{q_2a_i})^{n-1}}e^{-w}\,dw \\
&= \frac{c(a)q_2^{n-2}e^{q_2\Sigma a_i}}{(\sum_{i=1}^{n}e^{q_2a_i})^{n}}\Gamma(n), \qquad q_2 \geq 0.
\end{aligned}
\qquad (7.41)
$$

If $q_{2,\frac{\alpha}{2}}(a)$ and $q_{2,1-\frac{\alpha}{2}}(a)$ are the $\frac{\alpha}{2} \times 100$-th quantile and $(1 - \frac{\alpha}{2}) \times 100$-th quantile of the distribution of q_2 given a, obtained by numerically integrating (7.41), a conditional confidence interval for σ, with level $(1 - \alpha)$, is

$$\left(\frac{\hat{\sigma}}{q_{2,1-\frac{\alpha}{2}}(a)}, \frac{\hat{\sigma}}{q_{2,\frac{\alpha}{2}}(a)}\right)$$

and $\left(\hat{\nu}q_{2,\frac{\alpha}{2}}(a), \hat{\nu}q_{2,1-\frac{\alpha}{2}}(a)\right)$ is a conditional confidence interval for $\nu = 1/\sigma$. See Lawless (1982, Example 4.1.2) for a numerical example. $\triangle$

7.7 Inference in Composite Group Families

7.7.1 Data and Model Reduction

Let us consider first a special subclass of composite group families, that is models which depend on scale, location and shape parameters. Let $y_1, \ldots, y_n$

be independent realizations of a random variable Y with density

$$p(y; \mu, \sigma, \tau) = \frac{1}{\sigma} p_0 \left(\frac{y - \mu}{\sigma}; \tau \right), \qquad (7.42)$$

where $\mu \in \mathbb{R}$, $\sigma > 0$, $\tau \in T$ and $(y - \mu)/\sigma \in \mathcal{Y} \subseteq \mathbb{R}$. Let $(\hat{\mu}, \hat{\sigma})$ be the maximum likelihood estimate of (μ, σ) which is assumed to be finite and unique with probability one. Consider the one-to-one transformation of the data

$$(y_1, \ldots, y_n) \longleftrightarrow (\hat{\mu}, \hat{\sigma}, a),$$

with a still defined by (7.32). Because of the equivariance property of $(\hat{\mu}, \hat{\sigma})$, the statistic a is partially distribution constant for (μ, σ) (see Definition 4.3). Hence the density of $(\hat{\mu}, \hat{\sigma}, a)$ can be factorized in the form

$$p_{\hat{\mu}, \hat{\sigma}, A}(\hat{\mu}, \hat{\sigma}, a; \mu, \sigma, \tau) = p_{\hat{\mu}, \hat{\sigma} | A = a}(\hat{\mu}, \hat{\sigma}; \mu, \sigma, \tau, a) \, p_A(a; \tau). \qquad (7.43)$$

On the basis of (7.43), for inference about τ with (μ, σ) as a nuisance parameter, it is possible to use the marginal likelihood based on a. In other words, since the group $\bar{\mathcal{G}}$, induced in the parameter space by the scale and location transformations in the sample space, only acts on (μ, σ) (see (7.26)), the problem of inference about τ is *invariant* with respect to transformations of the data according to the actions of the group $\mathcal{G}^n$ with elements (7.23); hence it is natural to require that inference about τ should be based on invariant statistics, that is (see Theorem 7.1) on functions of the maximal invariant a.

The above considerations can clearly be extended to any composite group family: the maximal invariant with respect to the group of transformations which act on the sample space is the natural statistic on which a marginal likelihood can be based for inference about a shape parameter τ. Now, with reference to the proportional hazards model, Example 7.22, we are able to understand the reason why the marginal likelihood based on inverse ranks should be used (see Example 4.9). As was seen in Example 7.22, the model is invariant under increasing monotone transformations and (see Example 7.12) the set of ranks (in one-to-one correspondence with the set of inverse ranks) is a maximal invariant with respect to this group of transformations. There is one further important example of inference problems where invariance reduction leads to statistics based on ranks: the nonparametric two-sample problem, which is illustrated in the following example.

Example 7.27 *Nonparametric tests for comparing two populations*
Let $y_1, \ldots, y_m$ be independent realizations of a continuous r.v. with support

$\mathcal{Y} \subseteq \mathbb{R}$ and density $p_1(y)$; furthermore, let $y_{m+1}, \ldots, y_n$, with $n > m$, be independent realizations of a continuous r.v., with support $\mathcal{Y}$ and density $p_2(y)$; suppose that the first m observations are independent of the last $n - m$. Indicate the corresponding distribution functions by $F_1(\cdot)$ and by $F_2(\cdot)$. Take, as a statistical model $\mathcal{F}$, the set of all the joint densities of $y_1, \ldots, y_n$ such that $F_1(y) \leq F_2(y)$, for every $y \in \mathcal{Y}$, or $F_1(y) \geq F_2(y)$, for every $y \in \mathcal{Y}$.

Consider the problem of testing $H_0 : F_1(y) \leq F_2(y)$, for each $y \in \mathcal{Y}$ versus $H_1 : F_1(y) \geq F_2(y)$, for each $y \in \mathcal{Y}$, with $F_1(y) \neq F_2(y)$ for some $y \in \mathcal{Y}$. Thus, we can take $F_1(y) - F_2(y)$ as the parameter of interest τ; this is an infinite-dimensional parameter. Because a continuous and monotone increasing transformation maintains stochastic ordering, the family $\mathcal{F}$ can be considered to be a composite group family generated by two fixed joint densities in H_0 and in H_1, respectively, under the action of the group $\mathcal{G}$, with generic element (7.18).

Preliminary reduction by sufficiency leads to base inference on the two vectors of ordered observation $y_{1\text{ord}}$ and $y_{2\text{ord}}$ of the two samples, with joint density $m!(n - m)! \prod_{i=1}^{m} p_1(y_i) \prod_{i=m+1}^{n} p_2(y_i)$. On the basis of Example 7.12, the set of ranks $(v_1, \ldots, v_n)$ of the vector $(y_{1\text{ord}}, y_{2\text{ord}})$, in increasing order, is a maximal invariant with respect to $\mathcal{G}$. Equally, we could consider only the set of ranks of $y_{2\text{ord}}$, $v = (v_{m+1}, \ldots, v_n)$. The density of v (see Problem 7.32) is

$$p_V(v; p_1, p_2) = P(V_{m+1} = v_{m+1}, \ldots, V_n = v_n)$$

$$= m!(n - m)! \int_{w_1 < \cdots < w_n} \prod_{i=1}^{m} p_1(w_{v_i}) \prod_{i=m+1}^{n} p_2(w_{v_i}) dw_1 \cdots dw_n$$

$$= \binom{n}{m}^{-1} \int_{w_1 < \cdots < w_n} n! \prod_{i=1}^{n} p_1(w_{v_i}) \frac{\prod_{i=m+1}^{n} p_2(w_{v_i})}{\prod_{i=m+1}^{n} p_1(w_{v_i})} dw_1 \cdots dw_n$$

$$= \binom{n}{m}^{-1} E_1 \left\{ \prod_{i=m+1}^{n} \frac{p_2(Y_{(v_i)})}{p_1(Y_{(v_i)})} \right\}, \tag{7.44}$$

where the expectation in the last equality is evaluated with respect to the distribution of the order statistic $(Y_{(1)}, \ldots, Y_{(n)})$, based on n i.i.d. observations with density $p_1(y)$; here we assume that $p_1(y) = 0$ implies $p_2(y) = 0$.

Note that if $F_1(y) = F_2(y)$ the density (7.44) is reduced to $1/\binom{n}{m}$; in fact all the attributions of ranks to the second sample are equally likely under H_0. This result is the basis for the construction of tests for H_0 versus H_1.

Tests with optimal properties may exist if the statistical model is restricted to a stochastically ordered parametric family. In particular, if we consider a one-parameter exponential family with density $p(y; \theta) = \exp\{\theta y - K(\theta)\}h(y)$ and let $p_1(y) = p(y; \theta_1)$, $p_2(y) = p(y; \theta)$, with $\theta > \theta_1$, density (7.44) becomes

$$p_V(v; \theta) = \binom{n}{m}^{-1} e^{(n-m)\{K(\theta_1) - K(\theta)\}} E_1 \left\{ \prod_{i=m+1}^{n} e^{Y_{(v_i)}(\theta - \theta_1)} \right\}.$$

The locally most powerful test for $H_0 : \theta = \theta_1$ versus $H_1 : \theta > \theta_1$ based on ranks rejects the null hypothesis for large values of the statistic

$$\left(\frac{\partial}{\partial \theta} \log p_V(v; \theta) \right)\Big|_{\theta = \theta_1},$$

that is, for large values of

$$\sum_{i=m+1}^{n} E_1(Y_{(v_i)}). \tag{7.45}$$

If $p(y; \theta)$ is the $N(\theta, 1)$ density and $\theta_1 = 0$, we obtain the **Fisher–Yates test**, which is based on the sum of the expected values of the standard normal order statistics whose position corresponds to that of the ranks of the observations of the second sample. If $p(y; \theta)$ is the density of the $\mathcal{F}_{ne}^1$ generated by the $U(0, 1)$ distribution (see Example 5.1), the test based on (7.45) is equivalent to the **Wilcoxon test**, which rejects the null hypothesis for large values of the sum of the ranks of the second group; in fact, $E_1(Y_{(i)}) = i/(n+1)$.

Locally most powerful tests based on (7.44) can, alternatively, be obtained by assuming that $p_1(y)$ is a fixed density and that $p_2(y)$ is an element of the location family generated by $p_1(y)$, that is, that $p_2(y) = p_1(y - \theta)$, with $\theta \geq 0$. In this case, the locally most powerful test for $H_0 : \theta = 0$ versus $H_1 : \theta > 0$ rejects the null hypothesis for large values of the statistic

$$- \sum_{i=m+1}^{n} E_1 \left\{ \frac{p_1'(Y_{(v_i)})}{p_1(Y_{(v_i)})} \right\}. \tag{7.46}$$

We obtain the Fisher–Yates test again if $p_1(y)$ is the $N(0, 1)$ density, while we have the Wilcoxon test if $p_1(y)$ is the logistic density $Lo(0, 1)$ (see Table 7.1). The observed significance levels of the tests (7.45) and (7.46) should be referred to the null density $p_V(v; p_1, p_1) = \binom{n}{m}^{-1}$. $\triangle$

The various possibilities for data and model reduction in simple and composite group families are summarized in Table 7.2.

Table 7.2: *Data and model reduction in group families.*

statistical model	parameter of interest	criterion for reduction	reduced model
group family	$g \leftrightarrow \theta$	conditioning	$p_{\hat{\theta}\mid A=a}(\hat{\theta}; \theta, a)$
composite group family	τ, shape parameter	marginalizing	$p_V(v; \tau)$, v maximal invariant

7.7.2 Marginal Likelihood

The following result gives the general expression of marginal likelihood for inference about τ in composite group families, where $\mathcal{G}$ is the group of scale and location transformations.

Theorem 7.5 Let $y_1, \ldots, y_n$ be independent observations of a random variable Y with density (7.42). The marginal likelihood for τ based on the maximal invariant is

$$
\begin{aligned}
L_M(\tau) &= \int_{-\infty}^{+\infty} \int_0^{+\infty} \prod_{i=1}^n p_0\left(\frac{y_i - \mu}{\sigma}; \tau\right) \frac{1}{\sigma^{n+1}} d\sigma\, d\mu \\
&= \int_{-\infty}^{+\infty} \int_0^{+\infty} \left(\prod_{i=1}^n p(y_i; \mu, \sigma, \tau)\right) \frac{1}{\sigma} d\sigma\, d\mu.
\end{aligned}
\tag{7.47}
$$

Proof. Let us first carry out a sufficiency reduction to the order statistic $y_{\text{ord}} = (y_{(1)}, \ldots, y_{(n)})$ with density

$$
p_{Y_{\text{ord}}}(y_{(1)}, \ldots, y_{(n)}; \mu, \sigma, \tau) = n! \prod_{i=1}^n \sigma^{-1} p_0\left(\frac{y_{(i)} - \mu}{\sigma}; \tau\right).
$$

The family with this density is, in its turn, a composite group family generated by $n! \prod_{i=1}^n p_0(y_{(i)}; \tau)$, $\tau \in T$, under scale and location transformations (7.16) on the support of Y_{ord}. Here, it is convenient to adopt the representation

$$
v(y) = (v_3, \ldots, v_n) = \left(\frac{y_{(3)} - y_{(1)}}{y_{(2)} - y_{(1)}}, \ldots, \frac{y_{(n)} - y_{(1)}}{y_{(2)} - y_{(1)}}\right)
\tag{7.48}
$$

of the maximal invariant under scale and location transformations. The statistic v is partially distribution constant with respect to (μ, σ), hence, without loss of generality, we may assume $\mu = 0$ and $\sigma = 1$. For $\mu = 0$ and $\sigma = 1$, the order statistic $y_{\text{ord}} = (y_{(1)}, \ldots, y_{(n)})$ has density

$$p_{Y_{\text{ord}}}(y_{(1)}, \ldots, y_{(n)}; 0, 1, \tau) = n! \prod_{i=1}^{n} p_0(y_{(i)}; \tau) \,.$$

Consider the one-to-one transformation

$$\begin{cases} z = y_{(1)} \\ w = y_{(2)} - y_{(1)} \\ v_3 = (y_{(3)} - y_{(1)})/(y_{(2)} - y_{(1)}) \\ \vdots \\ v_n = (y_{(n)} - y_{(1)})/(y_{(2)} - y_{(1)}) \end{cases}$$

with inverse

$$\begin{cases} y_{(1)} = z \\ y_{(2)} = z + w \\ y_{(3)} = wv_3 + z \\ \vdots \\ y_{(n)} = wv_n + z \end{cases}$$

and Jacobian

$$J = \begin{pmatrix} 1 & 0 & 0 & \cdots & \cdots & 0 \\ 1 & 1 & 0 & \cdots & \cdots & 0 \\ 1 & v_3 & w & \cdots & \cdots & 0 \\ 1 & v_4 & 0 & w & \cdots & 0 \\ \vdots & \vdots & \vdots & \vdots & \vdots & \vdots \\ 1 & v_n & 0 & \cdots & \cdots & w \end{pmatrix} = \begin{pmatrix} A & B \\ C & D \end{pmatrix},$$

with $A = \begin{pmatrix} 1 & 0 \\ 1 & 1 \end{pmatrix}$, B a $2 \times (n-2)$ matrix of zeroes, $C = (\mathbf{1}_{n-2} \; v)$, $D = wI_{n-2}$. Therefore

$$|J| = |D|\,|A - CD^{-1}B| = |D|\,|A| = w^{n-2} \,,$$

from which

$$p_{z,w,v}(z, w, v_3, \ldots, v_n; \tau) = n! p_0(z; \tau) p_0(z + w; \tau) \left\{ \prod_{i=3}^{n} p_0(wv_i + z; \tau) \right\} w^{n-2} \,,$$

with $z \in \mathbb{R}$, $w \in \mathbb{R}^+$, $v_i \in \mathbb{R}^+$, for $i = 3, \ldots, n$. Thus, the marginal density of $v = (v_3, \ldots, v_n)$ is

$$p_V(v; \tau) = n! \int_{-\infty}^{+\infty} \int_0^{+\infty} p_0(z; \tau) p_0(z + w; \tau) \prod_{i=3}^n p_0(w v_i + z; \tau) w^{n-2} dw \, dz \, .$$

(7.49)

In the marginal likelihood the constant $n!$ can be neglected; furthermore, consider the change of variable

$$(u, x) = (u(z, w), x(z, w)) \, ,$$

with inverse $z = u + x y_{(1)}$, $w = x(y_{(2)} - y_{(1)})$ and Jacobian

$$J = \begin{pmatrix} 1 & y_{(1)} \\ 0 & y_{(2)} - y_{(1)} \end{pmatrix} ,$$

whose determinant $|J| = y_{(2)} - y_{(1)}$ does not depend on τ and thus does not contribute to the marginal likelihood. This gives the expression

$$L_M(\tau) = \int_{-\infty}^{+\infty} \int_0^{+\infty} p_0(u + x y_{(1)}; \tau) p_0(u + x y_{(2)}; \tau) \prod_{i=3}^n p_0(u + x y_{(i)}; \tau) x^{n-2} dx \, du$$

$$= \int_{-\infty}^{+\infty} \int_0^{+\infty} \left(\prod_{i=1}^n p_0(u + x y_i; \tau) \right) x^{n-2} dx \, du \, .$$

With the further change of variable

$$(\sigma, \mu) = \left(\frac{1}{x}, -\frac{u}{x} \right) , \qquad \text{with inverse} \qquad (u, x) = \left(-\frac{\mu}{\sigma}, \frac{1}{\sigma} \right)$$

and Jacobian

$$J = \begin{pmatrix} -\dfrac{1}{\sigma} & \dfrac{\mu}{\sigma^2} \\ 0 & -\dfrac{1}{\sigma^2} \end{pmatrix} ,$$

which has determinant $|J| = 1/\sigma^3$ up to the sign, lastly, one obtains

$$L_M(\tau) = \int_{-\infty}^{+\infty} \int_0^{+\infty} \prod_{i=1}^n p_0 \left(\frac{y_i - \mu}{\sigma}; \tau \right) \frac{1}{\sigma^{n+1}} d\sigma \, d\mu \, .$$

$\square$

The representation of the marginal likelihood obtained in Theorem 7.5 has a Bayesian interpretation: indeed the quantity on the right-hand side of (7.47) could be read as a posterior density (up to a normalizing constant) or as an integrated likelihood with respect to the improper prior distribution

$$\pi(\mu, \sigma) = \frac{1}{\sigma} d\mu \, d\sigma \,, \qquad \mu \in \mathbb{R}, \, \sigma > 0 \,. \qquad (7.50)$$

According to the prior distribution (7.50), the components μ and σ are independent, with μ uniformly distributed on $\mathbb{R}$ and σ with density proportional to σ^{-1} on $\mathbb{R}^+ \setminus \{0\}$. This is an uninformative prior distribution which can be intuitively justified as follows (Jeffreys, 1961, section 3.1; Cox and Hinkley, 1974, p. 377). The composition of the two location transformations is still a location transformation hence, the composition also defines an action of the group on itself, with respect to which it would seem reasonable to require that an uninformative prior distribution should be invariant. This then leads to the requirement that, for a location parameter, an uninformative prior distribution assigns the same probability to all the intervals $(\nu, \nu + \xi)$, for each $\nu \in \mathbb{R}$, with $\xi \in \mathbb{R}^+$ fixed; i.e. that it is a uniform distribution. Analogous reasoning for scale transformations adds the requirement that, for $\Delta > 1$ fixed, all the intervals $(\lambda, \Delta \lambda)$ must have the same prior probability for σ, for each $\lambda \in \mathbb{R}^+$. A prior distribution for σ, which satisfies this condition will have density proportional to $1/\sigma$; note that this density corresponds to an improper uniform prior for $\log \sigma$. Behind this there lies a new definition of an uninformative distribution on the parameter space Θ: if Θ has a group structure, an uninformative prior distribution must be invariant with respect to the group that acts on Θ. In general, for a given transformation group, one can define many actions of the group onto itself and, if such a group can be identified with the parameter space, it is possible to formulate different invariance requirements for a prior distribution. As regards the relations with frequentist inference procedures, the two most important actions are the left action and the right action; the former transforms an element g_0 of $\mathcal{G}$ into $g \circ g_0$, the latter associates the element $g_0 \circ g^{-1}$ to g_0. The corresponding invariant prior distributions are called, respectively, the left and right invariant prior distributions. According to these definitions, the distribution (7.50) is the right-invariant prior for scale and location transformations (see, e.g., Barndorff-Nielsen and Cox, 1994, Example 2.31). In a composite group family, the marginal likelihood for the shape parameter τ, based on the maximal invariant, can always be represented as an integrated likelihood with respect to the right invariant measure defined on the group of transformations $\mathcal{G}$ (Barndorff-Nielsen, 1988, section 2.2).

Example 7.28 *Testing for separate scale and location families*
One immediate application of Theorem 7.5 is to some problems of testing hypotheses on separate families (see Problem 5.35). Given n independent observations $y_1, \ldots, y_n$, from an element of a parametric family $\mathcal{F}$, consider the problem of testing $H_0 : \mathcal{F} = \mathcal{F}_0$ versus $H_1 : \mathcal{F} = \mathcal{F}_1$, where $\mathcal{F}_0$ and $\mathcal{F}_1$ are separate scale and location families generated by two continuous densities on $\mathbb{R}$. Suppose that the elements of $\mathcal{F}_0$ have density $(1/\sigma)p_0((y - \mu)/\sigma)$ and that the elements of $\mathcal{F}_1$ have density $(1/\sigma)p_1((y - \mu)/\sigma)$. This is a problem of testing hypotheses about the shape parameter τ in the composite group family $\mathcal{F}_0 \bigcup \mathcal{F}_1$, where τ plays the role of indicator of either $\mathcal{F}_0$ or $\mathcal{F}_1$. Because $\mathcal{F}_0$ and $\mathcal{F}_1$ are invariant (hence, also the union family is invariant) with respect to scale and location transformations, the inference problem here, is also invariant under scale and location transformations (see Example 7.18) in the sample space. Thus it would be sensible to require that the conclusions should be invariant too; that is, that the test statistic should be a function of the data only through the maximal invariant (see Theorem 7.1). The parametric model induced by the maximal invariant contains only two elements: the density of the maximal invariant referred to $\mathcal{F}_0$ and to $\mathcal{F}_1$, respectively. According to the Neyman–Pearson lemma (Theorem 3.3), there exists an optimum level α test in the reduced model, that is, there exists a **most powerful invariant test**. This test considers large values of the ratio of the marginal likelihoods to be significant against H_0,

$$\frac{(L_M)_1}{(L_M)_0} = \frac{\int_{-\infty}^{+\infty} \int_0^{+\infty} \prod_{i=1}^n p_1\left(\frac{y_i - \mu}{\sigma}\right) \frac{1}{\sigma^{n+1}} d\sigma\, d\mu}{\int_{-\infty}^{+\infty} \int_0^{+\infty} \prod_{i=1}^n p_0\left(\frac{y_i - \mu}{\sigma}\right) \frac{1}{\sigma^{n+1}} d\sigma\, d\mu}. \tag{7.51}$$

See Uthoff (1970) for some examples of tests based on the statistic (7.51). $\triangle$

Example 7.29 *Marginal likelihood for the shape parameter of the gamma distribution*
Let $y_1, \ldots, y_n$ be independent realizations of $Y \sim Ga(\nu, \sigma)$. One convenient representation of the maximal invariant with respect to scale transformations is

$$v(y) = \left(\frac{y_{(2)}}{y_{(1)}}, \ldots, \frac{y_{(n)}}{y_{(1)}}\right) = (v_2, \ldots, v_n).$$

The marginal likelihood (see Problem 7.34) is

$$L_M(\nu) = \int_0^{+\infty} \left\{ \prod_{i=1}^n \frac{(\frac{y_i}{\sigma})^{\nu-1} e^{-y_i/\sigma}}{\sigma \Gamma(\nu)} \right\} \frac{1}{\sigma} d\sigma$$

$$= \frac{\Pi y_i^{\nu-1}}{(\Gamma(\nu))^n} \int_0^{+\infty} \frac{1}{\sigma^{n\nu+1}} e^{-n\bar{y}_n/\sigma} d\sigma .$$

Consider the change of variable $z = n\bar{y}_n/\sigma$ with inverse $\sigma = n\bar{y}_n/z$ and $d\sigma = -(n\bar{y}_n/z^2)dz$. We obtain

$$L_M(\nu) = \frac{\Pi y_i^{\nu-1}}{(\Gamma(\nu)(n\bar{y}_n)^\nu)^n} \int_0^{+\infty} z^{n\nu-1} e^{-z} dz = \frac{\Gamma(n\nu)}{(\Gamma(\nu))^n} \frac{\Pi y_i^{\nu-1}}{(n\bar{y}_n)^{n\nu}} . \qquad (7.52)$$

It is easy to check that (7.52) depends on $(y_1, \ldots, y_n)$ only through $v(y)$. Indeed, we can write the marginal likelihood in the equivalent form

$$L_M(\nu) = \frac{\Gamma(n\nu)}{(\Gamma(\nu))^n} \left(\frac{\Pi y_i}{(\Sigma y_i)^n} \right)^\nu$$

with

$$\frac{\Pi y_i}{(\Sigma y_i)^n} = \frac{\Pi y_{(i)}}{(\Sigma y_{(i)})^n} = \prod_{i=1}^n \frac{y_{(i)}}{y_{(1)} + \ldots + y_{(n)}}$$

$$= \left(1 - \frac{y_{(2)} + \ldots y_{(n)}}{y_{(1)} + \ldots + y_{(n)}} \right) \prod_{i=2}^n \frac{y_{(i)}}{y_{(1)} + \ldots + y_{(n)}}$$

$$= \left(1 - \frac{v_2 + \ldots + v_n}{1 + v_2 + \ldots + v_n} \right) \prod_{i=2}^n \frac{v_i}{1 + v_2 + \ldots + v_n} .$$

The marginal likelihood can also be written as

$$L_M(\nu) = \exp \left\{ \nu \left(\Sigma \log y_i - n \log(n\bar{y}_n) \right) - n \left(\log \Gamma(\nu) - \frac{1}{n} \log \Gamma(n\nu) \right) \right\},$$

which corresponds to the likelihood for a one-parameter exponential family, with natural parameter ν and natural observation $\Sigma \log y_i - n \log(n\bar{y}_n)$. Therefore, a uniformly most powerful level α test for $H_0 : \nu = \nu_0$ versus $H_1 : \nu > \nu_0$ exists in the model reduced by invariance; this test rejects H_0 for large values of the statistic

$$\sum_{i=1}^n \log y_i - n \log(\sum_{i=1}^n y_i) = n \log \frac{\prod_{i=1}^n y_i^{1/n}}{n^{-1} \sum_{i=1}^n y_i} .$$

The test thus obtained is uniformly most powerful invariant; see also Lehmann (1986, section 6.3).   △

Recourse to marginal likelihood based on a maximal invariant could also offer a way of carrying out inference in a simple group family $\mathcal{F}_\mathcal{G}$, if the parameter of interest is a (not one-to-one) function $\tau(\theta)$ of the global parameter θ, where θ is in one-to-one correspondence with the elements of $\mathcal{G}$. Indeed, if for every fixed value of τ, a group $\mathcal{G}_1$ can be found for the corresponding subfamily $\mathcal{F}_\tau$, under which $\mathcal{F}_\tau$ is a simple group family, then a marginal likelihood for τ will be based on the maximal invariant with respect to $\mathcal{G}_1$.

Example 7.30 *Inference on the normal distribution function*
Let $y_1, \ldots, y_n$ be independent observations from $N(\mu, \sigma^2)$ and let

$$\pi = P(Y \leq y_0) = \Phi((y_0 - \mu)/\sigma)$$

be the parameter of interest, where y_0 is a fixed value. Consider the transformation $z_i = y_i - y_0$, $i = 1, \ldots, n$. The values z_i are independent realizations of $N(\nu, \sigma^2)$, with $\nu = \mu - y_0$; since $\pi = \Phi(-\nu/\sigma)$, the parameter of interest π is a one-to-one function of $\tau = \nu/\sigma$. The subfamily of normal distributions with ν/σ fixed is a simple group family with respect to the group of transformations $\mathcal{G}_1$ with generic element $g(z) = \xi z$, $\xi \in \mathbb{R} \setminus \{0\}$; here we could consider, for example, $N(1, \tau^{-2})$ as a generating element. Preliminary reduction by sufficiency leads to $(\bar{z}_n, s_n^2)$. The function $g(z_i) = \xi z_i$, $i = 1, \ldots, n$, transforms the sufficient statistic into $(\xi \bar{z}_n, \xi^2 s_n^2)$ and a maximal invariant is given by the ratio $\bar{z}_n / \sqrt{s_n^2}$ or, more usefully, by $t = \sqrt{n} \bar{z}_n / \sqrt{s_n^2}$ which has a non-central Student's t distribution with $n - 1$ degrees of freedom and non-centrality parameter τ. Since this is a family with a monotone likelihood ratio in τ (see, for example, Lehmann, 1986, p. 295), the uniformly most powerful invariant test (under $\mathcal{G}_1$) for $H_0 : \tau = \tau_0$ versus $H_1 : \tau > \tau_0$ rejects the null hypothesis for large values of t.   △

Example 7.31 *Testing hypotheses in the normal linear regression model*
Let $y = (y_1, \ldots, y_n)^{\mathrm{T}}$ be a realization of a random vector Y with distribution $N_n(\mu, \sigma^2 I_n)$, where $\mu = X\beta$, with X an $n \times k$ matrix of known constants and $\beta \in \mathbb{R}^k$ a k-dimensional vector of unknown regression coefficients; the error variance $\sigma^2 > 0$ is also unknown. Assume that the matrix X has full rank k, with $k < n$.

Many problems of testing hypotheses under a linear model can be reduced to the general form $H_0 : Q\beta = 0$, with Q a $q \times k$ ($q \leq k$) matrix of known constants, with full rank q. The model without the restrictions imposed by H_0 assumes that μ is an element of the k-dimensional linear subspace $\mathcal{V}_k$ of $\mathbb{R}^n$ generated by the columns of X. Under H_0 there are q further linear constraints, that is, μ is an element of a $(k - q)$-dimensional linear subspace $\mathcal{V}_{k-q}$ of $\mathcal{V}_k$.

The likelihood ratio test W for H_0 versus $H_1 : Q\beta \neq 0$ rejects the null hypothesis for large values of the statistic

$$F = \frac{(SSE_0 - SSE)/q}{SSE/(n - k)} , \qquad (7.53)$$

where SSE and SSE_0 indicate the residual sums of squares under the general hypothesis $\mu \in \mathcal{V}_k$ and the null hypothesis, respectively (see, for example, Azzalini, 1996, section 5.3.3). The null distribution of the statistic F is $F_{q,n-k}$; the non-null distribution is a non-central F with the same degrees of freedom.

We can show that the test based on the F statistic is optimal under a suitable reduced model, defined on the basis of invariance restrictions. Firstly, let us express the model in a canonical form, such that H_0 states that certain components of the mean vector of a multivariate normal distribution vanish. To do this, let A be an n-dimensional orthogonal matrix, whose first k rows form a basis of $\mathcal{V}_k$ and whose rows from the $(q+1)$-th to the k-th form a basis of $\mathcal{V}_{k-q}$. Hence, by construction, for $y \in \mathbb{R}^n$, the vector Ay has the last $n - k$ components equal to zero if, and only if, $y \in \mathcal{V}_k$; the first q components are also zero if, and only if, $y \in \mathcal{V}_{k-q}$. The random vector $Z = AY$ has a $N_n(A\mu, \sigma^2 I_n)$ distribution, because A is orthogonal. Let $\eta = A\mu$; since $\mu \in \mathcal{V}_k$, we have $\eta_{k+1} = \ldots = \eta_n = 0$; under H_0, we have in addition $\eta_1 = \ldots = \eta_q = 0$.

The testing problem in the transformed model concerns $(\eta_1, \ldots, \eta_q)$. However, it is usually only possible to find optimal tests, even in a reduced model, in problems where the parameter of interest is a scalar. For example, optimality can be reached in one-sided problems when the one-parameter family which corresponds to the reduced model has a monotone likelihood ratio (see Examples 7.29 and 7.30). In the case here, this can be done if we take

$$\tau = \frac{\sum_{i=1}^q \eta_i^2}{\sigma^2}$$

as the parameter of interest. Note that $\tau = 0$ if, and only if, H_0 is true and the alternatives which have the same distance as η from the origin, have the same value of τ. The density of Z is

$$p_z(z_1, \ldots, z_n; \eta, \sigma^2) = \frac{1}{(2\pi\sigma)^{n/2}} \exp\left\{ -\frac{1}{2\sigma^2} \left(\sum_{i=1}^k (z_i - \eta_i)^2 + \sum_{i=k+1}^n z_i^2 \right) \right\}.$$

For every fixed value of τ, this density can be regarded as being an element of a simple group family $\mathcal{F}_{\mathcal{G}}^{\tau}$, generated by the density $p_0(z)$ of an n-variate normal with mean vector $(\sqrt{\tau}, 0, \ldots, 0)$ and with covariance matrix I_n. The group $\mathcal{G}$ is given by the composition of three groups of transformations on $\mathbb{R}^n$, that is, $\mathcal{G} = \mathcal{G}_1 \circ \mathcal{G}_2 \circ \mathcal{G}_3$ (see Problem 7.6). The group $\mathcal{G}_3$ is made up of scale transformations $g_3(z_1, \ldots, z_n) = (\sigma z_1, \ldots, \sigma z_n)$, $\sigma > 0$; the group $\mathcal{G}_2$ is made up of the orthogonal transformations of the first q components of z; the group $\mathcal{G}_1$ has generic element

$$g_1(z_1, \ldots, z_q, z_{q+1}, \ldots, z_k, z_{k+1}, \ldots, z_n)$$
$$= (z_1, \ldots, z_q, z_{q+1} + c_1, \ldots, z_k + c_{k-q}, z_{k+1}, \ldots, z_n),$$

with $c_i \in \mathbb{R}$, $i = 1, \ldots, k - q$. The action of $\mathcal{G}_3$ on $p_0(z)$ gives marginal distributions with constant variance, σ^2; with the next action of $\mathcal{G}_2$ the mean vector of the first q components rotates, leaving τ unchanged; lastly, under the action of $\mathcal{G}_1$ the expected value of the components with indices from $q + 1$ to k is not necessarily equal to zero.

A preliminary sufficiency reduction leads to $(z_1, \ldots, z_k, \sum_{i=k+1}^{n} z_i^2)$. The statistic $(z_1, \ldots, z_q, \sum_{i=k+1}^{n} z_i^2)$ is a maximal invariant with respect to $\mathcal{G}_1$ (see Problem 7.11); a further invariance reduction with respect to $\mathcal{G}_2$ suggests only considering $(\sum_{i=1}^{q} z_i^2, \sum_{i=k+1}^{n} z_i^2)$ (see Example 7.13); lastly, for this statistic, the maximal invariant with respect to $\mathcal{G}_3$, that is the overall maximal invariant (see Problem 7.14), can be expressed as

$$F = \frac{\sum_{i=1}^{q} z_i^2 / q}{\sum_{i=k+1}^{n} z_i^2 / (n-k)} \, . \tag{7.54}$$

Because $y = A^{\mathrm{T}} z$ is an orthogonal transformation, it preserves distances and, thus, the statistics (7.54) and (7.53) coincide (Problem 7.37). On the basis of (7.54), the null and non-null distributions of the statistic F can immediately be recognized; more specifically, the non-centrality parameter is τ, with respect to which the distribution has a monotone likelihood ratio and the test obtained is uniformly most powerful invariant. $\triangle$

7.8 Bibliographic Note

Group theory has traditionally been attributed to Galois, Lie and Klein (see, for example Struik, 1966, sections 8.8 and 8.23). Fisher (1934) in his analysis of inference under scale and location models was the first to recognize the particular role in inference of the symmetry structures of group models. Systematic surveys of recent developments can be found in Fraser (1968, 1979), Barndorff-Nielsen, Blæsild and Eriksen (1989) and Eaton (1989). Introductory accounts of inference for group families can be found in Barndorff-Nielsen and Cox (1994, sections 1.4 and 2.8) and, from a frequency-decision viewpoint, in Lehmann (1986, Chapter 6) and Ferguson (1967, Chapter 4 and section 5.6).

The introduction to group families in sections 7.1 and 7.2 has, in part, been inspired by Lehmann (1983, section 1.3). The distinction between simple and composite group families, and the consequences of this for data and model reduction, have been developed in Barndorff-Nielsen, Blæsild and Eriksen (1989). Burridge (1981) proves a more general version of Theorem 7.3, which takes into account the possibility of grouping and, also, shows empirically that the reparameterization used in the proof is useful for accelerating the convergence of the algorithms for numerical maximization. Silvapulle and Burridge (1986) give a necessary and sufficient condition for the existence of the maximum likelihood estimate for the same class of models considered in Burridge (1981); the condition is similar to that of Silvapulle (1981) described in section 6.5. The discussion on the exact distribution of the maximum likelihood estimator in scale and location families (section 7.6.2) owes much to Lawless (1982, section 4.1.2 and Appendix G). For an extension to general simple group families, see Barndorff-Nielsen and Cox (1994, section 2.8). Severini (1994) and Small and Murdoch (1993) propose extensions of conditional inference to semiparametric scale and location families where the density $p_0(y)$ of the generating element is not specified. For further applications of invariance for obtaining nonparametric tests see Lehmann (1986, sections 6.8-6.10). Further references to testing for separate families can be found in Pace and Salvan (1990).

7.9   Problems

7.1 Show that if Y has a distribution which belongs to a location family, $\exp\{Y\}$ will have a distribution that belongs to a scale family.

[Section 7.1]

7.2 Check whether the class of location transformations with the restriction $\mu \geq 0$ is closed under inversion.

[Section 7.2]

7.3 Let $\mathcal{G}$ be a transformation group on $\mathcal{Y}$ and let $g_0 \in \mathcal{G}$. Show that, as g varies in $\mathcal{G}$, so too does $g \circ g_0$ vary throughout $\mathcal{G}$, that is, for every $g^* \in \mathcal{G}$ there is a transformation $g \in \mathcal{G}$ such that $g \circ g_0 = g^*$.

[Section 7.2]

7.4 Define suitable generalizations of the groups of transformations (7.5), (7.6) and (7.7) for the case where the space $\mathcal{Y}$ is $\mathbb{R}^n$. For each case, check in detail the group structure.

[Section 7.2]

7.5 Check whether the group of orthogonal transformations on $\mathbb{R}^n$ defined in Example 7.7 is commutative.

[Section 7.2]

7.6 Let $\mathcal{G}_1$ and $\mathcal{G}_2$ be two groups of transformations on $\mathcal{Y}$. Consider the class of transformations

$$\mathcal{G} = \mathcal{G}_1 \circ \mathcal{G}_2 \;=\; \left\{ g_1^1 \circ g_2^1 \circ \ldots \circ g_1^m \circ g_2^m, \; g_1^i \in \mathcal{G}_1, g_2^i \in \mathcal{G}_2, \right.$$
$$\left. i = 1, \ldots, m, m = 1, 2, \ldots \right\}$$

Show that $\mathcal{G}$ is a group and that $\mathcal{G}_1$ and $\mathcal{G}_2$ are subgroups of $\mathcal{G}$.

[Section 7.2]

7.7 Let $\mathcal{C}$ be a class of one-to-one transformations of $\mathcal{Y}$. Consider the class of transformations $\mathcal{G}$ of $\mathcal{Y}$ obtained through the composition of a finite number of elements of $\mathcal{C}$ or of their inverses. Show that $\mathcal{G}$ is a group (this is called the minimal group generated by $\mathcal{C}$).

[Section 7.2]

7.8 Show that the group of scale and location transformations (7.7) on $\mathbb{R}$ is transitive, but not free.

[Section 7.3]

7.9 Consider the transformation group on $\mathbb{R}^n \setminus N$, where $N = \{y \in \mathbb{R}^n : y_1 = y_2 = \cdots = y_n\}$, with generic element

$$g(y_1, \ldots, y_n) = (\mu + \sigma y_1, \ldots, \mu + \sigma y_n), \qquad \mu \in \mathbb{R}, \quad \sigma > 0.$$

Show that this group is transitive only for $n = 1$ and is free only for $n > 1$.

[Section 7.3]

7.10 With reference to Example 7.8, show that $u_2(y) = u_2(y')$ implies $u(y) = u(y')$.

[Section 7.3]

7.11 Let $\mathcal{Y} = \mathbb{R}^n$ and let $\mathcal{G}$ be the class of transformations on $\mathcal{Y}$ with generic element

$$g(y_1, \ldots, y_n) = (y_1 + c_1, \ldots, y_m + c_m, y_{m+1}, \ldots, y_n),$$

where $1 \le m < n$; the function $g(\cdot)$ is a location transformation, from y_i to $y_i + c_i$, on the first m components of y and leaves the last $n - m$ unchanged. Show that $\mathcal{G}$ is a group and obtain the maximal invariant under $\mathcal{G}$.

[Section 7.3]

7.12 Show that, if $\mathcal{F}$ is an invariant family with respect to a group $\mathcal{G}$ and if $\mathcal{F}$ satisfies the identifiability condition, the class $\bar{\mathcal{G}}$ of transformations induced in the parameter space is a group.

[Section 7.4]

7.13 Show that the family of distributions generated by the transformations (7.22) with the constraint $\mu = \beta_1 x_1 + \cdots + \beta_k x_k$ is a group family, that is, show that the assigned class of transformations will identify a group. Specify the action $\bar{g}$ induced by g on the parameter space $\Theta = \{(\beta_1, \ldots, \beta_k) \in \mathbb{R}^k, \quad \sigma > 0\}$.

[Section 7.4]

7.14 Let $\mathcal{G}$ be a transformation group on $\mathcal{Y}$ generated by two subgroups $\mathcal{G}_1$ and $\mathcal{G}_2$ (see Problem 7.6). Let $u(y)$ be a maximal invariant with respect to $\mathcal{G}_2$ and suppose that $u(y)$ is equivariant with respect to $\mathcal{G}_1$. Let $\bar{g}_1(\cdot)$ indicate the elements of the group $\tilde{\mathcal{G}}_1$ defined by (7.29) with $g = g_1$. Let $v(u)$ be a maximal invariant with respect to $\tilde{\mathcal{G}}_1$. Show that $v(u(y))$ is a maximal invariant with respect to $\mathcal{G}$.

[Sections 7.3 and 7.4; Lehmann, 1986, Theorem 6.2]

7.15 Let $\mathcal{F}_\mathcal{G} = \{p(y; \theta), \theta \in \Theta, y \in \mathcal{Y}\}$ be a simple group family with respect to $\mathcal{G}$. Let $f(y)$ be a measurable real function defined on $\mathcal{Y}$ with finite expectation for every $\theta \in \Theta$. Show that for every $g \in \mathcal{G}$,

$$E_\theta\{f(g(Y))\} = E_{\bar{g}\theta}\{f(Y)\}.$$

[Section 7.4]

7.16 Let

$$\mathcal{F}_0 = \left\{ p(y; \alpha, \beta) = \frac{y^{\alpha-1}(1-y)^{\beta-1}}{B(\alpha, \beta)}, \quad \alpha, \beta > 0, \ y \in [0, 1] \right\},$$

with $B(\alpha, \beta) = \Gamma(\alpha)\Gamma(\beta)/\Gamma(\alpha+\beta)$. Obtain the composite group family generated by $\mathcal{F}_0$ under the action of the group of scale and location transformations (7.7) on $\mathbb{R}$.

[Section 7.5]

7.17 Show that the family of log-normal distributions $Y = \exp(Z)$, where $Z \sim N(\mu, \sigma^2)$, is a composite group family, with respect to the group $\mathcal{G}$ of scale transformations, with shape parameter σ and scale parameter e^μ. Identify the generating class $\mathcal{F}_0$.

[Section 7.5]

7.18 Show that if Y has a distribution which belongs to a scale and location family $\mathcal{F}$, then e^Y will have a distribution belonging to a composite group family generated by the group of scale transformations on $\mathbb{R}^+ \setminus \{0\}$. Identify the generating family $\mathcal{F}_0$.

[Section 7.5]

7.19 Let $\mathcal{F}_0$ be the $N(\mu, 1)$ family, with $\mu > 0$. Consider the class $\mathcal{G}$ of transformations on $\mathbb{R}$ which is composed of just two elements: $g_1(y) = y$ and $g_2(y) = -y$. Show that $\mathcal{G}$ is a group. Identify the composite group family generated by $\mathcal{F}_0$ under the action of $\mathcal{G}$.

[Section 7.5]

7.20 Let $\mathcal{F}_{ne}^1$ be a natural exponential family of order 1 of continuous distributions. Construct the composite group family generated by $\mathcal{F}_{ne}^1$ under the action of the group of scale transformations on $\mathbb{R}$. Say whether the parametric family obtained is an exponential dispersion family.

[Section 7.5]

7.21 Let $\mathcal{F}_\mathcal{G} = \{p(y; \theta), \theta \in \Theta, y \in \mathcal{Y}\}$ be a composite group family with respect to the group $\mathcal{G}$. Let $\bar{\mathcal{G}}$ be the group induced by $\mathcal{G}$ on Θ and let $\tau(\theta)$ be the maximal invariant with respect to the action of $\bar{\mathcal{G}}$ on Θ. Show that the distribution of the maximal invariant, under the action of $\mathcal{G}$ on $\mathcal{Y}$, depends on θ only through $\tau(\theta)$. Illustrate this property with reference to the statistical model generated by random sampling from $N_2(\mu, \Sigma)$, under the group of scale and location transformations on $\mathbb{R}^2$, which has (7.16) as its generic element.

[Section 7.5]

7.22 Show that the family $\mathcal{F}_{\mathcal{G}}^n$ in (7.27) is a simple group family with respect to the action of the group $\mathcal{G}^n$, with generic element $g^n(y_1, \ldots, y_n) = (g(y_1), \ldots, g(y_n))$, $g \in \mathcal{G}$. (See Example 7.18.)

[Section 7.6.1]

7.23 Show that the statistic $c(y_1, y_2)$ defined in Example 7.23 is not invariant under location changes. Construct an analogous counter-example for a scale family.

[Section 7.6.1]

7.24 Obtain the expressions for the score vector and for the observed information matrix from log-likelihood (7.33).

[Section 7.6.2]

7.25 Identify a representation $(\hat{\theta}, a)$, analogous to that of (7.31), under random sampling of size n from a location family.

[Section 7.6.2]

7.26 Show that if $(\hat{\mu}, \hat{\sigma})$ is a sufficient statistic for (μ, σ) in a scale and location family, then $(\hat{\mu}, \hat{\sigma})$ and the configuration a are independent. Illustrate this with reference to random sampling from a $N(\mu, \sigma^2)$ distribution.

[Section 7.6.2]

7.27 Solve the question raised in Problem 7.25 with reference to a scale family. Identify a scale family where $\hat{\sigma}$ and a (with suitable a) are independent.

[Section 7.6.2]

7.28 Show that, when a is suitably redefined, Theorem 7.4 holds for any equivariant estimator $(\tilde{\mu}, \tilde{\sigma})$, that is, such that

$$(\tilde{\mu}, \tilde{\sigma})(\nu + \xi y_1, \ldots, \nu + \xi y_n) = (\nu + \xi \tilde{\mu}_0, \xi \tilde{\sigma}_0),$$

where $(\tilde{\mu}_0, \tilde{\sigma}_0)$ is the estimate $(\tilde{\mu}, \tilde{\sigma})$ calculated on $(y_1, \ldots, y_n)$.

[Section 7.6.2]

7.29 Let $y_1, \ldots, y_n$ $(n > 1)$ be independent observations from a distribution in the location family (7.1), whose elements have a finite second moment. Let $a = (a_2, \ldots, a_n) = (y_2 - y_1, \ldots, y_n - y_1)$, show that an equivariant estimator (see (7.30)) of μ has the form $\tilde{\mu}(y) = y_1 + b(a)$. Show that the mean squared error of an equivariant estimator does not depend on μ and that if an estimator minimizes the mean squared error conditionally on a, then it will also minimize the unconditional mean squared error. Verify that such an optimal estimator is

$$\tilde{\mu}^P(y) = y_1 - E_{\mu=0}(Y_1 \mid a),$$

and that it could be written in the form

$$\tilde{\mu}^P(y) = \frac{\displaystyle\int_{-\infty}^{+\infty} \mu \prod_{i=1}^{n} p_0(y_i - \mu) d\mu}{\displaystyle\int_{-\infty}^{+\infty} \prod_{i=1}^{n} p_0(y_i - \mu) d\mu}.$$

The estimator $\tilde{\mu}^P(y)$ is called the **Pitman estimator** of the location parameter μ (Pitman, 1938). Give a Bayesian interpretation of $\tilde{\mu}^P(y)$.

[Section 7.6]

7.30 Let $y_1, \ldots, y_n$ $(n > 1)$ be independent realizations of a r.v. with density belonging to a scale and location family (7.3) whose elements have finite second moment. Obtain the estimator with the minimum mean squared error within the class of equivariant estimators with respect to the transformations $\bar{g}$ (see Example 7.15), both when μ is the parameter of interest and when σ is the parameter of interest.

[Section 7.6; Zacks, 1981, section 5.6.3]

7.31 Say which of the marginal likelihoods in Examples 4.6, 4.7 and 4.8 can be interpreted as marginal likelihoods based on suitable maximal invariants.

[Section 7.7.1]

7.32 Let $Y_1, \ldots, Y_n$ be continuous and independent random variables, with support $\mathcal{Y} = \mathbb{R}$ and density $p_{Y_i}(y)$, $i = 1, \ldots, n$. Let $V = (V_1, \ldots, V_n)$ be the vector of ranks of $(Y_1, \ldots, Y_n)$; V_i is the position occupied by Y_i in the ordered vector $(Y_{(1)}, \ldots, Y_{(n)})$. Show that V has probability function

$$P(V_1 = v_1, \ldots, V_n = v_n) = \int_{w_1 < w_2 < \cdots < w_n} p_{Y_1}(w_{v_1}) \cdots p_{Y_n}(w_{v_n}) dw_1 \cdots dw_n.$$

(7.55)

(*Hint*: see Example 4.9.)

[Section 7.7.1]

7.33 Obtain the Jeffreys' uninformative prior distribution (2.12) for the parameters of a scale and location model and compare it with the uninformative invariant prior distribution (7.50).

[Section 7.7.2]

7.34 Obtain the marginal likelihood for the shape parameter, under random sampling of size n from an element of a composite group family which depends on a scale parameter σ and on a shape parameter τ.

[Section 7.7.2]

7.35 Obtain the most powerful invariant test based on n independent observations of the r.v. Y for $H_0 : Y \sim U(\mu, \mu + \sigma)$ versus $H_1 : Y \sim Exp(\mu, \sigma)$ (see Table 7.1).

[Section 7.7.2]

7.36 Compare the uniformly most powerful invariant test obtained in Example 7.29, with the uniformly most powerful similar test for the same problem, obtained in Example 5.14.

[Section 7.7.2]

7.37 Show in detail that the statistics (7.53) and (7.54), considered in Example 7.31, coincide.

[Section 7.7.2; Lehmann, 1986, section 7.2.2]

CHAPTER 8

ASYMPTOTIC METHODS:
INTRODUCTION AND ELEMENTARY
TECHNIQUES

8.1 Introduction

The first four chapters reviewed basic concepts of the theory of statistics. Chapters 5-7 looked at broad classes of models where the criteria for data and model reduction and the inference procedures outlined in the first part can, immediately, be applied. However, there is still some mismatch between exact inference procedures, which fully exploit the particular features of special classes of models, and approximations based on first-order theory, which work for any sufficiently smooth parametric model. Think, for example, of model reduction according to the conditionality principle. Even though this principle would seem to have general validity, its applications appear to be limited to group families and exponential families but, even here, only when the parameter of interest is a linear combination of canonical parameters. Furthermore, the problems of distribution which arise with reduction by conditioning can only be solved exactly in special cases, while the first-order approximations examined so far (see section 5.9) completely neglect the peculiar structure of the reference model.

Higher-order asymptotic techniques, to be presented in both this and subsequent chapters, make continuity of application of the diverse principles of data and model reduction and of exact and approximate inference possible. One early example is the approximate conditional likelihood, introduced in section 4.7, which illustrates how reduction by conditioning could be extended, and applied, beyond exact cases through the introduction of correction terms into the usual likelihood procedures. Some aspects of approximate conditional likelihood have yet to be clearly explained, e.g. the validity of the approximations considered, and the link with exact conditional likelihood, when the latter exists.

Any discussion of methods of approximation that are useful for statistical inference requires, first of all, a study of the techniques for approximating distributions. This study is of independent interest, and here we will concentrate on methods of approximation which arise from higher-order asymptotic considerations. Briefly, this entails refining the central limit theorem (or related results on convergence in distribution) by introducing correction terms which, in addition to the mean and the variance, take further information about the distribution into account. The approximations obtained in this way are, on the one hand, homogeneous with exact results, should these latter exist and, on the other, homogeneous with first-order asymptotic procedures. These methods have the advantage of being analytical rather than numerical, thus they facilitate the definition of inference procedures which combine accuracy with generality of application.

Some comments have already been made, in section 3.3, regarding the widespread presence of asymptotic arguments in statistics. These can be useful in specification problems (section 1.3), in inference problems (think e.g. of locally optimal tests, section 3.5.3, or of asymptotic efficiency of the maximum likelihood estimator, section 3.4.1), and in distribution problems.

Example 8.1 *Testing hypotheses: aspects of optimality and of distribution*
Suppose that the data $(y_1, \ldots, y_n)$ have been generated, by random sampling, from a r.v. Y with probability density $p_Y(y; \tau, \zeta)$. When testing hypotheses on τ we may encounter the crossings between identification of the null distribution and properties of optimality set out in Table 8.1. Cells 1, 4 and 9 of the table can be illustrated as follows.

1. If $Y \sim N(\tau, \zeta)$, with $\zeta > 0$, and we wish to test $H_0 : \tau \leq \tau_0$ versus $H_1 : \tau > \tau_0$, the usual Student's t test (Gosset, 1908), whose distribution is known, is uniformly most powerful both in the class of unbiased tests and in that of scale invariant tests (see for example Lehmann, 1986, sections 5.3 and 6.6).

4. If Y has a density which belongs to an $\mathcal{F}_{ne}^p$ with canonical parameter (τ, ζ), where τ is a scalar, there exists a uniformly most powerful similar test for the same problem of hypothesis testing considered in the point above.

Table 8.1: *Distribution and optimality of a test statistic.*

Null distribution of the test	Properties of the test		
	Optimal	Approx. optimal	*Ad hoc*
Known and usual	1	2	3
Known in principle, but ...	4	5	6
Not known (nuisance parameters)	7	8	9

Usually it is anything but easy to calculate the exact null distribution of the test statistic (see Example 5.13).

9. Consider τ as an indicator in the infinite-dimensional space of parametric families for continuous distributions and ζ as the parameter that identifies the elements of a given family. Both the dimension and the parameter space of ζ may depend on τ. The problem of testing for goodness-of-fit to a given parametric family is thus formalized as a problem of testing $H_0 : \tau = \tau_0$ versus $H_1 : \tau \neq \tau_0$; the value τ_0 may, for example, represent the family of normal distributions. For such a problem we could use the chi-squared test of goodness-of-fit (Pearson, 1900); this does not have optimality properties and the reference distribution is, typically, the asymptotic null distribution; the exact null distribution usually depends on the nuisance parameter ζ. △

Both this, and the following chapters, will deal, almost exclusively, with asymptotic methods for distribution problems, both with the aim of improving the first-order approximations of sampling distributions and, indirectly, with the aim of defining statistics and combinants on the basis of which inference is both simple and accurate. For an account of the role of asymptotic theory in problems of optimality, see Skovgaard (1989).

This chapter will present some further topics in first-order asymptotic theory and will discuss criteria for evaluating the accuracy of approximations for distributions. Some examples will highlight the possibility that first-order approximations may be inadequate. We will then describe some early ideas in the development of higher-order asymptotic methods, so as to give a gradual introduction to the subject.

Chapter 9 introduces the basic technical tools that are useful for asymptotic theory. Chapter 10 is an introduction to asymptotic methods for approximating distributions. Chapter 11 presents some improved inference procedures based on likelihood quantities, using the results of the previous two chapters.

8.2   Evaluating the Accuracy of an Approximation

Many statistics have distributions which, because of central limit results, can be approximated by a normal or chi-squared distribution when the sample size n, or, more generally, an index of information, is sufficiently large. For applications, it is important to obtain indications about when n is sufficiently large or, at least whether, for the value of n that is of interest, the approximation is satisfactory.

8.2.1   Berry–Esséen Inequality

It is unusual for an asymptotic result to be accompanied by a precise evaluation of the approximation error. Useful information about the speed of convergence of the distribution of a standardized sum of random variables to the normal distribution is offered by the following result, which is known as the Berry–Esséen inequality.

Theorem 8.1 Let Y be a random variable with $E(Y) = 0$, $\mathrm{Var}(Y) = \sigma^2$, $E(|Y|^3) = \rho$ (σ^2 and ρ positive and finite). Let $Y_1, \ldots, Y_n$ be independent copies of Y, let $F_n(\cdot)$ be the distribution function of $\sum Y_i / (\sqrt{n}\sigma)$ and let $\Phi(\cdot)$ be the standard normal distribution function. Then, for every $n \in \mathbb{N} \setminus \{0\}$,

$$\sup_{t \in \mathbb{R}} |F_n(t) - \Phi(t)| \leq \frac{3\rho}{\sigma^3 \sqrt{n}} \, . \tag{8.1}$$

Proof. See Feller (1971, p. 543), or Field and Ronchetti (1990, p. 8).   □

Theorem 8.1 establishes that convergence to the normal distribution, according to the central limit theorem, will be at a rate of order $O(n^{-1/2})$ and will be uniform. Observe that the upper bound given by the inequality (8.1) depends only on the third standardized absolute moment and not on any other characteristics of the distribution. Naturally, this runs counter to accuracy. If Y has a normal distribution, the upper bound (8.1) is not zero. The constant 3, which appears in the inequality, is not optimal. It has been improved by van Beek (1972), who proved that the value 3 in (8.1) can be replaced by 0.7975. A counterexample by Esséen (1956) showed that the optimal constant must be greater than 0.4.

The Berry–Esséen inequality is essentially different from the central limit theorem: indeed it is an approximation result which is valid for every n. Upper bounds like (8.1) are rarely used in statistical applications. The example below offers one explanation.

Example 8.2 *Uniform distribution*
If $Y \sim U(-0.5, 0.5)$, then

$$E(Y) = 0, \qquad \mathrm{Var}(Y) = \frac{1}{12}, \qquad E|Y|^3 = 2 \int_0^{1/2} y^3 \, dy = \frac{1}{32}.$$

On the basis of (8.1), the absolute error in the normal approximation of the distribution of $\sum Y_i \sqrt{12/n}$ will not be greater than a fixed $\varepsilon > 0$ if

$$n \geq \left(\frac{3/32}{12^{-3/2}\varepsilon} \right)^2 = n_\varepsilon^A$$

or, at least, using the van Beek improvement, if

$$n \geq \left(\frac{0.7975/32}{12^{-3/2}\varepsilon} \right)^2 = n_\varepsilon^B.$$

We obtain $n_\varepsilon^A = 15.1875/\varepsilon^2$, $n_\varepsilon^B = 1.0733/\varepsilon^2$, results which are illustrated in the table below.

ε	n_ε^A	n_ε^B
0.01	15 1875	10 733
0.05	6075	430
0.10	1519	108

Observe that, although the van Beek improvement is important, the inequality would suggest very large sample sizes, certainly far greater than those that are usually of interest in applications. On the other hand, already when

$n = 12$, the exact distribution of the sum of independent uniforms is almost indistinguishable from a normal distribution, at least for the purposes of many applications, including the generation of pseudo-random observations from a normal distribution.   △

8.2.2   Other Methods for Evaluating the Accuracy of a Normal Approximation for a Fixed n

Often the indices of skewness and kurtosis, ρ_3 and ρ_4 (see (3.20)), give useful indications about closeness to the normal distribution. Let Y be a random variable with a finite fourth moment and let $\mu = E(Y)$, $\sigma^2 = \text{Var}(Y)$, $\rho_3 = E(Y - \mu)^3/\sigma^3$ and $\rho_4 = E(Y - \mu)^4/\sigma^4 - 3$. Remember that for $Y \sim N(\mu, \sigma^2)$, $\rho_3 = \rho_4 = 0$. Thus, one can expect that the distribution of a statistic which has small values of ρ_3 and ρ_4 may be approximated fairly well by a normal distribution. For instance, if $Y_1, \ldots, Y_n$ are independent copies of Y and if $\bar{Y}_n = \sum Y_i/n$ is the sample mean, then from (3.12) and (3.19) we obtain

$$\rho_3\left(\bar{Y}_n\right) = \frac{\kappa_3(\bar{Y}_n)}{\kappa_2(\bar{Y}_n)^{3/2}} = \frac{\rho_3}{\sqrt{n}} = O\left(n^{-1/2}\right),$$

$$\rho_4\left(\bar{Y}_n\right) = \frac{\kappa_4(\bar{Y}_n)}{\kappa_2(\bar{Y}_n)^2} = \frac{\rho_4}{n} = O\left(n^{-1}\right).$$

Note that $\rho_4\left(\bar{Y}_n\right)$ tends to zero faster than does $\rho_3\left(\bar{Y}_n\right)$. In one sense, which will be explained better below, asymmetry of the parent distribution is the main indicator of the inadequacy of the normal approximation for the distribution of the sample mean.

A more thorough analysis is possible if the exact distribution of a statistic is known for every value of n. It is then possible to make a direct comparison between the true distribution and the asymptotic nominal distribution. In this way, we can identify the set of values of n where it is important to use the exact distribution, whereas for values of n where the error in the approximation is negligible, it might be simpler to use the asymptotic distribution. Exact distributions are usually difficult to calculate. However, evaluation of the adequacy of an asymptotic approximation can be carried out through simulation, so as to obtain a Monte Carlo estimate of the exact distribution of the statistic of interest.

The examples below illustrate various possibilities.

Example 8.3 *Gamma distribution*
Let $Y \sim Ga(\nu, 1)$. Because of closure under convolution of the gamma family

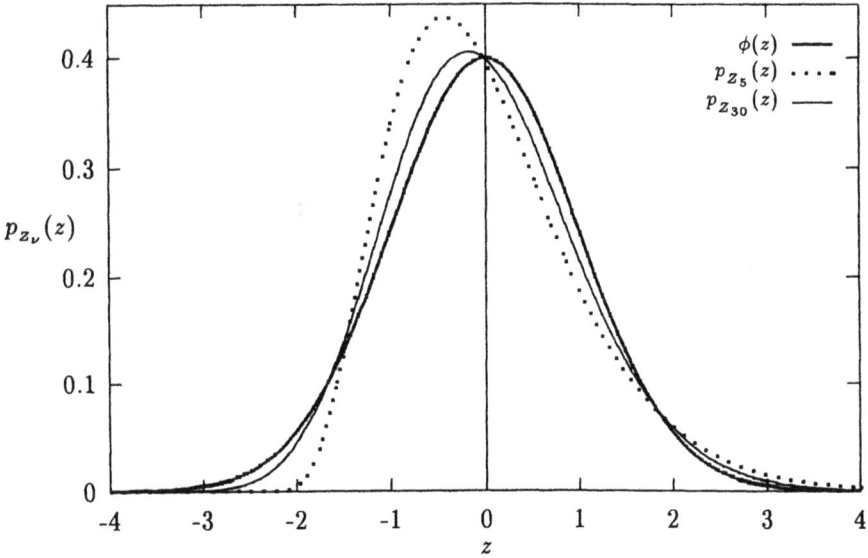

Figure 8.1: *Density of the standardized sum of ν i.i.d. exponential r.v.'s, Z_ν, $\nu = 5,30$, and of $Z \sim N(0,1)$.*

(cf. Problem 3.3), for large values of ν the distribution of $Z_\nu = (Y - \nu)/\sqrt{\nu}$ can be approximated with a standard normal distribution, both in terms of distribution function and in terms of density. Figure 8.1 compares the density functions of Z_ν, for $\nu = 5$ and 30, with the $N(0,1)$ density. The support of Z_ν is $[-\sqrt{\nu}, +\infty)$ because Y is positive. For small values of ν, for example $\nu < 20$, the approximation is particularly inaccurate in the lower tail. However, even for larger values of ν, for example for $\nu = 30$, the asymmetry of the distribution of Z_ν is still clear and the normal approximation is poor in the tails. $\triangle$

Example 8.4 *Uniform distribution*
Let Z_n be the standardized sum of n i.i.d. copies of $Y \sim U(0,1)$. The support of Z_n is $[-\sqrt{3n}, \sqrt{3n}]$; for the exact density of Z_n, see e.g. Field and Ronchetti (1990, section 3.5). Because the distribution of Z_n is symmetrical, we would expect that the normal approximation would be accurate even for relatively small values of n. This is confirmed by Figure 8.2. $\triangle$

Example 8.5 *Binomial distribution*
Let $Y \sim Bi(10, 0.4)$. Table 8.2 compares the exact values of the d.f. of Y

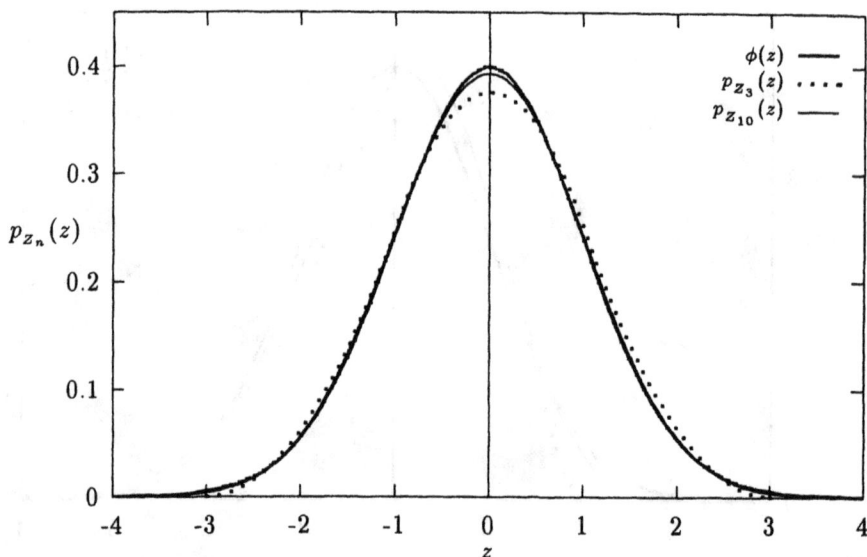

Figure 8.2: *Density of the standardized sum of n i.i.d. $U(0,1)$, Z_n, $n = 3,10$ and of $Z \sim$* *$N(0,1)$.*

with the approximate values which have been obtained both by crude normal approximation and using a continuity correction. Indeed, here we are approximating a discrete distribution having support $\{0, 1, \ldots, 10\}$ with a continuous distribution. A simple idea, in order to take the information about the support into account, would be to approximate the probability of each point, $y = 0, 1, \ldots, 10$, with the integral of the approximating continuous density over the interval $(y - 1/2, y + 1/2)$. In order to maintain normalization, we should also attribute the probability on the adjacent tail to the extreme points of the support. Note the surprising accuracy of the normal approximation with continuity correction.   △

Example 8.6 *Testing for separate exponential families*

Let $Y_1, \ldots, Y_n$ be independent realizations of a r.v. with p.d.f. $p^0(y)$ belonging to a parametric family. Consider the problem of testing $H_0 : p^0(y) \in \mathcal{F}_0$ versus $H_1 : p^0(y) \in \mathcal{F}_1$, with $\mathcal{F}_0$ and $\mathcal{F}_1$ separate exponential families with a common support. In some cases (see Problem 5.35) there exists a most powerful similar test. This is the case e.g. when $\mathcal{F}_0$ is the family of two-parameter log-normal

Table 8.2: *Approximations of the d.f. of* $Y \sim Bi(10, 0.4)$.

y	$F_Y(y)$	$\Phi((y-4)/\sqrt{2.4})$	$\Phi((y+0.5-4)/\sqrt{2.4})$
0	0.006	0.005	0.012
1	0.046	0.026	0.053
2	0.167	0.098	0.166
3	0.382	0.259	0.373
4	0.633	0.500	0.627
5	0.834	0.741	0.834
6	0.945	0.902	0.947
7	0.988	0.974	0.988
8	0.998	0.995	0.998
9	1.000	0.999	1.000
10	1.000	1.000	1.000

distributions (defined in Problem 3.1), while $\mathcal{F}_1$ is the family of two-parameter gamma distributions. The most powerful similar test for this problem rejects log-normality for large values of $\sum y_i$ in the conditional distribution given the observed value of $(\sum \log y_i, \sum (\log y_i)^2)$, which is a sufficient statistic for $\mathcal{F}_0$.

The actual levels of three tests have been evaluated through simulation (1000 Monte Carlo samples), using a nominal level $\alpha = 0.05$. The three tests are: the optimal test $T_{L,G}^{cond}$, with rejection region determined through simulation of the exact conditional null distribution (see Problem 2.10 for a description of the simulation algorithm); the test $T_{L,G}^{as}$, defined by referring the optimal test to the conditional asymptotic null normal distribution (see Theorem 5.11 and Pace and Salvan, 1990, section 4); the test $T_{L,G}^{C}$, based on the likelihood ratio, whose rejection region is obtained from the asymptotic normal distribution (Jackson, 1968).

The simulation results are given in Table 8.3; three log-normal distributions with different values of the shape parameter σ^2 (variance of the corresponding normal distribution) are considered with sample size $n = 20, 50, 100$. The tests which refer to an asymptotic distribution have actual levels which depend on the nuisance parameter and which are much smaller than 0.05 even for $n = 100$.

$\triangle$

Table 8.3: *Probability of rejection of log-normality, $\alpha = 0.05$.*

Test	n	$\sigma^2 = 0.5$	$\sigma^2 = 1.0$	$\sigma^2 = 2.0$
$T_{L,G}^{cond}$	20	0.05	0.06	0.05
	50	0.05	0.05	0.06
	100	0.06	0.05	0.06
$T_{L,G}^{as}$	20	0.01	0.00	0.00
	50	0.03	0.00	0.00
	100	0.03	0.01	0.00
$T_{L,G}^{C}$	20	0.02	0.01	0.00
	50	0.03	0.01	0.00
	100	0.04	0.02	0.00

8.2.3 Chi-Squared Approximations

Examples 8.3 and 8.6 in the previous section show that normal approximations may be unsatisfactory even with a moderate or large sample size.

Normal approximations are used in likelihood inference about a scalar parameter for the null distribution of Z, Z_u, Z_e (defined by formulae (3.44)-(3.46)), the signed versions of the likelihood ratio test W and of its asymptotically equivalent variants. In the presence of nuisance parameters, signed versions of the profile log-likelihood ratio statistic may be used (see formulae (4.33)-(4.35)). Chi-squared approximations for the null distribution of W, W_u or W_e are usually more precise than the normal approximations for their one-sided versions. In fact, while it can be shown that, under regularity conditions, Z has null distribution function $F_z(z) = \Phi(z) + O(n^{-1/2})$, in the one-parameter case the null distribution of W is given by $F_W(w) = \Phi(\sqrt{w}) - \Phi(-\sqrt{w}) + O(n^{-1})$, so that W has a χ_1^2 distribution with a smaller asymptotic error.

Below, we offer a preliminary explanation of the differences in the rate of asymptotic convergence to the normal and to the chi-squared distributions. As will be shown in section 10.3, if T_n is an asymptotically normal statistic, then, under regularity conditions, its distribution function may be expanded as

$$F_{T_n}(t) = \Phi(t) + \frac{1}{\sqrt{n}}h(t) + O(n^{-1}), \qquad (8.2)$$

where $h(t)$ is an even function. Therefore, the expansion

$$F_{U_n}(u) = \Phi(\sqrt{u}) - \Phi(-\sqrt{u}) + O(n^{-1}),$$

holds for the distribution function of $U_n = T_n^2$. Note that the χ_1^2 distribution function evaluated at u appears on the left-hand side. On the other hand, if we assume that $F_{U_n}(u) = \Phi(\sqrt{u}) - \Phi(-\sqrt{u}) + O(n^{-1})$ and that the distribution of T_n is symmetrical, then, necessarily, $h(t) = 0$ in (8.2) hence, the normal approximation for T_n has the same order of error as does the chi-squared approximation for U_n. To see this, from

$$F_{U_n}(u) = F_{T_n}(\sqrt{u}) - F_{T_n}(-\sqrt{u}) = \Phi(\sqrt{u}) - \Phi(-\sqrt{u}) + O(n^{-1}),$$

because of the symmetry of the distribution of T_n and of the normal distribution, we obtain

$$F_{U_n}(u) = 2F_{T_n}(\sqrt{u}) - 1 = 2\Phi(\sqrt{u}) - 1 + O(n^{-1}),$$

so that

$$F_{T_n}(\cdot) = \Phi(\cdot) + O(n^{-1}).$$

Here is a first explanation of why deviation from normality is, in the first instance, due to asymmetry. The error in normal approximations and in chi-squared approximations is of the same order, as will be seen in section 10.4, even if $\rho_3(T_n) = 0$ or, more generally, if $\rho_3(T_n) = o(n^{-1/2})$.

The example below illustrates the quality of the chi-squared approximation for the null distribution of W_p in a case where the exact distribution is not explicitly known.

Example 8.7 *Likelihood ratio test for the shape parameter of a gamma distribution*

Let $y_1, \ldots, y_n$ be independent observations from a $Ga(\nu, \phi)$ distribution. The profile log-likelihood ratio (4.25) for ν is

$$W_P(\nu) = 2n\left\{(\hat{\nu} - \nu)(\psi(\hat{\nu}) - 1) + \nu \log\left(\frac{\hat{\nu}}{\nu}\right) - \log\frac{\Gamma(\hat{\nu})}{\Gamma(\nu)}\right\}, \qquad (8.3)$$

where $\hat{\nu}$ is the m.l.e. of ν, the solution of the equation

$$\sum_{i=1}^{n} \log y_i / n - \log \bar{y}_n = \psi(\hat{\nu}) - \log \hat{\nu},$$

with $\psi(x) = d\log\Gamma(x)/dx$, the digamma function. For $n = 5$ and 10, Figure 8.3 shows graphs of the approximate χ_1^2 d.f. of $W_P(1)$ plotted against the exact null d.f., which has been estimated through 10 000 Monte Carlo samples from a $Ga(1,1)$ distribution. The straight line represents the simulated null distribution. Note the clear improvement in the approximation from $n = 5$ to $n = 10$; for example, when the nominal level is 0.10 the exact level is 0.17 for $n = 5$ and 0.13 for $n = 10$; analogously, with a nominal level equal to 0.01, the actual level is 0.03 for $n = 5$ and 0.02 for $n = 10$.   △

8.3   Improvements on First-Order Approximations: Historical Notes

Various authors have taken up the problem of how to improve on first-order asymptotic normal or chi-squared approximations when, with the n given for the specific application, there is reason to believe that the approximation is not satisfactory. This section describes some of the most important developments.

Edgeworth (1905) obtained an improvement in the normal approximation for the distribution of standardized sums through an asymptotic representation of the distribution function: this is known as the Edgeworth expansion. Formula (8.2) is one example of this.

Fisher (1921) calculated the exact distribution of the sample correlation coefficient r under random sampling from a bivariate normal distribution. The distribution of r depends heavily on the population value ρ of the correlation coefficient. As one would expect, r is almost centred at ρ. However, r has a bounded support and this translates into large changes of variance and skewness as ρ varies. Since there is a high level of non-normality for values of $|\rho|$ relatively far from zero, Fisher considered a monotonic transformation of r, with the dual aim of stabilizing variance and reducing skewness in the transformed scale. Stabilizing variance means rendering the variance of the transformed statistic asymptotically insensitive to variations in ρ. Reduction of skewness improves the adequacy of the normal approximation. Because the transformation considered is monotonic, inference procedures are still based on r. See Example 8.9 for more details on Fisher's transformation.

Later, many authors proposed transformations which sought to improve the numerical adequacy of asymptotic approximations of distributions. Wilson and Hilferty (1931) observed that if $T_n \sim \chi_n^2$, then $T_n^\dagger = 3\sqrt{n/2}\left(\sqrt[3]{T_n/n} - 1\right)$ is $N(0,1)$ with error of order $O(n^{-1})$. Note, in Figure 8.4, the normalizing effect of the transformation $T_n^\dagger$, for small sample sizes, when compared

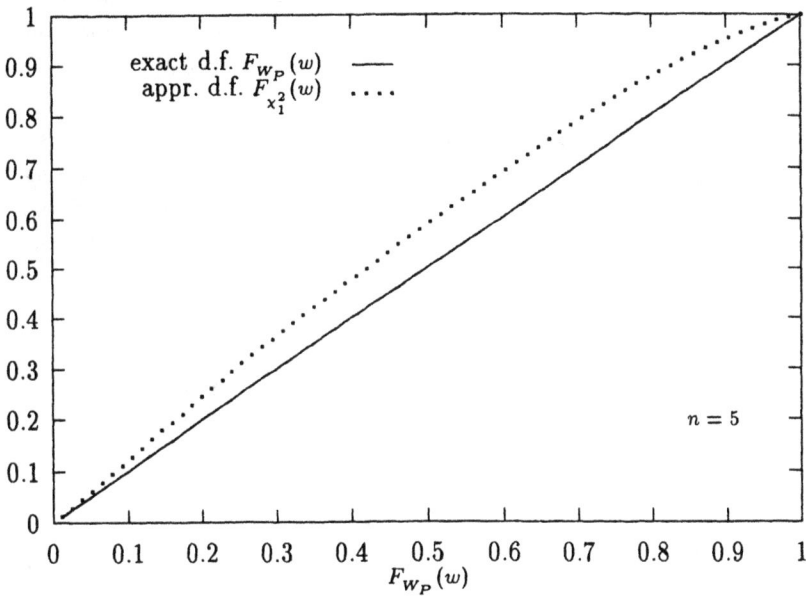

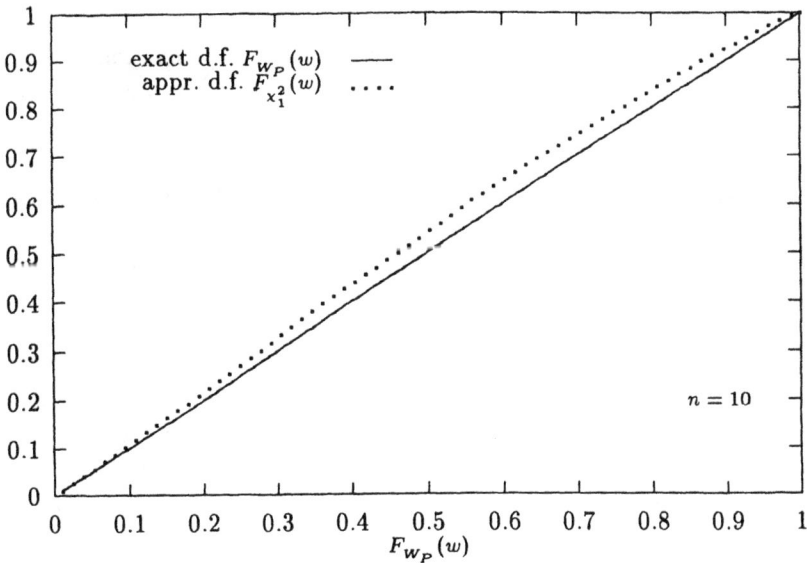

Figure 8.3: *Null distribution of* $W_P(\nu)$, $\nu = 1$, *with i.i.d.* $Ga(\nu, \phi)$ *observations,* $n = 5, 10$.

with $T_n^* = (T_n - n)/\sqrt{2n}$. Bartlett (1936b) found that the square root transformation asymptotically stabilizes the variance of a Poisson distribution; if $Y \sim P(\lambda)$, the distribution of $\sqrt{Y}$ is approximately $N(\sqrt{\lambda}, 1/4)$, with error of order $O(\lambda^{-1/2})$.

Bartlett (1937) proposed a correction to the log-likelihood ratio test W_P for the homogeneity of k variances when sampling from normal populations, this is called the Bartlett test. In general terms, the Bartlett correction is as follows. If W is a log-likelihood ratio statistic with asymptotic null χ_q^2 distribution, the transformation

$$\frac{q}{E(W)}W \, ,$$

renders the null expectation exactly equal to the nominal asymptotic value. This usually produces a more satisfactory numerical match with the reference χ_q^2 distribution. In order to apply the correction, the exact expectation $E(W)$ can be substituted with an asymptotic expansion $q + b/n$, which may be easier to calculate. In later studies, starting from that of Lawley (1956), it has been shown that, under regularity conditions, the Bartlett correction to W offers an overall improvement in the adequacy of the asymptotic distribution, in the sense that the order of error in the approximation of all null moments is reduced.

Cornish and Fisher (1937) inverted the Edgeworth expansion and obtained the general expression of a polynomial normalizing transformation. Haldane (1938) studied the family of transformations $Z_h = (Y/E(Y))^h$, with h chosen so as to minimize asymmetry or kurtosis. Anscombe (1948) offered further examples of transformations which stabilize variance, for the Poisson, binomial, negative binomial families.

Daniels (1954) obtained the saddlepoint expansion for the distribution of sums of independent random variables. This marks a considerable advance on the Edgeworth expansion for three reasons. Firstly, it improves qualitative adequacy since the density of a sum is approximated with a non-negative function. Secondly, the approximation error changes from being uniformly bounded as absolute error to being uniformly bounded as relative error; this is particularly important for approximations in the tails. Lastly, in many examples, the numerical accuracy of the approximation is often much better than expected on the basis of theoretical analysis.

Anscombe (1964) considered reparameterizations of statistical models in such a way as to render the log-likelihood around the maximum likelihood estimate approximately quadratic to a higher order, forcing third-order partial

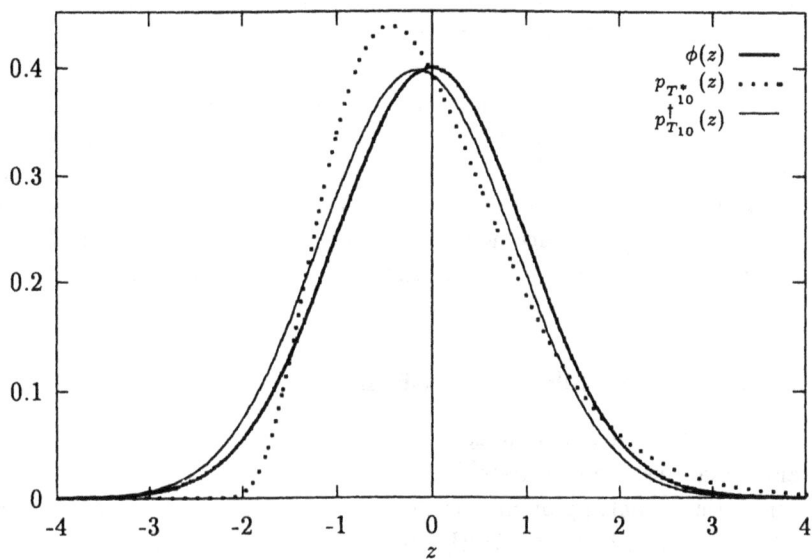

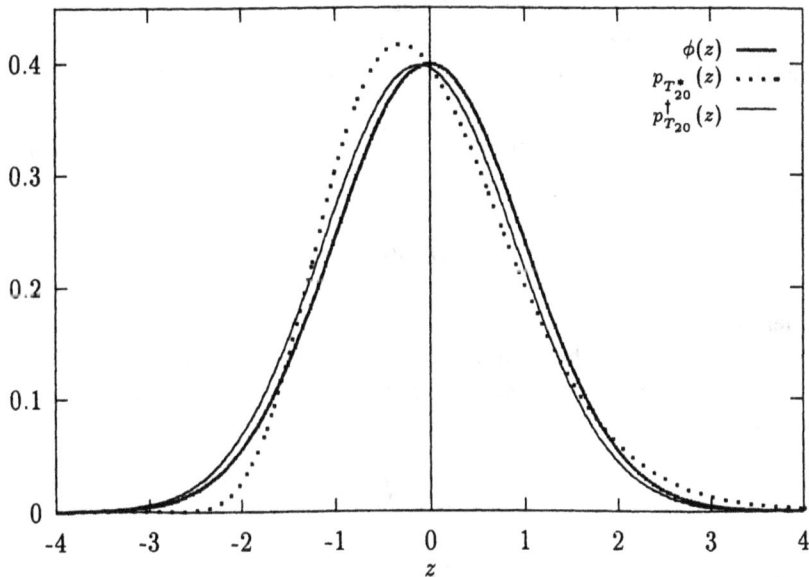

Figure 8.4: *Normalizing transformations for χ_n^2; densities of T_n^* and $T_n^\dagger$ compared with $\phi(z)$, $n = 10, 20$.*

derivatives to be zero. Box and Cox (1964) studied the parametric family of normalizing transformations $T = \left\{(Y+c)^h - 1\right\}/h$.

The contributions described above show that there have been two main lines of strategy. The first, and earliest, follows Edgeworth (1905) and leads to refinements in the approximation of the distribution, offering corrections, usually of order $O(n^{-1/2})$ and $O(n^{-1})$ when the limiting distribution is normal. The second, in line with Fisher (1921), suggests transforming the statistic in such a way as to reduce non-normality or, more generally, reduce the distance from the asymptotic distribution; the key idea here, is often to correct discrepancies in supports and moments.

8.4   Variance Stabilizing Transformations

By Theorems A.14 and A.15, asymptotic normality is preserved under smooth transformations. This suggests that it is possible to improve the adequacy of an asymptotic normal approximation through a smooth non-linear transformation of the statistic of interest. On the new scale, either the transformed statistic could have an almost symmetrical distribution, or the asymptotic variance could be independent of unknown parameters.

Let T_n be a scalar statistic calculated on a sample of size n from a random variable Y whose distribution depends on a scalar parameter θ. If, asymptotically, $T_n \dot\sim N(\theta,\ \sigma^2(\theta)/n)$, and $g(\cdot)$ is a differentiable function, with $g'(\theta) \neq 0$, then $g(T_n) \dot\sim N\left(g(\theta),\ \sigma^2(\theta)\,(g'(\theta))^2/n\right)$ (cf. Theorem A.14). The parameterization $\psi = g(\theta)$ is called a **variance stabilizing parameterization** if

$$\sigma(\theta)\,g'(\theta) = c\,, \tag{8.4}$$

where c is a constant. For inference about $\psi = g(\theta)$ based on the statistic $g(T_n)$ we could use the approximation $g(T_n) \dot\sim N(\psi, c^2/n)$, so that confidence intervals for ψ have constant width, as in location models. Solving (8.4) with respect to $g(\cdot)$ shows that a variance stabilizing parameterization is

$$\psi(\theta) = c \int_{\theta_0}^{\theta} \sigma(t)^{-1}\,dt\,.$$

Table 8.4 gives some examples of variance stabilizing transformations.

Example 8.8   *Variance stabilization in an* $\mathcal{F}_{ne}^1$

Consider n independent observations from a distribution in an $\mathcal{F}_{ne}^1$ with the

Table 8.4: *Examples of variance stabilizing transformations.*

Model	T_n	$\dfrac{\sigma^2(\theta)}{n}$	$\psi(\theta)$	$\dfrac{\sigma^2(\theta)}{n}\left(\psi'(\theta)\right)^2$	Author
Y_i, i.i.d., $i = 1,\ldots,n$, $P(\theta)$	$\bar{Y}_n$	$\dfrac{\theta}{n}$	$\sqrt{\theta}$	$\dfrac{1}{4n}$	Bartlett (1936b) Anscombe (1948)
Y_i, i.i.d., $i = 1,\ldots,n$, $Bi(1,\theta)$	$\bar{Y}_n$	$\dfrac{\theta(1-\theta)}{n}$	$\arcsin\sqrt{\theta}$	$\dfrac{1}{4n}$	Anscombe (1948)
Y_i, i.i.d., $i = 1,\ldots,n$, $Ga(1,\theta)$	$\bar{Y}_n$	$\dfrac{\theta}{n}$	$\sqrt{\theta}$	$\dfrac{1}{4n}$	Fisher (1922b)
(X_i,Y_i) i.i.d., $i = 1,\ldots,n$, bivariate normal with correlation coefficient ρ	r	$\dfrac{(1-\rho^2)^2}{n}$	$\mathrm{arctgh}\rho = \dfrac{1}{2}\log\dfrac{1+\rho}{1-\rho}$	$\dfrac{1}{n}$	Fisher (1921)

natural parameterization and let $T_n = \hat{\theta}$, the m.l.e. of θ. Asymptotically, $T_n \dot\sim N(\theta, 1/(nK''(\theta)))$. A variance stabilizing parameterization is

$$\psi(\theta) = c \int_{\theta_0}^{\theta} (K''(t))^{1/2}\, dt\,. \tag{8.5}$$

In this parameterization, the Fisher information is constant, as in a location model. The sample mean of natural observations, $\bar{Y}_n = \sum Y_i/n$, is a smooth increasing function of $\hat{\theta}$. Since $\bar{Y}_n \dot\sim N(K'(\theta), K''(\theta)/n)$, using the mean value parameterization, $\bar{Y}_n \dot\sim N(\mu, K''(\theta(\mu))/n)$, the variance stabilizing transformation is

$$\phi(\mu) = c \int_{\mu_0}^{\mu} (K''(\theta(m)))^{-1/2}\, dm\,.$$

With the change of variable $t = t(m)$, with inverse $m = m(t) = K'(t)$, and taking into account that $dm = K''(t)dt$, we obtain $\phi(\mu) = \psi(\theta(\mu))$, with $\psi(\cdot)$ defined by (8.5). $\triangle$

Obviously, simply stabilizing variance, does not make any direct contribution to improving asymptotic normality. The main objectives of variance stabilization are to make the (asymptotic) accuracy of an estimator independent of the parameter and to approach a situation of equal response variance if T_n plays the role of response variable in a regression model. However, in some specific examples a variance stabilizing transformation also improves the adequacy of the normal approximation. This is true in the case of Fisher's z transformation (Fisher, 1921), which is given in the last row of Table 8.4 and is described below.

Example 8.9 *Fisher's z transformation*
Let (X_i, Y_i), $i = 1, \ldots, n$, be i.i.d. with bivariate normal distribution and correlation coefficient ρ. If r is the sample correlation coefficient, the z transformation is defined by

$$z = \frac{1}{2} \log \frac{1+r}{1-r}\,.$$

With both r and z, we have departed from the usual convention of using a capital letter to indicate a random variable. Gayen (1951) obtained the following expansions for the first four cumulants of z:

$$E(z) = \frac{1}{2} \log \frac{1+\rho}{1-\rho} + \frac{\rho^2}{2(n-1)} \left\{ 1 + \frac{5+\rho^2}{4(n-1)} + O(n^{-2}) \right\},$$

$$\text{Var}(z) = \frac{1}{n-1}\left\{1+\frac{4-\rho^2}{2(n-1)}+O(n^{-2})\right\},$$

$$\rho_3(z) = \frac{\rho^3}{(n-1)^{3/2}}+O(n^{-2}),$$

$$\rho_4(z) = \frac{2}{(n-1)}+O(n^{-2}).$$

Note that $\rho_3(z)$ is order $O(n^{-3/2})$ instead of $O(n^{-1/2})$ as one would generally expect for a sample mean, or for a smooth function of sample means, and as happens for r. Furthermore, note that the leading term in the expansion of $\rho_4(z)$ does not depend on ρ. The usual first-order approximation for the distribution of z is

$$N\left(\frac{1}{2}\log\frac{1+\rho}{1-\rho},\ \frac{1}{n-3}\right),$$

where the value $1/(n-3)$ for asymptotic variance is based on the fact that, from

$$\text{Var}(z) \doteq \frac{1}{n-1}+\frac{4-\rho^2}{2(n-1)^2} = \frac{1}{n-3}\left\{1-\rho^2\frac{n-3}{2(n-1)^2}+O(n^{-2})\right\},$$

we have $\text{Var}(z) = 1/(n-3)\{1+O(n^{-2})\}$ if $\rho=0$.

The normal approximation for z is much more satisfactory than the usual one for r

$$r \overset{\sim}{\cdot} N\left(\rho, n^{-1}(1-\rho^2)^2\right).$$

In other words, it is numerically more adequate to evaluate

$$P(r \le r_0; \rho) \doteq \Phi\left(\sqrt{n-3}\left(\log\frac{1+r_0}{1-r_0}-\log\frac{1+\rho}{1-\rho}\right)/2\right)$$

than to evaluate

$$P(r \le r_0; \rho) \doteq \Phi\left(\sqrt{n}(r_0-\rho)/(1-\rho^2)\right).$$

Furthermore, the asymptotic error is of order $O(n^{-1})$ instead of $O(n^{-1/2})$ because asymmetry is almost entirely corrected. A graphic illustration of the normalizing effect of the Fisher's z transformation is given in Mudholkar (1983).

$\triangle$

Example 8.10 *Fisher's transformation for the χ_n^2 distribution*
If $T_n \sim \chi_n^2$, then $E(T_n) = n$, $\text{Var}(T_n) = 2n$, $\rho_3(T_n) = \sqrt{8/n}$, $\rho_4(T_n) = 12/n$.

One transformation that stabilizes variance (see Table 8.4) is $\sqrt{2T_n}$, wherein the approximation

$$\sqrt{2T_n} \;\dot\sim\; N\left(\sqrt{2n}, 1\right)$$

holds. This transformation also reduces asymmetry and the order of the fourth cumulant, since

$$\rho_3\left(\sqrt{2T_n}\right) = \frac{1}{\sqrt{2n}} \,,$$

$$\rho_4\left(\sqrt{2T_n}\right) = O(n^{-2}) \,.$$

However, the order of error in the normal approximation does not diminish.

$$\Delta$$

Holland (1973) showed that variance stabilization, i.e. making the covariance matrix proportional to the identity matrix, is not possible in the multi-parameter case.

8.5   Skewness Reducing Transformations

Let $T_n = \bar{Y}_n$ be the sample mean of n independent copies of a random variable Y with finite fourth moment. Let, as usual, $\mu = E(Y)$, $\sigma^2 = \mathrm{Var}(Y)$, $\bar{\mu}_3 = E\left\{(Y-\mu)^3\right\}$ and $\bar{\mu}_4 = E\left\{(Y-\mu)^4\right\}$. Then

$$E\left\{(\bar{Y}_n - \mu)^3\right\} \;=\; \bar{\mu}_3/n^2, \tag{8.6}$$
$$E\left\{(\bar{Y}_n - \mu)^4\right\} \;=\; 3\sigma^4/n^2 + (\bar{\mu}_4 - 3\sigma^4)/n^3 \,. \tag{8.7}$$

Theorem 8.2 Let $f(\cdot)$ be a smooth real-valued function defined on $\mathbb{R}$ with $f'(\mu) \neq 0$, $f''(\mu) \neq 0$. Then

$$E\left\{(f(\bar{Y}_n) - E\left(f(\bar{Y}_n)\right))^3\right\} = \frac{\bar{\mu}_3\left(f'(\mu)\right)^3 + 3\sigma^4 f'(\mu)^2 f''(\mu)}{n^2} + O(n^{-3}) \,. \tag{8.8}$$

Proof. Consider the expansion according to the delta method

$$f(\bar{Y}_n) = f(\mu) + f'(\mu)(\bar{Y}_n - \mu) + \frac{1}{2}f''(\mu)(\bar{Y}_n - \mu)^2 + O_p(\bar{Y}_n - \mu)^3$$

(see section A.3 and Azzalini, 1996, section A.8.4). Remember that $\bar{Y}_n - \mu = O_p(n^{-1/2})$. Thus, we have

$$f(\bar{Y}_n) = f(\mu) + f'(\mu)(\bar{Y}_n - \mu) + \frac{1}{2}f''(\mu)(\bar{Y}_n - \mu)^2 + O_p(n^{-3/2}),$$

and, taking expectations,

$$E(f(\bar{Y}_n)) = f(\mu) + \frac{1}{2}f''(\mu)\frac{\sigma^2}{n} + O(n^{-2}),$$

which gives

$$f(\bar{Y}_n) - E(f(\bar{Y}_n)) = f'(\mu)(\bar{Y}_n - \mu) + \frac{1}{2}f''(\mu)\left\{(\bar{Y}_n - \mu)^2 - \frac{\sigma^2}{n}\right\} + O_p(n^{-3/2}).$$

Hence, neglecting terms of order $O(n^{-3})$,

$$E\left\{(f(\bar{Y}_n) - E\left(f(\bar{Y}_n)\right))^3\right\}$$

$$= E\left\{\left(f'(\mu)(\bar{Y}_n - \mu) + \frac{1}{2}f''(\mu)\left((\bar{Y}_n - \mu)^2 - \frac{\sigma^2}{n}\right)\right)^3\right\}$$

$$= E\left\{(f'(\mu))^3(\bar{Y}_n - \mu)^3 + \frac{1}{8}(f''(\mu))^3\left((\bar{Y}_n - \mu)^2 - \frac{\sigma^2}{n}\right)^3\right.$$

$$+ \frac{3}{2}f'(\mu)^2 f''(\mu)(\bar{Y}_n - \mu)^2\left((\bar{Y}_n - \mu)^2 - \frac{\sigma^2}{n}\right)$$

$$\left. + \frac{3}{4}f'(\mu)f''(\mu)^2(\bar{Y}_n - \mu)\left((\bar{Y}_n - \mu)^2 - \frac{\sigma^2}{n}\right)^2\right\}$$

$$= f'(\mu)^3\frac{\bar{\mu}_3}{n^2} - \frac{3}{2}f'(\mu)^2 f''(\mu)\frac{\sigma^4}{n^2} + \frac{3}{2}f'(\mu)^2 f''(\mu)\frac{3\sigma^4}{n^2} + O(n^{-3}),$$

from which (8.8) follows by observing that

$$E\left\{(Y_n - \mu)^2 - \frac{\sigma^2}{n}\right\}^3 \quad \text{and} \quad E\left\{(\bar{Y}_n - \mu)\left((\bar{Y}_n - \mu)^2 - \frac{\sigma^2}{n}\right)^2\right\}$$

are of order $O(n^{-3})$. $\qquad\qquad\Box$

A transformation $f(\cdot)$ which reduces the asymptotic order of the third central moment is such that

$$E\left[\{f(\bar{Y}_n) - E\left(f(\bar{Y}_n)\right)\}^3\right] = O(n^{-3}),$$

so that, by Theorem 8.2, $f(\cdot)$ satisfies the differential equation

$$\bar{\mu}_3(f'(\mu))^3 + 3\sigma^4(f'(\mu))^2 f''(\mu) = 0. \qquad (8.9)$$

Finding a solution is simplified if $f(\cdot)$ is restricted to belong to a parametric class of functions. For example, consider the parametric class

$$f(y) = y^h \qquad h > 0 . \tag{8.10}$$

By imposing (8.9), we obtain

$$h = 1 - \frac{\mu\,\bar{\mu}_3}{3\sigma^4} . \tag{8.11}$$

Example 8.11 *The transformation Y^h for some parametric families*
i)   Poisson. If $Y \sim P(\theta)$, then $\mu = \sigma^2 = \bar{\mu}_3 = \theta$; from (8.11) we have $h = 1 - 1/3 = 2/3$. Note that the value obtained for h does not depend on θ.
ii) Chi-squared. If $Y \sim \chi_1^2$, then $\mu = 1$, $\sigma^2 = 2$, $\bar{\mu}_3 = 8$ and from (8.11) we obtain $h = 1/3$. If $T_n \sim \chi_n^2$ the transformation

$$\left(\frac{T_n}{n}\right)^{1/3} \overset{\cdot}{\sim} N\left(1, \frac{2}{9n}\right)$$

has standardized third cumulant of order $O(n^{-3/2})$. This is the **Wilson–Hilferty** transformation mentioned in section 8.3. Compare this result with that of Example 8.10. Figure 8.4 in section 8.3 illustrates the efficacy of the transformation for moderate values of n.
iii) Binomial. If $Y \sim Bi(1, \theta)$, then $\mu = \theta$, $\sigma^2 = \theta(1-\theta)$, $\bar{\mu}_3 = \theta(1-\theta)(1-2\theta)$ and from (8.11) we obtain $h = 1 - (1 - 2\theta)/(3(1 - \theta)) = 2/3 + \theta/(3(1 - \theta))$. Note that if $\theta \to 0$, $h \to 2/3$, which agrees with the result for the Poisson distribution. If $\theta = 1/2$, $h = 1$, which confirms that $\theta = 1/2$ is the parameter value for which the normal approximation for $Bi(n, \theta)$ is most accurate. $\triangle$

Another interesting parametric class is that of quadratic transformations of the sample mean, $f(\bar{Y}_n) = a\bar{Y}_n^2 + b\bar{Y}_n + c$. Here we wish to determine the values of a, b, c such that the transformed variable $f(\bar{Y}_n)$ has mean zero and approximately unit variance and $\rho_3 = 0$. More precisely, a, b and c should be such that

$$\begin{aligned}
E\left(f(\bar{Y}_n)\right) &= 0, \\
\text{Var}\left(f(\bar{Y}_n)\right) &= 1 + O(n^{-2}), \\
\rho_3\left(f(\bar{Y}_n)\right) &= O(n^{-3/2}).
\end{aligned} \tag{8.12}$$

If the first two relations are satisfied, the last will be satisfied if

$$E\left\{(f(\bar{Y}_n) - E(f(\bar{Y}_n)))^3\right\} = O(n^{-3/2}) .$$

By solving the equations imposed by the conditions (8.12) and using (8.8), two solutions are obtained: (a, b, c) and $(-a, -b, -c)$ with

$$a = -\frac{1}{6}\sqrt{n}\,\frac{\bar{\mu}_3}{\sigma^5}\,,$$

$$b = \sqrt{n}\left(\frac{1}{\sigma} + \frac{1}{3}\frac{\mu\,\bar{\mu}_3}{\sigma^5}\right)\,,$$

$$c = -\sqrt{n}\left(\frac{\mu}{\sigma} + \frac{1}{6}\frac{\mu^2\,\bar{\mu}_3}{\sigma^5}\right) + \frac{1}{6\sqrt{n}}\frac{\bar{\mu}_3}{\sigma^3}\,.$$

More compactly, let us define the standardized variable

$$Z_n = \frac{\bar{Y}_n - \mu}{\sigma/\sqrt{n}}\,,$$

then

$$f(\bar{Y}_n) = Z_n - \frac{\rho_3}{6\sqrt{n}}(Z_n^2 - 1)\,, \tag{8.13}$$

where $\rho_3 = \bar{\mu}_3/\sigma^3$ is the third standardized cumulant of the parent distribution. Transformation (8.13) is a special case of the Cornish–Fisher polynomial normalizing transformation which will be examined in section 10.6.

Example 8.12 *Quadratic normalizing transformation for the score function*
Let $\mathcal{F}$ be a one-parameter statistical model specified as $(\mathcal{Y}, p(y; \theta), \Theta)$, $\Theta \subseteq \mathbb{R}$. The quadratic normalizing transformation (8.13) can be applied to the score function $l_*(\theta)$ when this is the sum of n i.i.d. summands. With $Z_n = l_*(\theta)/\sqrt{i(\theta)}$ and $\rho_3(\theta) = E_\theta(l_*^3)/i(\theta)^{3/2}$, (8.13) takes the form

$$f(l_*(\theta)) = \frac{l_*(\theta)}{\sqrt{i(\theta)}} - \frac{\rho_3(\theta)}{6\sqrt{n}\,i(\theta)}(l_*(\theta)^2 - i(\theta))\,. \tag{8.14}$$

(See, also, Cox and Hinkley, 1974, p. 341, formula (85).) $\triangle$

8.6 Bibliographic Note

Even though most recent research has concentrated upon the development of general methods, there is still some interest in specific problems of finding normalizing transformations such as Fisher's z. For example, see Konishi (1978, 1981) and Smyth (1994). For a survey of the non-linear transformations, either of the parameters or of the variables, used in statistics, see Atkinson and Cox

(1988). Sakia (1992) offers an extensive bibliography about Box–Cox transformations.

Various monographs on asymptotic methods have appeared in recent years. In order to draw up a rough guide to these it is, again, convenient to distinguish between problems of inference and problems of distribution (see section 1.3). If the emphasis is on inference problems with a decision theoretic approach (see section 3.5), asymptotic techniques may be used both to approximate the risk function (variance of estimators, power of tests, ...) and to identify decision procedures that minimize an asymptotic expansion of the risk function. Pfanzagl (1982, 1985), Amari (1985, 1987), Le Cam (1986), Le Cam and Yang (1990), Ghosh (1994) mainly follow this approach; see also Skovgaard (1989) for a summary of the main results.

There are also many works which concentrate more specifically on inference and focus on the use of asymptotic methods for solving problems of distribution in relation to given statistics or combinants. The monographs by Barndorff-Nielsen and Cox (1989, 1994), Barndorff-Nielsen (1988), McCullagh (1987), Kolassa (1994) all deal mainly with this second class of problems, particularly in relation to likelihood methods. The application of asymptotic methods to distribution problems which arise in the area of robust and nonparametric methods are dealt with in Field and Ronchetti (1990).

8.7 Problems

8.1 Identify some examples of statistical procedures that could be classified into cells 3, 5, 6 and 8 of Table 8.1.

[Section 8.1]

8.2 Let $Y_1, \ldots, Y_n$ be i.i.d. random variables, with the same distribution as $Y = W - 1$, where W is exponential with mean 1. Construct an analogous table to that given in Example 8.2.

[Section 8.2.1]

8.3 Let W be a combinant with a null asymptotic χ^2_1 distribution. Assume that the expansion

$$E_\theta(W) = 1 + \frac{b(\theta)}{n} + O(n^{-2})$$

holds. Show that the Bartlett corrected combinant

$$W^\star = \frac{W}{1 + \dfrac{b(\theta)}{n}}$$

has null expectation $E_\theta(W^\star) = 1 + O(n^{-2})$.

[Section 8.3]

8.4 Check that the transformations $\psi(\theta)$ in Table 8.4 are variance stabilizing.

[Section 8.4]

8.5 Make a numerical comparison between some exact quantiles of the χ_n^2 distribution and their approximations, based on the asymptotic result in Example 8.10, for some values of n.

[Section 8.4]

8.6 Check relations (8.6) and (8.7) in detail.

[Section 8.5]

8.7 Using the same notation and assumptions as in section 8.5, obtain an asymptotic expansion for $\text{Var}(f(\bar{Y}_n))$ with error of order $O(n^{-3})$.

[Section 8.5]

8.8 Make the detailed calculations needed to obtain the quadratic normalizing transformation (8.13) as a solution to (8.12).

[Section 8.5]

8.9 Under the same assumptions as in section 8.5, show that if $\bar{\mu}_3(f(\bar{Y}_n)) = O(n^{-3})$, then $\rho_3(f(\bar{Y}_n)) = O(n^{-3/2})$.

[Section 8.5]

8.10 Under the same assumptions as in Example 8.8, show that the parameterization

$$\psi(\mu) = c \int_{\mu_0}^{\mu} (K''(\theta(m)))^{-1/3} \, dm \qquad (8.15)$$

makes the leading term of the third standardized cumulant of $\psi(\bar{Y}_n)$, that is of the m.l.e. of $\psi(\mu)$, equal to zero. Verify that the transformations $\mu^{2/3}$ for the Poisson distribution and $\mu^{1/3}$ for the gamma distribution with fixed shape parameter $\nu \geq 1$, are special cases of (8.15). Obtain $\psi(\mu)$ for the inverse Gaussian $IG(\phi, \lambda)$, with $\lambda > 0$, fixed.

[Section 8.5; Barndorff-Nielsen, 1978, p. 179; McCullagh and Nelder, 1989, p. 38]

8.11 With the same assumptions as in Problem 8.10, identify the parameterization $\psi(\theta)$ for which the leading term of the third standardized cumulant of $\psi(\hat{\theta})$ vanishes.

[Section 8.5; Hougaard, 1982, section 1]

8.12 Specify the expression of the quadratic normalizing transformation (8.14) under random sampling of size n from:

$i)$ a generic one-parameter exponential family;

$ii)$ the one-parameter exponential families considered in Table 5.1;

$iii)$ a generic location family;

$iv)$ the location families considered in Table 7.1, with $\sigma = 1$.

(In points $iii)$ and $iv)$, evaluate the moments required conditionally on the suitable configuration statistic and use the delta method.)

[Section 8.5]

CHAPTER 9

ASYMPTOTIC EXPANSIONS FOR STATISTICS

9.1 Index Notation

The notation $a = [a_r]$, where r is an index whose range, $\{1, \ldots, p\}$, is understood, was introduced in section 1.4.2, to indicate a vector a in $\mathbb{R}^p$ as, for example, $[l_r]$ for the score vector. Analogously, we denote by $\theta = [\theta^r]$ the parameter of a parametric statistical model. The convention that allows us to place the indices of elements of arrays both as upper and lower indices is the basis of index notation. If the indices $r, s, t, \ldots$ all have range $\{1, \ldots, p\}$, the symbols i_{rs}, i^{rs}, a_r^s indicate elements of $p \times p$ square matrices, while l_{rst}, b^{rst}, b_{rs}^t, b_r^{st} represent elements of arrays with dimensions $p \times p \times p$. An upper index, such as r in θ^r, is called a **contravariant** index, a lower index, such as i in a_i, is called a **covariant** index. Both these terms and the possibility, offered by the notation, of placing the index in either position, come from differential geometry. In the theory of statistical inference, concepts and notation from differential geometry are important for the study of transformation rules of combinants, and of statistics derived from combinants, under reparameterization of the statistical model. As well as this direct connection with transformation rules, which will be taken up in section 9.5, index notation introduces a markedly simplified representation of multivariate objects, which can be dealt with in the same way as the corresponding univariate quantities.

Example 9.1 *Index notation*
If $x = [x^i]$ and $y = [y^j]$ are two column vectors in $\mathbb{R}^p$, the $p \times p$ matrix of the products of their elements is $A = [a_{ij}] = xy^T = [x^i y^j]$. Index notation allows this to be written as the simple identity $a_{ij} = x^i y^j$, leaving the range of the indices as implicit and focussing attention on the definition of the elements of a given array instead of on the whole array. In this way, all the quantities are

dealt with algebraically as if they were scalars.   △

Usefulness of index notation is enhanced if Einstein's summation convention
is adopted. This says that when an index appears two or more times in a
product of elements of arrays, then summation over the range of that index is
understood. For example, let

$$l^r = \sum_{s=1}^{p} i^{rs} l_s .$$

The index s on which summation is performed is repeated and its range could
well be implicit. Even the summation symbol could be implicit and the expression becomes

$$l^r = i^{rs} l_s . \tag{9.1}$$

Example 9.2 *Summation convention*
Using index notation and the summation convention, some quantities which
are usually represented in matrix notation can be written more concisely in
index notation. For example, if $x = [x^r]$ and $y = [y^r]$ are column vectors in
$\mathbb{R}^p$ and $A = [a_{ij}]$ is a $p \times p$ matrix, the scalar product $x \cdot y$ and the quadratic
form $x^T A x$ can be expressed as

$$\begin{aligned} x \cdot y &= x^r y^r , \\ x^T A x &= a_{ij} x^i x^j . \end{aligned}$$

Furthermore, with the usual notation for likelihood quantities,

$$\begin{aligned} W_u &= l_*^T i^{-1} l_* = i^{rs} l_r l_s , \\ W_e &= (\hat{\theta} - \theta)^T i(\hat{\theta}) (\hat{\theta} - \theta) = \hat{\imath}_{rs} (\hat{\theta} - \theta)^r (\hat{\theta} - \theta)^s . \end{aligned}$$

These preliminary examples do not highlight all the possible uses of index
notation as they all concern quantities which could easily be expressed using
matrix notation. However, the cubic form

$$y = a_{ijk} x^i x^j x^k ,$$

with $A = [a_{ijk}]$ an array with dimension $p \times p \times p$, could not be written in
a compact form using matrix notation. For example, one possible expression
requires us first to write

$$z = [x^T A_k x] , \quad \text{where} \quad A_k = [a_{ijk}] \quad \text{with } k \text{ fixed} ,$$

then put $y = z^T x$, obtaining

$$y = \sum_{k=1}^{p} (x^T A_k \, x) x^k.$$

△

From now on, unless otherwise stated, index notation and the summation convention will be used.

Example 9.3 *Multivariate normal density*
The p.d.f. of $Y \sim N_p(\mu, \Sigma)$ is

$$p_Y(y; \mu, \Sigma) = \frac{1}{(2\pi)^{p/2}|\Sigma|^{1/2}} \exp\left\{-\frac{1}{2}(y - \mu)^r (y - \mu)^s \sigma^{rs}\right\},$$

where $\Sigma = [\sigma_{rs}]$, $\Sigma^{-1} = [\sigma^{rs}]$.

△

Example 9.4 *Kronecker delta*
The symbol δ_s^r, called the **Kronecker delta**, is defined by

$$\delta_s^r = \begin{cases} 1 & \text{if } r = s \\ 0 & \text{if } r \neq s. \end{cases} \tag{9.2}$$

Thus, $[\delta_s^r] = I_p$, where I_p indicates the identity matrix of order p. Note that

$$\delta_s^r x^s = x^r. \tag{9.3}$$

The fact that $[i^{rs}]$ is the inverse matrix of $[i_{rs}]$ is described by the equations

$$i^{rs} i_{st} = \delta_t^r, \quad r, t = 1, \ldots, p.$$

△

In a product of elements of arrays, where the summation convention applies, the repeated indices are called **dummy indices**; the non-dummy indices are called **free indices**.

If all the indices have range $\{1, \ldots, p\}$, the product of elements of arrays gives an array with dimension p^h, where h is the number of free indices. For example, the expression $a_{ijk} b^{il} c_{lj}$, where the indices i, j and l are dummy, and the index k is free, defines a vector as k varies in its own range. It is possible to redefine the names of the dummy indices without altering the result. For example, we have the identity $a_{ijk}b^{il}c_{lj} = a_{lmk}b^{ln}c_{nm}$; note that i has been replaced by l, j by m and l by n, and the identity is satisfied because the pairs of indices which occupy the same positions on the right-hand side and on the left-hand side have the same range. If we have expressions where indices with different ranges appear, in order to make the distinction clearer, we usually use indices which belong to separate subsets of the alphabet, according to the range. For example, $\{a, b, c, \ldots\}$, $\{i, j, k, \ldots\}$, $\{r, s, t, \ldots\}$.

We have a **symmetric array** when the generic element is invariant with respect to arbitrary permutations of the (free) indices.

Given two arrays with the same dimensions, we could, in a natural way, define their sum as the array obtained by adding the corresponding elements. For example, the sum of the arrays $[b_t^{rs}]$ and $[c_t^{rs}]$ is the array $[a_t^{rs}] = [b_t^{rs} + c_t^{rs}]$. The "product" of arrays can give different results depending on whether there are dummy indices present or not and, also, on the location of any such indices. For example, from the arrays $[b_t^{rs}]$ and $[c_t^{rs}]$, we can define the arrays $[a_{vw}^{rstu}] = [b_v^{rs}c_w^{tu}]$, $[a_v^{rtu}] = [b_v^{rs}c_s^{tu}]$, $[a^{rt}] = [b_v^{rw}c_w^{tv}]$, etc.. We will speak of **multiplication of arrays** only when, as in the first example, there are no dummy indices; hence, the dimension of the resulting array is given by the product of the dimensions of the component arrays. On the other hand, we will call **contraction of arrays** a product where dummy indices appear. The usual row by column matrix product is a contraction, since the matrix product of $[d_{ij}]$ and $[e_{ij}]$, $i, j = 1, \ldots, d$, has $f_{ij} = d_{ik}e_{kj}$ as its generic element.

The following Examples 9.5 and 9.6 translate some well-known theorems into index notation.

Example 9.5 *Differentiation of composite functions*
Let $g : \mathbb{R}^p \to \mathbb{R}^q$ and $f : \mathbb{R}^q \to \mathbb{R}$ be functions whose partial derivatives of every order exist, and indicate the generic arguments of g and f by $\theta = [\theta^r] \in \mathbb{R}^p$ and by $\phi = [\phi^a] \in \mathbb{R}^q$, respectively. The composite function $f(g(\theta))$ will have partial derivatives

$$\frac{\partial}{\partial \theta^r} f(g(\theta)) = f(g(\theta))_r = f_a\, g_r^a , \qquad (9.4)$$

where $f_a = \partial f(\phi)/\partial \phi^a \big|_{\phi = g(\theta)}$ and $g_r^a = \partial g^a(\theta)/\partial \theta^r$. Furthermore, the partial derivatives of order two, three and four are concisely expressed by the relations

$$f(g(\theta))_{r_1 r_2} = f_{a_1 a_2}\, g_{r_1}^{a_1} g_{r_2}^{a_2} + f_{a_1}\, g_{r_1 r_2}^{a_1} , \qquad (9.5)$$

$$f(g(\theta))_{r_1 r_2 r_3} = f_{a_1 a_2 a_3}\, g_{r_1}^{a_1} g_{r_2}^{a_2} g_{r_3}^{a_3} + f_{a_1 a_2}\, g_{r_1 r_2}^{a_1} g_{r_3}^{a_2} [3] + f_{a_1}\, g_{r_1 r_2 r_3}^{a_1} , \qquad (9.6)$$

$$f(g(\theta))_{r_1r_2r_3r_4} = f_{a_1a_2a_3a_4}\, g^{a_1}_{r_1} g^{a_2}_{r_2} g^{a_3}_{r_3} g^{a_4}_{r_4} + f_{a_1a_2a_3}\, g^{a_1}_{r_1r_2} g^{a_2}_{r_3} g^{a_3}_{r_4}\,[6]$$
$$+\ f_{a_1a_2}\, \{ g^{a_1}_{r_1r_2} g^{a_2}_{r_3r_4}\,[3] + g^{a_1}_{r_1r_2r_3}\, g^{a_2}_{r_4}\,[4] \}$$
$$+\ f_{a_1}\, g^{a_1}_{r_1r_2r_3r_4}\ , \tag{9.7}$$

where

$$f(g(\theta))_{r_1\dots r_k} = \partial^k f(g(\theta))/(\partial\theta^{r_1}\cdots\partial\theta^{r_k}),$$
$$f_{a_1\dots a_h} = \partial^h f(\phi)/(\partial\phi^{a_1}\cdots\partial\phi^{a_h})\Big|_{\phi=g(\theta)}$$

and

$$g^a_{r_1\dots r_k} = \partial^k g^a(\theta)/(\partial\theta^{r_1}\cdots\partial\theta^{r_k})\ .$$

The symbol $[k]$ indicates the sum of k summands which are similar to that written and are obtained by all possible permutations of the free indices. $\triangle$

Example 9.6 *Differentiation of the inverse function*
Let $\psi = \psi(\theta)$, with $\psi(\cdot)$ a one-to-one smooth function from $\Theta \subseteq \mathbb{R}^p$ into $\Psi \subseteq \mathbb{R}^p$. Indicate its inverse function by $\theta(\psi)$. Then, by definition,

$$\theta^r = \theta^r(\psi(\theta))\,, \qquad r = 1,\dots,p\,.$$

Formula (9.4) gives

$$\theta^r_s = \theta^r_a\,\psi^a_s\ .$$

On the other hand,

$$\theta^r_s = \frac{\partial}{\partial\theta^s}\,\theta^r = \begin{cases} 1 & \text{if } r = s \\ 0 & \text{if } r \neq s \end{cases} = \delta^r_s\,,$$

so that

$$\delta^r_s = \theta^r_a\,\psi^a_s\ . \tag{9.8}$$

Hence the Jacobian matrix of the inverse transformation $\theta(\psi)$ is the inverse of the Jacobian matrix of the transformation $\psi(\theta)$. $\triangle$

Often it is useful to express arrays such as $a_{r_1\dots r_m}$ or $b^{s_1\dots s_n}$ using an even more compact notation. The finite sequence of m indices $R_m = (r_1,\dots,r_m)$ is called a multiindex. Consider the multiindices R_m and $S_n = (s_1,\dots,s_n)$. We define

$$a_{R_m} = a_{r_1\dots r_m}$$
$$b^{S_n} = b^{s_1\dots s_n}$$
$$c^{S_n}_{R_m} = c^{s_1\dots s_n}_{r_1\dots r_m}\ .$$

Multiindices allow the multivariate Taylor formula to be expressed in a very simple form.

Example 9.7 *Multivariate Taylor formula*

Let $f : \mathbb{R}^p \rightarrow \mathbb{R}$ be a function which has continuous partial derivatives up to order $k + 1$ in $\mathbb{R}^p$ and let $\theta, \theta_0 \in \mathbb{R}^p$; then

$$f(\theta) = f(\theta_0) + \sum_{m=1}^{k} \frac{1}{m!} (\theta - \theta_0)^{R_m} f_{R_m}(\theta_0) + \frac{1}{(k+1)!} (\theta - \theta_0)^{R_{k+1}} f_{R_{k+1}}(\tilde{\theta}), \quad (9.9)$$

where

$$(\theta - \theta_0)^{R_m} = (\theta - \theta_0)^{r_1} \cdots (\theta - \theta_0)^{r_m}, \quad (9.10)$$

$$f_{R_m}(\theta) = \frac{\partial^m}{\partial \theta^{r_1} \dots \partial \theta^{r_m}} f(\theta), \quad (9.11)$$

and $\tilde{\theta} = \alpha \theta_0 + (1 - \alpha)\theta$, for some $\alpha \in (0, 1)$. $\triangle$

Furthermore, using multiindices allows the generic derivative of order m of a composite function to be expressed in a more compact form hence relations (9.4)-(9.7) are special cases. The general formula is

$$f(g(\theta))_{R_m} = \sum_{h=1}^{m} f_{A_h} \, G_{R_m}^{A_h}. \quad (9.12)$$

The factors f_{A_h} are immediately interpretable. The quantities $G_{R_m}^{A_h}$ in (9.12) are sums of products of derivatives of $g(\cdot)$, where summation extends over all the distinct partitions of R_m in h non-empty subsets. For example, for $m = 4$ the summand in (9.12) which corresponds to $h = 2$ is (see (9.7))

$$f_{a_1 a_2} G_{r_1 r_2 r_3 r_4}^{a_1 a_2} = f_{a_1 a_2} \left\{ g_{r_1 r_2}^{a_1} g_{r_3 r_4}^{a_2} [3] + g_{r_1 r_2 r_3}^{a_1} g_{r_4}^{a_2} [4] \right\}.$$

The general expression of $G_{R_m}^{A_h}$ is

$$G_{R_m}^{A_h} = \sum_{R_m/h} g_{R_{m_1}}^{a_1} \cdots g_{R_{m_h}}^{a_h}, \quad (9.13)$$

where, for $h \leq m$, the symbol $\sum_{R_m/h}$ indicates summation over all the partitions of R_m into h non-empty subsets $R_{m_1}, \dots R_{m_h}$. So as to write all these partitions in an orderly manner, we assume that the order of the indices in each subset is matched by the order in R_m and that, for $k = 1, \dots, h - 1$, the first index in R_{m_k} precedes the first index in $R_{m_{k+1}}$, with respect to the order in R_m. For example,

$$\sum_{R_3/2} g^{a_1}_{R_{3_1}} g^{a_2}_{R_{3_2}} = g^{a_1}_{r_1 r_2} g^{a_2}_{r_3} + g^{a_1}_{r_1 r_3} g^{a_2}_{r_2} + g^{a_1}_{r_1} g^{a_2}_{r_2 r_3} .$$

The number $N_{m,h}$ of summands in $\sum_{R_m/h}$ when $h \le m$ can be deduced from combinatorial rules. Since $N_{m,h}$ expresses the number of possible subdivisions of m objects in h non-empty subsets, we have immediately

$$N_{m,m} = N_{m,1} = 1 \qquad (9.14)$$

and, for $h \ge 2$, the recurrence relations

$$N_{m+1,h} = h N_{m,h} + N_{m,h-1} . \qquad (9.15)$$

The numbers $N_{m,h}$ are known as **Bernoulli numbers of second kind**. For example, $N_{5,2}$ is obtained on the basis of (9.14) and (9.15) from $N_{5,2} = 2N_{4,2} + N_{4,1} = 2N_{4,2} + 1$, $N_{4,2} = 2N_{3,2} + N_{4,1} = 2N_{3,2} + 1$, $N_{3,2} = 2N_{2,2} + N_{2,1} = 3$, $N_{4,2} = 7$, giving $N_{5,2} = 2 \times 7 + 1 = 15$. Observe that $N_{5,2}$, which is the number of possible subdivisions of 5 objects in 2 non-empty subsets, can be obtained directly as the sum of 5 and 10, where 5 is the number of subdivisions with 4 objects in one subset and 1 object in the other, 10 is the number of subdivisions with 2 objects in one subset and 3 in the other. If $q = 1$ (9.12) is simplified as

$$f(g(\theta))_{R_m} = \sum_{h=1}^{m} f^{(h)} \sum_{R_m/h} g_{R_{m_1}} \cdots g_{R_{m_h}} , \qquad (9.16)$$

where $f^{(h)} = (d^h f(\phi)/d\phi^h)\big|_{\phi = g(\theta)}$ and $g_{r_1 \dots r_k} = \partial^k g(\theta)/(\partial \theta^{r_1} \dots \partial \theta^{r_k})$.

Lastly, it is useful to have some differentiation rules for quantities expressed in terms of matrices of functions. Specifically, let $A = [a_{ij}] = [a_{ij}(\theta)]$ be a $d \times d$ matrix each of whose elements is a function of a p-dimensional variable $\theta \in \mathbb{R}^p$. Assume that A is non-singular for every $\theta \in \mathbb{R}^p$ and let $A^{-1} = [a^{ij}] = [a^{ij}(\theta)]$ be the inverse matrix of A. The two lemmas below offer differentiation rules for the elements of A^{-1} and for the logarithm of the determinant of A with respect to the components of θ.

Lemma 9.1 The partial derivative with respect to θ^r, $r = 1, \dots, p$, of the generic element of A^{-1} is

$$\frac{\partial}{\partial \theta^r} a^{ij}(\theta) = -a^{ik}(\theta)\, a^{lj}(\theta) \frac{\partial}{\partial \theta^r} a_{kl}(\theta) . \qquad (9.17)$$

Proof. By definition we have $a^{ik}a_{kl} = \delta^i_l$. Differentiating both sides of this equality with respect to θ^r, we obtain

$$a_{kl}\frac{\partial}{\partial\theta^r}a^{ik}(\theta) + a^{ik}\frac{\partial}{\partial\theta^r}a_{kl}(\theta) = 0 \, ,$$

from which

$$a_{kl}\frac{\partial}{\partial\theta^r}a^{ik}(\theta) = -a^{ik}\frac{\partial}{\partial\theta^r}a_{kl}(\theta) \, ,$$

and, after multiplying both sides by a^{lj},

$$\delta^j_k\frac{\partial}{\partial\theta^r}a^{ik}(\theta) = -a^{ik}a^{lj}\frac{\partial}{\partial\theta^r}a_{kl}(\theta) \, ,$$

which, using (9.3), gives (9.17). $\qquad\qquad\qquad\qquad\qquad\qquad\qquad\qquad\qquad\square$

Lemma 9.2 The partial derivative with respect to θ^r, $r = 1,\ldots,p$, of the logarithm of the determinant of A is

$$\frac{\partial}{\partial\theta^r}\log|A| = a^{ji}\frac{\partial}{\partial\theta^r}a_{ij}(\theta) \, . \qquad\qquad (9.18)$$

Proof. Note that

$$\frac{\partial}{\partial\theta^r}\log|A| = |A|^{-1}\frac{\partial}{\partial\theta^r}|A| = |A|^{-1}\frac{\partial|A|}{\partial a_{ij}}\frac{\partial a_{ij}}{\partial\theta^r} \, .$$

Let c_{ij} be the co-factor of a_{ij}, that is, the determinant of the $(p-1) \times (p-1)$ matrix obtained from A by suppressing the i-th row and the j-th column, multiplied by $(-1)^{i+j}$. Taking into account that $|A| = a_{ij}c_{kj}$, for $i = k$, and that $a^{ij} = |A|^{-1}c_{ji}$, we have

$$\frac{\partial|A|}{\partial a_{ij}} = c_{ij} \, ,$$

from which (9.18) is obtained, because $|A|^{-1}c_{ij} = a^{ji}$. $\qquad\qquad\qquad\qquad\square$

9.2   Likelihood Quantities

In the multiparameter case, index notation is useful for writing stochastic expansions of likelihood quantities compactly and, moreover, for checking invariance of combinants under reparameterization. In order to acquire the preliminary technical elements for these applications, first, we use index notation to express the multiparameter generalizations of some results which are familiar in the one-parameter case.

9.2.1 Null Moments

Let $l(\theta)$ be the log-likelihood associated with a statistical model with density $p_Y(y; \theta)$, $\theta \in \Theta \subseteq \mathbb{R}^p$, and denote the generic partial derivative of order m of $l(\theta)$ by

$$l_{R_m} = l_{R_m}(\theta) = l_{r_1 \dots r_m} .$$

Note that, under regularity conditions, the array of the log-likelihood derivatives of order m is symmetric. Let us introduce the following notation for null expected values of the log-likelihood derivatives and of their products:

$$
\begin{aligned}
\nu_{R_m} &= \nu_{R_m}(\theta) = E_\theta \left\{ l_{R_m}(\theta) \right\} , \\
\nu_{R_m, S_n} &= \nu_{R_m, S_n}(\theta) = E_\theta \left\{ l_{R_m}(\theta) l_{S_n}(\theta) \right\} , \\
\nu_{R_m, S_n, T_q} &= \nu_{R_m, S_n, T_q}(\theta) = E_\theta \left\{ l_{R_m}(\theta) l_{S_n}(\theta) l_{T_q}(\theta) \right\} ,
\end{aligned}
\tag{9.19}
$$

etc.. These quantities are symmetric both with respect to permutations of the multiindices and with respect to permutations within each multiindex. From (1.8), in a regular parametric model,

$$\nu_r = 0 . \tag{9.20}$$

From this basic relation we can obtain a sequence of balance relations known as **Bartlett relations** (Bartlett, 1953, section 2). In particular, by differentiating (9.20) with respect to θ^s, we obtain the information identity (1.9), which can be expressed, concisely, as

$$\nu_{rs} + \nu_{r,s} = 0 . \tag{9.21}$$

Differentiating (9.21) with respect to θ^t we obtain the third balance relation:

$$\nu_{rst} + \nu_{rs,t} + \nu_{rt,s} + \nu_{r,st} + \nu_{r,s,t} = 0 . \tag{9.22}$$

In fact, under regularity conditions which allow interchange of the order of differentiation and integration, we have

$$\frac{\partial}{\partial \theta^t} \nu_{rs} = \frac{\partial}{\partial \theta^t} \int l_{rs}(\theta) \exp\{\log p_Y(y; \theta)\} d\mu^* = \nu_{rst} + \nu_{rs,t}$$

and

$$\frac{\partial}{\partial \theta^t} \nu_{r,s} = \frac{\partial}{\partial \theta^t} \int l_r(\theta) l_s(\theta) \exp\{\log p_Y(y; \theta)\} d\mu^* = \nu_{rt,s} + \nu_{r,st} + \nu_{r,s,t} .$$

The derivative of a null moment ν of the type (9.19) with respect to one component of θ, for example, θ^t, is the sum of the null moments which are obtained from ν by inserting t into ν in every possible way as a covariant index. For example, $\partial \nu_{R_m, S_n} / \partial \theta^t = \nu_{R_m t, S_n} + \nu_{R_m, S_n t} + \nu_{R_m, S_n, t}$. The third Bartlett relation (9.22) can be written more concisely as

$$\nu_{rst} + \nu_{rs,t}[3] + \nu_{r,s,t} = 0 \,, \tag{9.23}$$

where [3] indicates the sum of three summands analogous to $\nu_{rs,t}$ obtained by all suitable permutations of the indices.

The fourth balance relation is

$$\nu_{rstu} + \nu_{rst,u}[4] + \nu_{rs,tu}[3] + \nu_{rs,t,u}[6] + \nu_{r,s,t,u} = 0 \,. \tag{9.24}$$

Index notation makes regular patterns clear in multivariate quantities. Numerical constants which appear as coefficients in the one-parameter case, often assume a combinatorial interpretation in the multivariate case. This can be seen if (9.24) is compared to the solution of Problem 9.5.

In a one-parameter model, index notation for likelihood quantities is not particularly useful. It is better to use the usual **power notation**, with definitions such as,

$$l_k = l_k(\theta) = \frac{\partial^k}{\partial \theta^k} l(\theta) \,, \quad k = 1, 2, \ldots \,, \tag{9.25}$$

$$\nu_k = \nu_k(\theta) = E_\theta(l_k(\theta)) \,, \tag{9.26}$$

$$\nu_{h,k} = \nu_{h,k}(\theta) = E_\theta \{l_h(\theta) l_k(\theta)\} \,, \tag{9.27}$$

and so on.

9.2.2 Asymptotic Orders

Below, it will often prove necessary to be able to identify the asymptotic order of likelihood quantities such as those in (9.19) under random sampling of size n. While it is immediately clear that the asymptotic order of ν_{R_m} is $O(n)$, for $m > 1$, the evaluation of the asymptotic order of quantities such as ν_{R_m, S_h} or ν_{R_m, S_h, T_q}, for $m, h, q = 1, 2, \ldots$, is not so straightforward. Hence we would like to have a simple criterion which makes this automatic.

It is best to take this problem up in the more general context of evaluating the asymptotic order of joint moments of sums of n independent copies, $Y_1, \ldots, Y_n$, of a d-dimensional random variable $Y = (Y^1, \ldots, Y^d)$. Let $\mu^{I_r} =$

$\mu^{i_1 i_2 \ldots i_r}$ be the generic joint moment of order r of Y about the origin and $\kappa^{I_r} = \kappa^{i_1, i_2, \ldots, i_r}$ be the generic joint cumulant of order r of Y (see section 3.2.7). Furthermore, let $\bar{\mu}^{I_r} = \bar{\mu}^{i_1 i_2 \ldots i_r}$ be a generic central moment of order r of Y. The joint cumulants of the d-dimensional sum $S_n = \sum Y_i$ are of order $O(n)$; this makes it easier to establish the order of moments, both central and about the origin, of S_n, once the relations which link moments and cumulants of order $r = 1, 2, \ldots$ are known. Special cases of these relations have already been introduced in (3.27)-(3.29). Consider the Taylor series expansions of the moment generating function and of the cumulant generating function of Y,

$$M_Y(t) = 1 + \sum_{r=1}^{\infty} \frac{1}{r!} \mu^{I_r} t_{I_r}$$

and

$$K_Y(t) = \sum_{r=1}^{\infty} \frac{1}{r!} \kappa^{I_r} t_{I_r},$$

where $t_{I_r} = t_{i_1} \cdots t_{i_r}$. Consider, also, the Taylor series expansion, around the origin, of $M_Y(t)$ expressed as $M_Y(t) = \exp(K_Y(t))$; the coefficients in this expansion are the moments about the origin of Y expressed in terms of cumulants of Y. Using (9.16) with $f(\cdot) = \exp(\cdot)$ and evaluating the result at the origin we obtain

$$\mu^{I_r} = \sum_{h=1}^{r} \sum_{I_r/h} \kappa^{I_{r_1}} \cdots \kappa^{I_{r_h}}. \tag{9.28}$$

Likewise, the inverse representation $K_Y(t) = \log M_Y(t)$ gives the general expression of cumulants in terms of moments about the origin, which is

$$\kappa^{I_r} = \sum_{h=1}^{r} (-1)^{h-1} (h-1)! \sum_{I_r/h} \mu^{I_{r_1}} \cdots \mu^{I_{r_h}}. \tag{9.29}$$

The relations (9.28) and (9.29) are also called **exlog relations** (see Barndorff-Nielsen and Cox, 1989, section 5.4), in that they give the coefficients of the series expansion of a function whose series expansion is given for logarithm, or exponential, transformation, respectively.

When relation (9.28) is applied to S_n it follows immediately that $\mu^{I_r}(S_n) = O(n^r)$, so long as $\kappa^i \neq 0$, for $i = 1, \ldots, d$. Relation (9.28) when referred to the centred variable $S_n - [\kappa^i]$ gives $\bar{\mu}^{I_r}(S_n) = O(n^k)$, where k is the maximum number of subsets into which $I_r = (i_1, \ldots, i_r)$ can be subdivided such that each subset contains at least two elements. Since $k = r/2$ if r is even and $k = (r-1)/2$ if r is odd, we have

$$\bar{\mu}^{I_r}(S_n) = \begin{cases} O(n^{r/2}) & \text{if } r \text{ is even} \\ O(n^{(r-1)/2}) & \text{if } r \text{ is odd}. \end{cases} \tag{9.30}$$

Now, consider likelihood quantities under random sampling of size n from an element of the statistical model $(\mathcal{Y}, p_Y(y; \theta), \Theta)$, with $\theta \in \mathbb{R}^p$. The above remarks make it possible to say, for example, that $E_\theta\{l_{r_1} \cdots l_{r_h}\}$, with $h = 2, 3, \ldots$ is of order $O(n^{h/2})$ if h is even, and of order $O(n^{(h-1)/2})$ if h is odd. Analogously, ν_{r,S_m}, the covariance between l_r and l_{S_m}, is a cumulant and is of order $O(n)$. Furthermore, the centred variables

$$H_{R_m} = l_{R_m} - \nu_{R_m}, \tag{9.31}$$

are of order $O_p(n^{1/2})$, for $m > 1$. On the basis of (9.30) (see also (3.28) and (3.29)), we get, for example, the relations

$$E_\theta(l_r l_s l_t H_{uv}) = O(n^2) = \nu_{r,uv} i_{st}[3] + O(n) \tag{9.32}$$

and

$$E_\theta(l_r l_s l_t H_{uvw}) = O(n^2) = \nu_{r,uvw} i_{st}[3] + O(n), \tag{9.33}$$

where [3] refers to the permutations of the indices r, s and t. Other relations of the same kind which will be useful later are

$$E_\theta(l_r l_s l_t l_u) = i_{rs} i_{tu}[3] + O(n) \tag{9.34}$$

and

$$E_\theta(l_r l_s H_{tu} H_{vw}) = \nu_{r,tu} \nu_{s,vw} + \nu_{r,vw} \nu_{s,tu} + i_{rs}(\nu_{tu,vw} - i_{tu} i_{vw}) + O(n). \tag{9.35}$$

The relations (9.32)-(9.35) give an evaluation of central moments in terms of cumulants where the terms which give a contribution of order $o(n^2)$ are ignored.

For ease of reference, the orders in probability under random sampling of size n of some likelihood quantities are summarized in Table 9.1. The same asymptotic orders hold under the assumption that joint cumulants of log-likelihood derivatives are of order $O(n)$.

Table 9.1: *Asymptotic order of likelihood quantities.*

Quantity		Order
l_r	sum of r.v.'s with zero mean	$O_p(n^{1/2})$
l_{rs}	sum of r.v.'s with non-zero mean	$O_p(n)$
i_{rs}		$O(n)$
i^{rs}		$O(n^{-1})$
ν_{rst}		$O(n)$
l_{rst}		$O_p(n)$
$\nu_{rs,t}$		$O(n)$
$E_\theta(l_r l_s H_{tu})$		$O(n)$
$E_\theta(l_r l_s l_t H_{uv})$		$O(n^2)$
$i^{rs} l_r l_s$		$O_p(1)$
l_{R_m}	$m \geq 2$	$O_p(n)$
ν_{R_m}	$m \geq 2$	$O(n)$
$H_{R_m} = l_{R_m} - \nu_{R_m}$		$O_p(n^{1/2})$

9.3   Some Basic Tools

Many asymptotic approximations in statistics are derived using one or more of the following basic tools: the stochastic Taylor formula, the inversion formula for asymptotic series and the Laplace expansion for evaluating an integral. These formulae are all consequences of the multivariate Taylor formula (given in Example 9.7).

9.3.1   The Stochastic Taylor Formula

Let $\{Y_n\}$ be a sequence of d-dimensional random variables, such that

$$Y_n = c + O_p(n^{-\alpha}), \quad \alpha > 0,$$

where $c \in \mathbb{R}^d$. Then, if $f : \mathbb{R}^d \to \mathbb{R}$ is a smooth function, that is, if it admits continuous partial derivatives up to order $k + 1$, at least in a neighbourhood of c, then the expansion

$$f(Y_n) = f(c) + \sum_{m=1}^{k} \frac{1}{m!}(Y_n - c)^{I_m} f_{I_m}(c) + O_p\left(n^{-(k+1)\alpha}\right) \qquad (9.36)$$

holds, where $(Y_n - c)^{I_m} = (Y_n - c)^{i_1} \cdots (Y_n - c)^{i_m}$, $m = 1, 2, \ldots$, and the notation in (9.11) is used for the partial derivatives. If $f(\cdot)$ does not depend on n and, consequently, $f_{I_m}(c) = O(1)$, $m = 1, 2, \ldots$, the first term that is neglected, and the remainder, (see (9.9)), is of order $O_p\left(n^{-(k+1)\alpha}\right)$.

Typical sequences where the stochastic Taylor formula is applied are: $\{\bar{Y}_n\}$, the sequence of the sample means of n independent random variables, and $\{\hat{\theta}_n\}$, the sequence of maximum likelihood estimators, when the observations y are a random sample of size n. Since, under regularity conditions,

$$\bar{Y}_n = \mu + O_p(n^{-1/2}),$$
$$\hat{\theta}_n = \theta + O_p(n^{-1/2}),$$

for both sequences $\alpha = 1/2$, and the subsequent terms of the expansion (9.36) are of the order of the positive integer powers of $n^{-1/2}$, that is, $O_p(n^{-1/2})$, $O_p(n^{-1})$, $O_p(n^{-3/2})$, etc..

If Y_n has joint moments up to order $k + 1$, expansion (9.36) allows us to approximate the moments of $f(Y_n)$ using the moments of Y_n. This is a natural extension of the delta method, mentioned in section 8.5 (see also section A.3). Observe that if $Y_n = \bar{Y}_n$ and $c = \mu$, then on the basis of (9.30), we have

$E((\bar{Y}_n - \mu)^{I_m}) = O(n^{-m/2})$ if m is even and $E((\bar{Y}_n - \mu)^{I_m}) = O(n^{-(m+1)/2})$ if m is odd. Consequently, in the asymptotic expansion of the expectation of $f(\bar{Y}_n)$ successive terms are of the order of the powers $0, 1, \ldots$ of n^{-1}.

9.3.2 Inversion of Asymptotic Series

Here, for the sake of clarity, we will distinguish between the univariate and the multivariate cases. Let $y = f(x)$, $x \in \mathbb{R}$ be a real smooth function which admits a power series expansion

$$y = a_0 + a_1 x + a_2 x^2 + a_3 x^3 + \ldots .$$

Without loss of generality, assume that $a_0 = 0$ and $a_1 = 1$, so that

$$y = x + a_2 x^2 + a_3 x^3 + \ldots . \tag{9.37}$$

Furthermore, assume that the terms in the series (9.37) depend on an asymptotic parameter n; more specifically, let $x = O(n^{-\alpha})$, $\alpha > 0$, while $a_i = O(1)$, $i = 2, 3, \ldots$.

Note that, as a first approximation, $x = y + O(n^{-2\alpha})$. Because $f'(0) = 1$, the function $y = f(x)$ can be inverted in a neighbourhood of $y = 0$, with inverse $x = g(y)$. We wish to express $g(y)$, in a neighbourhood of $y = 0$, as a power series in the form

$$x = y + b_2 y^2 + b_3 y^3 + \ldots ,$$

with b_2 and b_3 yet to be determined as functions of a_2 and a_3, ignoring terms of order $O(n^{-4\alpha})$. The solution is known as the Lagrange inversion formula (De Bruijn, 1961, p. 22).

One direct approach to this problem is first to write (9.37) as

$$x = y - a_2 x^2 - a_3 x^3 + \ldots \tag{9.38}$$

and iteratively substitute x, which appears on the right-hand side, according to its expression given by (9.38), until the only terms that cannot be neglected (for example y up to, and including, the third power) are functions only of y. This procedure will be called the iterative substitution method. The first substitution gives

$$\begin{aligned} x &= y - a_2(y - a_2 x^2)^2 - a_3 y^3 + O(n^{-4\alpha}) \\ &= y - a_2 y^2 + 2a_2^2 x^2 y - a_3 y^3 + O(n^{-4\alpha}) . \end{aligned} \tag{9.39}$$

The second step simply requires that x^2 is substituted by y^2 on the right-hand side of (9.39). This gives

$$
\begin{aligned}
x &= y - a_2 y^2 + 2a_2^2 y^3 - a_3 y^3 + O(n^{-4\alpha}) \\
&= y - a_2 y^2 - (a_3 - 2a_2^2) y^3 + O(n^{-4\alpha}) ,
\end{aligned}
$$

which is in the desired form

$$
x = y + b_2 y^2 + b_3 y^3 + O(n^{-4\alpha}), \qquad\qquad (9.40)
$$

with

$$
\begin{aligned}
b_2 &= -a_2 , \\
b_3 &= -(a_3 - 2a_2^2) .
\end{aligned}
$$

An alternative way for obtaining the inversion formula (9.40) is to use the differentiation rule for an implicit function. Consider the equation

$$
f(x, y) = y - x - a_2 x^2 - a_3 x^3 + \ldots = 0 . \qquad\qquad (9.41)
$$

This defines the function $x = g(y)$, such that $f(g(y), y) = 0$. An expansion of $g(y)$ as a power series in a neighbourhood of $y = 0$ gives

$$
x = g(0) + g'(0)y + \frac{1}{2} g''(0) y^2 + \frac{1}{3} g'''(0) y^3 + \ldots . \qquad\qquad (9.42)
$$

Since $f(0, 0) = 0$, $g(0) = 0$. The coefficients in the expansion (9.42) may be calculated by exploiting the differentiation rule

$$
g'(y) = -\frac{\partial f(x, y)}{\partial y} \bigg/ \frac{\partial f(x, y)}{\partial x} \bigg|_{x=x(y)} .
$$

From (9.41),

$$
\frac{\partial f(x, y)}{\partial y} = 1 , \qquad \frac{\partial f(x, y)}{\partial x} = -\left(1 + 2a_2 x + 3a_3 x^2 + \ldots\right)
$$

hence

$$
g'(y) = \frac{1}{1 + 2a_2 x + 3a_3 x^2 + \ldots} \bigg|_{x=x(y)} ;
$$

this gives $g'(0) = 1$, since $x(0) = 0$. Furthermore,

$$
g''(y) = -\frac{(2a_2 + 6a_3 x + \ldots) g'(y)}{(1 + 2a_2 x + 3a_3 x^2 + \ldots)^2} \bigg|_{x=x(y)} ,
$$

so that $g''(0) = -2a_2$. Lastly,

$$g'''(y) = -\frac{1}{(1 + 2a_2 x + 3a_3 x^2 + \ldots)^4}\{(6a_3 + 24a_4 x + \ldots)g'(y)$$

$$+(2a_2 + 6a_3 x + \ldots)g''(y)$$

$$-2(2a_2 + 6a_3 x + \ldots)^2(g'(y))^2(1 + 2a_2 x + \ldots)\}\Big|_{x=x(y)},$$

which gives $g'''(0) = (-1)\{6a_3 - 4a_2^2 - 8a_2^2\} = -6(a_3 - 2a_2^2)$. By substituting $g'(0)$, $g''(0)$ and $g'''(0)$ in (9.42) we obtain (9.40).

If the coefficients a_i, $i = 2, 3, \ldots$, in (9.37), are of order $O(n^{\beta_i})$, then the terms that correspond to successive powers of x do not necessarily correspond also to decreasing powers of n. If, for example, $\alpha = 1/2$, $\beta_2 = -1/2$, and $\beta_3 = 0$, then both $a_2 x^2$ and $a_3 x^3$ are of order $O(n^{-3/2})$. We shall often seek to obtain expansions whose successive terms are of the order of decreasing powers of n. It is then useful, already in the inversion phase, to take into account the fact that the coefficients are of different order in n: this is done in a simple way by following the iterative substitution method.

It is not difficult to use index notation to obtain the multivariate version of (9.40). Let $f : \mathbb{R}^p \to \mathbb{R}^p$ and

$$x^{R_m} = x^{r_1} x^{r_2} \cdots x^{r_m} \, .$$

If in a neighbourhood of $x = 0$ the expansion

$$y^a = x^a + A^a_{rs} x^{rs} + A^a_{rst} x^{rst} + A^a_{rstu} x^{rstu} + \ldots \, ,$$

holds for $y = f(x)$, implying that, in a neighbourhood of the origin, $f(x)$ is the identity function plus a perturbation, then $f(x)$ can be inverted in a neighbourhood of the origin. We wish to determine the coefficients $B^a_{R_m}$ in the power series expansion of the inverse function

$$x^a = y^a + B^a_{rs} y^{rs} + B^a_{rst} y^{rst} + B^a_{rstu} y^{rstu} + \ldots \, . \tag{9.43}$$

As in the previous case, the starting point is

$$x^a = y^a - A^a_{rs} x^{rs} - A^a_{rst} x^{rst} - A^a_{rstu} x^{rstu} + \ldots \, .$$

With three substitution steps we obtain

$$
\begin{aligned}
B^a_{rs} &= -A^a_{rs} \\
B^a_{rst} &= -(A^a_{rst} - 2A^a_{rv}A^v_{st}) \\
B^a_{rstu} &= -(A^a_{rstu} - 2A^a_{rv}A^v_{stu} - 3A^a_{rsv}A^v_{tu} + A^a_{vw}A^v_{rs}A^w_{tu} + 4A^a_{rv}A^v_{sw}A^w_{tu}) .
\end{aligned}
\tag{9.44}
$$

9.3.3 Laplace Expansion

Here too, it is convenient to take the univariate case first. Consider the integral

$$
I(n) = \int_{\mathbf{R}} \exp\{-ng(y)\}\, dy , \tag{9.45}
$$

where $g(\cdot)$ is a smooth real function with a unique absolute minimum at $\tilde{y}$, so that $g'(\tilde{y}) = 0$ and $g''(\tilde{y}) > 0$. Under these assumptions, the asymptotic behaviour of $I(n)$ is entirely determined by the local behaviour of $g(\cdot)$ in a neighbourhood of $\tilde{y}$. In particular, we have

$$
I(n) = \frac{\exp\{-n\tilde{g}\}\sqrt{2\pi}}{\sqrt{n\tilde{g}''}} \left\{ 1 - \frac{1}{8}\frac{\tilde{g}^{IV}}{n(\tilde{g}'')^2} + \frac{5}{24n}\frac{(\tilde{g}''')^2}{(\tilde{g}'')^3} + O(n^{-2}) \right\} , \tag{9.46}
$$

with $\tilde{g} = g(\tilde{y})$, $\tilde{g}'' = g''(\tilde{y})$, $\tilde{g}''' = g'''(\tilde{y})$ and $\tilde{g}^{IV} = g^{IV}(\tilde{y})$. To obtain (9.46), consider the Taylor expansion of $g(y)$ around $\tilde{y}$

$$
\begin{aligned}
g(y) &= \tilde{g} + \frac{1}{2}(y - \tilde{y})^2 \tilde{g}'' + \frac{1}{6}(y - \tilde{y})^3 \tilde{g}''' \\
&\quad + \frac{1}{24}(y - \tilde{y})^4 \tilde{g}^{IV} + O((y - \tilde{y})^5) ,
\end{aligned}
$$

which gives

$$
I(n) = e^{-n\tilde{g}} \int_{\mathbf{R}} e^{-\frac{n}{2}(y-\tilde{y})^2 \tilde{g}''} e^{-\frac{n}{6}(y-\tilde{y})^3 \tilde{g}''' - \frac{n}{24}(y-\tilde{y})^4 \tilde{g}^{IV} + nO(y-\tilde{y})^5}\, dy ,
$$

where the $N\left(\tilde{y}, (n\tilde{g}'')^{-1}\right)$ density (up to the normalizing constant) is isolated as the first factor under the integral. If we multiply and divide by the normalizing constant $\sqrt{n\tilde{g}''/(2\pi)}$ and change the integration variable to $z = (y - \tilde{y})\sqrt{n\tilde{g}''}$, we obtain

$$
\begin{aligned}
I(n) &= \frac{e^{-n\tilde{g}}\sqrt{2\pi}}{\sqrt{n\tilde{g}''}} \int_{\mathbf{R}} \exp\left\{ -\frac{z^3 \tilde{g}'''}{6\sqrt{n}(\tilde{g}'')^{3/2}} - \frac{z^4 \tilde{g}^{IV}}{24n(\tilde{g}'')^2} + O(n^{-3/2}) \right\} \phi(z)\, dz \\
&= \frac{e^{-n\tilde{g}}\sqrt{2\pi}}{\sqrt{n\tilde{g}''}} \int_{\mathbf{R}} \left(1 - \frac{1}{6\sqrt{n}}\frac{\tilde{g}'''}{(\tilde{g}'')^{3/2}}z^3 - \frac{1}{24n}\frac{\tilde{g}^{IV}}{(\tilde{g}'')^2}z^4 \right. \\
&\quad \left. + \frac{1}{72n}\frac{(\tilde{g}''')^2}{(\tilde{g}'')^3}z^6 + O(n^{-3/2}) \right) \phi(z)\, dz ,
\end{aligned}
$$

where the expansion $e^x = 1 + x + x^2/2 + \ldots$ is used and the summands of order $O(n^{-3/2})$ are neglected; $\phi(\cdot)$ is the standard normal density. In order to evaluate the required integral, remember that, if $Z \sim N(0,1)$, then $E(Z^k)$ is zero if k is odd and $(k-1)(k-3) \cdots 3 \cdot 1$ if k is even. In the end, we obtain (9.46). Note that the relative error is of order $O(n^{-2})$ and not $O(n^{-3/2})$, because the term of order $O(n^{-3/2})$ only involves expectations of odd powers of Z.

The asymptotic expansion (9.46) holds even when the integral (9.45) is restricted to an interval $[a,b]$, where $\tilde{y}$ is an inner point. In fact, proceeding as above, after the change of variable $z = (y - \tilde{y})\sqrt{n\tilde{g}''}$, the integration interval becomes $[c_1\sqrt{n}, c_2\sqrt{n}]$, with $c_1 < 0 < c_2$. For $c > 0$, $\int_c^{+\infty} \phi(z)dz = 1 - \Phi(c) = O(\phi(c)/c)$ as $c \to +\infty$, moreover, $\int_c^{+\infty} z\phi(z)dz = \phi(c)$, and iterative integration by parts shows that $\int_c^{+\infty} z^r \phi(z)dz$ with $r \in \mathbb{N}$, is of order $o(c^{-k})$, for every $k > 0$. Hence, the contribution made by the intervals $(-\infty, c_1\sqrt{n})$ and $(c_2\sqrt{n}, +\infty)$ to the integral can be neglected, because it is of order $o(n^{-k})$ for every $k > 0$.

Example 9.8 *Stirling's approximation*
Consider the expression of $n!$ through the gamma function

$$n! = \Gamma(n+1) = \int_0^{+\infty} x^n e^{-x} dx = \int_0^{+\infty} \exp\{n \log x - x\} dx$$

$$= \int_0^{+\infty} \exp\{-n(\frac{x}{n} - \log x)\} dx.$$

With the change of variable $y = x/n$ we have

$$\Gamma(n+1) = n \int_0^{+\infty} \exp\{-n(y - \log y - \log n)\} dy$$

$$= n e^{n \log n} \int_0^{+\infty} \exp\{-n(y - \log y)\} dy$$

$$= n^{n+1} I(n),$$

where

$$I(n) = \int_0^{+\infty} \exp\{-ng(y)\} dy, \quad g(y) = y - \log y.$$

Now, $g'(y) = 1 - 1/y$, $g'(1) = 0$, $g''(y) = y^{-2} > 0$, hence, $\tilde{y} = 1$, $\tilde{g} = 1$, $\tilde{g}'' = 1$, $\tilde{g}''' = -2$, $\tilde{g}^{IV} = 6$. Consequently,

$$I(n) = \frac{e^{-n}\sqrt{2\pi}}{\sqrt{n}} \left\{ 1 - \frac{6}{8n} + \frac{20}{24n} + O(n^{-2}) \right\}$$

and, in conclusion,

$$\Gamma(n+1) = n^{n+1/2}\sqrt{2\pi}e^{-n}\left\{1 + \frac{1}{12n} + O(n^{-2})\right\}. \qquad (9.47)$$

$$\triangle$$

Using index notation, it is easy to obtain the multivariate version of the Laplace expansion. Let $g : \mathbb{R}^p \to \mathbb{R}$ be a smooth function, with a unique absolute minimum at $\tilde{y}$. Indicate by $\tilde{g}_i$ the partial derivative $\partial g(y)/\partial y^i$ calculated at $\tilde{y}$ and by $\tilde{g}_{ij} = (\partial^2 g(y)/\partial y^i \partial y^j)\big|_{y=\tilde{y}}$ the elements of the Hessian matrix calculated at $\tilde{y}$; the matrix $[\tilde{g}_{ij}]$ is positive definite. Higher-order derivatives of g will be indicated by $g_{I_m}(y)$, with $\tilde{g}_{I_m} = g_{I_m}(\tilde{y})$. Using the same steps as in the univariate case we see that

$$I(n) = \int_{\mathbb{R}^p} \exp\{-ng(y)\}dy \qquad (9.48)$$

admits the asymptotic expansion

$$I(n) = \frac{e^{-n\tilde{g}}(2\pi)^{p/2}}{n^{p/2}\left|[\tilde{g}_{ij}]\right|^{1/2}}\left\{1 + \frac{C_1}{n} + O(n^{-2})\right\}, \qquad (9.49)$$

where

$$\begin{aligned}
C_1 &= \frac{1}{24}\left\{3\tilde{g}^{ij}\tilde{g}^{hk}\tilde{g}^{lm}\tilde{g}_{ijh}\tilde{g}_{klm} + 2\tilde{g}^{ik}\tilde{g}^{jl}\tilde{g}^{hm}\tilde{g}_{ijh}\tilde{g}_{klm}\right. \\
&\quad \left. - 3\tilde{g}^{ij}\tilde{g}^{hk}\tilde{g}_{ijhk}\right\}.
\end{aligned}$$

In statistical applications it is often useful to have asymptotic approximations of integrals such as

$$I(n) = \int_{\mathbb{R}^p} \exp\{-ng(y)\}b(y)dy, \qquad (9.50)$$

where $b(y)$ is a function of order $O(1)$ such that $b(\tilde{y}) \neq 0$. Along the same line of reasoning as above, by expanding both $g(\cdot)$ and $b(\cdot)$ in a Taylor series around $\tilde{y}$, and by re-ordering the terms according to their order in n, we obtain the expansion

$$I(n) = \frac{e^{-n\tilde{g}}(2\pi)^{p/2}\tilde{b}}{n^{p/2}\left|[\tilde{g}_{ij}]\right|^{1/2}}\left\{1 + \frac{C_1^{(b)}}{n} + O(n^{-2})\right\}, \qquad (9.51)$$

with

$$C_1^{(b)} = \frac{1}{24}\{3\tilde{g}^{ij}\tilde{g}^{hk}\tilde{g}^{lm}\tilde{g}_{ijh}\tilde{g}_{klm} + 2\tilde{g}^{ik}\tilde{g}^{jl}\tilde{g}^{hm}\tilde{g}_{ijh}\tilde{g}_{klm}$$

$$-3\tilde{g}^{ij}\tilde{g}^{hk}\tilde{g}_{ijhk} - 12\tilde{g}^{ij}\tilde{g}^{hk}\frac{\tilde{b}_k}{\tilde{b}}\tilde{g}_{ijh} + 12\tilde{g}^{ij}\frac{\tilde{b}_{ij}}{\tilde{b}}\},$$

where $\tilde{b} = b(\tilde{y})$, $\tilde{b}_i = \partial b(y)/\partial y^i\big|_{y=\tilde{y}}$ and $\tilde{b}_{ij} = \partial^2 b(y)/(\partial y^i y^j)\big|_{y=\tilde{y}}$.

A change of variable which does not depend on n either in (9.48) or in (9.50) will produce another integral as in (9.50). It is possible to check that expansion (9.51) respects the rule of change of variable of integrals (see Problem 9.12).

Example 9.9 *Approximation of marginal likelihood in scale, location and shape families*

The asymptotic expansion (9.51) allows a simple approximation of the marginal likelihood (7.47) to be obtained. $L_M(\tau)$ can be re-written as

$$L_M(\tau) = \int_{-\infty}^{+\infty} \int_0^{+\infty} \exp\{l(\mu,\sigma,\tau)\}\frac{1}{\sigma}d\sigma\,d\mu\,, \tag{9.52}$$

where $l(\mu,\sigma,\tau) = -n\log\sigma + \sum\log p_0((y_i - \mu)/\sigma;\tau)$. Suppose that $q_0(\cdot;\tau) = -\log p_0(\cdot;\tau)$ has a strictly positive second derivative for every τ. We can obtain an approximation for $L_M(\tau)$ with error of order $O(n^{-1})$ by evaluating the leading term in (9.51) with $p = 2$, $-ng(y_1, y_2) = l(y_1, y_2, \tau)$ and $b(y_1, y_2) = 1/y_2$. Here, $\tilde{y} = (\hat{\mu}_\tau, \hat{\sigma}_\tau)$ is the maximum likelihood estimate of (μ,σ) for fixed τ. The approximation sought is

$$L_M(\tau) = L_P(\tau)(\hat{\sigma}_\tau)^{-1}|j_{\mu\sigma}(\hat{\mu}_\tau,\hat{\sigma}_\tau,\tau)|^{-1/2}\{1 + O(n^{-1})\}\,, \tag{9.53}$$

where $j_{\mu\sigma}(\hat{\mu}_\tau,\hat{\sigma}_\tau,\tau)$ is the block (μ,σ) of the observed information matrix, evaluated at $(\hat{\mu}_\tau,\hat{\sigma}_\tau,\tau)$. If we exploit the representation (7.33) for the log-likelihood $l(\mu,\sigma,\tau)$ with fixed τ, it is easy to verify that

$$|j_{\mu\sigma}(\hat{\mu}_\tau,\hat{\sigma}_\tau,\tau)| = (\hat{\sigma}_\tau)^{-4}D_\tau(a)\,,$$

where

$$D_\tau(a) = \left\{\left(\sum_{i=1}^n q_0''(a_i;\tau)\right)\left(n + \sum_{i=1}^n a_i^2 q_0''(a_i;\tau)\right) - \left(\sum_{i=1}^n a_i q_0''(a_i;\tau)\right)^2\right\},$$

with $a_i = a_i(\tau) = (y_i - \hat{\mu}_\tau)/\hat{\sigma}_\tau$, $i = 1,\ldots,n$.

In terms of log-likelihood,

$$l_M(\tau) = l_P(\tau) + \log(\hat{\sigma}_\tau) - \frac{1}{2}\log D_\tau(a) + O(n^{-1}) \ .$$

This expansion confirms that the profile likelihood is an approximation, with
error of order $O(1)$, of the marginal likelihood, as was mentioned in section 4.3.
The correction to $l_P(\tau)$ is of order $O(1)$, because $\log D_\tau(a)$ makes the same
contribution to log-likelihood as does $\log(D_\tau(a)/n^2)$.   △

Example 9.10 *Bayesian interpretation of approximate conditional likelihood*
Consider a parametric model with likelihood function $l(\tau, \zeta)$ of order $O_p(n)$.
Let τ be the parameter of interest, while ζ is treated as a nuisance parameter.
In the Bayesian view, inference about τ is based on the posterior marginal
distribution for τ which is proportional to the integrated likelihood

$$L_I(\tau) = \int \exp\{l(\tau, \zeta)\}\pi(\tau, \zeta)d\zeta \ , \qquad (9.54)$$

where $\pi(\tau, \zeta)$ is a prior density for (τ, ζ). Suppose that this prior density has the
form $\pi(\tau, \zeta) \propto \pi_\zeta(\zeta)$, that is, equivalently, that τ and ζ are independent in the
prior distribution, and that τ has a uniform improper marginal distribution
while ζ is distributed according to the density $\pi_\zeta(\zeta)$. Furthermore, assume
that, for each fixed τ, $l(\tau, \zeta)$ admits only one maximum at $\hat{\zeta}_\tau$. Then, from
(9.51),

$$L_I(\tau) = \frac{e^{l_P(\tau)}\pi_\zeta(\hat{\zeta}_\tau)}{|j_{\zeta\zeta}(\tau, \hat{\zeta}_\tau)|^{1/2}}\{1 + O(n^{-1})\} \ , \qquad (9.55)$$

where $l_P(\tau)$ is the profile log-likelihood. If τ and ζ are orthogonal, then
$\pi_\zeta(\hat{\zeta}_\tau) = \pi_\zeta(\hat{\zeta}) + O_p(n^{-1})$ because of (4.38). Hence, in (9.55), the contribution of the prior distribution is negligible in comparison to that of both the
profile likelihood and the observed information. Thus

$$\begin{aligned}
l_I(\tau) &= \log L_I(\tau) \\
&= l_P(\tau) - \frac{1}{2}\log|j_{\zeta\zeta}(\tau, \hat{\zeta}_\tau)| + O_p(n^{-1}) = l_{AC}(\tau) + O_p(n^{-1}) \ ,
\end{aligned}$$

where $l_{AC}(\tau)$ is the approximate conditional log-likelihood introduced in section 4.7, formula (4.42).

Consequently, with this choice of the prior, inference about τ based on
$l_{AC}(\tau)$ can be justified using asymptotic arguments both according to the Fisherian approach and in the non-personalistic Bayesian context. Observe that
the result holds for any specification of the prior density $\pi_\zeta(\zeta)$.   △

9.4 Fundamental Asymptotic Expansions

By using the notation and tools introduced in the previous sections, it is not difficult to obtain asymptotic expansions for likelihood quantities such as $\hat{\theta} - \theta$, $l(\hat{\theta}) - l(\theta)$, and for the profile score function $\partial l_P(\tau)/\partial \tau$ (see formula (4.36)). For applications, the most convenient asymptotic expansions are power series in the elements of the score vector, with coefficients that depend on higher-order log-likelihood derivatives and on their null expectations. These expansions are the basis for the study of higher-order asymptotic properties. For example, one can obtain an asymptotic approximation for the bias and for the covariance matrix of $\hat{\theta}$. It is also possible to obtain an expansion for the Bartlett correction $b(\theta)$, where $b(\theta)$ is such that $E_\theta \left\{ 2 \left(l(\hat{\theta}) - l(\theta) \right) \right\} = p \left\{ 1 + n^{-1} b(\theta) + O(n^{-2}) \right\}$. Evaluation of the null expectation of the profile score is important in order to introduce modifications of the profile likelihood that improve inferential accuracy when nuisance parameters are present.

9.4.1 Expansion of $\hat{\theta} - \theta$

Consider a parametric statistical model $\mathcal{F}$ with density $p_Y(y; \theta)$, $\theta \in \Theta \subseteq \mathbb{R}^p$. Assume that the maximum likelihood estimate $\hat{\theta}$ is a consistent solution of the likelihood equation $l_r(\theta) = 0$, $r = 1, \ldots, p$. Under regularity conditions, $l_r(\hat{\theta})$ can be Taylor expanded around θ using formula (9.36), which gives

$$0 = l_r(\hat{\theta}) = l_r + (\hat{\theta} - \theta)^s l_{rs} + \frac{1}{2}(\hat{\theta} - \theta)^{st} l_{rst}$$
$$+ \frac{1}{6}(\hat{\theta} - \theta)^{stu} l_{rstu} + O_p(n^{-1}), \qquad (9.56)$$

where $(\hat{\theta} - \theta)^{R_m} = (\hat{\theta} - \theta)^{r_1} \cdots (\hat{\theta} - \theta)^{r_m}$.

It is useful to identify the order in n of the single terms that appear in (9.56). Table 9.1 can help. Specifically, l_r and $(\hat{\theta} - \theta)^s l_{rs}$ are of order $O_p(n^{1/2})$, $(\hat{\theta} - \theta)^{st} l_{rst} = O_p(1)$ and $(\hat{\theta} - \theta)^{stu} l_{rstu} = O_p(n^{-1/2})$. The order of the remainder is that of the first neglected term, $(\hat{\theta} - \theta)^{stuv} l_{rstuv}$.

The second step is to separate the stochastic part of the coefficients l_{R_m}, $m > 1$, from the non-stochastic part. To this end, consider the decomposition $l_{R_m} = \nu_{R_m} + H_{R_m}$, where $\nu_{R_m} = O(n)$ while the quantities H_{R_m}, defined by (9.31), are of order $O_p(n^{1/2})$ for $m > 1$. Remembering that $\nu_{rs} = -i_{rs}$, (9.56) can be re-written as

$$0 = l_r + (\hat{\theta} - \theta)^s(-i_{rs} + H_{rs}) + \frac{1}{2}(\hat{\theta} - \theta)^{st}(\nu_{rst} + H_{rst})$$

$$+\frac{1}{6}(\hat{\theta}-\theta)^{stu}(\nu_{rstu}+H_{rstu})+O_p(n^{-1})\,,$$

which, when re-ordered, gives

$$(\hat{\theta}-\theta)^s i_{rs} = l_r \overset{\bullet}{+} (\hat{\theta}-\theta)^s H_{rs} + \frac{1}{2}(\hat{\theta}-\theta)^{st}\nu_{rst}$$

$$\overset{\bullet}{+} \frac{1}{2}(\hat{\theta}-\theta)^{st}H_{rst} + \frac{1}{6}(\hat{\theta}-\theta)^{stu}\nu_{rstu} \overset{\bullet}{+} O_p(n^{-1}). \quad (9.57)$$

In order to make it easier to read the formula, both in (9.57) and in the rest of this chapter, we will use the symbol $\overset{\bullet}{+}$ every time the terms which follow are asymptotically smaller for an order $O_p(n^{-1/2})$ than are the preceding terms. In the same way, the symbol $\overset{\bullet\bullet}{+}$ will be used to indicate a fall of order $O_p(n^{-1})$ in the asymptotic order.

Now, it is possible to isolate $(\hat{\theta}-\theta)^{\bar{r}}$ on the left-hand side by multiplying both sides of (9.57) by $i^{r\bar{r}}$; this gives

$$(\hat{\theta}-\theta)^s i_{rs} i^{r\bar{r}} = l_r i^{r\bar{r}} \overset{\bullet}{+} (\hat{\theta}-\theta)^s H_{rs} i^{r\bar{r}} + \frac{1}{2}(\hat{\theta}-\theta)^{st}\nu_{rst} i^{r\bar{r}}$$

$$\overset{\bullet}{+} \frac{1}{2}(\hat{\theta}-\theta)^{st} H_{rst} i^{r\bar{r}} + \frac{1}{6}(\hat{\theta}-\theta)^{stu}\nu_{rstu} i^{r\bar{r}} \overset{\bullet}{+} O_p(n^{-2})\,.$$

Let

$$H_s^r = i^{rt} H_{st}, \ldots, H_{S_n}^r = i^{rt} H_{S_n t}\,, \quad (9.58)$$

$$\nu_s^r = i^{rt}\nu_{st}, \ldots, \nu_{S_n}^r = i^{rt}\nu_{S_n t}\,, \quad (9.59)$$

and remember the definition of l^r given by (9.1). Thus, for $r = 1,\ldots,p$, we have

$$(\hat{\theta}-\theta)^r = l^r \overset{\bullet}{+} H_s^r(\hat{\theta}-\theta)^s + \frac{1}{2}\nu_{st}^r(\hat{\theta}-\theta)^{st} \overset{\bullet}{+} \frac{1}{2}H_{st}^r(\hat{\theta}-\theta)^{st}$$

$$+\frac{1}{6}\nu_{stu}^r(\hat{\theta}-\theta)^{stu} \overset{\bullet}{+} O_p(n^{-2})\,, \quad (9.60)$$

which is the implicit form of the expansion sought.

Expansion (9.60) does not allow the relations (9.44) for the inversion of asymptotic series to be applied directly. However, the iterative substitution method (see section 9.3.2) can be applied. It is convenient to use the compact notation $l^{st} = l^s l^t$, $l^{stu} = l^s l^t l^u$, etc.. After some simple algebra the result is

$$(\hat{\theta}-\theta)^r = l^r \overset{\bullet}{+} \frac{1}{2}\nu_{st}^r l^{st} + H_s^r l^s \overset{\bullet}{+} \frac{1}{6}\left(\nu_{stu}^r + 3\nu_{sv}^r \nu_{tu}^v\right) l^{stu}$$

$$+\frac{1}{2}\left(H_{st}^r + 2\nu_{sv}^r H_t^v + H_v^r \nu_{st}^v\right)l^{st} + H_v^r H_s^v l^s \overset{\bullet}{+} O_p(n^{-2})\,. \quad (9.61)$$

Here, the proof of (9.61) will only be touched upon. Starting from (9.60), and neglecting the terms $O_p(n^{-3/2})$, with one substitution step, we get

$$(\hat{\theta} - \theta)^r = l^r \overset{\bullet}{+} \frac{1}{2}\nu_{st}^r\left(l^{st} \overset{\bullet}{+} O_p(n^{-3/2})\right) + H_s^r\left(l^s \overset{\bullet}{+} O_p(n^{-1})\right) \overset{\bullet}{+} O_p(n^{-3/2})$$

$$= l^r \overset{\bullet}{+} \frac{1}{2}\nu_{st}^r l^{st} + H_s^r l^s \overset{\bullet}{+} O_p(n^{-3/2})\,.$$

This is expansion (9.61) up to and including terms of order $O_p(n^{-1})$. For obtaining further terms, one can proceed in the same way.

On the basis of (9.61),

$$(\hat{\theta} - \theta)^r = l^r \overset{\bullet}{+} O_p(n^{-1}) = O_p(n^{-1/2})\,.$$

The asymptotic normality of $\sqrt{n}(\hat{\theta} - \theta)$ is a direct consequence of the representation $\sqrt{n}(\hat{\theta} - \theta)^r = \sqrt{n}i^{rs}l_s + O_p(n^{-1/2})$ and of the central limit theorem for the standardized score vector.

9.4.2 Asymptotic Bias of $\hat{\theta}$

Expansion (9.61) makes it possible to obtain an asymptotic expansion for the bias of $\hat{\theta}$. We have

$$E_\theta(\hat{\theta} - \theta)^r = 0 \overset{\bullet\bullet}{+} \frac{1}{2}\nu_{st}^r E_\theta(l^{st}) + E_\theta(H_s^r l^s) \overset{\bullet\bullet}{+} O(n^{-2})\,,$$

where the expectation of the term of order $O_p(n^{-3/2})$ in (9.61) is of order $O(n^{-2})$; in fact, the expectations $E_\theta(l_r l_s l_t)$, $E_\theta(l_r l_s H_{tu})$ and $E_\theta(l_r H_{st} H_{uv})$, as third-order central moments of sums, are of order $O(n)$. In addition, from the identities

$$E_\theta(l^{st}) = E_\theta(l^s l^t) = i^{s\bar{s}}i^{t\bar{t}}E_\theta(l_{\bar{s}}l_{\bar{t}}) = i^{s\bar{s}}i^{t\bar{t}}i_{\bar{s}\bar{t}} = i^{s\bar{s}}\delta_{\bar{s}}^t = i^{st}\,,$$

$$E_\theta(H_s^r l^s) = i^{r\bar{r}}i^{s\bar{s}}E_\theta\left\{(l_{\bar{r}s} + i_{\bar{r}s})l_{\bar{s}}\right\} = i^{r\bar{r}}i^{s\bar{s}}(\nu_{\bar{r}s,\bar{s}}+0) = i^{r\bar{r}}i^{s\bar{s}}\nu_{\bar{r}s,\bar{s}} = i^{rs}i^{tu}\nu_{st,u}$$

and

$$i^{st}\nu_{st}^r = i^{st}i^{r\bar{r}}\nu_{\bar{r}st} = i^{rs}i^{tu}\nu_{stu}\,,$$

we obtain

$$E_\theta(\hat{\theta} - \theta)^r = \frac{1}{2}i^{rs}i^{tu}(\nu_{stu} + 2\nu_{st,u}) \overset{\bullet\bullet}{+} O(n^{-2})\,. \quad (9.62)$$

Thus the first-order bias corrected maximum likelihood estimator is

$$\hat{\theta}_I = \hat{\theta} - d(\hat{\theta}),\tag{9.63}$$

where $d(\theta) = [d^r(\theta)]$, with

$$d^r(\theta) = \frac{1}{2}i^{rs}i^{tu}(\nu_{stu} + 2\nu_{st,u}).$$

Since

$$d^r(\hat{\theta}) = d^r(\theta) \overset{\bullet}{+} d^r_v(\hat{\theta} - \theta)^v \overset{\bullet}{+} O_p(n^{-2}),$$

with

$$
\begin{aligned}
d^r_v &= \frac{\partial d^r}{\partial \theta^v} = i^{rw}i^{sy}i^{tu}(\nu_{stu} + 2\nu_{st,u})(\nu_{vwy} + \nu_{v,wy}) \\
&\quad + \frac{1}{2}i^{rs}i^{tu}(\nu_{stuv} + \nu_{stu,v} + 2\nu_{stv,u} + 2(\nu_{st,uv} + \nu_{st,u,v})) \\
&= O(n^{-1}),
\end{aligned}\tag{9.64}
$$

the corrected estimator $\hat{\theta}_I$ has bias of order $O(n^{-2})$.

Example 9.11 *One-parameter exponential families*
Consider an $\mathcal{F}^1_{ne}$ with density

$$p_Y(y;\theta) = \exp\{\theta y - nK(\theta)\}h(y),\tag{9.65}$$

where $\theta \in \tilde{\Theta} \subseteq \mathbb{R}$. Using power notation (see (9.25)-(9.27)), expansion (9.62) becomes

$$E_\theta(\hat{\theta} - \theta) = \frac{1}{2}i^{-2}(\nu_3 + 2\nu_{2,1}) \overset{\bullet\bullet}{+} O(n^{-2}).\tag{9.66}$$

Since $l_1 = l_* = y - nK'(\theta)$, $l_2 = -nK''(\theta)$, $l_3 = -nK'''(\theta)$, the quantities in (9.66) are $i(\theta) = nK''(\theta)$, $\nu_3 = -nK'''(\theta)$ and $\nu_{2,1} = E_\theta(l_2l_1) = -nK''(\theta)E_\theta(l_*) = 0$. Hence, the asymptotic expansion for the bias of $\hat{\theta}$ is

$$E_\theta(\hat{\theta} - \theta) = -\frac{1}{2}\frac{K'''(\theta)}{n(K''(\theta))^2} \overset{\bullet\bullet}{+} O(n^{-2}).\tag{9.67}$$

Consider the particular case of random sampling of size n from a $Ga(\nu, 1)$ distribution with data $y = (y_1, \ldots, y_n)$. The natural observation $s = \sum \log y_i$ has density of the form $p_s(s;\nu) = \exp\{\nu s - n\Gamma(\nu)\}h(s)$. The maximum likelihood estimate $\hat{\nu}$ is the solution to the equation $s/n = \Gamma'(\nu)$ and, from (9.67),

$$E_\nu(\hat{\nu} - \nu) = -\frac{1}{2n}\frac{\Gamma'''(\nu)}{(\Gamma''(\nu))^2} \overset{\bullet\bullet}{+} O(n^{-2}).$$

Table 9.2: *Estimates of bias and mean squared error of $\hat{\nu}$ and $\hat{\nu}'_I$.*

n	$E(\hat{\nu} - \nu)$	$E(\hat{\nu}'_I - \nu)$	$E(\hat{\nu} - \nu)^2$	$E(\hat{\nu}'_I - \nu)^2$
5	1.21	−0.003	13.78	1.98
10	0.33	−0.01	0.58	0.24
20	0.15	0.007	0.18	0.12

The first-order bias corrected m.l.e. is

$$\hat{\nu}_I = \hat{\nu} + \frac{1}{2n} \frac{\Gamma'''(\hat{\nu})}{(\Gamma''(\hat{\nu}))^2} , \qquad (9.68)$$

which has bias of order $O(n^{-2})$.

If the scale parameter is unknown, that is if the reference distribution is a two-parameter gamma, $Ga(\nu, \phi)$, the maximum likelihood estimate of the ν component corrected for bias of order $O(n^{-1})$ is

$$\hat{\nu}'_I = \hat{\nu} + \frac{1}{2n} \frac{\hat{\nu}^2 \psi''(\hat{\nu}) - \hat{\nu}\psi'(\hat{\nu}) + 2}{\{\hat{\nu}\psi'(\hat{\nu}) - 1\}^2} ,$$

where $\psi'(\cdot)$ and $\psi''(\cdot)$ are the first two derivatives of the digamma function $\psi(\nu) = d\log(\Gamma(\nu))/d\nu$. See Abramowitz and Stegun (1965, sections 6.3 and 6.4) for an account of these special functions. Table 9.2 shows the estimates of the bias and of the mean squared error of $\hat{\nu}$ and $\hat{\nu}'_I$ based on a simulation with 2000 samples for $n = 5, 10, 20$ and $\nu = \phi = 1$.   △

Unbiasedness (exact or approximate) depends on the parameterization. Hence, corrections of the kind described above are of limited interest. On the other hand, any bias corrections of estimating equations, for example, of equations based on the profile score function, obey the principle of parameterization invariance. This type of correction will be examined in more detail below (see sections 9.4.6 and 11.6).

9.4.3 Variance and Other Cumulants of $\hat{\theta}$

The asymptotic expansion (9.61) can also be used to obtain approximations of moments of $\hat{\theta}$ of order higher than one. Here we will restrict ourselves to the asymptotic expansion for $\text{Var}_\theta(\hat{\theta})$ in the one-parameter case. Using power notation, expansion (9.61) becomes

$$\hat{\theta} - \theta = i^{-1}l_* \overset{\bullet}{+} \frac{1}{2}i^{-3}\nu_3 l_*^2 + i^{-2}H_2 l_*$$
$$\overset{\bullet}{+} \frac{1}{6}i^{-4}\nu_4 l_*^3 + \frac{1}{2}i^{-5}\nu_3^2 l_*^3 + \frac{1}{2}i^{-3}H_3 l_*^2$$
$$+\frac{3}{2}i^{-4}\nu_3 H_2 l_*^2 + i^{-3}H_2^2 l_* \overset{\bullet}{+} O_p(n^{-2}) .$$

Hence,

$$(\hat{\theta} - \theta)^2 = i^{-2}l_*^2 \overset{\bullet}{+} i^{-4}\nu_3 l_*^3 + 2i^{-3}H_2 l_*^2$$
$$\overset{\bullet}{+} \frac{5}{4}i^{-6}\nu_3^2 l_*^4 + 3i^{-4}H_2^2 l_*^2 + 4i^{-5}\nu_3 H_2 l_*^3$$
$$+\frac{1}{3}i^{-5}\nu_4 l_*^4 + i^{-4}H_3 l_*^3 \overset{\bullet}{+} O_p(n^{-5/2}) .$$

Exploiting relations (9.32)-(9.35), we have

$$E_\theta(\hat{\theta} - \theta)^2 = i^{-1} \overset{\bullet\bullet}{+} i^{-3} \left\{ 2\nu_{1,1,2} + (\nu_{2,2} - i^2) + 2\nu_{2,2} + \nu_4 + 3\nu_{3,1} \right\}$$
$$+i^{-4} \left\{ \frac{11}{4}\nu_3^2 + 9\nu_3\nu_{1,2} + 6\nu_{1,2}^2 \right\} \overset{\bullet\bullet}{+} O(n^{-3}) . \tag{9.69}$$

Lastly, from (9.69) and (9.66), we get

$$\text{Var}_\theta(\hat{\theta}) = i^{-1} \overset{\bullet\bullet}{+} i^{-3} \left\{ 2\nu_{1,1,2} + (\nu_{2,2} - i^2) + 2\nu_{2,2} + \nu_4 + 3\nu_{3,1} \right\}$$
$$+\frac{i^{-4}}{2} \left\{ 5\nu_3^2 + 16\nu_3\nu_{1,2} + 10\nu_{1,2}^2 \right\} \overset{\bullet\bullet}{+} O(n^{-3}) . \tag{9.70}$$

We know that $E_\theta(\hat{\theta} - \theta) = d(\theta) \overset{\bullet\bullet}{+} O(n^{-2})$ with $d(\theta) = (2i^2)^{-1}(\nu_3 + 2\nu_{1,2})$. From (9.64) for $p = 1$ we have

$$d'(\theta) = i^{-3}(\nu_3^2 + 2\nu_{1,2}^2 + 3\nu_3\nu_{1,2}) + \frac{1}{2}i^{-2}\{\nu_4 + 3\nu_{1,3} + 2(\nu_{2,2} + \nu_{1,1,2})\} .$$

It is easy to check that the asymptotic expansion (9.70) could be rewritten as

$$\text{Var}_\theta(\hat{\theta}) = i^{-1} \overset{\bullet\bullet}{+} 2i^{-1}d'(\theta) + 2\{d(\theta)\}^2 + i^{-1}\gamma^2 \overset{\bullet\bullet}{+} O(n^{-3}) , \tag{9.71}$$

where γ is the curvature defined by (5.91). Expansion (9.71) also immediately offers an asymptotic expansion with error of order $O(n^{-3})$ for the asymptotic variance of the bias corrected estimator $\hat{\theta}_r = \hat{\theta} - d(\hat{\theta})$. Since

$$\hat{\theta} - \theta = i^{-1}l_* \overset{\bullet}{+} O_p(n^{-1})$$

and

$$d(\hat{\theta}) = d(\theta) \stackrel{\bullet}{+} d'(\theta)(\hat{\theta} - \theta) \stackrel{\bullet}{+} O_p(n^{-2}) \,,$$

then

$$
\begin{aligned}
\mathrm{Var}_\theta(\hat{\theta}_I) &= \mathrm{Var}_\theta(\hat{\theta} - d(\hat{\theta})) = \mathrm{Var}_\theta(\hat{\theta}) + \mathrm{Var}_\theta(d(\hat{\theta})) - 2\,\mathrm{Cov}_\theta(\hat{\theta}, d(\hat{\theta})) \\
&= \mathrm{Var}_\theta(\hat{\theta}) - 2\,\mathrm{Cov}_\theta(\hat{\theta}, d(\hat{\theta})) \stackrel{\bullet\bullet}{+} O(n^{-3}) \\
&= \mathrm{Var}_\theta(\hat{\theta}) - 2i^{-1}d'(\theta) \stackrel{\bullet\bullet}{+} O(n^{-3}) \,.
\end{aligned}
$$

Consequently, up to order $O(n^{-2})$, $\hat{\theta}_I$ has asymptotic variance equal to (9.71) except for the term $2i^{-1}d'(\theta)$, that is

$$\mathrm{Var}_\theta(\hat{\theta}_I) = i^{-1} \stackrel{\bullet\bullet}{+} 2\{d(\theta)\}^2 + i^{-1}\gamma^2 \stackrel{\bullet\bullet}{+} O(n^{-3}) \,. \tag{9.72}$$

Note that in (9.72) the term of order $O(n^{-2})$ is non-negative; in particular, it is zero in a full exponential family (where $\gamma = 0$) if $\hat{\theta}$ is unbiased ($d(\theta) = 0$), i.e., whenever the mean value parameterization is considered. This confirms the well-known exact result that the maximum likelihood estimator reaches the Cramér–Rao lower bound in this same situation (see section 5.8). In a regular problem, it can be shown (see, e.g. Efron, 1975) that if $\tilde{\theta}$ is an estimator of θ with bias of order $O(n^{-2})$, then

$$\mathrm{Var}_\theta(\tilde{\theta}) = \mathrm{Var}_\theta(\hat{\theta}_I) + \Delta^2(\theta) \stackrel{\bullet\bullet}{+} O(n^{-3}) \,,$$

where $\Delta^2(\theta) \geq 0$; furthermore $\Delta^2(\theta) = 0$ if $\tilde{\theta} = \hat{\theta}_I$. Thus, $\hat{\theta}_I$ is said to be second-order efficient.

For $p > 1$, Peers and Iqbal (1985) obtained the following asymptotic expansions for the cumulants of $\hat{\theta}$ of order two, three and four, indicated, for simplicity, by $\kappa^{r,s}$, $\kappa^{r,s,t}$, $\kappa^{r,s,t,u}$:

$$
\begin{aligned}
\kappa^{r,s} ={}& \mathrm{Cov}_\theta(\hat{\theta}^r \hat{\theta}^s) = i^{rs} \\
&\stackrel{\bullet\bullet}{+} i^{rv} i^{sw} i^{tu} \{\nu_{vwtu} + \nu_{v,wtu} + \nu_{t,vwu} \\
&\qquad + 2E_\theta(l_v l_u H_{wt}) + 3E_\theta(H_{vt} H_{wu})\} \\
&+ i^{rv} i^{sw} i^{tu} i^{zy} \{(3/2)\nu_{vtz}\nu_{wuy} + \nu_{wuy}(\nu_{z,vt} + \nu_{v,tz}) \\
&\quad + \nu_{z,vt}(2\nu_{w,uy} + \nu_{u,wy}) + \nu_{v,wz}(2\nu_{t,uy} + \nu_{tuy}) \\
&\quad + \nu_{vwz}(2\nu_{t,uy} + \nu_{tuy})\} \stackrel{\bullet\bullet}{+} O(n^{-3})
\end{aligned}
\tag{9.73}
$$

$$\kappa^{r,s,t} = i^{rv}i^{sw}i^{tu}(2\nu_{vwu} + 3\nu_{v,wu}) \overset{\bullet\bullet}{+} O(n^{-3}) \tag{9.74}$$

$$\kappa^{r,s,t,u} = i^{rv}i^{sw}i^{ty}i^{uz}\{8\nu_{v,wyz} + 3\nu_{v,wyz} + 9E_\theta(H_{vw}H_{yz}) + 6E_\theta(l_v l_w H_{yz})\}$$
$$+ i^{rv}i^{sw}i^{ty}i^{uz}i^{y'z'}(\nu_{v,wy'} + \nu_{y',vw} + \nu_{vwy'})(\nu_{y,z,z'} + \nu_{yzz'})$$
$$\overset{\bullet\bullet}{+} O(n^{-4}). \tag{9.75}$$

The central moments of log-likelihood derivatives $E_\theta(l_v l_u H_{wt})$, $E_\theta(H_{vt}H_{wu})$, etc., have simply been indicated, as this renders evaluation of the asymptotic order of each term of the expansions more immediate.

Expansion (9.73) allows one to obtain a multiparameter generalization of expansion (9.72) and to extend the second-order efficiency result for $\hat{\theta}_I$ to the multiparameter case. See section 9.6 for the relevant references.

9.4.4 Expansion of $l(\hat{\theta}) - l(\theta)$

Direct application of the stochastic Taylor formula gives

$$l(\hat{\theta}) - l(\theta) = (\hat{\theta} - \theta)^r l_r + \frac{1}{2}(\hat{\theta} - \theta)^{rs}l_{rs}$$
$$+ \frac{1}{6}(\hat{\theta} - \theta)^{rst}l_{rst} \overset{\bullet}{+} \frac{1}{24}(\hat{\theta} - \theta)^{rstu}l_{rstu} \overset{\bullet}{+} O_p(n^{-3/2}).$$

Using the quantities H_{R_m} defined by (9.31) we have

$$l(\hat{\theta}) - l(\theta) = (\hat{\theta} - \theta)^r l_r - \frac{1}{2}(\hat{\theta} - \theta)^{rs}i_{rs}$$

$$\overset{\bullet}{+} \frac{1}{2}(\hat{\theta} - \theta)^{rs}H_{rs} + \frac{1}{6}(\hat{\theta} - \theta)^{rst}\nu_{rst}$$

$$\overset{\bullet}{+} \frac{1}{6}(\hat{\theta} - \theta)^{rst}H_{rst} + \frac{1}{24}(\hat{\theta} - \theta)^{rstu}\nu_{rstu} \overset{\bullet}{+} O_p(n^{-3/2}). \tag{9.76}$$

By substituting (9.61) in (9.76) and neglecting contributions of order $o_p(n^{-1})$, we obtain the expansion

$$l(\hat{\theta}) - l(\theta) = \frac{1}{2}i_{rs}l^{rs} \overset{\bullet}{+} \frac{1}{6}\left(\nu_{rst}l^t + 3H_{rs}\right)l^{rs}$$
$$\overset{\bullet}{+} \frac{1}{24}\left(\nu_{rstu} + 3\nu_{rsv}\nu_{tu}^v\right)l^{rstu} + \frac{1}{6}\left(H_{rst} + 3\nu_{rsv}H_t^v\right)l^{rst}$$
$$+ \frac{1}{2}H_{rv}H_s^v l^{rs} \overset{\bullet}{+} O_p(n^{-3/2}). \tag{9.77}$$

For a partial proof of (9.77), up to, and including, the terms of order $O_p(n^{-1/2})$, substitute the expansion (9.61) of $\hat{\theta} - \theta$ into (9.76) and ignore the terms of order $o_p(n^{-1/2})$, obtaining

$$
\begin{aligned}
l(\hat{\theta}) - l(\theta) \;=\; & \{l^r + \tfrac{1}{2}\nu_{st}^r l^{st} + H_s^r l^s\} l_r \\
& - \tfrac{1}{2}\{l^{rs} + \nu_{tu}^s l^{rtu} + H_t^s l^{rt}\} i_{rs} \\
& + \tfrac{1}{2} l^{rs} H_{rs} + \tfrac{1}{6} l^{rst} \nu_{rst} + O_p(n^{-1}) .
\end{aligned} \tag{9.78}
$$

Note that $l^r l_r = i_{rs} l^{rs}$, $H_s^r l_r = H_{rs} l^r$, $i_{rs} \nu_{tu}^s = \nu_{tru}$ and $i_{rs} H_t^s = H_{rt}$. Hence, (9.78) gives (9.77) directly, up to and including the term of order $O_p(n^{-1/2})$.

Since l_* has a null asymptotic $N_p(0, [i_{rs}])$ distribution, the leading term of (9.77) gives the well-known result of null asymptotic χ_p^2 distribution for $W(\theta) = 2(l(\hat{\theta}) - l(\theta))$.

9.4.5 *Expansion of $E_\theta(W)$*

It is not difficult to obtain an asymptotic expansion for $E_\theta(W(\theta))$ of the form

$$
E_\theta(W(\theta)) = p\{1 + n^{-1}b(\theta) + O(n^{-2})\} , \tag{9.79}
$$

which can be used for the Bartlett correction. The Bartlett-corrected combinant $W^\star(\theta) = W(\theta)/(1 + n^{-1}b(\theta))$, has asymptotic null χ_p^2 distribution with error of order $O(n^{-2})$, instead of order $O(n^{-1})$, as it would be the case for $W(\theta)$. For further details on this point, see sections 11.3 and 11.4. Here we shall only deal with the calculation of the Bartlett correction.

Example 9.12 *One-parameter exponential families (cont.)*
If

$$
p(y; \theta) = \exp\{\theta y - nK(\theta)\} h(y), \quad \theta \in \tilde{\Theta} \subseteq \mathbb{R} ,
$$

then (see also Example 9.11)

$$
\begin{aligned}
W(\theta) \;=\; & 2(l(\hat{\theta}) - l(\theta)) = \frac{1}{nK''(\theta)}(l_*)^2 \overset{\bullet}{-} \frac{1}{3}\frac{K'''(\theta)}{n^2(K''(\theta))^3}(l_*)^3 \\
& \overset{\bullet}{-} \frac{1}{12n^3(K''(\theta))^4}\left\{ K^{IV}(\theta) - 3\frac{(K'''(\theta))^2}{K''(\theta)} \right\} (l_*)^4 \overset{\bullet}{+} O_p(n^{-3/2}) .
\end{aligned} \tag{9.80}
$$

Note that l_* is the only stochastic ingredient in (9.80). By using (9.21), (9.23), (9.24) and (9.34), we obtain

$$
\begin{aligned}
E_\theta(l_*^2) &= -E_\theta(l_2) = nK''(\theta) \\
E_\theta(l_*^3) &= \nu_{1,1,1} = -\nu_3 - 3\nu_{2,1} = -\nu_3 = nK'''(\theta) \\
E_\theta(l_*^4) &= \nu_{1,1,1,1} = -\nu_4 - 4\nu_{3,1} - 3\nu_{2,2} - 6\nu_{1,1,2} \\
&= nK^{IV}(\theta) - 0 - 3n^2(K''(\theta))^2 + 6nK''(\theta)E_\theta(l_*^2) \\
&= nK^{IV}(\theta) + 3n^2(K''(\theta))^2 = 3n^2(K''(\theta))^2 + O(n) .
\end{aligned}
$$

Hence, writing

$$
\rho_3 = \frac{K'''(\theta)}{(K''(\theta))^{3/2}}, \quad \rho_4 = \frac{K^{IV}(\theta)}{(K''(\theta))^2} ,
$$

if we take the expectation of both sides of expansion (9.80) we get

$$
E_\theta(W(\theta)) = 1 - \frac{1}{3n}\rho_3^2 - \frac{3}{12n}(\rho_4 - 3\rho_3^2) + O(n^{-2}) ,
$$

since the expectation of the term of order $O_p(n^{-3/2})$ in (9.80) is of order $O(n^{-2})$. Therefore,

$$
E_\theta(W(\theta)) = 1 - \frac{1}{12n}(3\rho_4 - 5\rho_3^2) + O(n^{-2}). \tag{9.81}
$$

From (9.80) we can also obtain an expansion for $Z = \text{sgn}(\hat{\theta} - \theta)\sqrt{W}$. Let $Y = l_*/\sqrt{nK''(\theta)}$ be the standardized score. We can write

$$
Z = Y\sqrt{1 - \frac{\rho_3}{3\sqrt{n}}Y - \frac{\rho_4 - 3\rho_3^2}{12n}Y^2} + O_p(n^{-3/2}) ,
$$

and, using $\sqrt{1+x} = 1 + x/2 - x^2/8 + O(x^3)$, we get

$$
Z = Y - \frac{\rho_3}{6\sqrt{n}}Y^2 - \frac{\rho_4}{24n}Y^3 + \frac{\rho_3^2}{9n}Y^3 + O_p(n^{-3/2}). \tag{9.82}
$$

$$\triangle$$

The general expression for the quantity $b(\theta)/n$ which appears in (9.79) can be obtained from direct term by term evaluation of the expectation of the expansion (9.77), by using the approximations (9.32)-(9.35) and the identity

$$
E_\theta(H_{rs}l_tl_u) = \nu_{rs,tu} - i_{rs}i_{tu} .
$$

By re-ordering the result and redefining the dummy indices, the final expression is

$$B(\theta) = \frac{b(\theta)}{n} = \frac{1}{12p} \left\{ i^{rs} i^{tu} B_{rstu} + i^{rs} i^{tu} i^{vw} B_{rstuvw} \right\} , \qquad (9.83)$$

where

$$
\begin{aligned}
B_{rstu} &= 3\nu_{rstu} + 12(\nu_{rt,su} + \nu_{r,t,su} + \nu_{r,stu}) \\
B_{rstuvw} &= 3\nu_{rst}\nu_{uvw} + 6\nu_{rtv}\nu_{suw} \\
&\quad + 12(\nu_{r,st}\nu_{uvw} + \nu_{r,st}\nu_{uv,w} + \nu_{r,tv}\nu_{su,w}) \\
&\quad + 4\nu_{r,t,v}\nu_{suw} + 24\nu_{r,tv}\nu_{suw} .
\end{aligned}
$$

If we consider a partition of θ such as $\theta = (\tau, \zeta)$, with τ a k-dimensional parameter of interest, we immediately obtain the expression of the Bartlett correction for the profile log-likelihood ratio

$$W_P(\tau) = 2 \left\{ l(\hat{\tau}, \hat{\zeta}) - l(\tau, \hat{\zeta}_\tau) \right\} . \qquad (9.84)$$

Indeed,

$$E_\theta(W_P(\tau)) = E_\theta \left\{ 2(l(\hat{\tau}, \hat{\zeta}) - l(\tau, \zeta)) \right\} - E_\theta \left\{ 2(l(\tau, \hat{\zeta}_\tau) - l(\tau, \zeta)) \right\} . \qquad (9.85)$$

The first expectation on the right-hand side can be evaluated directly on the basis of (9.79) and (9.83); the second, requires adapting these formulae to the $(p - k)$-dimensional model which corresponds to a fixed value of τ. This gives an expression of the form

$$E_\theta \left\{ 2 \left(l(\tau, \hat{\zeta}_\tau) - l(\tau, \zeta) \right) \right\} = (p - k) \left\{ 1 + n^{-1} \tilde{b}(\theta) \right\} + O(n^{-2})$$

so that

$$E_\theta(W_P(\tau)) = k + n^{-1} \left\{ pb(\theta) + (p - k)\tilde{b}(\theta) \right\} + O(n^{-2}) .$$

9.4.6 Expansion of the Profile Score

Consider a parametric statistical model with parameter $\theta = (\tau, \zeta)$, where $\tau = [\tau^r]$ is a k-dimensional parameter of interest, $r = 1, \ldots, k$, and $\zeta = [\zeta^a]$, $a = 1, \ldots, p - k$ is a nuisance component. Use the indices $r, s, t, \ldots$ for the components of τ and the indices $a, b, c, \ldots$ for the components of ζ. Let $[i_{ab}]$ be the block (ζ, ζ) of the overall information matrix and let κ^{ab} be a generic element of $[i_{ab}]^{-1}$. We want to obtain an asymptotic expansion for the profile score (see formula (4.36))

$$l_P(\tau)_* = [\tilde{l}_r] = [l_r(\tau, \hat{\zeta}_\tau)].$$

We have

$$\tilde{l}_r = l_r + l_{ra}(\hat{\zeta}_\tau - \zeta)^a \overset{\bullet}{+} \frac{1}{2} l_{rab}(\hat{\zeta}_\tau - \zeta)^{ab} \overset{\bullet}{+} O_p(n^{-1/2}). \qquad (9.86)$$

Furthermore, from (9.61)

$$(\hat{\zeta}_\tau - \zeta)^a = \kappa^{ab} l_b \overset{\bullet}{+} \frac{1}{2} \kappa^{ab} \kappa^{ce} \kappa^{df} \nu_{bcd} l_e l_f + \kappa^{ab} \kappa^{cd} H_{bc} l_d \overset{\bullet}{+} O_p(n^{-3/2}). \qquad (9.87)$$

Writing $l_{ra} = H_{ra} - i_{ra}$ and substituting (9.87) into (9.86) we obtain

$$\begin{aligned}
\tilde{l}_r = {} & l_r - i_{ra} \kappa^{ab} l_b \overset{\bullet}{+} H_{ra} \kappa^{ab} l_b - \frac{1}{2} i_{ra} \kappa^{ab} \kappa^{ce} \kappa^{df} \nu_{bcd} l_e l_f \\
& - i_{ra} \kappa^{ab} \kappa^{cd} H_{bc} l_d + \frac{1}{2} \nu_{rab} \kappa^{ac} \kappa^{bd} l_c l_d \overset{\bullet}{+} O_p(n^{-1/2}). \qquad (9.88)
\end{aligned}$$

Therefore, by elementary calculations,

$$E_\theta(\tilde{l}_r) = \kappa^{ab}(\nu_{ra,b} - i_{rc} \kappa^{cd} \nu_{b,ad}) + \frac{1}{2} \kappa^{ab}(\nu_{rab} - i_{rc} \kappa^{cd} \nu_{abd}) \overset{\bullet\bullet}{+} O(n^{-1}). \qquad (9.89)$$

If ζ has a fixed dimension, the null expectation of the profile score is of order $O(1)$. By (9.22) the identities

$$\kappa^{ab}(\nu_{rab} + 2\nu_{ra,b}) = -\kappa^{ab}(\nu_{r,a,b} + \nu_{r,ab})$$

and

$$\kappa^{ab} i_{rc} \kappa^{cd}(\nu_{abd} + 2\nu_{b,ad}) = -\kappa^{ab} i_{rc} \kappa^{cd}(\nu_{a,b,d} + \nu_{d,ab})$$

hold. Thus, (9.89) could also be written as

$$E_\theta(\tilde{l}_r) = -\frac{1}{2} \kappa^{ab} \left\{ (\nu_{r,ab} - i_{rc} \kappa^{cd} \nu_{d,ab}) + (\nu_{r,a,b} - i_{rc} \kappa^{cd} \nu_{a,b,d}) \right\} \overset{\bullet\bullet}{+} O(n^{-1}). \qquad (9.90)$$

The quantities inside round brackets represent, respectively, $\text{Cov}_\theta(l_{ab}, l_r - i_{rc} \kappa^{cd} l_d)$ and $\text{Cov}_\theta(l_a l_b, l_r - i_{rc} \kappa^{cd} l_d)$, where $l_r - i_{rc} \kappa^{cd} l_d$ is the residual of the linear least squares regression of l_r on $l_\zeta = [l_a]$. Expansion (9.90) allows for direct checking that the term of order $O(1)$ of $E_\theta(\tilde{l}_r)$ behaves in regular manner under interest-respecting reparameterizations (see section 9.5.3).

9.5 Parameterization Invariance and Asymptotic Expansions

Consider a parametric statistical model $\mathcal{F}$ with two distinct parameterizations θ and ψ, where the function $\psi(\theta)$ is one-to-one and smooth, with smooth inverse $\theta(\psi)$. It has already been seen in section 2.11 that the combinant $W = W(\theta)$ is parameterization invariant. Furthermore, expansion (9.77) gives $W = W_u + O_p(n^{-1/2})$, with $W_u = i_{rs}l^{rs}$ and, from (9.76), we obtain $W = W_e + O_p(n^{-1/2})$, with $W_e = (\hat{\theta}-\theta)^{rs}\hat{i}_{rs}$ thus, the relations (3.42) and (3.43) are verified. However, as in the one-parameter case, while W_u is parameterization invariant, W_e is not. It is easy to check this using index notation.

Transformation rules for basic likelihood quantities were given in section 1.4.3: see (1.13), (1.17)-(1.20) and Problem 1.5. Further likelihood quantities have been defined in this chapter, such as ν_{R_m}, ν_{R_m,S_n}, and H_{R_m}. It is useful to give the transformation rules for these new quantities. In particular,

$$\bar{\nu}_{abc} = \nu_{rst}\theta_a^r\theta_b^s\theta_c^t - i_{rs}\theta_{ab}^r\theta_c^s[3] , \tag{9.91}$$

$$\bar{\nu}_{abcd} = \nu_{rstu}\theta_a^r\theta_b^s\theta_c^t\theta_d^u + \nu_{rst}\theta_a^r\theta_b^s\theta_{cd}^t[6]$$
$$-i_{rs}(\theta_{ab}^r\theta_{cd}^s[3] + \theta_a^r\theta_{bcd}^s[4]) , \tag{9.92}$$

$$\bar{H}_{ab} = \bar{l}_{ab} + \bar{i}_{ab} = H_{rs}\theta_a^r\theta_b^s + l_r\theta_{ab}^r , \tag{9.93}$$

$$\bar{H}_{abc} = \bar{l}_{abc} - \bar{\nu}_{abc} = H_{rst}\theta_a^r\theta_b^s\theta_c^t + H_{rs}\theta_{ab}^r\theta_c^s[3] + l_r\theta_{abc}^r , \tag{9.94}$$

and so on. Furthermore, for the generic element of the inverse of the expected information matrix the transformation rule

$$\bar{i}^{ab} = i^{rs}\psi_r^a\psi_s^b \tag{9.95}$$

holds; (9.95) is a direct consequence of (9.8). Indeed, starting from

$$\bar{i}_{ab}\bar{i}^{bc} = \delta_a^c$$

and using (1.20) we get

$$i_{rs}\theta_a^r\theta_b^s\,\bar{i}^{bc} = \delta_a^c .$$

If both sides of this identity are multiplied in sequence by ψ_t^a, by i^{ut} and by ψ_u^a, we obtain

$$i_{ts}\theta_b^s\,\bar{i}^{bc} = \psi_t^c ,$$
$$\theta_b^u\,\bar{i}^{bc} = i^{ut}\psi_t^c ,$$
$$\bar{i}^{ac} = i^{ut}\psi_t^c\psi_u^a .$$

Re-labelling indices, we get (9.95).

9.5.1   Tensors

Transformations (1.13), (1.20) and (9.95) are the simplest of all those obtained. They only require that the quantity in the original parameterization and the Jacobian of the reparameterization be known; these ingredients combine in a multilinear form. Quantities such as these are of special interest.

Definition 9.1 The array $T^{S_n}_{R_m}(\theta)$ will be called a tensor with covariant degree m and contravariant degree n, or, more concisely, an (n,m)-tensor, if, under reparameterization, it obeys the transformation rule

$$\bar{T}^{B_n}_{A_m}(\psi) = T^{S_n}_{R_m}(\theta(\psi))\,\theta^{r_1}_{a_1}\ldots\theta^{r_m}_{a_m}\psi^{b_1}_{s_1}\ldots\psi^{b_n}_{s_n}\ . \tag{9.96}$$

Often, for the sake of brevity, we use the term tensor with reference to one individual element of an array rather than with reference to the entire array. According to the definition above, l_r is a $(0,1)$ tensor, i_{rs} is a $(0,2)$ tensor, i^{rs} is a $(2,0)$ tensor and $\hat{\jmath}_{rs}$ is a $(0,2)$ tensor.

Note the analogy in notation for the covariant degree and the cardinality of the lower multiindex and, that between the contravariant degree and cardinality of the upper multiindex. This possibility of indicating the covariant or contravariant nature of a tensor through the notation was the original reason for the convention, according to which the indices of arrays can be placed either as lower or as upper indices. Often, even when an array does not behave like a tensor, the transformation rule of a tensorial component is highlighted, as for example, $H_{rs}\theta^r_a\theta^s_b$ or $H_{rst}\theta^r_a\theta^s_b\theta^t_c$ in (9.93) and (9.94). Usually, the tensorial properties of this component suggest positioning the indices of the corresponding array as either upper or lower indices.

A $(0,0)$ tensor is a parameterization invariant quantity. Below, for simplicity, it will be called an invariant quantity. The following result is fundamental: if T_{R_m} is an $(0,m)$ tensor and U^{S_m} is an $(m,0)$ tensor, then the quantity obtained from the contraction $T_{R_m}U^{R_m}$ is an invariant quantity. Indeed, exploiting (9.96)

$$
\begin{aligned}
\bar{T}_{A_m}\bar{U}^{A_m} &= T_{R_m}\theta^{r_1}_{a_1}\cdots\theta^{r_m}_{a_m}\,U^{\bar{R}_m}\psi^{a_1}_{\bar{r}_1}\cdots\psi^{a_m}_{\bar{r}_m} \\
&= T_{R_m}U^{\bar{R}_m}\,\delta^{r_1}_{\bar{r}_1}\cdots\delta^{r_m}_{\bar{r}_m} \\
&= T_{R_m}U^{R_m}\ .
\end{aligned}
$$

If one defines a quantity through a multiplication of tensors, or through a contraction where, unlike the previous case, not all the indices are dummy indices, one will obtain, as the result, another tensor which has a covariant degree equal to the number of covariant free indices and a contravariant degree equal to the number of contravariant free indices. For example, if T^{rs} is a (2,0) tensor and U_t is a (0,1) tensor, the product $T^{rs} U_t$ is a (2,1) tensor; the contraction $T^{rs} U_s$ is a (1,0) tensor; we obtain an invariant quantity by the contraction $T^{rs} U_r U_s$.

Example 9.13 *Contraction of score and inverse of the information matrix* Since i^{rs} is a (2,0) tensor and l_r is a (0,1) tensor, the quantity $l^r = i^{rs} l_s$, obtained by contraction, is a (1,0) tensor, that is, it satisfies the transformation rule

$$\bar{l}^a = l^r \psi_r^a . \tag{9.97}$$

In addition, the contraction $W_u = i^{rs} l_r l_s$ is an invariant quantity. $\triangle$

The notion of a tensor is particularly useful for checking the parameterization invariance of an asymptotic expansion, truncated at a certain order, of a parameterization invariant combinant. If invariance is not satisfied, there may be two combinants that are asymptotically equivalent to a fixed order but are not, necessarily, both invariant.

Take, for example, the Bartlett correction. If the expansion (9.77) truncated after the term of order $O_p(n^{-1})$ does not define an invariant quantity, then the quantity $B(\theta)$ given by (9.83) and the modified combinant $W^\star(\theta)$ may not be invariant either. Except for the leading term, and possibly the next one, direct checking for the invariance of an asymptotic expansion is usually rather cumbersome. However, if one is able to express the asymptotic expansion of an invariant quantity only in terms of tensors, checking becomes immediate since it is reduced to a question of balance of covariant and contravariant degrees in contractions.

Analogously, if a combinant, whose asymptotic expansion is being considered, behaves like a tensor, then it is natural to demand that the expansion which has been truncated at the desired asymptotic order should also define a tensor of the same order. Tensorial representation of the expansion is, clearly, very useful for checking this.

9.5.2 Invariance of the Expansion of $l(\hat{\theta}) - l(\theta)$

Consider the expansion (9.77) of $l(\hat{\theta}) - l(\theta)$. The leading term $i_{rs} l^r l^s = i^{rs} l_r l_s$ was found to be invariant in Example 9.13. Now, the aim is to examine the behaviour of the term of order $O_p(n^{-1/2})$, that is $(\nu_{rst}l^t + 3H_{rs})l^{rs}$. From (9.91), (9.93) and (9.97) we have

$$
\begin{aligned}
\bar{\nu}_{abc}\bar{l}^{abc} + 3\bar{H}_{ab}\bar{l}^{ab} = {} & \left(\nu_{rst}\theta^r_a\theta^s_b\theta^t_c - i_{rs}\theta^r_{ab}\theta^s_c[3]\right) l^{uvw}\psi^a_u\psi^b_v\psi^c_w \\
& +3\left(H_{rs}\theta^r_a\theta^s_b + l_r\theta^r_{ab}\right) l^{tu}\psi^a_t\psi^b_u \;,
\end{aligned}
$$

from which, by isolating the quantities of interest,

$$
\begin{aligned}
\bar{\nu}_{abc}\bar{l}^{abc} + 3\bar{H}_{ab}\bar{l}^{ab} = {} & \nu_{rst}l^{uvw}\theta^r_a\theta^s_b\theta^t_c\psi^a_u\psi^b_v\psi^c_w + 3H_{rs}l^{tu}\theta^r_a\theta^s_b\psi^a_t\psi^b_u \\
& -i_{rs}l^{tuv}(\theta^r_{ab}\theta^s_c[3])\psi^a_t\psi^b_u\psi^c_v + 3l_r l^{st}\theta^r_{ab}\psi^a_s\psi^b_t \\
= {} & \nu_{rst}l^{rst} + 3H_{rs}l^r l^s + \text{REMAINDER}\,.
\end{aligned}
$$

It is relatively easy to show that REMAINDER $= 0$. To do this, consider that

$$
\begin{aligned}
i_{rs}l^{tuv}(\theta^r_{ab}\theta^s_c & + \theta^r_{ac}\theta^s_b + \theta^r_a\theta^s_{bc})\psi^a_t\psi^b_u\psi^c_v \\
& = i_{rs}l^s l^{tu}\theta^r_{ab}\psi^a_t\psi^b_u + i_{rs}l^s l^{tv}\theta^r_{ac}\psi^a_t\psi^c_v + i_{rs}l^r l^{uv}\theta^s_{bc}\psi^b_u\psi^c_v \\
& = l_r l^{tu}\theta^r_{ab}\psi^a_t\psi^b_u + l_r l^{tv}\theta^r_{ac}\psi^a_t\psi^c_v + l_s l^{uv}\theta^s_{bc}\psi^b_u\psi^c_v \\
& = 3\, l_r l^{st}\theta^r_{ab}\psi^a_s\psi^b_t \;.
\end{aligned}
$$

Analogously, but with more complex calculations, it is possible to show that the term of order $O_p(n^{-1})$ in (9.77) is also invariant.

One explicitly tensorial representation of expansion (9.77) is given by

$$
\begin{aligned}
l(\hat{\theta}) - l(\theta) = {} & \frac{1}{2}i_{rs}l^{rs} \overset{\bullet}{+} \frac{1}{2}U_{rs}l^{rs} + \frac{1}{6}\xi_{rst}l^{rst} \overset{\bullet}{+} \frac{1}{2}T_{rv}i^{vw}T_{ws}l^{rs} + \frac{1}{6}U_{rst}l^{rst} \\
& +\frac{1}{24}\xi_{rstu}l^{rstu} \overset{\bullet}{+} O_p(n^{-3/2})\,,
\end{aligned} \tag{9.98}
$$

where:

$$
\begin{aligned}
T_{rs} &= H_{rs} - \nu_{t,rs}l^t \;, \\
U_{rs} &= H_{rs} - \nu_{t;rs}l^t \;,
\end{aligned}
$$

with $\nu_{t;rs} = \nu_{t,rs} + \nu_{t,r,s}$, are $(0,2)$ tensors;

$$
\begin{aligned}
U_{rst} &= H_{rst} - i^{vw}\nu_{v;rs}H_{wt}[3] - (\nu_{u;rst} - i^{vw}\nu_{v;rs}\nu_{u;tw}[3])l^u \;, \\
\xi_{rst} &= 2\nu_{r,s,t} \;,
\end{aligned}
$$

with

$$
\nu_{r;stu} = \nu_{r,stu} + \nu_{r,s,tu}[3] + \nu_{r,s,t,u}\,,
$$

are $(0,3)$ tensors;

$$\xi_{rstu} = \nu_{rstu} + \nu_{r;stu}[4] + i^{vw}\nu_{v;rs}\nu_{w;tu}[3] - i^{vw}\xi_{rsv}\nu_{w;tu}[6]$$

is a $(0,4)$ tensor. The algebraic check for equality between the terms up order $O_p(n^{-1})$ of expansions (9.77) and (9.98) and for the transformation rules of the quantities which appear in (9.98) is, relatively, direct.

As a consequence of the invariance of the expansion (9.77), the Bartlett correction $b(\theta)$ expressed by (9.83) is invariant too. The corrected combinant $W^\star = W(\theta)/(1 + n^{-1}b(\theta))$ is also invariant. This means that the adequacy of the approximation based on the null asymptotic χ_p^2 distribution of $W^\star$ for given values of n and of θ is an intrinsic property of the statistical model $\mathcal{F}$. For an explicitly tensorial expression of the Bartlett correction, see Barndorff-Nielsen and Cox (1994, formulae (5.33)-(5.37)).

9.5.3 *Tensorial Behaviour of the Expansion of the Expected Value of the Profile Score*

The discussion in the preceding subsections can be extended to the case where there are nuisance parameters present and, thus, one is interested in obtaining asymptotic expansions that obey the principle of invariance under interest-respecting reparameterizations. With reference to expansions of the profile score and its expectation obtained in section 9.4.3, the hope is that, by cutting off the expansion at a fixed order, the quantities obtained will transform under interest-respecting reparameterizations just as does $\tilde{l}_r$.

Let $\psi(\theta) = (\phi, \chi) = (\phi(\tau), \chi(\tau, \zeta))$ be an interest-respecting reparameterization (see section 4.2.4). Adopt the convention, as in section 9.4.3, of using the letters $r, s, \ldots$ as indices for the components of τ and the letters $a, b, \ldots$ for those of ζ. Under reparameterization the profile score function $\tilde{l}_r = l_r(\tau, \hat{\zeta}_\tau)$ transforms according to the rule

$$\bar{\tilde{l}}_{\tilde{r}} = l_r\big(\tau(\phi), \hat{\zeta}_{\tau(\phi)}\big)\tau_{\tilde{r}}^r , \tag{9.99}$$

with $\tau_{\tilde{r}}^r = (\partial/\partial\phi^{\tilde{r}})\tau^r(\phi)$. Indeed,

$$\bar{\tilde{l}}_{\tilde{r}} = l_r\tau_{\tilde{r}}^r + l_a\zeta_{\tilde{r}}^a ,$$

with $\zeta_{\tilde{r}}^a = (\partial/\partial\phi^{\tilde{r}})\zeta^a(\phi, \chi)$, and the second summand is zero when it is evaluated at the partial maximum likelihood estimate. On the basis of (9.99), the

profile score behaves like a $(0,1)$ tensor under interest-respecting reparameterizations. Furthermore, it is easy to check that each of the two summands in
the leading term of (9.90),

$$T_r = \kappa^{ab}(\nu_{r,ab} - i_{rc}\kappa^{cd}\nu_{d,ab}) \qquad (9.100)$$

and

$$U_r = \kappa^{ab}(\nu_{r,a,b} - i_{rc}\kappa^{cd}\nu_{a,b,d}) , \qquad (9.101)$$

behaves like a $(0,1)$ tensor under interest-respecting reparameterizations. For
T_r, check is based on the following transformation rules

$$\bar{\kappa}^{\bar{a}\bar{b}} = \kappa^{ab}\chi^{\bar{a}}_a\chi^{\bar{b}}_b ,$$

where $\chi^{\bar{a}}_a = (\partial/\partial\zeta^a)\chi^{\bar{a}}(\tau,\zeta)$,

$$\bar{i}_{\bar{r}\bar{c}} = (i_{ra}\tau^r_{\bar{r}} + i_{ab}\zeta^b_{\bar{r}})\zeta^a_{\bar{c}} ,$$

$$\bar{\nu}_{\bar{r},\bar{a}\bar{b}} = \nu_{r,ab}\tau^r_{\bar{r}}\zeta^a_{\bar{a}}\zeta^b_{\bar{b}} + i_{ra}\tau^r_{\bar{r}}\zeta^a_{\bar{a}\bar{b}} + \nu_{c,ab}\zeta^c_{\bar{r}}\zeta^a_{\bar{a}}\zeta^b_{\bar{b}} + i_{ab}\zeta^a_{\bar{a}\bar{b}}\zeta^b_{\bar{r}} ,$$

$$\bar{\nu}_{\bar{d},\bar{a}\bar{b}} = \nu_{d,ab}\zeta^a_{\bar{a}}\zeta^b_{\bar{b}}\zeta^d_{\bar{d}} + i_{ad}\zeta^d_{\bar{d}}\zeta^a_{\bar{a}\bar{b}} .$$

If the parameter of interest is a scalar, the integral with respect to τ of $T_r + U_r$,
gives an invariant quantity which may be used in order to define an adjustment
for the profile likelihood whose associated score has an expected value of order
$O(n^{-1})$. This will be examined in section 11.6.

9.6   Bibliographic Note

Index notation is used in applied mathematics, mathematical physics and differential geometry. Although it is useful in multivariate statistical calculations,
it only appeared sporadically before the 1980's (see for example, Bartlett, 1953;
Lawley, 1956). The publication of the monograph by McCullagh (1987) (see,
also, McCullagh, 1984b) greatly encouraged the adoption of this notation. See
section 1.8 of McCullagh (1987) for further references and for the links with
notational conventions in physics and differential geometry.

Bartlett relations for moments of log-likelihood derivatives offer a tool for
simplifying asymptotic calculations for likelihood quantities. Skovgaard (1986)
shows that analogous relations are also satisfied for cumulants of log-likelihood
derivatives. McCullagh (1987, section 7.2.2) gives asymptotic expansions for
non-null cumulants of log-likelihood derivatives. Mykland (1994) shows that
Bartlett relations are also satisfied by martingales, where the higher-order
derivatives are replaced by measures of variation of the process.

Almost all asymptotic statistical calculations are based on the Taylor formula. Here we have confined the discussion to smooth functions with finite-dimensional domain. For extensions of the Taylor formula to functionals, which would be useful, for example, for obtaining expansions such as (3.79) which include higher-order terms too, see von Mises (1947), Fernholz (1983) and Withers (1983).

The Laplace expansion offers a tool for approximating integrals. It is used in Bayesian theory in order to obtain approximations for moments and posterior densities of functions of the parameter. For this, see Tierney and Kadane (1986), Tierney, Kass and Kadane (1989a,b), Wong and Li (1992). Shun and McCullagh (1995) obtain a modified Laplace approximation for multiple integrals on a domain with dimension proportional to the square-root of the asymptotic parameter n. The extension of the approximation of marginal likelihood, considered in Example 9.9, to general composite group families has been obtained by Barndorff-Nielsen and Jupp (1987). Evans and Swartz (1995) survey the major techniques for the approximation of integrals useful in statistics.

The asymptotic expansions presented in section 9.4 and other formulae which are useful for asymptotic calculations are collected in Barndorff-Nielsen, Blæsild, Pace and Salvan (1991). The expansion of the profile score in section 9.4.3 was obtained by McCullagh and Tibshirani (1990); the tensorial expression explained in section 9.5.3 is given in Barndorff-Nielsen (1994). Cordeiro and McCullagh (1991) specialize formula (9.62) to generalized linear models. The expansions considered in section 9.4 are expected/observed likelihood expansions. For a detailed presentation, see Barndorff-Nielsen and Cox (1994, Chapter 5). In particular, Barndorff-Nielsen and Cox (1994, section 5.6) offers an introduction to invariant Taylor expansions, that is, to methods, based on geometric concepts, which automatically allow asymptotic expansions, that fit with the principle of parameterization invariance, to be obtained. One special case is observed likelihood expansions which, whenever an ancillary is available, make it possible to obtain invariant expansions that are useful for approximate evaluations of moments, cumulants, etc., conditionally on the observed value of the ancillary: for one example, see Problem 11.4. Pace and Salvan (1994a) show that when the procedure followed in section 9.4.2, in order to obtain the asymptotic expansion of the log-likelihood ratio, is applied to a generic invariant function $f(\hat{\theta})$, it will usually produce an invariant expansion, at least for the first four asymptotic orders. A general tensorial expression is also given for this expansion.

The complexity of carrying out asymptotic calculations, which are essentially based on the repeated application of the multivariate Taylor formula, makes the use of computer algebra attractive. For this, see Kendall (1993) and Andrews and Stafford (1993).

For further study of the results of higher-order efficiency of the maximum likelihood estimator see, for example, the monographs by Akahira and Takeuchi (1981), Bhattacharya and Denker (1990), Ghosh (1994).

9.7 Problems

9.1 Interpret the relation $c_{R_m}^{S_n} = a_{R_m}^{T_i} b_{T_i}^{S_n}$ in the light of the summation convention.

[Section 9.1]

9.2 Show that the number $N_{m,h}$ of ways in which it is possible to subdivide a set of m objects in h non-empty subsets is given by relations (9.14) and (9.15).

[Section 9.1]

9.3 For $i, j = 1, \ldots, n$, let $u_{ij} = 1$ and let $\delta_{ij} = 1$ if $i = j$, otherwise $\delta_{ij} = 0$. Consider the covariance matrix $\Sigma = [\sigma_{ij}]$, where $\sigma_{ij} = \delta_{ij} + cu_{ij}$, for c a suitable scalar. This corresponds to equicorrelated observations. Show that the matrix inverse of Σ has elements $\sigma^{ij} = \delta_{ij} + bu_{ij}$, where $b = -c/(1+nc)$, $c \neq -1/n$. Use this result to show that $\det \Sigma = 1 + nc$ and, therefore, that Σ is a covariance matrix if $c > -1/n$. (*Hint*: remember that $\det(\Sigma) \det(\Sigma^{-1}) = 1$.)

[Section 9.1]

9.4 Interpret and prove the fourth balance relation (9.24).

[Section 9.2.1]

9.5 Write the relation (9.24) specializing it for $\Theta \subseteq \mathbb{R}$, using power notation.

[Section 9.2.1]

9.6 Balance relations for null moments of log-likelihood derivatives have been given in section 9.2.1 up to the fourth order. Suggest the form of the k-th relation in the multiparameter case.

[Section 9.2.1; Barndorff-Nielsen and Cox, 1994, section 5.2]

9.7 Obtain the asymptotic order under random sampling of size n of the quantities: $l(\hat{\theta}) - l(\theta)$; $i^{rs} l_r$; $l_t l_s + i^{rs} l_{rt}$; $i^{rs} i^{tu}(l_{su} + i_{su})$.

[Section 9.2.2]

9.8 Obtain the asymptotic order under random sampling of size n of the quantities: $\nu_{rs,tuv}$; $\nu_{r,s,tu}$; $E_\theta\{l_r(l_{st}+i_{st})l_{uvw}\}$; $i^{rs}\nu_{s,tu}\nu_{r,vw}$.

[Section 9.2.2]

9.9 Prove formula (9.44) for the coefficient $B^a_{rs}, B^a_{rst}, B^a_{rstu}$ of the inversion formula (9.43).

[Section 9.3.2]

9.10 Check that the Laplace expansion (9.46) has error of order $O(n^{-2})$.

[Section 9.3.3]

9.11 Consider the Bessel function (2.8), $I_0(\lambda)$, which appears as a normalizing constant of the von Mises distribution in Example 2.14. For

$$I_0(\lambda) = \frac{1}{2\pi} \int_{-\pi}^{\pi} e^{\lambda \cos t} dt$$

obtain the asymptotic expansion

$$I_0(\lambda) = \frac{e^\lambda}{\sqrt{2\pi\lambda}} \left\{ 1 + \frac{1}{8\lambda} + O(\lambda^{-2}) \right\} .$$

[Section 9.3.3]

9.12 Apply expansion (9.51) to the case $p = 1$ and check that the Laplace approximation with error of order $O(n^{-2})$ respects the identity

$$\int_{\mathbb{R}} e^{-ng(y)}b(y)dy = \int_{\mathbb{R}} e^{-ng(y(w))}b(y(w))|y'(w)|dy ,$$

where $y(w)$ is a real function defined on $\mathbb{R}$ which is one-to-one and differentiable together with its inverse.

[Section 9.3.3; Pace and Salvan, 1994b]

9.13 Consider an integral of the form

$$I(n) = \int_0^{+\infty} \exp\{-ng(y)\}b(y)\, dy ,$$

where $g(\cdot)$ is a smooth function from $\mathbb{R}^+$ to $\mathbb{R}$ with finite absolute minimum $\tilde{g}$ at the boundary point $y = 0$ and with right-hand derivatives $\tilde{g}'_+, \tilde{g}''_+, \ldots$, at $y = 0$, finite and with $\tilde{g}'_+ \neq 0$. Assume that $b(y)$ is a real function defined on

$\mathbb{R}^+$, such that $\tilde{b} = b(0) \neq 0$ and with finite right-hand derivative $\tilde{b}'_+$ at $y = 0$. Obtain the asymptotic expansion

$$I(n) = \frac{\exp\{-n\tilde{g}\}\tilde{b}}{n\tilde{g}'_+}\left\{1 + \frac{1}{n}\left(\frac{\tilde{b}'_+}{\tilde{b}}\frac{1}{\tilde{g}'_+} - \frac{\tilde{g}''_+}{(\tilde{g}'_+)^2}\right) + O(n^{-2})\right\}.$$

[Section 9.3.3]

9.14 Consider the approximation of marginal likelihood obtained in Example 9.9 and discuss its application to the problem of testing for separate scale and location families (see section 7.7). Obtain an approximation for the most powerful invariant test based on n independent observations of the r.v. Y, for $H_0 : Y \sim N(\mu, \sigma^2)$ versus $H_1 : Y \sim EV(\mu, \sigma)$.

[Section 9.3.3; Ventura, 1994]

9.15 Complete the proof of (9.61), obtaining the term of order $O_p(n^{-3/2})$.

[Section 9.4.1]

9.16 Let $\mathcal{F}$ be a parametric statistical model with p-dimensional parameter θ and data y. Let $q(y; \theta)$ be a combinant of order $O_p(n^{1/2})$ when θ is the true value of the parameter and of order $O(n)$ otherwise. Obtain a stochastic expansion for $(\tilde{\theta} - \theta)^r$, where $\tilde{\theta}$ is the unique solution to the estimating equation $q(y; \theta) = 0$.

[Section 9.4.1]

9.17 Let $\psi(\theta)$ be a one-to-one smooth function $\psi : \Theta \to \Psi \subseteq \mathbb{R}^p$. Obtain an expansion for $E_\theta(\psi(\hat{\theta}))$ up to and including terms of order $O(n^{-1})$, both when $p = 1$ and in the multiparameter case $p > 1$.

[Section 9.4.2]

9.18 Let Y_i, $i = 1, \ldots, n$ be independent r.v.'s with a $N(\mu, \sigma^2)$ distribution. Obtain the correction of order $O(n^{-1})$ for the bias of the maximum likelihood estimator of σ^2, $\hat{\sigma}^2 = n^{-1}\sum_{i=1}^{n}(Y_i - \bar{Y})^2$.

[Section 9.4.2]

9.19 Obtain an expression for the bias of $\hat{\theta}$ with error of order $O(n^{-2})$ under random sampling of size n from a distribution in a regular exponential family of order 2 with density

$$p(y; \theta) = \exp\{\theta^1 y_1 + \theta^2 y_2 - K(\theta^1, \theta^2)\} h(y). \tag{9.102}$$

Apply the result to the case where (9.102) is the density of the natural observations of the model induced by n independent observations Y_i with density

$$p_{Y_i}(y; \theta) = (\theta^1 + \theta^2 z_i)e^{-(\theta^1+\theta^2 z_i)y}, \qquad y > 0,$$

where (θ^1, θ^2) belongs to the natural parameter space, and the z_i are positive and not all equal, known constants.

[Section 9.4.2]

9.20 Apply expansions (9.73)-(9.75) to the one-parameter case and check that (9.73) gives (9.70).

[Section 9.4.3]

9.21 Adapt expansions (9.74) and (9.75) to a natural exponential family of order p.

[Section 9.4.3]

9.22 Adapt (9.77) to the one-parameter case and calculate, directly, its expectation up to and including the terms of order $O(n^{-1})$. Compare the result obtained with the relation (84) on p. 339 of Cox and Hinkley (1974). (*Hint*: for the definition of the quantities in this formula see *ibidem* p. 309.)

[Sections 9.4.4 and 9.4.5]

9.23 Under random sampling of size n from a regular one-parameter model with parameterization θ, consider the three combinants $(\hat{\theta} - \theta)i(\theta)^{1/2}$, $(\hat{\theta} - \theta)i(\hat{\theta})^{1/2}$, and $l_*(\theta)i(\hat{\theta})^{-1/2}$. Show that for all three the null expected value and the null third standardized cumulants are proportional to

$$i(\theta)^{-3/2} \left(\nu_3(\theta)(1 + c) + 3c\nu_{1,2}(\theta) \right) + O(n^{-3/2}), \qquad (9.103)$$

where power notation has been used and c is a constant that can take the values -1, 0, $1/2$, 1 and 2. Show that in the parameterization $\psi = \psi(\theta)$, with

$$\psi(\theta) = \alpha + \beta \int_{\theta_1}^{\theta} \exp\left\{ -\frac{1}{3} \int_{\theta_2}^{t} h(u)\, du \right\} dt,$$

where α and β are arbitrary constants, θ_1 and θ_2 are arbitrarily chosen values in the parameter space and

$$h(\theta) = i(\theta)^{-1} \left(\nu_3(\theta)(1 + c) + 3c\nu_{1,2}(\theta) \right),$$

the leading term of (9.103) is zero.

[Section 9.4; DiCiccio, 1984]

9.24 Show that the quantity U_r defined by equation (9.101) behaves like a (0,1) tensor under interest-respecting reparameterizations.

[Section 9.5.3]

CHAPTER 10

ASYMPTOTIC EXPANSIONS FOR DISTRIBUTIONS

10.1 Generating Functions for a Standardized Sum

Let Y be a one-dimensional random variable with cumulant generating function $K_Y(t)$ and let $Y_1, \ldots, Y_n$ be independent copies of Y. Let $S_n = \sum_{i=1}^n Y_i$, and let $S_n^* = (S_n - n\mu)/\sqrt{n\sigma^2}$ be the standardized sum, where, as usual, $\mu = \kappa_1 = E(Y)$ and $\sigma^2 = \kappa_2 = \text{Var}(Y)$. Because of standardization, $\kappa_1(S_n^*) = 0$ and $\kappa_2(S_n^*) = 1$. Furthermore, cumulants of order higher than two are unaffected by location transformations. From the relations (3.19) for scale transformations and (3.12) for sums of independent random variables, for $r = 3, 4, \ldots$, we obtain

$$\kappa_r(S_n^*) = \kappa_r \left(\frac{S_n}{\sqrt{n\sigma^2}} \right) = \frac{\kappa_r(S_n)}{\left(\sqrt{n\sigma^2} \right)^r} = \frac{n\kappa_r}{n^{r/2}\sigma^r} = n^{1-r/2}\rho_r , \qquad (10.1)$$

where κ_r and ρ_r are, respectively, the cumulants and the standardized cumulants of Y. Because $\kappa_2(S_n^*) = 1$, for $r = 3, 4 \ldots$ under random sampling of size n

$$\rho_r(S_n^*) = \kappa_r(S_n^*) = O\left(n^{-\frac{r}{2}+1} \right) .$$

As n diverges, the cumulants of S_n^* draw closer to those of the $N(0,1)$ distribution.

For fixed t, as n varies, the expansion

$$K_{S_n^*}(t) = \kappa_1(S_n^*)t + \kappa_2(S_n^*)\frac{t^2}{2!} + \kappa_3(S_n^*)\frac{t^3}{3!} + \kappa_4(S_n^*)\frac{t^4}{4!} + O(n^{-3/2})$$

holds. Therefore

$$K_{S_n^*}(t) = \frac{1}{2}t^2 + \frac{\rho_3}{6\sqrt{n}}t^3 + \frac{\rho_4}{24n}t^4 + O(n^{-3/2}) . \qquad (10.2)$$

381

For t fixed as n diverges, the moment generating function of S_n^*. has the asymptotic expansion

$$M_{S_n^*}(t) = e^{\frac{1}{2}t^2} \left\{ 1 + \frac{\rho_3}{6\sqrt{n}}t^3 + \frac{\rho_4}{24n}t^4 + \frac{\rho_3^2}{72n}t^6 + O(n^{-3/2}) \right\}. \qquad (10.3)$$

Indeed, from (10.2)

$$M_{S_n^*}(t) = \exp\left\{K_{S_n^*}(t)\right\} = e^{\frac{1}{2}t^2} \exp\left\{ \frac{1}{6\sqrt{n}}\rho_3 t^3 + \frac{1}{24n}\rho_4 t^4 + O(n^{-3/2}) \right\}$$

and, since $e^x = 1 + x + x^2/2 + O(x^3)$, we can immediately see that $x^2/2$ gives only one term which is not of order $O(n^{-3/2})$, that is, $\left(\rho_3 t^3/(6\sqrt{n})\right)^2/2$. As n diverges for fixed t, (10.3) offers the usual elementary justification of the central limit theorem. Furthermore, it explicitly gives correction terms of order $O(n^{-1/2})$ and $O(n^{-1})$ for the moment generating function of the $N(0,1)$ distribution.

We could consider obtaining an inversion of (10.3), which would express the density of the standardized sum, $p_{S_n^*}(\cdot)$, according to an expansion which has the $N(0,1)$ density as its leading term and explicit correction terms of order $O(n^{-1/2})$ and $O(n^{-1})$. This inversion can be carried out thanks to the properties of Hermite polynomials, which will be introduced in the section below; this leads to the Edgeworth expansion which will be presented in section 10.3.

10.2   Hermite Polynomials

For $r = 0, 1, \ldots$, consider the sequence of functions $H_r(y)$ defined by

$$H_r(y)\phi(y) = (-1)^r \frac{d^r}{dy^r}\phi(y), \qquad (10.4)$$

where $\phi(y)$ is the standard normal density and the zero-order derivative of a function is taken to be the function itself. For $r = 0$

$$H_0(y) = 1. \qquad (10.5)$$

Furthermore, by differentiating both sides of (10.4), we obtain

$$\begin{aligned}
\frac{d^{r+1}}{dy^{r+1}}\phi(y) &= (-1)^r \left\{ H_r'(y)\phi(y) + H_r(y)\phi'(y) \right\} \\
&= (-1)^r \left\{ H_r'(y) - yH_r(y) \right\} \phi(y) \\
&= (-1)^{r+1} \left\{ yH_r(y) - H_r'(y) \right\} \phi(y),
\end{aligned}$$

from which we have

$$(-1)^{r+1} \frac{d^{r+1}}{dy^{r+1}} \phi(y) \;=\; \{yH_r(y) - H'_r(y)\}\, \phi(y)$$
$$=\; H_{r+1}(y)\phi(y),$$

still using (10.4). Hence, the functions $H_r(y)$ satisfy the recurrence relationship

$$H_{r+1}(y) = yH_r(y) - H'_r(y)\,. \tag{10.6}$$

Because $H_0(y) = 1$ and $H'_0(y) = 0$, from (10.6) we can see, by induction on r, that $H_r(y)$ is a polynomial of degree r: this is called the Hermite polynomial of degree r. For the polynomials $H_r(y)$ up to the sixth degree, we obtain, recursively from (10.5) and (10.6), the expressions:

$$
\begin{aligned}
H_1(y) &= y \\
H_2(y) &= y^2 - 1 \\
H_3(y) &= y(y^2 - 1) - 2y = y^3 - 3y \\
H_4(y) &= y(y^3 - 3y) - (3y^2 - 3) = y^4 - 6y^2 + 3 \\
H_5(y) &= yH_4(y) - H'_4(y) = y^5 - 10y^3 + 15y \\
H_6(y) &= yH_5(y) - H'_5(y) = y^6 - 15y^4 + 45y^2 - 15\,.
\end{aligned}
$$

Note the following important regularities:
i) the coefficient of y^r in $H_r(\cdot)$ is 1;
ii) the even degree polynomials are even functions, the odd degree polynomials are odd functions, so that, in particular, $H_{2r+1}(0) = 0$.
    One immediate consequence of the definition is that the function $H_r(y)\phi(y)$ has indefinite integral $-H_{r-1}(y)\phi(y)$ for $r \geq 1$; in fact

$$- H_{r-1}(y)\phi(y) = (-1)^r \frac{d^{r-1}}{dy^{r-1}} \phi(y)$$

$$= (-1)^r \int \left(\frac{d^r}{dy^r} \phi(y) \right) dy = \int H_r(y)\phi(y)dy\,. \tag{10.7}$$

Using (10.7) it is possible to show that Hermite polynomials are a sequence of orthogonal polynomials with respect to the scalar product with weight function $\phi(y)$. More precisely, for $r, s = 0, 1, \ldots, r \neq s$, we have

$$\int_{-\infty}^{+\infty} H_r(y)H_s(y)\phi(y)dy = 0\,. \tag{10.8}$$

The key result for the inversion of expansion (10.3) for $M_{S_n^*}(t)$ is the identity

$$\int_{-\infty}^{+\infty} e^{ty} H_r(y)\phi(y)dy = t^r e^{\frac{1}{2}t^2},$$
(10.9)

for $r \in \mathbb{N}$. Relation (10.9) can be proved by induction on r. Clearly, (10.9) holds for $r = 0$. Assume that (10.9) is satisfied for $r - 1$. Using (10.4), and integrating by parts, we have

$$\begin{aligned}
\int_{-\infty}^{+\infty} e^{ty} H_r(y)\phi(y)dy &= (-1)^r \int_{-\infty}^{+\infty} e^{ty} \left(\frac{d^r}{dy^r}\phi(y)\right) dy \\
&= -(-1)^r \int_{-\infty}^{+\infty} te^{ty} \left(\frac{d^{r-1}}{dy^{r-1}}\phi(y)\right) dy \\
&= \int_{-\infty}^{+\infty} te^{ty}(-1)^{r-1} \left(\frac{d^{r-1}}{dy^{r-1}}\phi(y)\right) dy \\
&= t \int_{-\infty}^{+\infty} e^{ty} H_{r-1}(y)\phi(y)dy = t^r e^{\frac{1}{2}t^2}.
\end{aligned}$$

10.3   Edgeworth Expansion for Density Functions

Relation (10.9) allows the asymptotic expansion (10.3) of $M_{S_n^*}(t)$ to be written as

$$\begin{aligned}
M_{S_n^*}(t) &= \int_{-\infty}^{+\infty} e^{ty} \left\{ 1 + \frac{1}{6\sqrt{n}}\rho_3 H_3(y) + \frac{1}{24n}\rho_4 H_4(y) \right. \\
&\qquad \left. + \frac{1}{72n}\rho_3^2 H_6(y) + O(n^{-3/2}) \right\} \phi(y)dy.
\end{aligned}$$

Comparison with the definition

$$M_{S_n^*}(t) = \int_{-\infty}^{+\infty} e^{ty} p_{S_n^*}(y)dy,$$

where $p_{S_n^*}(y)$ is the density of S_n^*, would suggest the approximation

$$p_{S_n^*}(y) = \overset{E}{p}_{S_n^*}(y) + O(n^{-3/2}),$$
(10.10)

with

$$\overset{E}{p}_{S_n^*}(y) = \phi(y)\left\{ 1 + \frac{1}{6\sqrt{n}}\rho_3 H_3(y) + \frac{1}{24n}\rho_4 H_4(y) + \frac{1}{72n}\rho_3^2 H_6(y) \right\}.$$
(10.11)

Formula (10.10) is called the **Edgeworth expansion**. It agrees with the first-order local limit result established by Theorem A.6. The validity of expansion (10.10), of which only a heuristic justification has been offered here, can be rigorously proved. Specifically, if Y has finite moments up to the fifth order, then the validity of (10.10) can be shown:

i) in the continuous case (Feller, 1971, section 16.2; Jensen, 1995a, section 1.5), with the additional assumption that there exists a value $\bar{n} \in \mathbb{N} \setminus \{0\}$ such that the density $p_{S_{\bar{n}}^*}(y)$ is continuous and bounded (or, equivalently, such that the characteristic function of S_n^* is absolutely integrable) for every $n > \bar{n}$;

ii) in the discrete case, for non-degenerate **lattice variables**, that is, those which have support

$$S_Y = \{a + hj; \quad a \in \mathbb{R}, h > 0, j \in \mathcal{J} \subseteq \mathbb{Z}\}, \qquad (10.12)$$

if the characteristic function of Y satisfies the condition

$$\sup_{\varepsilon < |t| < \pi} |C_Y(t)| < 1$$

for every $\varepsilon > 0$; however, as in the local limit theorem, (10.11) should be considered for y values which belong to the support of S_n^* and $p_{S_n^*}^{E}(y)$ gives an approximation for $(\sigma\sqrt{n}/h)P(S_n^* = y)$ (see Jensen, 1995a, section 1.5).

In both cases, it can be shown that the absolute error in the approximation can be bounded by a constant which is independent of y. In other words, the absolute error is of order $O(n^{-3/2})$ uniformly in y.

It is important to stress that the error, kept under control in (10.10), is absolute; hence we would expect there to be degradation in the approximation for values of y in the tails, where $p_{S^*}(y)$ becomes comparable to the error. The leading term of the relative error is a linear combination of Hermite polynomials; with fixed n it could explode as y diverges.

Example 10.1 *Gamma distribution (cont.)*
Under the same assumptions as in Example 8.3, we wish to find the Edgeworth expansion for the density of the standardized variable Z_ν. For $Y \sim Ga(\nu, 1)$, $\rho_3(Y) = 2/\sqrt{\nu}$, $\rho_4(Y) = 6/\nu$. Hence,

$$p_{Z_\nu}^{E}(z) = \phi(z)\left\{1 + \frac{1}{3\sqrt{\nu}}H_3(z) + \frac{1}{4\nu}H_4(z) + \frac{1}{18\nu}H_6(z)\right\}.$$

Figure 10.1 compares the density functions of Z_ν, for $\nu = 5, 10$, with $\phi(z)$ and $p_{Z_\nu}^E(z)$. Passing from the normal approximation to the Edgeworth approximation offers a marked improvement at the centre of the distribution. However, the approximation is qualitatively unsatisfactory in the lower tail where it gives negative values. Figures 10.2 and 10.3 illustrate the behaviour of absolute and relative errors. Note that while the absolute error remains bounded, the relative error diverges in the tails. $\triangle$

For the discussion below, it is useful to consider the Edgeworth expansion up to, and including, terms of order $O(n^{-3/2})$. This can be obtained through steps analogous to those which give (10.10). If we assume, among the regularity conditions, that moments of Y exist up to the sixth order, then

$$p_{S_n^*}(y) = p_{S_n^*}^E(y) + \frac{1}{n^{3/2}}\phi(y)\left\{\frac{1}{120}\rho_5 H_5(y)\right.$$

$$\left. + \frac{1}{144}\rho_3\rho_4 H_7(y) + \frac{1}{1296}\rho_3^3 H_9(y)\right\} + O(n^{-2}). \quad (10.13)$$

10.4 Edgeworth Expansion for Distribution Functions

In the continuous case formula (10.7) allows us to explicitly integrate the Edgeworth expansion (10.10) for the density of S_n^*. This gives the Edgeworth expansion for the distribution function of S_n^*,

$$F_{S_n^*}(y) = F_{S_n^*}^E(y) + O(n^{-3/2}), \quad (10.14)$$

where

$$F_{S_n^*}^E(y) = \Phi(y) - \phi(y)\left\{\frac{1}{6\sqrt{n}}\rho_3 H_2(y) + \frac{1}{24n}\rho_4 H_3(y) + \frac{1}{72n}\rho_3^2 H_5(y)\right\}.$$

Note that the term of order $O(n^{-1/2})$ in $F_{S_n^*}^E(y)$ is an even function of y. This justifies expansion (8.2) and, consequently, the fact that the error in the chi-squared approximation of the distribution of $(S_n^*)^2$ is of order $O(n^{-1})$. By integrating (10.13), we can immediately see that the term of order $O(n^{-3/2})$ in (10.14) is also an even function of y. Consequently, the asymptotic expansion for the distribution function of $(S_n^*)^2$ has the form

$$F_{(S_n^*)^2}(t) = \Phi(\sqrt{t}) - \Phi(-\sqrt{t}) + \frac{g(\sqrt{t})}{n} + O(n^{-2}), \quad (10.15)$$

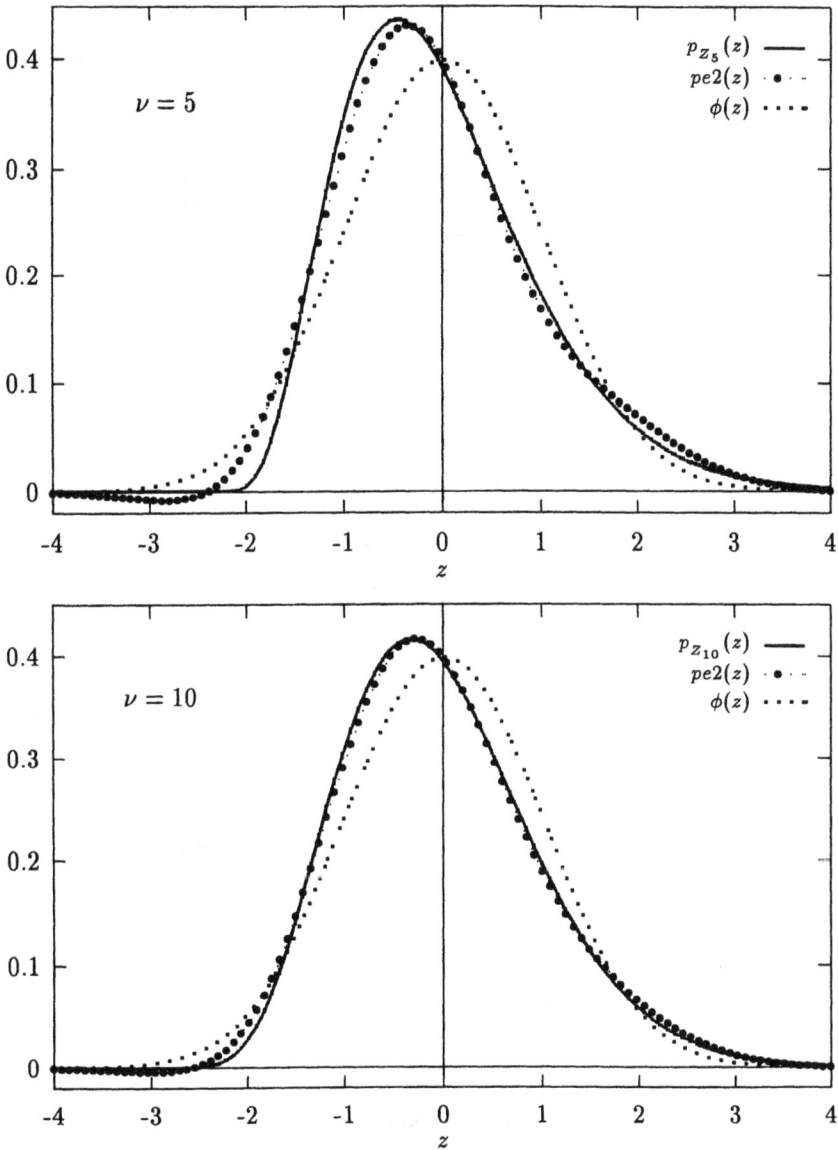

Figure 10.1: *Exact density, $p_{Z_\nu}(z)$, of the standardized sum Z_ν of ν i.i.d. exponential r.v.'s, $\phi(z)$ and $pe2(z) = p_{Z_\nu}^E(z)$; $\nu = 5, 10$.*

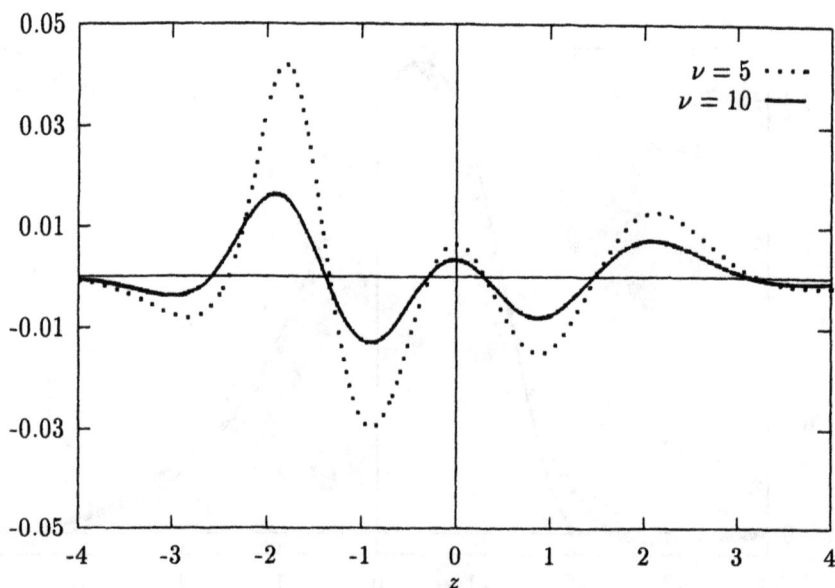

Figure 10.2: *Standardized sum of ν i.i.d. exponential r.v.'s: absolute error $p_{Z_\nu}^E(z) - p_{Z_\nu}(z)$.*

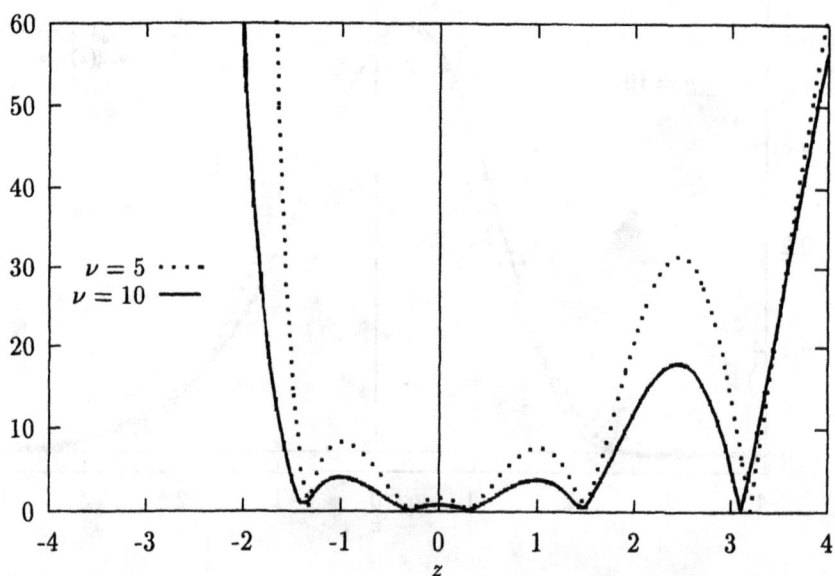

Figure 10.3: *Standardized sum of ν i.i.d. exponential r.v.'s: absolute value of the percentage relative error, $100|p_{Z_\nu}^E(z) - p_{Z_\nu}(z)|/p_{Z_\nu}(z)$.*

where the expression of the function $g(\cdot)$ can immediately be deduced from the expression of $F^{E}_{S^*_n}(y)$ in (10.14).

Expansion (10.14) for the distribution function of S^*_n holds when Y has a continuous distribution. If, instead, Y is a discrete variable, the distribution function of S^*_n will be discontinuous with jumps of order $O(n^{-1/2})$; since $F^{E}_{S^*_n}(y)$ is continuous, the error in (10.14) is usually of order $O(n^{-1/2})$. However, for lattice variables with support (10.12), it is possible to introduce corrections which, at specific points, maintain the order $O(n^{-3/2})$ of the error. In particular, Kolassa and McCullagh (1990) obtained the following fairly simple correction. Let y be a point in the support of S^*_n, and let

$$y^+ = y + \frac{h}{2\sigma\sqrt{n}}$$

be y with a continuity correction; furthermore, let

$$c = 1 + \frac{h^2}{24n\sigma^2}.$$

Then,

$$F_{S^*_n}(y) = F^{E}_{S^*_n}(cy^+) + O(n^{-3/2}). \tag{10.16}$$

Example 10.2 *Binomial distribution (cont.)*
Under the same assumptions as in Example 8.5, with $Y \sim Bi(10, 0.4)$, we wish to approximate $F_Y(y)$ with $F^{E}_{S^*_n}(cy^+)$ (see (10.16)), for $y = 0, 1, \ldots, 10$. We have $y^+ = (y + 0.5 - 4)/\sqrt{2.4}$ and $c = 1 + 1/57.6$. The standardized cumulants are $\rho_3 = 0.2/\sqrt{2.4}$ and $\rho_4 = -0.44/2.4$. In Table 10.1 the exact values of the d.f. of Y are compared with those approximated according to the Edgeworth expansion (10.16) with errors of order $O(n^{-1/2})$, $O(n^{-1})$ and $O(n^{-3/2})$. In the table these approximations are indicated by E0, E1, E2, respectively; E0 is the normal approximation with continuity correction, see Table 8.2. There is an overall improvement in the approximation from E0 to E1 and from E1 to E2. In this example the normal approximation is already fairly satisfactory; often, when the normal approximation is already qualitatively satisfactory, the Edgeworth expansion will give particularly accurate approximations. $\triangle$

Example 10.3 *Poisson distribution*
Because of the closure under convolution of the Poisson family, the Edgeworth expansion (10.16) offers an asymptotic approximation for the d.f. of $Y \sim P(\theta)$ for large values of θ. Using the same notation as in Example 10.2, Tables 10.2

Table 10.1: *Approximations of the d.f. of* $Y \sim Bi(10, 0.4)$.

y	$F_Y(y)$	E0	E1	E2
0	0.006047	0.011934	0.008148	0.006858
1	0.046357	0.053291	0.046538	0.046750
2	0.167290	0.166461	0.162455	0.166555
3	0.382281	0.373443	0.378579	0.381722
4	0.633104	0.626557	0.635934	0.632791
5	0.833762	0.833539	0.837858	0.833758
6	0.945239	0.946709	0.945899	0.945687
7	0.987706	0.988066	0.986613	0.987902
8	0.998323	0.998162	0.997595	0.998237
9	0.999896	0.999808	0.999696	0.999832
10	1.00000	0.999986	0.999974	0.999988

and 10.3 show the values of the exact d.f. and the Edgeworth approximations with various orders of error for $\theta = 1$ and for $\theta = 4$. For $y = 0, 1, \ldots$, $y^+ = (y+0.5-\theta)/\sqrt{\theta}$. Furthermore, $c = 1+1/(24\theta)$ and the standardized cumulants are $\rho_3 = \theta^{-1/2}$, $\rho_4 = \theta^{-1}$. The approximation E2 is quite accurate, even for $P(1)$, which is strongly asymmetric. $\triangle$

At the centre of the distribution, for $y = 0$, that is, correspondingly to $E(S_n^*)$, the Edgeworth expansions for density and distribution functions show important simplifications. Since $H_2(0) = -1$, $H_4(0) = 3$, $H_6(0) = -15$, we have

$$p_{S_n^*}^{E}(0) = \frac{1}{\sqrt{2\pi}}\left\{1 + \frac{1}{24n}(3\rho_4 - 5\rho_3^2)\right\},$$

$$F_{S_n^*}^{E}(0) = \frac{1}{2} + \frac{1}{\sqrt{2\pi}}\frac{1}{6\sqrt{n}}\rho_3.$$

At $y = 0$, the normal approximation for the density of S_n^* has error of order $O(n^{-1})$. This is also true for $y = \pm\sqrt{3}$. From (10.13), the Edgeworth expansion for the density has error of order $O(n^{-2})$ at $y = 0$: the error could be of order $O(n^{-2})$ also at other points whose position depends on the standardized cumulants up to the fifth order. Thus we would expect there to be oscillation for the absolute error, analogous to that shown in Figure 10.2. Similar ob-

Table 10.2: *Approximations of the d.f. of $Y \sim P(1)$.*

y	$F_Y(y)$	E0	E1	E2
0	0.367879	0.308537	0.343549	0.354069
1	0.735759	0.691463	0.741067	0.730547
2	0.919699	0.933193	0.912640	0.925742
3	0.981012	0.993791	0.982448	0.980251
4	0.996340	0.999767	0.998805	0.996453
5	0.999406	0.999996	0.999975	0.999828
6	0.999917	1.00000	1.000000	0.999998

Table 10.3: *Approximations of the d.f. of $Y \sim P(4)$.*

y	$F_Y(y)$	E0	E1	E2
0	0.018316	0.040059	0.023704	0.020569
1	0.091578	0.105650	0.094378	0.092312
2	0.238103	0.226627	0.234902	0.236722
3	0.433470	0.401294	0.430434	0.432317
4	0.628837	0.598706	0.629860	0.627976
5	0.785130	0.773373	0.786339	0.784520
6	0.889326	0.894350	0.887797	0.889863
7	0.948866	0.959941	0.946681	0.949816
8	0.978637	0.987776	0.978044	0.978867
9	0.991868	0.997020	0.992566	0.991628
10	0.997160	0.999423	0.998007	0.997038
11	0.999085	0.999912	0.999586	0.999139
12	0.999726	0.999989	0.999934	0.999806

servations can be made with reference to approximations for the distribution function.

The introduction of further terms into the expansion (10.3) of the moment generating function of S_n^* allows us to obtain Edgeworth expansions with errors of order $O(n^{-5/2})$, $O(n^{-3})$, etc., under the same regularity conditions, provided that Y admits moments up to the sixth, seventh order, etc.. Usually this possibility is ignored, firstly because non-normality is usually adequately described in terms of skewness and kurtosis; secondly, the introduction of polynomials of a higher degree could accentuate possible anomalies in tail behaviour; lastly, even when Y admits moments of every order, convergence of the sequence of Edgeworth expansions to $F_{S_n^*}(y)$ is only guaranteed under very restrictive conditions (see Feller, 1971, section 16.4).

10.5   Anomalies in Edgeworth Approximations

So far we have mainly discussed the Edgeworth expansion in terms of an asymptotic approximation for fixed y as n diverges. However, it is clear that, in statistical applications, the divergence of n is entirely hypothetical. What is of interest here, is the adequacy of the approximation for a given n as y varies. From this point of view the Edgeworth expansion is somewhat unsatisfactory in that $p_{S_n^*}^E(y)$ in (10.10) is not necessarily a density because it may be negative in some regions (see Figure 10.1). Consequently, $F_{S_n^*}^E(y)$ given by (10.14) is not necessarily a distribution function, as it may not be monotonic and may take values from outside the interval $[0, 1]$. However, it is true that $\lim_{y \to -\infty} F_{S_n^*}^E(y) = 0$ and $\lim_{y \to +\infty} F_{S_n^*}^E(y) = 1$. For fixed values of ρ_3 and ρ_4, the function $p_{S_n^*}^E(y)$ could be non-negative for every y for values of n which are sufficiently large, $n \geq \bar{n}(\rho_3, \rho_4)$. One numerical study of the region of values of $\rho_3/\sqrt{n}, \rho_4/n$ which ensure the positivity of $p_{S_n^*}^E(y)$ for each y can be found in Draper and Tierney (1972). In particular, $\bar{n}(\rho_3, \rho_4)$ depends on ρ_3 only through $|\rho_3|$, because ρ_3 is the coefficient of the odd function $H_3(y)$, while the other terms in the expansion are all even functions; furthermore, we cannot have $p_{S_n^*}^E(y) \geq 0$ for each y if $\rho_4 < 0$ or if $\rho_4 > 4n$, which is easy to verify for $\rho_3 = 0$.

Example 10.4 *Gamma distribution (cont.)*
Under the assumptions of Example 10.1, the Edgeworth approximation for the

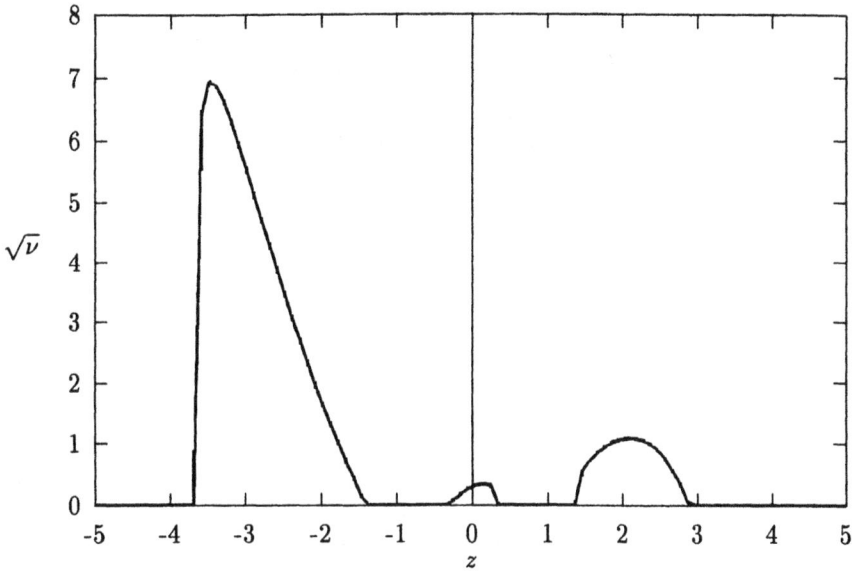

Figure 10.4: *Minimum value of $\nu^{1/2}$ such that $p^E_{Z_\nu}(z) > 0$, as a function of z.*

density of the standardized sum is positive if

$$1 + \frac{1}{3\sqrt{\nu}}H_3(z) + \frac{1}{4\nu}H_4(z) + \frac{1}{18\nu}H_6(z) > 0 \, ,$$

that is, if

$$g(z, \nu) = 2z^6 - 21z^4 + 12\sqrt{\nu}z^3 + 36z^2 - 36\sqrt{\nu}z + 36\nu - 3 > 0 \, .$$

For fixed z, $g(z, \nu)$ is a quadratic function of $\sqrt{\nu}$, positive for values of $\sqrt{\nu}$ greater than the value $\sqrt{\nu^*(z)}$ which represents the maximum between zero and the largest of the two real roots of the equation $g(z, \nu) = 0$ (if there are no real roots, then $g(z, \nu) > 0$ for every ν). Hence, the function $\nu^*(z)$ is the minimum value of the shape parameter ν where the Edgeworth approximation evaluated at z is positive. Figure 10.4 gives the graph of $\sqrt{\nu^*(z)}$. The maximum of $\nu^*(z)$ with respect to z is about 49; for $\nu > 49$, the Edgeworth approximation for the density of Z_ν is positive for every z. △

Example 10.5 *Laplace distribution*
Let $\bar{Y}_n$ be the sample mean of n i.i.d. copies of a r.v. $Y \sim La(\mu, \sigma)$ (see Table

3.1). The cumulant generating function of Y is

$$K_Y(t) = t\mu - \log(1 - (\sigma t)^2), \quad |t| < \sigma^{-1},$$

from which $\kappa_1 = \mu$, $\kappa_2 = 2\sigma^2$, $\kappa_3 = 0$, $\kappa_4 = 12\sigma^4$, $\rho_3 = 0$, $\rho_4 = 3$. The Edgeworth expansion for the distribution function of the standardized sample mean is

$$F^{\scriptstyle E}_{S^*_n}(y) = \Phi(y) - \frac{1}{8n}\phi(y)(y^3 - 3y) + O(n^{-2}).$$

$$\triangle$$

10.6 Cornish–Fisher Expansion and Polynomial Normalizing Transformation

The inversion of the Edgeworth approximation $F^{\scriptstyle E}_{S^*_n}(y)$ makes it possible to obtain an approximation for the quantiles of S^*_n in terms of the quantiles of a normal distribution. Let y_α be the α-th quantile of the distribution of S^*_n and let u_α be the α-th quantile of $U \sim N(0,1)$. The purpose here is to solve the equation

$$\alpha = F_{S^*_n}(y_\alpha)$$

with respect to y_α. From (10.14), since, by definition, $\Phi(u_\alpha) = \alpha$, we have

$$\Phi(u_\alpha) = \Phi(y_\alpha) - \phi(y_\alpha)\left\{\frac{\rho_3}{6\sqrt{n}}H_2(y_\alpha) + \frac{\rho_4}{24n}H_3(y_\alpha) + \frac{\rho_3^2}{72n}H_5(y_\alpha)\right\} + O(n^{-3/2}),$$

so that

$$\begin{aligned}
\Phi(u_\alpha) - \Phi(y_\alpha) &= -\phi(y_\alpha)\left\{\frac{\rho_3}{6\sqrt{n}}H_2(y_\alpha) + \frac{\rho_4}{24n}H_3(y_\alpha)\right.\\
&\qquad \left. + \frac{\rho_3^2}{72n}H_5(y_\alpha)\right\} + O(n^{-3/2}).
\end{aligned} \tag{10.17}$$

Substitution of the expansion

$$\Phi(u_\alpha) = \Phi(y_\alpha) + (u_\alpha - y_\alpha)\phi(y_\alpha) - (u_\alpha - y_\alpha)^2\frac{y_\alpha\phi(y_\alpha)}{2} + O(u_\alpha - y_\alpha)^3$$

into (10.17) gives

$$\begin{aligned}
&(u_\alpha - y_\alpha)\phi(y_\alpha) - (u_\alpha - y_\alpha)^2\frac{y_\alpha\phi(y_\alpha)}{2} + O(u_\alpha - y_\alpha)^3\\
&= -\phi(y_\alpha)\left\{\frac{\rho_3}{6\sqrt{n}}H_2(y_\alpha) + \frac{\rho_4}{24n}H_3(y_\alpha) + \frac{\rho_3^2}{72n}H_5(y_\alpha)\right\} + O(n^{-3/2}).
\end{aligned}$$

Let $\Delta_\alpha = u_\alpha - y_\alpha$; obvious simplifications give

$$\Delta_\alpha = -\left\{ \frac{\rho_3}{6\sqrt{n}} H_2(y_\alpha) + \frac{\rho_4}{24n} H_3(y_\alpha) + \frac{\rho_3^2}{72n} H_5(y_\alpha) \right\}$$
$$+ \frac{y_\alpha}{2} \Delta_\alpha^2 + O(\Delta_\alpha^3) + O(n^{-3/2}).$$

Note that $\Delta_\alpha = O(n^{-1/2})$. By the iterative substitution method (see section 9.3.2), neglecting the terms of order $o(n^{-1})$, we get

$$\Delta_\alpha = -\frac{\rho_3}{6\sqrt{n}} H_2(y_\alpha) - \frac{\rho_4}{24n} H_3(y_\alpha) - \frac{\rho_3^2}{72n} H_5(y_\alpha)$$
$$+ \frac{y_\alpha}{2} \frac{\rho_3^2}{36n} (H_2(y_\alpha))^2 + O(n^{-3/2}).$$

Hence

$$u_\alpha = y_\alpha - \frac{\rho_3}{6\sqrt{n}}(y_\alpha^2 - 1) - \frac{\rho_4}{24n}(y_\alpha^3 - 3y_\alpha) + \frac{\rho_3^2}{36n}(4y_\alpha^3 - 7y_\alpha) + O(n^{-3/2}) \quad (10.18)$$

which expresses the quantiles of the $N(0,1)$ distribution as a polynomial in the quantiles of S_n^*.

The relation sought is the inverse of (10.18). The result is obtained, once again, through the iterative substitution method. Since $y_\alpha = u_\alpha + O(n^{-1/2})$, we only have to substitute u_α for y_α in the summands of order $O(n^{-1})$ in (10.18). In the terms of order $O(n^{-1/2})$ it is best to use the representation

$$y_\alpha^2 = u_\alpha^2 + \frac{\rho_3}{3\sqrt{n}}(y_\alpha^2 - 1) + O(n^{-1})$$
$$= u_\alpha^2 + \frac{\rho_3}{3\sqrt{n}}(u_\alpha^2 - 1) + O(n^{-1}).$$

Hence

$$y_\alpha = u_\alpha + \frac{\rho_3}{6\sqrt{n}}(u_\alpha^2 - 1) + \frac{\rho_4}{24n}(u_\alpha^3 - 3u_\alpha) - \frac{\rho_3^2}{36n}(2u_\alpha^3 - 5u_\alpha) + O(n^{-3/2}). \quad (10.19)$$

Relation (10.19) is called the Cornish–Fisher expansion for inversion of the Edgeworth expansion and can be used in order to approximate the quantiles of S_n^* for the α values where it is monotonically increasing.

An interesting interpretation of (10.18) can be made in terms of random variables. Think of α as if it were a realization of the uniform distribution on $[0, 1]$: then, u_α is a realization of $U \sim N(0, 1)$ and, analogously, y_α is a realization of $Y = S_n^*$. Then (10.18) can be read as

$$U = u(Y) = Y - \frac{\rho_3}{6\sqrt{n}}(Y^2 - 1) - \frac{\rho_4}{24n}(Y^3 - 3Y)$$

$$+ \frac{\rho_3^2}{36n}(4Y^3 - 7Y) + O_p(n^{-3/2}), \qquad (10.20)$$

that is, as a polynomial normalizing transformation for $Y = S_n^*$. The numerical error $O(n^{-3/2})$ of (10.18) becomes $O_p(n^{-3/2})$, because the neglected terms are polynomials in $Y = O_p(1)$, with non-stochastic coefficients of order $O(n^{-3/2})$ or lower. Bearing in mind that $F_{u(Y)}(z) = \Phi(z) + O(n^{-3/2})$, (10.20) can be used to represent the Edgeworth expansion for the distribution function of S_n^* in the asymptotically equivalent form

$$F_{S_n^*}(y) = \Phi(u(y)) + O(n^{-3/2}).$$

Example 10.6 *Bartlett correction in a one-parameter exponential family*
Asymptotic expansions of the combinants W and Z in a one-parameter exponential family were obtained in Example 9.12. Expansion (9.82) for Z is rather similar to expansion (10.20) for the normalizing transformation $u(Y)$. More specifically, if, as in Example 9.12, the standardized score is indicated by Y, then

$$Z = u(Y) - \frac{\rho_3}{6\sqrt{n}} + \frac{1}{n}\left(\frac{7\rho_3^2}{36} - \frac{\rho_4}{8}\right)Y + O_p(n^{-3/2}).$$

It is easy to verify that, by ignoring the terms of order $O(n^{-3/2})$, $E(Z) = m = -\rho_3/(6\sqrt{n})$ and $\text{Var}(Z) = v^2 = 1 + 2(7\rho_3^2/36 - \rho_4/8)/n$. Hence

$$E(W) = E(Z^2) = m^2 + v^2 = 1 + \frac{1}{12n}(5\rho_3^2 - 3\rho_4) + O(n^{-2}),$$

which agrees with (9.81). We can immediately see that the variable Z, standardized through the use of the approximations m and v^2 for $E(Z)$ and $\text{Var}(Z)$, coincides with the normalizing transformation $u(Y)$ except for terms of order $O_p(n^{-3/2})$, so,

$$\frac{Z + \rho_3/(6\sqrt{n})}{1 + (7\rho_3^2/36 - \rho_4/8)/n} = U + O_p(n^{-3/2}),$$

with $U \sim N(0, 1)$. Equivalently,

$$F_z(z) = \Phi\left(\frac{z - m}{v}\right) + O(n^{-3/2}), \qquad (10.21)$$

which is a representation of the Edgeworth expansion for the distribution function of Z. From (10.21) we obtain

$$
\begin{aligned}
F_W(t) &= \Phi\left(\frac{\sqrt{t}-m}{v}\right) - \Phi\left(\frac{-\sqrt{t}-m}{v}\right) + O(n^{-3/2}) \\
&= \Phi\left(\frac{\sqrt{t}-m}{1+O(n^{-1})}\right) - \Phi\left(\frac{-\sqrt{t}-m}{1+O(n^{-1})}\right) + O(n^{-3/2}) \\
&= \Phi\left((\sqrt{t}-m)\right) - \Phi\left((-\sqrt{t}-m)\right) + O(n^{-1}) \\
&= \Phi\left(\sqrt{t}\right) - m\phi(\sqrt{t}) - \Phi\left(-\sqrt{t}\right) + m\phi(-\sqrt{t}) + O(n^{-1}) \\
&= \Phi\left(\sqrt{t}\right) - \Phi\left(-\sqrt{t}\right) + O(n^{-1}).
\end{aligned}
$$

This relation shows that the distribution of W is chi-squared with one degree of freedom, with an error of order $O(n^{-1})$.

Now, consider the Bartlett corrected combinant $W^\star = W/E(W)$. The distribution function of $W^\star$ is

$$
F_{W^\star}(t) = F_Z\left(\sqrt{t}\sqrt{E(W)}\right) - F_Z\left(-\sqrt{t}\sqrt{E(W)}\right).
$$

From (10.21) we obtain

$$
F_{W^\star}(t) = \Phi\left(\frac{\sqrt{t}\sqrt{m^2+v^2}-m}{v}\right) - \Phi\left(\frac{-\sqrt{t}\sqrt{m^2+v^2}-m}{v}\right) + O(n^{-2}),
$$

where the error becomes of order $O(n^{-2})$ because of cancellation of the contributions of order $O(n^{-3/2})$ in the Edgeworth expansion of the distribution function of $Z/\sqrt{E(W)}$ (see (10.15)). The existence of such an expansion is guaranteed by the fact that $Z/\sqrt{E(W)}$ is a smooth function of Y, whose distribution function, in its turn, admits an Edgeworth expansion. By letting $c = m/v$, with $c = O(n^{-1/2})$, we have

$$
\begin{aligned}
F_{W^\star}(t) &= \Phi\left(\sqrt{t}\sqrt{1+c^2}-c\right) - \Phi\left(-\sqrt{t}\sqrt{1+c^2}-c\right) + O(n^{-2}) \\
&= \Phi\left(\sqrt{t}(1+c^2/2)-c\right) - \Phi\left(-\sqrt{t}(1+c^2/2)-c\right) + O(n^{-2}).
\end{aligned}
$$

An elementary expansion gives

$$
\Phi(x - c + xc^2/2) = \Phi(x) - c\phi(x) + c^3\phi(x)/6 + O(n^{-2}).
$$

In conclusion,
$$F_{W^*}(t) = \Phi(\sqrt{t}) - \Phi(-\sqrt{t}) + O(n^{-2})\,.$$

The effect of the Bartlett correction is that W^* has a null χ_1^2 distribution with error of order $O(n^{-2})$. $\triangle$

10.7 Saddlepoint Expansion for Density Functions

As was stressed in section 10.3, the approximation for the distribution of the standardized sum S_n^* given by the Edgeworth expansion (10.10) is more accurate at the centre of the distribution of S_n^*, especially at $y = 0 = E(S_n^*)$, than it is in the tails. It is worth briefly mentioning the main reasons why this should be so. Above all, at $y = 0$ the term of order $O(n^{-1/2})$ in (10.10) is zero: hence, at $y = 0$, the error of the normal approximation becomes $O(n^{-1})$ instead of $O(n^{-1/2})$. Furthermore, the error of order $O(n^{-3/2})$ in (10.10) is an absolute error and its impact is felt less strongly in the regions where the density of S_n^* is large. Lastly, $p_{S_n^*}^{E}(y)$ is non-negative for values of n which are smaller at the centre than in the tails. For example, $\rho_3^2 \leq 12n - 3$ is a sufficient condition for $p_{S_n^*}^{E}(0) \geq 0$.

The Edgeworth approximation can be improved by exploiting its behaviour at $y = 0$. One way of doing this is to express the density of S_n^* at y in terms of the density of an auxiliary distribution. Then we apply the Edgeworth expansion to the density obtained, in such a way that the expansion is calculated in correspondence with the mean of the auxiliary distribution for every y.

Calculations are more straightforward if the sum, S_n, is considered rather than the standardized sum, S_n^*. Consider the $\mathcal{F}_{ne}^1$ generated by $p_{S_n}(s)$ through exponential tilting, whose elements have density

$$p_{S_n}(s; \theta) = e^{\theta s - n K_Y(\theta)} p_{S_n}(s)\,. \tag{10.22}$$

If S_n has density (10.22) it has moment generating function (see (5.6))

$$M_{S_n}(t; \theta) = \exp\left\{ n\left(K_Y(\theta + t) - K_Y(\theta) \right) \right\}\,,$$

cumulant generating function

$$K_{S_n}(t; \theta) = n\left(K_Y(\theta + t) - K_Y(\theta) \right)$$

and cumulants

$$E_\theta(S_n) = K_{S_n}'(0; \theta) = n K_Y'(\theta)\,, \tag{10.23}$$

$$\text{Var}_\theta(S_n) = K''_{S_n}(0;\theta) = nK''_Y(\theta) , \tag{10.24}$$

$$\kappa_r(S_n;\theta) = K^{(r)}_{S_n}(0;\theta) = nK^{(r)}_Y(\theta) . \tag{10.25}$$

The corresponding standardized cumulants are

$$\rho_r(S_n;\theta) = \frac{1}{n^{r/2-1}}\rho_r(\theta) , \tag{10.26}$$

where $\rho_r(\theta) = K^{(r)}_Y(\theta)/ (K''_Y(\theta))^{r/2}$. On the basis of (10.22), the density of S_n can be represented as

$$p_{S_n}(s) = p_{S_n}(s;0) = e^{-\theta s+nK_Y(\theta)}p_{S_n}(s;\theta) , \tag{10.27}$$

for each θ in the natural parameter space of the exponential family (10.22).

The saddlepoint approximation can be obtained, through the Edgeworth expansion, by evaluating the density $p_{S_n}(s;\theta)$ in (10.27) in correspondence with a suitably chosen value of θ. Specifically, θ is chosen as a function of s so that the expected value of S_n for the tilted distribution (10.22) is, exactly, s. In other words, the expansion is evaluated for $\theta = \hat\theta(s)$, where $\hat\theta(s)$ satisfies the equality (see (10.23))

$$E_{\hat\theta(s)}(S_n) = nK'_Y(\hat\theta(s)) = s . \tag{10.28}$$

The value $\hat\theta = \hat\theta(s)$ is called the **saddlepoint** and (10.28) the **saddlepoint equation**. Note that $\hat\theta$ can be interpreted as the maximum likelihood estimate of θ with reference to the auxiliary statistical model $\mathcal{F}^1_{ne}$ with density (10.22) and data s. Hence, if this $\mathcal{F}^1_{ne}$ is regular or, at least, steep, the saddlepoint $\hat\theta$ exists and is unique for each s within the closed convex hull of the support of S_n. Furthermore, note that in (10.27), $\hat\theta$ renders the factor that is evaluated exactly, minimum, and the factor that is evaluated approximately, maximum. It is easier to approximate a larger quantity.

The Edgeworth expansion for $p_{S_n}(s;\theta)$ is

$$p_{S_n}(s;\theta) = \frac{1}{\sqrt{\text{Var}_\theta(S_n)}} \phi\left(\frac{s - E_\theta(S_n)}{\sqrt{\text{Var}_\theta(S_n)}}\right)$$

$$\left\{1 + \frac{1}{6\sqrt{n}}\rho_3(\theta)H_3\left(\frac{s - E_\theta(S_n)}{\sqrt{\text{Var}_\theta(S_n)}}\right) + \frac{1}{24n}\rho_4(\theta)H_4\left(\frac{s - E_\theta(S_n)}{\sqrt{\text{Var}_\theta(S_n)}}\right)\right.$$

$$\left. + \frac{1}{72n}\rho_3^2(\theta)H_6\left(\frac{s - E_\theta(S_n)}{\sqrt{\text{Var}_\theta(S_n)}}\right) + O(n^{-3/2})\right\} ,$$

with $E_\theta(S_n)$, $\text{Var}_\theta(S_n)$ given by (10.23), (10.24) and $\rho_r(\theta)$ as in (10.26). With $\theta = \hat{\theta}$, defined by (10.28), all the Hermite polynomials are evaluated at zero and

$$
p_{S_n}(s;\hat{\theta}) = \frac{1}{\sqrt{nK_Y''(\hat{\theta})}}\phi(0)
$$
$$
\left\{1 + \frac{1}{24n}\rho_4(\hat{\theta})H_4(0) + \frac{1}{72n}\rho_3^2(\hat{\theta})H_6(0) + O(n^{-2})\right\},
$$

where the error is of order $O(n^{-2})$ because of (10.13). Since $H_4(0) = 3$ and $H_6(0) = -15$, with the notation $\hat{\rho}_r = \rho_r(\hat{\theta})$, $r = 3, 4$, we have

$$
p_{S_n}(s;\hat{\theta}) = \frac{1}{\sqrt{2\pi nK_Y''(\hat{\theta})}}\left\{1 + \frac{1}{24n}\left(3\hat{\rho}_4 - 5\hat{\rho}_3^2\right) + O(n^{-2})\right\}. \tag{10.29}
$$

Therefore, with $\theta = \hat{\theta}$ in (10.27) and taking (10.29) into account, we obtain

$$
p_{S_n}(s) = p_{S_n}^t(s)\left\{1 + R_n\right\}, \tag{10.30}
$$

where

$$
p_{S_n}^t(s) = \frac{1}{\sqrt{2\pi nK_Y''(\hat{\theta})}}\exp\{nK_Y(\hat{\theta}) - \hat{\theta}s\} \tag{10.31}
$$

and

$$
R_n = \frac{1}{24n}\left(3\hat{\rho}_4 - 5\hat{\rho}_3^2\right) + O(n^{-2}). \tag{10.32}
$$

Relation (10.30) is called the **saddlepoint expansion**. This name is connected to an alternative derivation by inversion of the characteristic function of $\bar{Y}_n = S_n/n$, that is, through the approximation of the integral (see (3.5))

$$
p_{\bar{Y}_n}(y) = \frac{n}{2\pi}\int_{-\infty}^{+\infty} e^{n\{K_Y(it) - ity\}}\, dt
$$

using a technique which is an extension of the Laplace expansion (see section 9.3.3) in order to deal with the integration of a complex-valued function. More specifically, the function to be integrated is expanded around the point $t = \hat{t}(y)$ which is the solution to the equation, equivalent to (10.28), $K_Y'(t) = y$. The point $(\hat{t}, 0)$ is a saddlepoint of the real part of the function $K_Y(z) - zy$, where $z = u + iv$ is a complex variable which takes values in the strip of the complex plane where $K_Y(u)$ is finite.

Sometimes (10.30) is also called a tilted expansion, with reference to its derivation from exponential tilting. The superscript t in $p^t_{S_n}(s)$ stands for "tilted". The term indirect Edgeworth expansion is also used.

Rigorous justification of (10.30) could be carried out under the same assumptions as those required for the validity of (10.10) for $\hat{\theta}(s) \in C$, with C a compact subset of $\tilde{\Theta}$, and $\tilde{\Theta}$ the natural parameter space of the auxiliary model with density (10.22).

There is one important difference between the two expansions (10.10) and (10.30). In (10.10) the relative error of approximation $p_{S^*_n}(y)/p^E_{S^*_n}(y) - 1$ is of order $O(n^{-3/2})$ uniformly in $y = \dfrac{s - n\mu}{\sqrt{n\sigma^2}}$, provided that $\dfrac{s - n\mu}{\sqrt{n\sigma^2}} = O(1)$, that is, provided that s belongs to a normal deviation region

$$|s - n\mu| \leq c\sqrt{n}.$$

On the other hand for (10.30) the relative error $p_{S_n}(y)/p^t_{S_n}(y) - 1$ is equal to R_n, which is given by (10.32), and depends on s only through s/n. Because of the continuity, in θ, both of cumulants and of the mean value mapping, R_n is of order $O(n^{-1})$ and is uniformly bounded in a large deviation region

$$\left|\frac{s}{n} - \mu\right| \leq c,$$

that is, for $|s - n\mu| \leq cn$. In some cases the relative error is uniformly bounded in all the support of $\bar{Y}_n = S_n/n$ (Daniels, 1954; Jensen, 1988).

One further advantage of the saddlepoint approximation is that $p^t_{S_n}(y)$ is always positive, like the normal density which is the leading term of the Edgeworth expansion. However, the approximation is, usually, more accurate overall. Furthermore, instead of $p^t_{S_n}(s)$, one could use the normalized saddlepoint approximation

$$\tilde{p}^t_{S_n}(s) = \frac{c_n}{\sqrt{2\pi}} \frac{e^{-\hat{\theta}s + nK_Y(\hat{\theta})}}{\sqrt{nK''_Y(\hat{\theta})}}, \tag{10.33}$$

with c_n a normalizing constant, that is, using the notation of the continuous case, $c_n^{-1} = \int p^t_{S_n}(s)ds$, where integration is over the support of S_n.

The normalized saddlepoint approximation is qualitatively more adequate than the Edgeworth approximation because it approximates a density using another density. There is also one further advantage. If the term of order $O(n^{-1})$ in R_n (see (10.32)) does not depend on $\hat{\theta}$, hence does not depend on s, then it could be incorporated into the normalizing constant and the relative error of approximation would be reduced to order $O(n^{-2})$. Suppose that $R_n = A/n + O_s(n^{-2})$, where possible dependency on s in the term of order $O(n^{-2})$ is highlighted. Integration of (10.30) gives

$$1 = \left(1 + \frac{A}{n}\right) c_n^{-1} + O(n^{-2}),$$

from which

$$
\begin{aligned}
p_{S_n}(s) &= c_n p_{S_n}^t(s) \left\{ c_n^{-1}\left(1 + \frac{A}{n}\right) + O_s(n^{-2}) \right\} \\
&= \tilde{p}_{S_n}^t(s) \left\{ 1 + O_s(n^{-2}) \right\}.
\end{aligned}
$$

If the remainder term of order $O(n^{-2})$ does not depend on s, the approximation (10.33) will match the exact density. Blæsild and Jensen (1985) show that this is true if, and only if, Y has a normal, a gamma, or an inverse Gaussian distribution. Usually, the term of order $O(n^{-1})$ in R_n depends on s through $\hat{\theta}$. It is possible to show that normalization does reduce the error to order $O(n^{-3/2})$, at least in a normal deviation region (Durbin, 1980). Indeed, the expansion of $R_n = R_n(\hat{\theta})$ around $\hat{\theta} = 0$ gives

$$R_n = \frac{A(\hat{\theta})}{n} + O_s(n^{-2}) = \frac{A(0)}{n} + \hat{\theta}\frac{A'(0)}{n} + O_s(n^{-2}).$$

Note that in a normal deviation region $\hat{\theta}$ is of order $O_s(n^{-1/2})$ because it can be interpreted as the maximum likelihood estimate of $\theta = 0$.

The normalized saddlepoint approximation is more accurate than the Edgeworth approximation but has the drawback that it requires more complex calculations. It should be stressed that the cumulant generating function or, at least its approximation, must be available in order to determine $\hat{\theta}$. This requirement is satisfied by exponential families.

Example 10.7 *Exponential distribution*
Let Y_i, $i = 1, \ldots, n$, be independent random variables with p.d.f. $p(y) = e^{-y}$, $y > 0$. From Table 3.1, $M_Y(\theta) = (1 - \theta)^{-1}$, for $\theta < 1$, hence, $K_Y(\theta) = -\log(1 - \theta)$, $K_Y'(\theta) = 1/(1 - \theta)$, $K_Y''(\theta) = 1/(1 - \theta)^2$. Let $s = \Sigma y_i$; the saddlepoint is the value $\hat{\theta}$ of θ such that $n/(1 - \theta) = s$, that is, $\hat{\theta} = 1 - n/s$. The ingredients of (10.31) are: $K_Y(\hat{\theta}) = -\log(1 - \hat{\theta}) = \log(s/n)$, $-\hat{\theta}s = n - s$ and $K_Y''(\hat{\theta}) = s^2/n^2$. We obtain

$$p_{S_n}^t(s) = \frac{\exp\{n\log(s/n) + n - s\}}{\sqrt{2\pi n(s^2/n^2)}} = d_n s^{n-1} e^{-s},$$

where $d_n = n^{-n+\frac{1}{2}} e^n / \sqrt{2\pi}$. This approximation matches the exact density except for the normalizing constant. The relations

$$
\begin{aligned}
\frac{1}{\Gamma(n)} &= \frac{e^{n-1}(n-1)^{-n+1/2}}{\sqrt{2\pi}} \left(1 + O(n^{-1})\right) \\
&= d_n e^{-1} \left(1 - \frac{1}{n}\right)^{-n+\frac{1}{2}} \left(1 + O(n^{-1})\right) = d_n \left(1 + O(n^{-1})\right)
\end{aligned}
$$

hold because of Stirling's approximation (9.47), so that it is immediate to check that $p_{S_n}(s) = p_{S_n}^t(s) \left\{1 + O(n^{-1})\right\}$ and that the relative error of order $O(n^{-1})$ is independent of s. In this example $\bar{p}_{S_n}^t(s)$ is exact. △

Example 10.8 *Exponential family of order 1*
Let Y_i, $i = 1, \ldots, n$, be independent random variables with p.d.f.

$$p(y; \phi) = \exp\left\{\phi y - K(\phi)\right\} h(y),$$

with $y \in \mathcal{Y}$, $\phi \in \tilde{\Phi} \subseteq \mathbb{R}$, and natural parameter space $\tilde{\Phi}$. The sum $S_n = \Sigma Y_i$ has density

$$p_{S_n}(s; \phi) = \exp\{\phi s - nK(\phi)\}\tilde{h}(s),$$

with, using the notation of the continuous case, $\tilde{h}(s) = \int \prod h(y_i)\, dy_i$, where integration is over the region $\{(y_1, \ldots, y_n) \in \mathcal{Y}^n : \sum y_i = s\}$. Usually, it is difficult to calculate $\tilde{h}(s)$. Because S_n has cumulant generating function $K_{S_n}(\theta; \phi) = nK_Y(\theta; \phi) = n(K(\phi + \theta) - K(\phi))$, the saddlepoint is the value $\hat{\theta}$ of θ such that $nK'(\phi + \hat{\theta}) = s$. The maximum likelihood estimate $\hat{\phi}$ satisfies the relation $nK'(\hat{\phi}) = s$, thus, $\hat{\theta} = \hat{\phi} - \phi$. The ingredients of (10.31) are: $K_Y(\hat{\theta}; \phi) = K(\phi + \hat{\phi} - \phi) - K(\phi) = K(\hat{\phi}) - K(\phi)$, $K_Y''(\hat{\theta}; \phi) = K''(\phi + \hat{\phi} - \phi) = K''(\hat{\phi})$ and the saddlepoint approximation is

$$p_{S_n}^t(s) = \frac{\exp\{n(K(\hat{\phi}) - K(\phi)) - (\hat{\phi} - \phi)s\}}{\sqrt{2\pi n K''(\hat{\phi})}}. \tag{10.34}$$

Thus, we obtain the approximation

$$\tilde{h}(s) = \frac{\exp\{nK(\hat{\phi}) - \hat{\phi}s\}}{\sqrt{2\pi n K''(\hat{\phi})}} \left\{1 + O(n^{-1})\right\}$$

for $\tilde{h}(s)$. △

Example 10.9 *Binomial distribution*
Let Y_i, $i = 1, \ldots, n$, be independent random variables with a $Bi(1, \phi)$ distribution. From (10.31), the saddlepoint approximation for the probability function of S_n, that is, for the binomial probabilities, is

$$p_{S_n}^t(s; \phi) = \frac{\exp\left\{ n\left(\log(1 - \phi) - \log(1 - \hat{\phi})\right) - \log\frac{\hat{\phi}(1-\phi)}{\phi(1-\hat{\phi})}s\right\}}{\sqrt{2\pi\hat{\phi}(1 - \hat{\phi})}}.$$

Note that $p_{S_n}^t(s; \phi) = p_{S_n}(s; \phi)r(s; n)$, where

$$r(s; n) = \frac{1}{\sqrt{2\pi n}}\hat{\phi}^{-s-1/2}\left(1 - \hat{\phi}\right)^{-n+s-1/2} / \binom{n}{s}.$$

One immediate application of Stirling's approximation gives $r(s; n) = 1 + O(n^{-1})$. △

Example 10.10 *Uniform distribution (cont.)*
Let $\bar{Y}_n = S_n/n$ be the mean of n i.i.d. copies of $Y \sim U(0, 1)$. The c.g.f. of Y is $K_Y(\theta) = \log((e^\theta - 1)/\theta)$. Hence, the saddlepoint equation is

$$\frac{e^\theta(\theta - 1) + 1}{\theta(e^\theta - 1)} = \bar{y}_n . \tag{10.35}$$

Furthermore,

$$K_Y''(\theta) = \frac{(e^\theta - 1)^2 - \theta^2 e^\theta}{\theta^2(e^\theta - 1)^2} .$$

By numerically solving (10.35) and calculating (10.31) for a grid of values $\bar{y}_n \in [0, 1]$ we obtain the density sketched in Figure 10.5 for $n = 3$ and $n = 5$. This graph shows an almost imperceptible improvement over the normal approximation. However, as Figure 10.6 shows, the difference between the two approximations, in terms of relative errors, is very marked in the tails. △

10.8   Lugannani–Rice Expansion for Distribution Functions

Unlike with the Edgeworth approximation, integration of (10.31), in order to obtain the corresponding approximation for the distribution function of S_n, is not immediate. Assume that S_n has a continuous distribution and let $p_{S_n}^t(s)$ indicate the saddlepoint approximation. The distribution function of S_n could

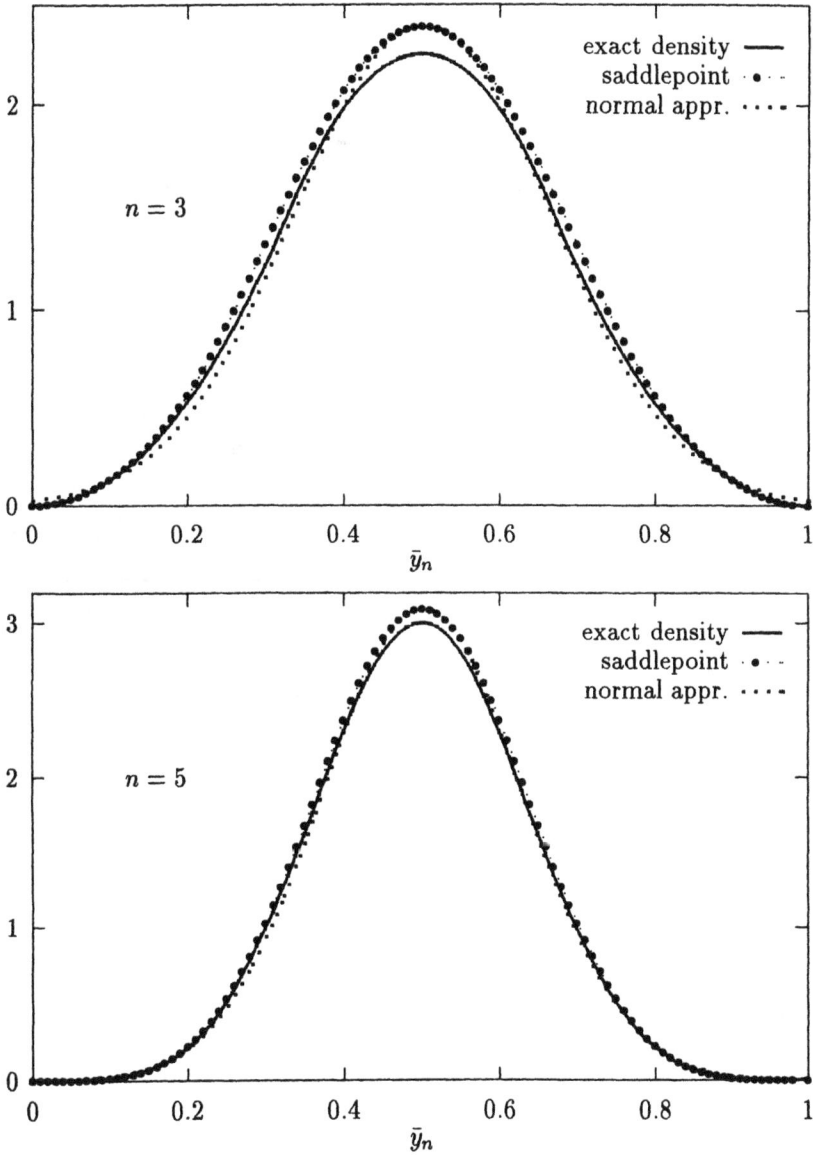

Figure 10.5: *Density of the sample mean of n i.i.d.* $U(0,1)$, $n = 3, 5$.

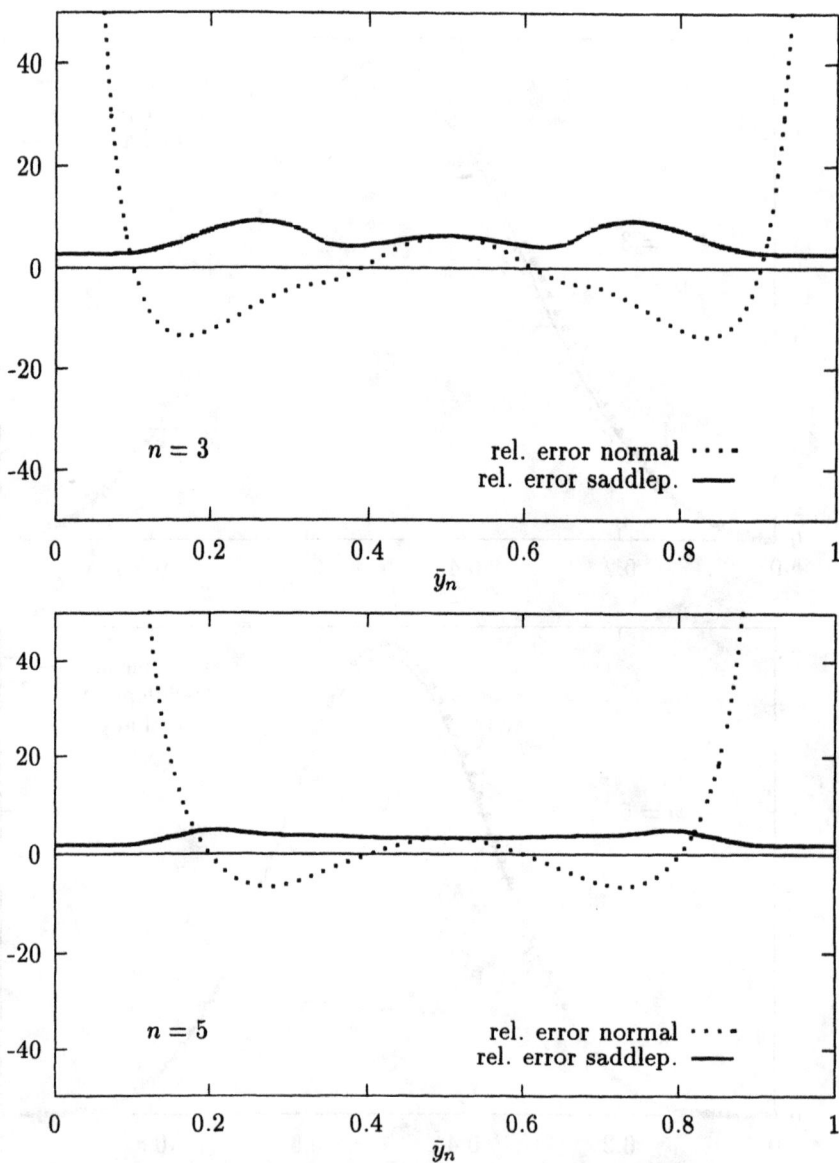

Figure 10.6: *Relative errors in approximations of the density of the sample mean of n i.i.d.* $U(0,1)$, $n = 3, 5$.

then be approximated on the basis of the relation

$$F_{S_n}(s) = \int_{-\infty}^{s} p_{S_n}(u)du = \int_{-\infty}^{s} p_{S_n}^t(u)\, du + O(n^{-1})$$

$$= \int_{-\infty}^{s} \frac{1}{\sqrt{2\pi n K_Y''(\hat\theta)}}\, e^{nK_Y(\hat\theta)-\hat\theta u}\, du + O(n^{-1}), \qquad (10.36)$$

where $nK_Y'(\hat\theta) = u$. Let

$$I = \int_{-\infty}^{s} \frac{1}{\sqrt{2\pi n K_Y''(\hat\theta)}}\, e^{nK_Y(\hat\theta)-\hat\theta u}\, du\,.$$

Consider the change of variable $\hat\theta = \hat\theta(u)$ in I. For simplicity, assume that both s and $\hat\theta$ vary on the whole real line. Since $du/d\hat\theta = nK_Y''(\hat\theta)$,

$$I = \int_{-\infty}^{\hat\theta(s)} \sqrt{\frac{nK_Y''(\hat\theta)}{2\pi}}\, e^{n(K_Y(\hat\theta)-\hat\theta K_Y'(\hat\theta))}\, d\hat\theta\,.$$

Consider the further change of variable

$$q = q(\hat\theta) = \mathrm{sgn}(\hat\theta)\sqrt{2(\hat\theta K_Y'(\hat\theta) - K_Y(\hat\theta))}\,.$$

Note that $\sqrt{n}q$ matches the test statistic

$$Z = \mathrm{sgn}(\hat\theta - \theta_0)\sqrt{2(l(\hat\theta) - l(\theta_0))}$$

for $H_0 : \theta = \theta_0 = 0$ under a model with density (10.22). We can immediately verify that

$$\frac{dq}{d\hat\theta} = \frac{|\hat\theta|K_Y''(\hat\theta)}{\sqrt{2(\hat\theta K_Y'(\hat\theta) - K_Y(\hat\theta))}} = \frac{\hat\theta K_Y''(\hat\theta)}{q(\hat\theta)}\,,$$

which is a positive quantity for each $\hat\theta \neq 0$; furthermore, $q(0) = 0$ and, consequently, at $\hat\theta = 0$, the derivative $dq/d\hat\theta$ has the indeterminate form $0/0$. Using the expansion

$$0 = K_Y(0) = K_Y(\hat\theta) - \hat\theta K_Y'(\hat\theta) + \frac{1}{2}\hat\theta^2 K_Y''(\hat\theta) + O(\hat\theta^3)\,,$$

we obtain

$$\lim_{\hat\theta \to 0} \frac{dq}{d\hat\theta} = \sqrt{K_Y''(0)}\,. \qquad (10.37)$$

Thus, $q(\hat{\theta})$ is a strictly increasing and continuous function. With the change of variable $q = q(\hat{\theta})$ the integral I could be rewritten as

$$I = \int_{-\infty}^{q_s} \sqrt{\frac{n}{2\pi K_Y''(\hat{\theta})}} \, \frac{q}{\hat{\theta}} \, e^{-nq^2/2} \, dq = \int_{-\infty}^{q_s} f(q) \, \phi(q; n^{-1}) \, dq \,,$$

where $q_s = q(\hat{\theta}(s))$, while $\phi(x; n^{-1})$ indicates the $N(0, n^{-1})$ density; furthermore,

$$f(q) = \frac{q}{\hat{\theta}\sqrt{K_Y''(\hat{\theta})}} \,,$$

with $\hat{\theta}$ expressed as a function of q. The integral I has now been rewritten in a form which makes it possible to apply the Laplace expansion (see section 9.3.3). However, this method offers three separate expressions for the approximation according to whether q_s is negative, equal to zero, or positive. A more elegant solution is obtained starting from the representation

$$\begin{aligned} I &= \int_{-\infty}^{q_s} (1 + f(q) - 1) \, \phi(q; n^{-1}) \, dq \\ &= \Phi(\sqrt{n}q_s) + \int_{-\infty}^{q_s} \frac{f(q) - 1}{q} \, q \, \phi(q; n^{-1}) \, dq \,. \end{aligned}$$

Integration by parts of the last summand, taking $q\phi(q; n^{-1})$ as the differential factor, since $\lim_{q \to -\infty} f(q)\phi(q; n^{-1})/q = 0$, gives

$$\begin{aligned} I &= \Phi(\sqrt{n}q_s) - \frac{1}{n}\phi(q_s; n^{-1})\frac{f(q_s) - 1}{q_s} \\ &\quad + \frac{1}{n} \int_{-\infty}^{q_s} \frac{d}{dq}\left(\frac{f(q) - 1}{q}\right) \phi(q; n^{-1}) \, dq \,, \end{aligned}$$

where the last integral is still of the form $\int_{-\infty}^{q_s} f(q)\phi(q; n^{-1}) \, dq$, with a different specification of the function $f(\cdot)$. This integral is multiplied by n^{-1}, so that its contribution is of order $O(n^{-1})$. This gives the asymptotic expansion

$$I = \left\{\Phi(\sqrt{n}q_s) + \phi(q_s; n^{-1})\frac{1 - f(q_s)}{nq_s}\right\} \left\{1 + O(n^{-1})\right\} \,.$$

Writing $r_s = \sqrt{n}q_s$, this is the approximation known as the **Lugannani–Rice** **expansion** (Lugannani and Rice, 1980)

$$F_{S_n}(s) = \Phi(r_s) + \phi(r_s)\left(\frac{1}{r_s} - \frac{1}{v_s}\right) + O(n^{-1}), \qquad (10.38)$$

where

$$r_s = \text{sgn}(\hat{\theta})\sqrt{2n\left\{\hat{\theta}K'_Y(\hat{\theta}) - K_Y(\hat{\theta})\right\}},$$

$$v_s = \hat{\theta}\sqrt{nK''_Y(\hat{\theta})},$$

and $\hat{\theta} = \hat{\theta}(s)$ is the saddlepoint. Observe that v_s is the test statistic

$$Z_e = (\hat{\theta} - \theta_0)\sqrt{nK''_Y(\hat{\theta})}$$

for $H_0 : \theta = \theta_0 = 0$ under the model with density (10.22).

In (10.38) the leading term $\Phi(r_s)$ is of order $O(1)$; the correction term is of order $O(n^{-1/2})$ if r_s is of order $O(1)$, that is, for $\hat{\theta}$ of order $O(n^{-1/2})$, since

$$\frac{1}{r_s} - \frac{1}{v_s} = \frac{1}{6\sqrt{n}}\frac{K'''_Y(\hat{\theta})}{(K''_Y(\hat{\theta}))^{3/2}} + O(n^{-1}).$$

This expansion also makes it possible to eliminate the singularity of (10.38) at $s = E(S_n)$, which corresponds to $\hat{\theta}(s) = 0$, giving

$$F_{S_n}(E(S_n)) = \frac{1}{2} + \frac{\rho_3}{6\sqrt{2\pi n}} + O(n^{-1}).$$

Comparison with the Edgeworth expansion shows that, at least in a normal deviation region, the error must be of order $O(n^{-3/2})$. Jensen (1995a, section 3.3) shows that the approximation given by (10.38) holds with relative error of order $O(n^{-3/2})$ for r_s bounded and with relative error of order $O(n^{-1})$ when r_s is of order $O(\sqrt{n})$.

Expansion (10.38) could be expressed in the asymptotically equivalent form (Jensen, 1995a, Theorem 5.1.1)

$$F_{S_n}(s) = \Phi(r_s^*)\left\{1 + O(n^{-1})\right\}, \tag{10.39}$$

with

$$r_s^* = r_s - \frac{1}{r_s}\log\frac{r_s}{v_s}. \tag{10.40}$$

A heuristic justification of (10.39) could be based on the expansions

$$\Phi(r_s^*) = \Phi(r_s) - \phi(r_s)\frac{1}{r_s}\log\frac{r_s}{v_s} + \dots$$

and

$$\log \frac{r_s}{v_s} = \frac{r_s}{v_s} - 1 + \dots .$$

Thus, in the continuous case, (10.40) is a normalizing transformation of S_n based on the saddlepoint expansion.

With suitable continuity corrections, (10.38) and (10.40) are also valid for lattice variables: see Daniels (1987), Pierce and Peters (1992), Kolassa (1994, p. 78) and Jensen (1995a, p. 79). The simplest correction is to substitute the quantity v_s in (10.38) and (10.40) with

$$v_s^D = \sqrt{n K_Y''(\hat{\theta})}(e^{\hat{\theta}} - 1) .$$

10.9   Multivariate Edgeworth and Saddlepoint Expansions

Using index notation it is easy to obtain the multivariate versions of expansions (10.10) and (10.30). Let $Y = (Y^1, \dots, Y^d)$ be a d-dimensional random vector with moment generating function $M_Y(t)$. Let us use the notation introduced in section 3.2.7 for the moments and the cumulants of Y. Let $Y_1, \dots, Y_n$ be independent copies of Y and let

$$S_n = \sum_{i=1}^{n} Y_i \quad \text{and} \quad S_n^* = \frac{S_n - n\mu}{\sqrt{n}} \qquad (10.41)$$

be the d-dimensional sum and standardized sum, where $\mu = [\kappa^i]$ is the mean vector. As in the univariate case, we have the relations

$$\kappa^{i_1, \dots, i_r}(S_n^*) = \frac{1}{n^{r/2-1}} \kappa^{i_1, \dots, i_r}, \qquad r \geq 2 ,$$

where $\kappa^{i_1, \dots, i_r}$ indicates the generic cumulant of order r of Y. Specifically,

$$\begin{aligned}
\kappa^i(S_n^*) &= 0 \\
\kappa^{i,j}(S_n^*) &= \kappa^{i,j} \\
\kappa^{i,j,k}(S_n^*) &= \frac{1}{\sqrt{n}} \kappa^{i,j,k} \\
\kappa^{i,j,k,l}(S_n^*) &= \frac{1}{n} \kappa^{i,j,k,l}
\end{aligned}$$

and, consequently, as n diverges, all the joint cumulants of S_n^* of order higher than two will tend to zero; the higher the order is, the faster convergence will be.

Sections 10.9.1-10.9.3 below, give the Edgeworth, mixed and saddlepoint expansions for the density of S_n^* or of S_n. The conditions for the validity of these approximations are not listed here as they are natural generalizations of the regularity conditions illustrated above for the univariate case. See e.g. Jensen (1995a, sections 1.5 and 2.2) for further study.

10.9.1 Multivariate Edgeworth Expansion

Let $M_{S_n^*}(t)$, where $t = (t_1, \ldots, t_d)$, be the moment generating function of S_n^*. With analogous steps to those which, in the univariate case, give (10.3), we obtain the expansion

$$
M_{S_n^*}(t) = e^{\frac{1}{2}\kappa^{i,j}t_{ij}} \left\{ 1 + \frac{1}{6\sqrt{n}}\kappa^{i,j,k}t_{ijk} + \frac{1}{24n}\kappa^{i,j,k,l}t_{ijkl} \right.
$$
$$
\left. + \frac{1}{72n}\kappa^{i,j,k}\kappa^{l,m,n}t_{ijklmn} + O(n^{-3/2}) \right\}, \qquad (10.42)
$$

where $t_{I_r} = t_{i_1} \cdots t_{i_r}$. Inversion of (10.42) is made easier by Hermite polynomials. However, the components of S_n^* are usually correlated and it is better to adopt a definition of Hermite polynomials which has coefficients which depend on the covariance matrix of Y, $\kappa = [\kappa^{i,j}]$. A generic Hermite polynomial $h_{I_r}(y; \kappa)$ of order r, $r = 1, 2 \ldots$, and with index I_r, is defined by

$$
\phi_d(y; \kappa) h_{I_r}(y; \kappa) = (-1)^r \phi_{/I_r}(y; \kappa),
$$

where $y = (y^1, \ldots, y^d) \in \mathbb{R}^d$ and $\phi_{/I_r}(y; \kappa) = (\partial^r/(\partial y^{i_1} \cdots \partial y^{i_r}))\phi_d(y; \kappa)$. Specifically, with the notations $[\kappa_{ij}] = \kappa^{-1}$, $y_i = \kappa_{ij}y^j$ and $y_{I_r} = y_{i_1} \cdots y_{i_r}$, we have

$$
\begin{aligned}
h_i(y; \kappa) &= y_i \\
h_{ij}(y; \kappa) &= y_{ij} - \kappa_{ij} \\
h_{ijk}(y; \kappa) &= y_{ijk} - y_i \kappa_{jk}[3] \\
h_{ijkl}(y; \kappa) &= y_{ijkl} - y_{ij}\kappa_{kl}[6] + \kappa_{ij}\kappa_{kl}[3] \\
h_{ijklm}(y; \kappa) &= y_{ijklm} - y_{ijk}\kappa_{lm}[10] + y_i\kappa_{jk}\kappa_{lm}[15] \\
h_{ijklmn}(y; \kappa) &= y_{ijklmn} - y_{ijkl}\kappa_{mn}[15] + y_{ij}\kappa_{kl}\kappa_{mn}[45] - \kappa_{ij}\kappa_{kl}\kappa_{mn}[15].
\end{aligned}
$$

Note that $h_{I_r}(0; \kappa) = 0$ if r is odd. Furthermore, the key result for the inversion of expansion (10.42) of the moment generating function of S_n^* is

$$
\int_{\mathbb{R}^d} e^{t \cdot y} h_{I_r}(y; \kappa)\phi_d(y; \kappa)dy = t_{I_r} \exp\left\{ \frac{1}{2}\kappa^{i,j}t_{ij} \right\}, \qquad (10.43)
$$

which generalizes (10.9).

By combining (10.42) and (10.43) we obtain the multivariate Edgeworth expansion for the density of S_n^*

$$p_{S_n^*}(y) = p_{S_n^*}^E(y) + O(n^{-3/2}), \tag{10.44}$$

where

$$p_{S_n^*}^E(y) = \phi_d(y; \kappa)\left\{1 + Q_3(y; \kappa^{(3)}) + Q_4(y; \kappa^{(4)})\right\},$$

with

$$Q_3(y; \kappa^{(3)}) = \frac{1}{6\sqrt{n}}\kappa^{i,j,k}h_{ijk}(y; \kappa)$$

and

$$Q_4(y; \kappa^{(4)}) = \frac{1}{24n}\kappa^{i,j,k,l}h_{ijkl}(y; \kappa) + \frac{1}{72n}\kappa^{i,j,k}\kappa^{l,m,n}h_{ijklmn}(y; \kappa) .$$

10.9.2 Multivariate Saddlepoint Expansion

The saddlepoint approximation for the density of S_n is obtained starting from

$$p_{S_n}(s) = p_{S_n}(s; 0) = \exp\{-\theta \cdot s + nK_Y(\theta)\}p_{S_n}(s; \theta) , \tag{10.45}$$

where $p_{S_n}(s; \theta)$ is the density of an element of the $\mathcal{F}_{ne}^d$ generated by $p_{S_n}(s)$. Let $s = (s^1, \ldots, s^d)$. The saddlepoint $\hat{\theta} = \hat{\theta}(s)$ is the solution in $\theta = (\theta^1, \ldots, \theta^d)$ to the equations $n(\partial/\partial\theta^i)K_Y(\theta) = s^i$, $i = 1, \ldots, d$. Approximation of the factor $p_{S_n}(s; \theta)$ in (10.45) using the Edgeworth expansion (10.44), and evaluation of the result at $\theta = \hat{\theta}$, gives the multivariate saddlepoint expansion

$$p_{S_n}(s) = p_{S_n}^t(s)\{1 + R_n\} , \tag{10.46}$$

where

$$p_{S_n}^t(s) = \frac{1}{(2\pi)^{d/2}\left\{\det[n\hat{\kappa}^{i,j}]\right\}^{1/2}}\exp\{nK_Y(\hat{\theta}) - \hat{\theta} \cdot s\} , \tag{10.47}$$

with

$$\hat{\kappa}^{i,j} = \left.\frac{\partial^2}{\partial\theta^i\partial\theta^j}K_Y(\theta)\right|_{\theta=\hat{\theta}}, \qquad i, j = 1, \ldots, d .$$

The term R_n is a relative error of order $O(n^{-1})$ and can be expressed as

$$R_n = \frac{1}{24n}\hat{\kappa}^{i,j,k,l}\hat{\kappa}_{ij}\hat{\kappa}_{kl}[3] - \frac{1}{72n}\hat{\kappa}^{i,j,k}\hat{\kappa}^{l,m,n}\hat{\kappa}_{ij}\hat{\kappa}_{kl}\hat{\kappa}_{mn}[15] + O(n^{-2}),$$

where

$$\hat{\kappa}^{i_1,\dots,i_r} = \frac{\partial^r}{\partial\theta^{i_1}\cdots\partial\theta^{i_r}}K_Y(\theta)\Big|_{\theta=\hat{\theta}} \quad \text{and} \quad [\hat{\kappa}_{ij}] = [\hat{\kappa}^{i,j}]^{-1}.$$

Example 10.11 *Exponential family of order p*
Let $Y = (Y^1,\dots,Y^p)$ be a r.v. belonging to an $\mathcal{F}_e^p$ with density

$$p_Y(y;\theta) = \exp\{\theta \cdot y - K(\theta)\}h(y),$$

where $y \in \mathcal{Y} \subseteq \mathbb{R}^p$, $\theta \in \tilde{\Theta} \subseteq \mathbb{R}^p$, with $\tilde{\Theta}$ the natural parameter space. Let $Y_1,\dots,Y_n$ be i.i.d. copies of Y. With analogous steps to those used in Example 10.8, the saddlepoint approximation (10.47) for the density of $S_n = \sum Y_i$ is

$$p_{S_n}^t(s) = \frac{\exp\{(\theta - \hat{\theta}) \cdot s - n(K(\theta) - K(\hat{\theta}))\}}{(2\pi)^{p/2}\left\{\det(n\hat{K}_{\theta\theta})\right\}^{1/2}}, \tag{10.48}$$

where $\hat{K}_{\theta\theta}$ is the matrix with generic element

$$(\partial^2/\partial\theta^i\partial\theta^j)K(\theta)\Big|_{\theta=\hat{\theta}}.$$

$\triangle$

While in the univariate case the normalized saddlepoint approximation is only exact for normal, gamma and inverse Gaussian distributions, in the multivariate case there are many more cases of exactness. One example is offered by the minimal sufficient statistic of a random sample from $N_p(\mu, \Sigma)$ with unknown μ and Σ (Barndorff-Nielsen and Cox, 1989, p. 185).

10.9.3 Mixed Expansion

In some cases, the cumulant generating function of Y may prove difficult to find or, it may not even exist. However, there may be a known marginal cumulant generating function of a block of components of Y. For example, think of the bivariate variable $Y = (X, X^2)$, where X has a Poisson distribution. The cumulant generating function of X is well known; Y admits cumulants of every order, but does not admit a cumulant generating function.

Let $Y = (Y^{(1)}, Y^{(2)})$ denote one subdivision of the d-dimensional variable Y into two blocks of components, with dimensions d_1 and d_2 respectively. Suppose that the cumulant generating function of $Y^{(1)}$, $K_{Y^{(1)}}(t)$, is known. Let $Y_1, \ldots, Y_n$ be independent copies of Y and let $S_n = (S_n^{(1)}, S_n^{(2)})$ be the sum random variable. We could approximate the density of S_n by combining a saddlepoint expansion for $S_n^{(1)}$ and an Edgeworth expansion for $S_n^{(2)}$. To this end, consider the **partial exponential tilting** which gives a family of densities with elements

$$p_{S_n}(s^{(1)}, s^{(2)}; \theta^{(1)}) = \exp\{s^{(1)} \cdot \theta^{(1)} - nK_{Y^{(1)}}(\theta^{(1)})\}p_{S_n}(s^{(1)}, s^{(2)}) . \quad (10.49)$$

The tilting parameter $\theta^{(1)}$ has dimension d_1. Assume that S_n admits finite joint cumulants up to the fifth order with respect to the distribution defined by the partial exponential tilting (10.49). For the Edgeworth expansion of the left-hand side of (10.49), the most convenient value of $\theta^{(1)}$ is $\hat{\theta}^{(1)}$, which is the solution of

$$n\frac{\partial}{\partial\theta^{(1)}}K_{Y^{(1)}}(\theta^{(1)}) = s^{(1)} .$$

Let us denote by $n\kappa^{i_1, \ldots, i_r}(\theta^{(1)})$ a generic cumulant of order r of S_n with respect to the density (10.49). With reference to the same distribution, let $n\mu^{(2)}(\theta^{(1)})$ be the expectation of $S_n^{(2)}$ and let $n\kappa(\theta^{(1)})$ be the covariance matrix of S_n. The mixed expansion (Edgeworth-saddlepoint) for the density of S_n

$$p_{S_n}(s^{(1)}, s^{(2)}) = p_{S_n}^m(s^{(1)}, s^{(2)})\left\{1 + O(n^{-3/2})\right\} , \quad (10.50)$$

where

$$p_{S_n}^m(s^{(1)}, s^{(2)}) = \exp\left\{nK_{Y^{(1)}}(\hat{\theta}^{(1)}) - s^{(1)} \cdot \hat{\theta}^{(1)}\right\}\frac{1}{\sqrt{n}}\phi_d(s^*; \kappa(\hat{\theta}^{(1)}))$$

$$\times \left\{1 + Q_3(s^*; \kappa^{(3)}(\hat{\theta}^{(1)})) + Q_4(s^*; \kappa^{(4)}(\hat{\theta}^{(1)}))\right\} ,$$

with

$$Q_3(s^*; \kappa^{(3)}(\hat{\theta}^{(1)})) = \frac{1}{6\sqrt{n}} \kappa^{i,j,k} h_{ijk}(s^*; \kappa(\hat{\theta}^{(1)})) ,$$

$$Q_4(s^*; \kappa^{(4)}(\hat{\theta}^{(1)})) = \frac{1}{24n} \kappa^{i,j,k,l}(\hat{\theta}^{(1)}) h_{ijkl}(s^*; \kappa(\hat{\theta}^{(1)}))$$

$$+ \frac{1}{72n} \kappa^{i,j,k}(\hat{\theta}^{(1)}) \kappa^{l,m,n}(\hat{\theta}^{(1)}) h_{ijklmn}(s^*; \kappa(\hat{\theta}^{(1)})) ,$$

and

$$s^* = \left(0, \frac{s^{(2)} - n\mu^{(2)}(\hat{\theta}^{(1)})}{\sqrt{n}}\right).$$

Example 10.12 *Poisson distribution*
Let $X_1, \ldots, X_n$ be independent copies of $X \sim P(e^\theta)$. Let $Y = (X, X^2) = (Y^{(1)}, Y^{(2)})$ and let $S_n^{(1)} = \sum X_i$, $S_n^{(2)} = \sum X_i^2$. Since

$$K_{Y^{(1)}}(\theta^{(1)}) = e^{\theta + \theta^{(1)}} - e^\theta,$$

then $\hat{\theta}^{(1)} = \hat{\theta} - \theta$, with $\hat{\theta} = \log \bar{x}_n$, so that $K_{Y^{(1)}}(\hat{\theta}^{(1)}) = e^{\hat{\theta}} - e^\theta$. The tilted family is made up of the densities of $(\sum T_i, \sum T_i^2)$, where the T_i are independent and distributed as $T \sim P(e^{\theta^{(1)} + \theta})$. Then, $\mu^{(2)}(\theta^{(1)})$ is equal to the second moment of T, $e^{\theta^{(1)} + \theta}(1 + e^{\theta^{(1)} + \theta})$, and the elements of $\kappa(\theta^{(1)})$ are $\mathrm{Var}(T)$, $\mathrm{Cov}(T, T^2)$ and $\mathrm{Var}(T^2)$. Therefore,

$$p_{S_n}^m(s^{(1)}, s^{(2)}) = \exp\{n(e^{\hat{\theta}} - e^\theta) - s^{(1)}(\hat{\theta} - \theta)\} \frac{1}{\sqrt{n}} \phi_2(s^*; \kappa(\hat{\theta}^{(1)})) \left\{1 + O(n^{-1/2})\right\},$$

where

$$s^* = \left(0, \frac{s^{(2)} - ne^{\hat{\theta}}(e^{\hat{\theta}} + 1)}{\sqrt{n}}\right)$$

and

$$\kappa(\hat{\theta}^{(1)}) = \begin{pmatrix} e^{\hat{\theta}} & e^{\hat{\theta}}\left(1 + 2e^{\hat{\theta}}\right) \\ e^{\hat{\theta}}\left(1 + 2e^{\hat{\theta}}\right) & e^{\hat{\theta}}\left(1 + 6e^{\hat{\theta}} + 4e^{\hat{\theta}}\right) \end{pmatrix}.$$

$\wedge$

10.10 Asymptotic Expansions for Conditional Distributions

10.10.1 Three Methods

Using the same notation as in section 10.9.3, we wish to obtain approximations for the conditional density of $S_n^{(2)}$ given $S_n^{(1)} = s^{(1)}$, which we assume can be expressed as the ratio between the joint density of $(S_n^{(1)}, S_n^{(2)})$ and the marginal density of $S_n^{(1)}$,

$$p_{S_n^{(2)}|S_n^{(1)} = s^{(1)}}(s^{(2)}; s^{(1)}) = \frac{p_{S_n^{(1)}, S_n^{(2)}}(s^{(1)}, s^{(2)})}{p_{S_n^{(1)}}(s^{(1)})}. \tag{10.51}$$

One way of obtaining an approximation is that of substituting the numerator
and denominator in (10.51) by suitable expansions: either Edgeworth, saddlepoint or mixed. Among all possible combinations, the most interesting, from
the point of view of applications, are those which use:

i) an Edgeworth expansion both for the numerator and for the denominator;

ii) a saddlepoint expansion both for the numerator and for the denominator;

iii) a mixed expansion for the numerator and a saddlepoint expansion for the
denominator.

Using the first of these, we obtain the **double Edgeworth expansion**, whose
leading term is a d_2-variate normal density; the correction term can be expressed as a function of the polynomials $Q_3(\cdot)$ and $Q_4(\cdot)$ (see (10.44)) which
appear in the Edgeworth expansion for numerator and denominator. It is
worth referring the expansion to the standardized sum $S_n^* = (S_n^{*(1)}, S_n^{*(2)})$ (see
(10.41)). Below, the indices $a, b, c, \ldots$ (with range $\{1, \ldots, d_1\}$) refer to the
components of $S_n^{*(1)}$, the indices $r, s, t, \ldots$ (with range $\{d_1 + 1, \ldots, d\}$) refer
to the components of $S_n^{*(2)}$, the indices $i, j, k, \ldots$ (with range $\{1, \ldots, d\}$) refer
to all the components of S_n^*. Let $\kappa_{ab}^{(1)}$ be the generic element of the inverse of
the covariance matrix of $Y^{(1)}$ and let $\beta_a^r = \kappa^{r,b} \kappa_{ab}^{(1)}$ be the least squares linear
regression coefficients of $S_n^{*(2)}$ on $S_n^{*(1)}$. Lastly, let $\bar{\kappa} = [\kappa^{r,s} - \kappa^{r,a} \kappa^{s,b} \kappa_{ab}^{(1)}]$ be
the covariance matrix of the residuals of this linear regression. With simple
algebra, we obtain

$$P_{S_n^{*(2)}|S_n^{*(1)}=y^{(1)}}\left(y^{(2)}; y^{(1)}\right) = P_{S_n^{*(2)}|S_n^{*(1)}=y^{(1)}}^{EE}\left(y^{(2)}; y^{(1)}\right) + O(n^{-3/2}), \quad (10.52)$$

where

$$P_{S_n^{*(2)}|S_n^{*(1)}=y^{(1)}}^{EE}\left(y^{(2)}; y^{(1)}\right) = \phi_{d_2}(\bar{y}^{(2)}; \bar{\kappa})$$
$$\times \left\{1 + Q_3^N - Q_3^D + Q_4^N - Q_4^D - Q_3^D(Q_3^N - Q_3^D)\right\},$$

with

$$\bar{y}^{(2)} = [y^r - \beta_a^r y^a],$$

and Q_3^N, Q_4^N are the correction terms of orders $O(n^{-1/2})$ and $O(n^{-1})$ in the
Edgeworth expansion for the numerator, while Q_3^D, Q_4^D are the analogous terms
for the denominator.

The second combination of expansions we mentioned gives the **double saddlepoint expansion**

$$P_{S_n^{(2)}|S_n^{(1)}=s^{(1)}}\left(s^{(2)}; s^{(1)}\right) = P_{S_n^{(2)}|S_n^{(1)}=s^{(1)}}^{tt}\left(s^{(2)}; s^{(1)}\right)\{1 + R_n\}, \quad (10.53)$$

where

$$p^{tt}_{S_n(2)|S_n(1)=s(1)}\left(s^{(2)};s^{(1)}\right) = \frac{\exp\left\{nK_Y(\hat{\theta}) - nK_Y(\hat{\theta}_0^{(1)},0) - \hat{\theta}\cdot s + \hat{\theta}_0^{(1)}\cdot s^{(1)}\right\}}{(2\pi)^{d_2/2}\left\{\det[n\hat{\kappa}^{i,j}]/\det[n\tilde{\kappa}^{a,b}]\right\}^{1/2}},$$

with $\hat{\theta}_0^{(1)}$ the solution in $\theta^{(1)}$ of $(\partial/\partial\theta^{(1)})K_Y(\theta^{(1)},0) = s^{(1)}$,

$$\tilde{\kappa}^{a,b} = \left.\frac{\partial^2}{\partial\theta^a\partial\theta^b}K_Y(\theta)\right|_{\theta=(\hat{\theta}_0^{(1)},0)}, \qquad a,b = 1,\ldots,d_1.$$

The term R_n of order $O(n^{-1})$ is

$$R_n = \frac{1}{24n}\hat{\kappa}^{i,j,k,l}\hat{\kappa}_{ij}\hat{\kappa}_{kl}[3] - \frac{1}{72n}\hat{\kappa}^{i,j,k}\hat{\kappa}^{l,m,n}\hat{\kappa}_{ij}\hat{\kappa}_{kl}\hat{\kappa}_{mn}[15]$$

$$-\frac{1}{24n}\tilde{\kappa}^{a,b,c,d}\tilde{\kappa}_{ab}^{(1)}\tilde{\kappa}_{cd}^{(1)}[3] + \frac{1}{72n}\tilde{\kappa}^{a,b,c}\tilde{\kappa}^{d,e,f}\tilde{\kappa}_{ab}^{(1)}\tilde{\kappa}_{cd}^{(1)}\tilde{\kappa}_{ef}^{(1)}[15] + O(n^{-2}),$$

where

$$\tilde{\kappa}^{A_h} = \left.\frac{\partial^h}{\partial\theta^{a_1}\cdots\partial\theta^{a_h}}K_Y(\theta)\right|_{\theta=(\hat{\theta}_0^{(1)},0)}.$$

The third possibility, that is, mixed approximation for the numerator and saddlepoint approximation for the denominator in (10.51), gives the mixed (Edgeworth-saddlepoint) expansion

$$p_{S_n(2)|S_n(1)=s(1)}\left(s^{(2)};s^{(1)}\right) = p^m_{S_n(2)|S_n(1)=s(1)}\left(s^{(2)};s^{(1)}\right)\left\{1+O(n^{-3/2})\right\},$$

$$(10.54)$$

with

$$p^m_{S_n(2)|S_n(1)=s(1)}\left(s^{(2)};s^{(1)}\right) = \frac{1}{\sqrt{n}}\phi_{d_2}(s^{*(2)};\tilde{\tilde{\kappa}})$$

$$\times\left\{1 + Q_3(s^*;\kappa^{(3)}) + Q_4(s^*;\kappa^{(4)}) - \tilde{Q}_4^{(1)}\right\}.$$

The d_2-variate normal density is evaluated at $s^{*(2)} = (s^{(2)} - n\mu^{(2)}(\hat{\theta}^{(1)}))/\sqrt{n}$, the covariance matrix $\tilde{\tilde{\kappa}}$ has the generic element $\tilde{\kappa}^{r,s} - \tilde{\kappa}^{r,a}\tilde{\kappa}^{s,b}\tilde{\kappa}_{ab}^{(1)}$, where $n\tilde{\kappa}^{r,s,\ldots,a,b,\cdots}$ is the generic joint cumulant of S_n with respect to the density (10.49), evaluated at $(\hat{\theta}_0^{(1)},0)$. The quantities s^*, $Q_3(s^*;\kappa^{(3)})$ and $Q_4(s^*;\kappa^{(4)})$ are the same as those which appear in (10.50), while

$$\tilde{Q}_4^{(1)} = \frac{1}{24n}\tilde{\kappa}^{a,b,c,d}\tilde{\kappa}_{ab}^{(1)}\tilde{\kappa}_{cd}^{(1)}[3] - \frac{1}{72n}\tilde{\kappa}^{a,b,c}\tilde{\kappa}^{d,e,f}\tilde{\kappa}_{ab}^{(1)}\tilde{\kappa}_{cd}^{(1)}\tilde{\kappa}_{ef}^{(1)}[15].$$

10.10.2 *Exponential Families and Approximate Conditional Inference*

Assume that the reference model is an exponential family of order p with density

$$p_{T,U}(t, u; \tau, \zeta) = e^{\tau \cdot t + \zeta \cdot u - nK(\tau, \zeta)} h(t, u), \qquad (10.55)$$

where $\theta = (\tau, \zeta) \in \tilde{\Theta}$, with $\tilde{\Theta}$ the natural parameter space; the component t of the p-dimensional natural observation (t, u) has dimension $k < p$. Note that the factor $h(t, u)$ may not be known in an explicit form. Generally, the double saddlepoint approximation (10.53) gives a fairly accurate approximation of the conditional density of T given $U = u$, whose structure is given by (5.30). This is interesting because, usually, it is difficult to obtain an exact expression of the term $K_u(\tau)$ which appears in (5.30). With an immediate adaptation of (10.48), the saddlepoint approximation for the joint density of (T, U) is

$$p_{T,U}^t(t, u) = \frac{\exp\{(\tau - \hat{\tau}) \cdot t + (\zeta - \hat{\zeta}) \cdot u - n(K(\tau, \zeta) - K(\hat{\tau}, \hat{\zeta}))\}}{(2\pi)^{p/2} \{\det[n\hat{\kappa}^{i,j}]\}^{1/2}},$$

where $[\hat{\kappa}^{i,j}]$ is the matrix of second partial derivatives of $K(\tau, \zeta)$ with respect to the components of (τ, ζ) evaluated at $(\hat{\tau}, \hat{\zeta})$. Use the indices i, j, with range $\{1, \ldots, p\}$ in order to refer to the components of $\theta = (\tau, \zeta)$ and to those of the sufficient statistic (t, u), whereas the indices a, b, with range $\{k+1, \ldots, p\}$, identify the components of ζ and of u. The saddlepoint approximation for the marginal density of U is

$$p_U^t(u) = \frac{\exp\{(\zeta - \hat{\zeta}_\tau) \cdot u - n(K(\tau, \zeta) - K(\tau, \hat{\zeta}_\tau))\}}{(2\pi)^{(p-k)/2} \{\det[n\tilde{\kappa}^{a,b}]\}^{1/2}},$$

where

$$\tilde{\kappa}^{a,b} = \frac{\partial^2}{\partial \zeta^a \partial \zeta^b} K(\tau, \zeta) \Bigg|_{(\tau, \zeta) = (\tau, \hat{\zeta}_\tau)}, \qquad a, b = k+1, \ldots, p.$$

The ratio $p_{T,U}^t(t, u)/p_U^t(u)$ gives the double saddlepoint approximation

$$
\begin{aligned}
p_{T|U=u}^{tt}(t; u) &= \exp\Big\{(\tau - \hat{\tau}) \cdot t + (\hat{\zeta}_\tau - \hat{\zeta}) \cdot u \\
&\qquad - nK(\tau, \hat{\zeta}_\tau) + nK(\hat{\tau}, \hat{\zeta})\Big\} \\
&\qquad \times \frac{\{\det[n\tilde{\kappa}^{a,b}]\}^{1/2}}{(2\pi)^{k/2} \{\det[n\hat{\kappa}^{i,j}]\}^{1/2}} \{1 + O(n^{-1})\}. \qquad (10.56)
\end{aligned}
$$

In particular, note that (10.56) does not depend on ζ and is, therefore, consistent with the partial sufficiency of u for ζ. Relation (10.56) immediately offers an approximation for the conditional likelihood $L_C(\tau)$ defined by (5.33). Ignoring the factors in (10.56) which only depend on (t, u), we have

$$L_C(\tau) = L_P(\tau)M(\tau)\left\{1 + O(n^{-1})\right\}, \qquad (10.57)$$

where

$$L_P(\tau) = \exp\left\{\tau \cdot t + \hat{\zeta}_\tau \cdot u - nK(\tau, \hat{\zeta}_\tau)\right\}$$

is the profile likelihood for τ and

$$M(\tau) = \left\{\det[\tilde{\kappa}^{a,b}]\right\}^{1/2}$$

is a modification factor. Therefore, $l_{AC}(\tau) = l_P(\tau) + \log M(\tau)$ gives an approximation of the conditional log-likelihood $l_C(\tau)$. Observe that this approximation is the same as the Cox and Reid approximate conditional log-likelihood (5.56). We can also deduce that the profile log-likelihood $l_P(\tau)$ offers an approximation of $l_C(\tau)$ with error of order $O(1)$.

Comparison between (10.57) and the exact conditional likelihood (5.33) suggests the approximation

$$K_u(\tau) \doteq nK(\tau, \hat{\zeta}_\tau) - \hat{\zeta}_\tau \cdot u - \frac{1}{2}\log\left(\det[\tilde{\kappa}^{a,b}]\right) \qquad (10.58)$$

for $K_u(\tau)$. Because of (5.30), T given $U = u$ has cumulant generating function $K_{T|U=u}(v) = K_u(\tau + v) - K_u(\tau)$, with $v \in \mathbb{R}^k$ such that $\tau + v$ belongs to the natural parameter space of the conditional family. Thus one can conjecture that the partial derivatives with respect to the components of τ of the right-hand side of (10.58) are approximations for the conditional cumulants of T given $U = u$. The validity of this conclusion is shown in Pace and Salvan (1992). In particular, if we denote by $\kappa_{2\cdot1}^{r,s,\cdots}$ the conditional cumulants of T given $U = u$, we obtain the expansions

$$\kappa_{2\cdot1}^r = \tilde{\kappa}^r - \frac{1}{2}\tilde{\kappa}_{ab}\,\tilde{\tilde{\kappa}}^{a,b,r} + O(n^{-1}) \qquad (10.59)$$

$$\kappa_{2\cdot1}^{r,s} = \tilde{\tilde{\kappa}}^{r,s} + \frac{1}{2}\tilde{\kappa}_{ab}\,\tilde{\kappa}_{cd}\left(\tilde{\tilde{\kappa}}^{a,c,r}\,\tilde{\tilde{\kappa}}^{b,d,s} + \tilde{\tilde{\kappa}}^{a,r,s}\,\tilde{\kappa}^{b,c,d} - \tilde{\kappa}^{a,d}\,\tilde{\tilde{\kappa}}^{b,c,r,s}\right)$$
$$+ O(n^{-1}) \qquad (10.60)$$

$$\kappa_{2\cdot1}^{r,s,t} = \tilde{\tilde{\kappa}}^{r,s,t} + O(1) \qquad (10.61)$$

$$\kappa_{2\cdot1}^{r,s,t,u} = \tilde{\tilde{\kappa}}^{r,s,t,u} - \tilde{\kappa}_{ab}\,\tilde{\tilde{\kappa}}^{a,r,s}\,\tilde{\tilde{\kappa}}^{b,t,u}[3] + O(1). \qquad (10.62)$$

Here, κ_{ab} denotes a generic element of $[\kappa^{a,b}]^{-1}$. The symbol $\bar{\kappa}^{r,s,\dots,a,b,\dots}$ indicates a residual cumulant, that is, a generic joint cumulant of $\bar{Z} = ([\bar{Z}_a], [\bar{Z}_r])$, where $\bar{Z}_a = U_a - \kappa^a$ and $\bar{Z}_r = T_r - \kappa^r - \beta_a^r(U_a - \kappa^a)$, with $\beta_a^r \doteq \kappa^{r,b}\kappa_{ab}$, the coefficients of linear least squares regression of T_r on U. The symbols $\tilde{\kappa}^{a,b,\dots}$ and $\tilde{\bar{\kappa}}^{r,s,\dots,a,b,\dots}$ indicate, as usual, evaluation at $(\tau, \hat{\zeta}_r)$. The residual cumulants which appear in (10.59)-(10.62) are given by the following formulae:

$$
\begin{aligned}
\bar{\kappa}^{a,b,r} &= \kappa^{a,b,r} - \beta_c^r \kappa^{a,b,c}, \\
\bar{\kappa}^{r,s} &= \kappa^{r,s} - \beta_a^r \beta_b^s \kappa^{a,b}, \\
\bar{\kappa}^{a,r,s} &= \kappa^{a,r,s} - \beta_b^s \kappa^{a,b,s}[2] + \beta_b^r \beta_c^s \kappa^{a,b,c}, \\
\bar{\kappa}^{a,b,r,s} &= \kappa^{a,b,r,s} - \beta_c^r \kappa^{a,b,c,s}[2] + \beta_c^r \beta_d^s \kappa^{a,b,c,d}, \\
\bar{\kappa}^{r,s,t} &= \kappa^{r,s,t} - \beta_a^r \kappa^{a,s,t}[3] + \beta_a^r \beta_b^s \kappa^{a,b,t}[3] - \beta_a^r \beta_b^s \beta_c^t \kappa^{a,b,c}, \\
\bar{\kappa}^{r,s,t,u} &= \kappa^{r,s,t,u} - \beta_a^r \kappa^{a,s,t,u}[4] + \beta_a^r \beta_b^s \kappa^{a,b,t,u}[6] \\
&\quad - \beta_a^r \beta_b^s \beta_c^t \kappa^{a,b,c,u}[4] + \beta_a^r \beta_b^s \beta_c^t \beta_d^u \kappa^{a,b,c,d}.
\end{aligned}
$$

The approximations (10.59)-(10.62) hold even if the joint distribution of (T, U) is not an exponential family (Pace and Salvan, 1992). If, for example, $Y_1, \dots, Y_n$ are independent observations with density

$$p(y; \theta) = \exp\{\theta \cdot s(y) - K(\theta)\}h(y), \tag{10.63}$$

for $\theta \in \tilde{\Theta} \subset \mathbb{R}^p$, with $\tilde{\Theta}$ the natural parameter space and if $T_n = T_n(Y_1, \dots, Y_n)$ is any statistic of order $O_p(n)$, then formulae (10.59)-(10.62) give approximations for the conditional cumulants of T_n given $S_n = \sum s(Y_i) = s$. Of course, the joint cumulants $\kappa^{a,b,\dots,r,s,\dots}$ of (S_n, T_n) must be available, and they must be evaluated at $\hat{\theta}$. This possibility is illustrated in the following example.

Example 10.13 *Approximation of the minimum variance unbiased estimator*
Let $Y_1, \dots, Y_n$ be independent observations with density (10.63) and let $T_n/n = T_n(Y_1, \dots, Y_n)/n$ be an unbiased estimator of $g(\theta)$, where $g(\cdot) : \tilde{\Theta} \to \mathbb{R}$ is a smooth function. The uniformly minimum variance unbiased estimator of $g(\theta)$ is $\hat{g}_{UMVU} = E(T_n/n | S_n = s)$. For this estimator (10.59) leads to the expansion

$$\hat{g}_{UMVU} = E_{\hat{\theta}}\left(\frac{T_n}{n}\right) - \frac{1}{2n}\hat{\kappa}_{ab}\left(\hat{\kappa}^{a,b,T} - \hat{\kappa}^{T,d}\hat{\kappa}_{dc}\hat{\kappa}^{a,b,c}\right) + O_p(n^{-2}), \tag{10.64}$$

where the indices $a, b, \dots = 1, \dots, p$ point to the components of S_n, while $\kappa^{a,b,T}$ and $\kappa^{T,d}$ indicate joint cumulants of S_n and T_n. All the cumulants indicated are evaluated at $\hat{\theta}$. The cumulant generating function of (S_n, T_n) is

$$K_{S_n, T_n}(z, v; \theta) = n\{K(\theta + z) - K(\theta)\} + \log E_{\theta+z}\left(e^{vT_n}\right).$$

On the other hand, because T_n is unbiased,

$$\frac{\partial}{\partial v} K_{S_n, T_n}(z, v; \theta)\bigg|_{z=0, v=0} = g(\theta).$$

Since $K_{S_n, T_n}(z, v; \theta)$ depends on v and θ only through $\theta + v$ then

$$\kappa^{T, d} = g(\theta)_{/d} = (\partial/\partial \theta^d) g(\theta)$$

and

$$\kappa^{a, b, T} = g(\theta)_{/ab} = (\partial^2/\partial \theta^a \partial \theta^b) g(\theta).$$

Therefore

$$\hat{g}_{UMVU} = g(\hat{\theta}) - \frac{1}{2n} \hat{\kappa}_{ab} \left(g(\hat{\theta})_{/ab} - g(\hat{\theta})_{/d} \, \hat{\kappa}_{dc} \hat{\kappa}^{a, b, c} \right) + O_p(n^{-2}). \qquad (10.65)$$

The UMVU estimator admits an asymptotic expansion which has the maximum likelihood estimator $g(\hat{\theta})$ as its leading term. This is hardly surprising since $\hat{\theta}$ is a minimal sufficient statistic. It would also seem natural to suppose that the correction term of order $O_p(n^{-1})$ in (10.65) is the same as the correction of $g(\hat{\theta})$ for the bias of order $O(n^{-1})$. This can be verified immediately using the expansion

$$E_\theta \left(g(\hat{\theta}) \right) = g(\theta) + g(\theta)_{/a} E_\theta(\hat{\theta} - \theta)^a + \frac{1}{2} g(\theta)_{/ab} E_\theta \left((\hat{\theta} - \theta)^a (\hat{\theta} - \theta)^b \right) + O(n^{-2})$$

and also using (9.62) and (9.73) specialized to a multiparameter exponential family. $\triangle$

If the reference model has density (10.55) with $k = 1$, integrating the approximate conditional density (10.56) gives very accurate approximations for observed significance levels of conditional tests based on the distribution of T given $U = u$ and for the confidence intervals for τ based on this conditional distribution. Skovgaard (1987) and Pierce and Peters (1992) obtained the following conditional versions of, respectively, (10.38) and (10.39):

$$P_\tau (T \leq t | U = u) = \Phi(r_P) + \phi(r_P) \left(\frac{1}{r_P} - \frac{1}{Cv_P} \right) + O(n^{-1}), \qquad (10.66)$$

$$P_\tau (T \leq t | U = u) = \Phi(r_P^*) + O(n^{-1}). \qquad (10.67)$$

The quantities which appear in (10.66) are defined as follows:

$$r_P = \text{sgn} (\hat{\tau} - \tau) \sqrt{2 \{l_P(\hat{\tau}) - l_P(\tau)\}}, \qquad (10.68)$$

with $l_p(\tau) = \tau \cdot t + \hat{\zeta}_\tau \cdot u - nK(\tau, \hat{\zeta}_\tau)$;

$$C = \left(\frac{\det[\hat{\kappa}^{a,b}]}{\det[\tilde{\kappa}^{a,b}]} \right)^{1/2} ; \qquad (10.69)$$

$$v_p = (\hat{\tau} - \tau)j_p(\hat{\tau})^{1/2}, \qquad (10.70)$$

with $j_p(\tau)$ given by (5.53). Lastly, in (10.67)

$$r_p^* = r_p + \frac{1}{r_p} \log\left(\frac{Cv_p}{r_p} \right). \qquad (10.71)$$

Pierce and Peters (1992) showed that the quantity $r_p^{-1} \log(C)$ controls the adjustment of r_p in order to take the presence of nuisance parameters into account, while $r_p^{-1} \log(v_p/r_p)$ effects a correction aimed at improving normal approximation. With suitable corrections, the approximations (10.66) and (10.67) are also valid in the discrete case (Pierce and Peters, 1992, section 4). If, for simplicity, we assume that, conditionally on $U = u$, T takes on successive integer values, one correction that offers satisfactory results is

$$P_\tau(T \le t | U = u) = \Phi\left(r_p^* + \frac{1}{r_p} \log h\{-(\hat{\tau} - \tau)\} \right) + O(n^{-1}), \qquad (10.72)$$

$$P_\tau(T \ge t | U = u) = 1 - \Phi\left(r_p^* + \frac{1}{r_p} \log h(\hat{\tau} - \tau) \right) + O(n^{-1}), \qquad (10.73)$$

with

$$h(\hat{\tau} - \tau) = \frac{1 - e^{-(\hat{\tau} - \tau)}}{\hat{\tau} - \tau}.$$

Example 10.14 *Conditional confidence intervals for the shape parameter of the gamma distribution*
With the same assumptions as in Example 5.13, we wish to obtain an approximation for the confidence interval of τ with level $1 - \alpha$, obtained as the intersection of the intervals based on the level $\alpha/2$ tests for $H_0 : \tau \le \tau_0$ versus $H_1 : \tau > \tau_0$ and for $H_0 : \tau \ge \tau_0$ versus $H_1 : \tau < \tau_0$. In other words, we wish to obtain an approximation of the interval $(\hat{\tau}_{inf}, \hat{\tau}_{sup})$, where $\hat{\tau}_{inf}$ satisfies the equation

$$P_{\hat{\tau}_{inf}}(T \le t | U = u) = 1 - \alpha/2, \qquad (10.74)$$

Table 10.4: *Coverage probabilities for* $\tau = 1$, $n = 5, 10$, $1 - \alpha = 0.9, 0.95$.

n	$1 - \alpha$	PC0	PC1
5	0.90	0.964	0.906
	0.95	0.978	0.954
10	0.90	0.931	0.898
	0.95	0.967	0.945

while $\hat{\tau}_{sup}$ satisfies the equation

$$P_{\hat{\tau}_{sup}}(T \leq t | U = u) = \alpha/2. \qquad (10.75)$$

Table 10.4 gives the results of a simulation (2000 trials) carried out in order to estimate the unconditional coverage probabilities of the intervals obtained from (10.74) and (10.75) using the approximation (10.67) and the coverage probabilities of the usual intervals based on the asymptotically pivotal quantity (10.70). The estimated coverage probabilities are indicated by PC1 and by PC0, respectively. The quantities required for calculating (10.71) are

$$l_p(\hat{\tau}) - l_p(\tau) = n\left\{(\hat{\tau} - \tau)(\psi(\hat{\tau}) - 1) + \tau \log\left(\frac{\hat{\tau}}{\tau}\right) - \log\left(\frac{\Gamma(\hat{\tau})}{\Gamma(\tau)}\right)\right\},$$

$$C = \sqrt{\tau/\hat{\tau}} \text{ and } j_p(\hat{\tau}) = n\left(\psi'(\hat{\tau}) - \hat{\tau}^{-1}\right). \qquad \triangle$$

Example 10.15 *Inference on the log-odds ratio*
In order to compare the behaviour of the approximate conditional log-likelihood function with the exact conditional log-likelihood and with the profile log-likelihood, consider the hypothetical data used in McCullagh and Nelder (1989, Table 7.1), given here in Table 10.5. As in McCullagh and Nelder (1989, section 7.4.1), we assume that t and v are independent realizations of $Bi(n, \phi_1)$ and $Bi(m, \phi_2)$, respectively. Using the same notation as in Example 5.12, let τ, the log-odds ratio, be the parameter of interest. On the basis of (5.35) the exact conditional log-likelihood is

$$l_C(\tau) = 2\tau - \log\left(4 + 18e^\tau + 12e^{2\tau} + e^{3\tau}\right).$$

Starting from (5.34), with simple calculations, the profile log-likelihood is seen to be

$$l_P(\tau) = 2\tau + 3\hat{\zeta}_\tau - \left\{3\log\left(1 + e^{\tau + \hat{\zeta}_\tau}\right) + 4\log\left(1 + e^{\hat{\zeta}_\tau}\right)\right\},$$

Table 10.5: *Hypothetical data from McCullagh and Nelder (1989, Table 7.1).*

	successes	failures	total
treatment	$t = 2$	1	$n = 3$
control	$v = 1$	3	$m = 4$
total	$u = 3$	4	7

with

$$e^{\hat{\zeta}_r} = \frac{\sqrt{(e^\tau + 1)^2 + 48e^\tau} - e^\tau - 1}{8e^\tau}.$$

The modification term required in order to obtain $l_{AC}(\tau)$ is

$$\log M(\tau) = \frac{1}{2} \log \left(3 \frac{e^{\tau + \hat{\zeta}_r}}{\left(1 + e^{\tau + \hat{\zeta}_r}\right)^2} + 4 \frac{e^{\hat{\zeta}_r}}{\left(1 + e^{\hat{\zeta}_r}\right)^2} \right).$$

The values of τ which maximize $l_c(\tau)$, $l_p(\tau)$ and $l_{AC}(\tau)$ are about equal to 1.493, 1.792 and 1.567, respectively. Figure 10.7 compares the behaviour of the three log-likelihoods by means of the graphs of the corresponding combinants W. This is the same as considering normalized log-likelihoods by eliminating the additive constants which are different for each of the three functions. The approximation $l_{AC}(\tau)$ of $l_c(\tau)$ is fairly accurate. △

10.11 Bibliographic Note

The Edgeworth expansion has been used in statistics for a long time. Recent presentations are in Barndorff-Nielsen and Cox (1989), Reid (1991) and Kolassa (1994). See any of these for further bibliographic references. Regularity conditions for the extension of the Edgeworth expansion to smooth functions of sample means are examined in depth in Bai and Rao (1991). Mykland (1995b) considers Edgeworth expansions for martingales. A brief summary of the properties of Hermite polynomials and of other orthogonal polynomials is given in Thisted (1988, section 5.3.2).

Barndorff-Nielsen and Cox (1979) underline the usefulness of Edgeworth and saddlepoint approximations in parametric inference. This paper is the starting point for many contributions in subsequent years. Reid (1988) offers an introduction to saddlepoint method in statistics. Jensen (1995c) gives

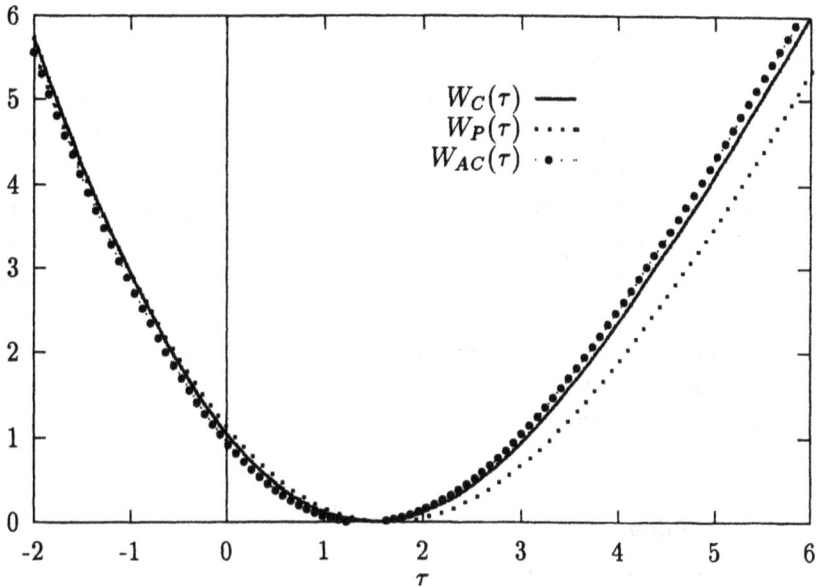

Figure 10.7: *Behaviour of* $2(\hat{l} - l)$ *for* $l_C(\tau)$, $l_P(\tau)$ *and* $l_{AC}(\tau)$ *with data from Table 10.5.*

a brief introduction to Edgeworth and saddlepoint approximations, with numerical examples and historical remarks. The exponential tilting method for obtaining asymptotic expansions, first introduced by Esscher (1932), is used in Daniels (1954) and in Barndorff-Nielsen and Cox (1989, Chapter 4); for a more recent examination, which pays particular attention to regularity conditions, see Jensen (1995a, Chapters 1 and 2). For an introduction to saddlepoint approximations through inversion of the characteristic function, see Daniels (1954), Field and Ronchetti (1990), Kolassa (1994). The monograph by Field and Ronchetti (1990) deals specifically with approximations for the distribution of solutions to estimating equations; see, also, Daniels (1983), Ronchetti (1990), Jensen (1995a, section 4.2), Gatto and Ronchetti (1996) and Strawderman, Casella and Wells (1996). For applications to time series, see Taniguchi (1991), Jensen (1995a, Chapter 9), Wang (1992a), Lieberman (1994a). For applications in multivariate statistics, see Jensen (1991), Butler, Huzurbazar and Booth (1992a, 1992b, 1993), Lieberman (1994b). In the nonparametric case, and also for approximating bootstrap distributions, see Robinson (1982), Davison and Hinkley (1988), Feuerverger (1989), Wang (1990, 1992b, 1993), Ronchetti and Welsh (1994), Monti and Ronchetti (1993) and DiCiccio, Martin and Young (1994). Harvill and Newton (1995) consider an application to order statistics. For the applications to conditional inference in exponential families and in generalized linear models considered in section 10.10.2, see Davison (1988). See Pierce and Peters (1992) for further numerical examples. Kolassa and Tanner (1994) and Forster, McDonald and Smith (1996) consider a Monte Carlo method based on the Gibbs sampler for estimating conditional distributions. A different approach to approximate conditional inference is outlined by Waterman and Lindsay (1996a,b). The main objective is the approximation of the conditional score through projection of the profile score in the linear space spanned by the derivatives of the log-likelihood function with respect to the nuisance parameter. Kolassa (1996) obtains a second-order saddlepoint approximation for conditional distribution functions that extends the results of Skovgaard (1987).

10.12   Problems

10.1 Let S_n be the sum of n independent, but not identically distributed random variables, with c.g.f. $K_{Y_i}(t)$, $i = 1, \ldots, n$. Obtain the first four cumulants of S_n^* and establish the order of $\rho_3(S_n^*)$ and of $\rho_4(S_n^*)$.

[Section 10.1]

10.2 Prove the orthogonality relations (10.8). (*Hint*: suppose that $s > r$ and integrate by parts $r + 1$ times.)

[Section 10.2]

10.3 Let S_n be a sum of independent and identically distributed random variables with mean μ and variance σ^2. Suppose that the Edgeworth expansion (10.10) holds for the density of the standardized sum S_n^*. Furthermore, let $g(\cdot)$ be a real-valued function defined on $\mathbb{R}$, with continuous first derivative g' such that $g'(\mu) \neq 0$ and which has derivatives up to the fourth order. Using asymptotic expansions for the cumulants of $g(S_n/n)$ up to the fourth order, write the Edgeworth expansion for the density of $g(S_n/n)$ so as to obtain a representation with leading term $\phi((g(S_n/n) - g(\mu))/\sqrt{\sigma^2 g'(\mu)^2/n})/\sqrt{\sigma^2 g'(\mu)^2/n}$. Show that the expansion obtained is asymptotically equivalent to that which one would have after transforming the approximate density (10.11) of S_n^* adopting the usual rules.

[Section 10.3; Barndorff-Nielsen and Cox, 1989, section 4.5]

10.4 Consider $\bar{Y}_n$, the sample mean of n i.i.d. copies of the r.v. Y with logistic distribution, $Lo(\mu, \sigma)$ (see Table 3.1). Obtain the Edgeworth expansion for the d.f. of $\bar{Y}_n$.

[Section 10.4]

10.5 Do detailed calculations which give the Edgeworth approximation for the distribution function of the standardized mean given in Example 10.5. Justify the order of error. Generalize the structure of the expansions to a generic symmetric distribution.

[Section 10.5]

10.6 With reference to Example 10.5, state whether the error in the normal approximation of S_n^* depends on (μ, σ). Investigate the values of n for which $F_{S_n^*}^E(\cdot)$ is a distribution function.

[Section 10.5]

10.7 Show that, for the mode y_{mod} of S_n^*, an immediate application of the expansion (10.10) gives the expansion

$$y_{\text{mod}} = -\frac{\rho_3}{2\sqrt{n}} + O(n^{-3/2}). \qquad (10.76)$$

Exploiting the Cornish–Fisher expansion, show that the median of S_n^* is

$$y_{0.5} = -\frac{\rho_3}{6\sqrt{n}} + O(n^{-3/2}). \qquad (10.77)$$

Verify that relations (10.76) and (10.77) make it possible to obtain the traditional approximate relation for unimodal distributions

$$(\text{mean - mode}) = 3(\text{mean - median}).$$

[Sections 10.3 and 10.6]

10.8 Under the same assumptions as in Example 10.5 obtain the saddlepoint approximations for the density and for the distribution function of S_n^*. Perform a small simulation study to compare the accuracy of the various approximations.

[Section 10.7]

10.9 Let $Y_1, \ldots, Y_n$ be independent random variables with an $IG(\phi, \lambda)$ distribution. Show that the normalized saddlepoint approximation for the density of $\bar{Y}_n$ gives the exact result (see Theorem 6.1).

[Section 10.7]

10.10 Let $\bar{Y}_n$ be the sample mean of n i.i.d. copies of the r.v. Y with logistic distribution $Lo(\mu, \sigma)$ (see Table 3.1). Obtain the saddlepoint approximations for both the density and the distribution function of $\bar{Y}_n$. Bearing in mind the results of Problem 10.4, perform a small simulation study in order to compare the accuracy of the various approximations.

[Sections 10.7 and 10.8]

10.11 Justify exactness of the normalized saddlepoint approximation for the density of $S_n = \sum Y_i$ when Y_i, $i = 1, \ldots, n$, are independent r.v.'s with a $N(\mu, \sigma^2)$ or a $Ga(\nu, 1)$ distribution.

[Section 10.7]

10.12 Obtain the expressions for the approximations of the conditional log-likelihood for τ based on (10.57) for the problems considered in Examples 5.11, 5.12 and 5.13. Compare these approximations with the exact expressions given in these three examples.

[Section 10.10.2]

10.13 Assume that the reference statistical model has density (10.55). Exploiting expansions for conditional cumulants, obtain an approximation for the conditional maximum likelihood estimating equation.

[Section 10.10.2]

10.14 Let Y_i, $i = 1, \ldots, n$, be independent $Ga(1, 1/\lambda_1)$ random variables for $i = 1, \ldots, m$ ($m < n$) and $Ga(1, 1/\lambda_2)$ for $i = m + 1, \ldots, n$. Let $\tau = \lambda_1 - \lambda_2$. Obtain an approximation for the observed significance level of the uniformly most powerful similar test for $H_0 : \tau = 0$ versus $H_1 : \tau < 0$. Calculate this approximation with $m = 16$, $n - m = 19$, $\sum_{i=1}^{n} y_i = 923$, $\sum_{i=m+1}^{n} y_i = 234$. Compare the approximate value with that based on first-order asymptotic theory, that is, on r_p and with the exact value calculated starting from the null distribution of $T = \sum_{i=m+1}^{n} Y_i$ given $S = \sum_{i=1}^{n} Y_i = s$.

[Section 10.10.2; Pace and Salvan, 1992]

10.15 Under the same assumptions as in Example 5.14, and exploiting the result of Problem 5.31, use the saddlepoint expansion to obtain an approximation for the distribution function of $-\log(R)$, where the statistic R is defined by (5.69). Discuss the relation between the confidence intervals for τ based on this approximation and those obtained in Example 10.14.

[Section 10.10.2]

10.16 With reference to the model (10.55), and exploiting the approximations (10.59)-(10.62) for conditional cumulants, obtain an asymptotic expansion for $\hat{\tau}_c$, the conditional maximum likelihood of τ (obtained by maximizing (5.33)), in terms of of $\hat{\tau}$. Obtain an asymptotic expansion with error of order $O(n^{-3})$ for the variance of $\hat{\tau}_c$.

[Section 10.10.2]

CHAPTER 11

LIKELIHOOD AND HIGHER-ORDER ASYMPTOTICS

11.1 Introduction

The central limit theorem allows us to obtain the usual first-order asymptotic approximations for the distribution of likelihood quantities. Analogously, the higher-order results, presented in Chapter 10, make it possible to develop higher-order asymptotic methods for inference based on the likelihood. These methods are often fairly recent, having first appeared in the 1980s. Research is far from being complete and these methods are not yet widely used in applications either. This chapter will present some of the main results and highlight the concepts, techniques and examples. Many results will only be stated, and the reader referred to the literature for proofs.

Most higher-order asymptotic theory for likelihood inference can be justified by means of the p^* formula of Barndorff-Nielsen (1980, 1983), which will be presented in section 11.2. This is an approximation for the density of the maximum likelihood estimator, conditionally on an ancillary statistic, which generalizes the saddlepoint expansion for the density of $\hat{\theta}$ in full exponential families. Applications of the p^* formula are usually not direct. Rather, it is a general tool for obtaining specific refinements of first-order results which incorporate the conditionality principle. These refinements particularly concern null distributions of W and $Z = \text{sgn}(\hat{\theta} - \theta)\sqrt{W}$, even in the presence of nuisance parameters, and suitable modifications of the profile likelihood. Results for the log-likelihood ratio statistic are illustrated in sections 11.3 and 11.4. Refinements of the distribution of one-sided versions, like Z, are in section 11.5. Some modifications of the profile likelihood are illustrated in section 11.6.

431

11.2   Approximation for the Distribution of $\hat{\theta}$: the $p^\star$ Formula

Let $y_1, \ldots, y_n$ be independent observations of a random variable with density in an $\mathcal{F}^1_{ne}$,

$$p(y; \theta) = \exp\{\theta y - K(\theta)\} p_0(y),$$

with $\theta \in \tilde{\Theta} \subseteq \mathbb{R}$, where $\tilde{\Theta}$ is the natural parameter space. If $y = (y_1, \ldots, y_n)$ and $s = \sum_{i=1}^n y_i$, then the log-likelihood is

$$l(\theta; y) = l(\theta; s) = \theta s - nK(\theta).$$

Because of the one-to-one correspondence which, in an $\mathcal{F}^1_{ne}$, binds s with $\hat{\theta}$, expressed by the likelihood equation $nK'(\hat{\theta}) = s$, the log-likelihood could equally well be written as (see relation (5.45))

$$l(\theta; s) = l(\theta; \hat{\theta}) = n\left\{\theta K'(\hat{\theta}) - K(\theta)\right\},$$

hence

$$n\left\{K(\hat{\theta}) - K(\theta)\right\} - (\hat{\theta} - \theta)s = l(\theta; \hat{\theta}) - l(\hat{\theta}; \hat{\theta}).$$

Thus, the saddlepoint expansion for the density of $S_n = \sum_{i=1}^n Y_i$, given by (10.34), can also be written as

$$p_{S_n}(s; \theta) = \frac{1}{\sqrt{2\pi n K''(\hat{\theta})}} \exp\left\{l(\theta; \hat{\theta}) - l(\hat{\theta}; \hat{\theta})\right\}\left\{1 + O(n^{-1})\right\}, \qquad (11.1)$$

where $\hat{\theta} = \hat{\theta}(s)$. Expansion (11.1) holds, both when the distribution of S_n is continuous and when it is discrete. In the continuous case, since

$$\frac{ds(\hat{\theta})}{d\hat{\theta}} = nK''(\hat{\theta}) = j(\hat{\theta}) = \hat{j},$$

from (11.1) the density of $\hat{\theta}$ has the expansion

$$p_{\hat{\theta}}(\hat{\theta}; \theta) = p^\star(\hat{\theta}; \theta)\left\{1 + O(n^{-3/2})\right\}, \qquad (11.2)$$

where

$$p^\star(\hat{\theta}; \theta) = c(\theta)|j(\hat{\theta})|^{1/2} \exp\{l(\theta; \hat{\theta}) - l(\hat{\theta}; \hat{\theta})\}. \qquad (11.3)$$

Note that $(2\pi)^{-1/2}$ has been replaced by $c(\theta)$ thinking that a renormalization has been carried out, which also has the effect of reducing the order of error from $O(n^{-1})$ to $O(n^{-3/2})$, at least for values of $\hat{\theta}$ in a normal deviation region $\sqrt{n}|\hat{\theta} - \theta| \leq c_0$.

Expression (11.3) is a special case of the p^* formula of Barndorff-Nielsen (1980, 1983). He showed that expansion (11.2) can be extended beyond full exponential families provided attention is paid to the distribution of $\hat{\theta}$ conditionally on an ancillary statistic. In order to introduce this result, it is worth examining how (11.3) could be interpreted within a generic one-parameter model $\mathcal{F}$ with minimal sufficient statistic s. Although (11.3) is expressed only in terms of likelihood quantities, we may write $l(\theta; s) = l(\theta; \hat{\theta})$ only if $\hat{\theta}$ is a sufficient statistic. Furthermore, outside of natural exponential families, $j(\theta)$ is $j(\theta; s)$. However, if a is ancillary and, thus, the correspondence between the minimal sufficient statistic s and $(\hat{\theta}, a)$ is one-to-one, then we may write

$$l(\theta; s) = l(\theta; \hat{\theta}, a)$$

and

$$j(\theta; s) = j(\theta; \hat{\theta}, a).$$

Hence, it would seem natural to think that, if (11.3) could be extended to a generic model $\mathcal{F}$, $l(\theta; \hat{\theta})$ should be replaced with $l(\theta; \hat{\theta}, a)$ and $j(\theta)$ with $j(\theta; \hat{\theta}, a)$. Since $\hat{\theta}$ is minimal sufficient in the conditional model of $\hat{\theta}$ given a, a reasonable conjecture is that the extension of (11.2) and (11.3) is in the conditional form

$$p_{\hat{\theta}|A=a}(\hat{\theta}; \theta, a) = p^*(\hat{\theta}; \theta|a)\{1 + O(n^{-3/2})\}, \tag{11.4}$$

where

$$p^*(\hat{\theta}; \theta|a) = c(\theta, a)\left|j(\hat{\theta}; \hat{\theta}, a)\right|^{1/2} \exp\{l(\theta; \hat{\theta}, a) - l(\hat{\theta}; \hat{\theta}, a)\}, \tag{11.5}$$

with $c(\theta, a)$ a normalizing constant,

$$c(\theta, a)^{-1} = \int_{\hat{\Theta}} \left|j(\hat{\theta}; \hat{\theta}, a)\right|^{1/2} e^{l(\theta; \hat{\theta}, a) - l(\hat{\theta}; \hat{\theta}, a)} d\hat{\theta},$$

where $\hat{\Theta}$ is the support of $\hat{\theta}$.

Before discussing the validity of (11.4), let us consider its formal multiparameter extension. Let $\mathcal{F}$ be a parametric statistical model with parameter space $\Theta \subseteq \mathbb{R}^p$, $p \geq 1$. The saddlepoint expansion (10.48) for the distribution of the minimal sufficient statistic in an $\mathcal{F}^p_{ne}$ differs from (11.1) only in the factor $\{\det(n\hat{K}_{\theta\theta})\}^{-1/2}$ instead of $(nK''(\hat{\theta}))^{-1/2}$; on the other hand, the factor

$$e^{(\theta-\hat{\theta})\cdot s - n(K(\theta) - K(\hat{\theta}))}$$

is again equal to

$$e^{l(\theta;\hat\theta)-l(\hat\theta;\hat\theta)}$$

because, taking the relation $s = nK_\theta(\hat\theta)$ into account, with $K_\theta(\theta) = (\partial/\partial\theta)K(\theta)$, we can write

$$l(\theta;\hat\theta) = n\{\theta \cdot K_\theta(\hat\theta) - K(\theta)\}.$$

These considerations suggest the general expansion

$$p_{\hat\theta|A=a}(\hat\theta;\theta,a) = p^\star(\hat\theta;\theta|a)\{1 + O(n^{-3/2})\}, \qquad (11.6)$$

where

$$p^\star(\hat\theta;\theta|a) = c(\theta,a)\left|j(\hat\theta;\hat\theta,a)\right|^{1/2}\exp\{l(\theta;\hat\theta,a) - l(\hat\theta;\hat\theta,a)\}. \qquad (11.7)$$

Expression (11.7) is known as Barndorff-Nielsen's $p^\star$ formula. It is simple to check directly that, for scale and location families, this gives the exact distribution of $\hat\theta = (\hat\mu,\hat\sigma)$ conditionally on $a = (a_1,\ldots,a_n)$, where a is the configuration statistic with $a_i = (y_i - \hat\mu)/\hat\sigma$, $i = 1,\ldots,n$. Indeed, if $y_1,\ldots,y_n$ are independent realizations of a random variable Y with density $p(y;\mu,\sigma) = \sigma^{-1}p_0((y-\mu)/\sigma)$, $\mu \in \mathbb{R}$, $\sigma > 0$, from (7.33) we have

$$l(\mu,\sigma;\hat\mu,\hat\sigma,a) - l(\hat\mu,\hat\sigma;\hat\mu,\hat\sigma,a)$$
$$= -n\log\sigma - \sum_{i=1}^n q_0\left(\frac{\hat\sigma}{\sigma}a_i + \frac{\hat\mu-\mu}{\sigma}\right) + n\log\hat\sigma + \sum_{i=1}^n q_0(a_i).$$

Furthermore, from Example 9.9, we have

$$\left|j(\hat\mu,\hat\sigma;\hat\mu,\hat\sigma,a)\right|^{1/2} = (\hat\sigma)^{-2}\sqrt{D(a)},$$

where

$$D(a) = \left\{\left(\sum_{i=1}^n q_0''(a_i)\right)\left(n + \sum_{i=1}^n a_i^2 q_0''(a_i)\right) - \left(\sum_{i=1}^n a_i q_0''(a_i)\right)^2\right\}.$$

Hence the $p^\star$ formula (11.7) becomes

$$p^\star(\hat\mu,\hat\sigma;\mu,\sigma|a) = c(\mu,\sigma,a)\frac{\hat\sigma^{n-2}}{\sigma^n}\prod_{i=1}^n p_0\left(\frac{\hat\sigma}{\sigma}a_i + \frac{\hat\mu-\mu}{\sigma}\right), \qquad (11.8)$$

which matches the exact density (7.38); furthermore, comparison with this latter shows that the normalizing constant $c(\mu, \sigma, a)$ in (11.7) does not depend on (μ, σ).

Barndorff-Nielsen, Blæsild and Eriksen (1989, Theorem 8.1) show that, for all group models that satisfy weak regularity conditions, (11.7) matches the exact formula for the density of the maximum likelihood estimator conditionally on the maximal invariant a. Furthermore, for such models the normalizing constant does not depend on the parameter θ, that is, $c(\theta, a) = c(a)$.

Other cases of exactness of the $p^\star$ formula can be found under families with $IG(\phi, \lambda)$ distribution (see Problem 11.1) and with generalized inverse Gaussian distribution (Jørgensen, 1982).

Under regularity conditions, the $p^\star$ formula also holds outside of the exponential or group families. It usually gives an approximation for the distribution of $\hat{\theta}$ given a with a relative error of order $O(n^{-3/2})$. See Skovgaard (1990) for an overall discussion and for regularity conditions. Heuristically, the argument to show the general validity of the $p^\star$ formula is as follows (see also Jensen, 1995a, section 4.2). Because $(\hat{\theta}, a)$ is sufficient, for $\theta, \theta_0 \in \Theta$, we can write

$$p_{\hat{\theta}, A}(\hat{\theta}, a; \theta_0) = \exp\{l(\theta_0; \hat{\theta}, a) - l(\theta; \hat{\theta}, a)\} p_{\hat{\theta}, A}(\hat{\theta}, a; \theta) \,.$$

Dividing both sides by $p_A(a)$, we obtain

$$\frac{p_{\hat{\theta}, A}(\hat{\theta}, a; \theta_0)}{p_A(a)} = \exp\{l(\theta_0; \hat{\theta}, a) - l(\theta; \hat{\theta}, a)\} \frac{p_{\hat{\theta}, A}(\hat{\theta}, a; \theta)}{p_A(a)}$$

and, hence

$$p_{\hat{\theta}|A=a}(\hat{\theta}; \theta_0, a) = \exp\{l(\theta_0; \hat{\theta}, a) - l(\theta; \hat{\theta}, a)\} p_{\hat{\theta}|A=a}(\hat{\theta}; \theta, a) \,. \tag{11.9}$$

Note that in order to draw this conclusion $p_A(a)$ must not depend on θ, at least not asymptotically in a neighbourhood of θ_0.

Now it is possible to approximate the right-hand side of (11.9) using the same technique as that adopted in section 10.7 to obtain the saddlepoint expansion. This means approximating $p_{\hat{\theta}|A=a}(\hat{\theta}; \theta, a)$ according to the Edgeworth expansion with θ chosen so as to satisfy the equation $E_\theta(\hat{\theta}|a) = \hat{\theta}$. As a value of θ, $\hat{\theta}$ does not exactly satisfy this relation, however, the asymptotic expansion

$$E_\theta(\hat{\theta}|a) = \theta + O(n^{-1}) \tag{11.10}$$

holds (see Problem 11.4). Furthermore, since

$$\phi_p(x; \kappa) = \phi_p(0; \kappa) \left\{ 1 + O(\|x\|^2) \right\} \,,$$

the approximation sought for the left-hand side of (11.9) is

$$p_{\hat{\theta}|A=a}(\hat{\theta};\theta_0,a) = e^{l(\theta_0;\hat{\theta},a)-l(\hat{\theta};\hat{\theta},a)}\phi_p(0;\text{Var}_{\hat{\theta}}(\hat{\theta}|a))\left\{1+O(n^{-1})\right\}. \quad (11.11)$$

The error is of order $O(n^{-1})$ since the argument of the Hermite polynomials of degree three is of order $O(n^{-1/2})$. Lastly, the approximation $\text{Var}_{\hat{\theta}}(\hat{\theta}|a) = \hat{\jmath}^{-1} + O(n^{-2})$ (see Problem 11.4) gives

$$p_{\hat{\theta}|A=a}(\hat{\theta};\theta_0,a) = e^{l(\theta_0;\hat{\theta},a)-l(\hat{\theta};\hat{\theta},a)}(2\pi)^{-p/2}\left|\hat{\jmath}\right|^{1/2}\left\{1+O(n^{-1})\right\}. \quad (11.12)$$

Relation (11.6) is obtained from (11.12) by means of renormalization. The technically delicate aspect of this, which will not be examined in depth here, concerns the validity of (11.6). Usually, the quality of the p^* approximation is comparable with that of saddlepoint approximations: the relative error of order $O(n^{-1})$ can be uniformly bounded in a large deviation region $\hat{\theta} - \theta = O(1)$, while in a normal deviation region the error is of order $O(n^{-3/2})$ (see Skovgaard, 1990).

The statistic a in (11.7) could also be ancillary only in an approximate sense, for example with density $p_A(a;\theta) = p_0(a) + O(n^{-1/2})$, as is the case for the Efron-Hinkley ancillary (5.90). Then, (11.6) holds with relative error of order $O(n^{-1})$, for a in a normal deviation region. See Barndorff-Nielsen (1984) and Skovgaard (1990) for further study.

The p^* formula also has the following properties:
i) it is invariant under one-to-one transformations of the data y;
ii) it transforms in a regular manner under a reparameterization; more precisely, if $\psi = \psi(\theta)$ is a reparameterization then

$$p^*_{\hat{\psi}|a}(\hat{\psi};\psi,a) = p^*_{\hat{\theta}|a}(\theta(\hat{\psi});\theta(\psi),a)|\partial\theta(\hat{\psi})/\partial\hat{\psi}|;$$

note that this is the transformation rule for densities with respect to the Lebesgue measure under smooth one-to-one transformations;
iii) the normalizing constant, $c(\theta,a)$, is parameterization invariant.

Calculating the normalizing constant $c(\theta,a)$ requires that a representation of s as $(\hat{\theta},a)$ should be available. Let

$$\bar{c}(\theta,a) = c(\theta,a)(2\pi)^{p/2}; \quad (11.13)$$

then, as is suggested by (11.12),

$$\bar{c}(\theta,a) = 1 + O(n^{-1}) \quad (11.14)$$

(see Barndorff-Nielsen and Cox, 1994, section 6.3). Thus we obtain the approximation with error of order $O(n^{-1})$

$$p^\dagger_{\hat{\theta}|a}(\hat{\theta}; \theta, a) = (2\pi)^{-p/2} \left| \hat{j} \right|^{1/2} \exp\{l(\theta; \hat{\theta}, a) - l(\hat{\theta}; \hat{\theta}, a)\}. \qquad (11.15)$$

The error remains of order $O(n^{-1})$ even if the statistic a used in the representation $l(\theta; \hat{\theta}, a)$ is ancillary in the approximate sense with density $p_A(a; \theta) = p_0(a) + O(n^{-1/2})$. Hence, the approximation given by (11.15) is stable in the sense defined in section 2.8 (see also Barndorff-Nielsen and Cox, 1994, p. 176).

Lastly, with the expansion

$$\begin{aligned}
l(\theta; \hat{\theta}, a) - l(\hat{\theta}; \hat{\theta}, a) &= -(1/2)(\hat{\theta} - \theta)^{rs} \hat{j}_{rs} + O_p(n^{-1/2}) \\
&= -(1/2)(\hat{\theta} - \theta)^{rs} i_{rs} + O_p(n^{-1/2}),
\end{aligned}$$

(11.15) gives the usual normal approximation for the null distribution of $\hat{\theta}$.

If $p = 1$ and if $\hat{\theta}$ has continuous distribution, integration of the $p^\star$ formula, adopting the same method of approximation as for the Lugannani–Rice expansion (10.38), gives (Barndorff-Nielsen and Cox, 1994, section 6.7; Reid, 1995b)

$$P_\theta(\hat{\theta} \le \hat{\theta}^0 | a) = \left\{ \Phi(r_{\hat{\theta}^0}) + \phi(r_{\hat{\theta}^0}) \left(\frac{1}{r_{\hat{\theta}^0}} - \frac{1}{v_{\hat{\theta}^0}} \right) \right\} \left\{ 1 + O(n^{-3/2}) \right\}, \qquad (11.16)$$

where

$$r_{\hat{\theta}} = \text{sgn}(\hat{\theta} - \theta) \left\{ 2 \left(l(\hat{\theta}; \hat{\theta}, a) - l(\theta; \hat{\theta}, a) \right) \right\}^{1/2} \qquad (11.17)$$

and

$$v_{\hat{\theta}} = \left\{ l_{;1}(\theta; \hat{\theta}, a) - l_{;1}(\hat{\theta}; \hat{\theta}, a) \right\} \left(j(\hat{\theta}) \right)^{-1/2} \qquad (11.18)$$

The symbol $l_{;1}(\theta; \hat{\theta}, a)$ indicates the partial derivative $\partial l(\theta; \hat{\theta}, a)/\partial \hat{\theta}$. Expression (11.16) could also be given in the asymptotically equivalent form (Jensen, 1995a, Theorem 5.1.1)

$$P_\theta(\hat{\theta} \le \hat{\theta}^0 | a) = \Phi(r^\star_{\hat{\theta}^0}) \left\{ 1 + O(n^{-3/2}) \right\}, \qquad (11.19)$$

with

$$r^\star_{\hat{\theta}} = r_{\hat{\theta}} + \frac{1}{r_{\hat{\theta}}} \log \frac{v_{\hat{\theta}}}{r_{\hat{\theta}}}. \qquad (11.20)$$

As for the Lugannani–Rice expansion, the singularity at $\hat{\theta}^0 = \theta$ in (11.16) and (11.19) can be removed. Let us define the mixed derivative

$$l_{m;n}(\theta; \hat{\theta}, a) = \frac{\partial^{m+n}}{\partial \theta^m \partial \hat{\theta}^n} l(\theta; \hat{\theta}, a). \qquad (11.21)$$

Then, let

$$t_3 = -\frac{l_{3;}(\theta; \theta, a) + 3l_{2;1}(\theta; \theta, a)}{(l_{1;1}(\theta; \theta, a))^{3/2}} ;$$

the quantities which enter (11.16) and (11.19) at $\hat{\theta}_0 = \theta$, due to continuity, become

$$\Phi(r_\theta) + \phi(r_\theta) \left(\frac{1}{r_\theta} - \frac{1}{v_\theta} \right) = \frac{1}{2} + \frac{t_3}{6\sqrt{2\pi}} \qquad (11.22)$$

and

$$r_\theta^\star = \frac{t_3}{6} .$$

See Barndorff-Nielsen and Cox (1994, section 6.7) for guidelines on how to prove these relations and for some numerical examples which compare the accuracy of (11.16) and of (11.19).

11.3 Null Distribution of W

Let us the discuss first the one-parameter case. Assume that $\hat{\theta}$ has a continuous distribution which can be approximated using the $p^\star$ formula (11.5). This approximation can be rewritten as

$$p_{\hat{\theta}|A=a}(\hat{\theta}; \theta, a) = c(\theta, a)| \hat{j} |^{1/2} e^{-w/2}\{1 + O(n^{-3/2})\}, \qquad (11.23)$$

where $w = 2(l(\hat{\theta}; \hat{\theta}, a) - l(\theta; \hat{\theta}, a))$. For fixed a, w is not a one-to-one function of $\hat{\theta}$. Specifically, taking into account the expansion $w = i(\theta)(\hat{\theta} - \theta)^2 + \dots$, locally at least, there are two values of $\hat{\theta}$ which give the same value of w. We assume that, for fixed a, the mapping $\hat{\theta} \to w$, restricted, respectively, to $\hat{\theta} < \theta$ and to $\hat{\theta} > \theta$ (the event $\hat{\theta} = \theta$ has zero probability), is one-to-one. Hence, $\hat{\theta}$ can be expressed in terms of w according to the functions $\hat{\theta}_1(w)$ for $\hat{\theta} < \theta$ and $\hat{\theta}_2(w)$ for $\hat{\theta} > \theta$. Thus, the approximation for the density of W is

$$p_{W|A=a}(w; \theta, a) = c(\theta, a)e^{-w/2} \sum_{h=1}^{2} \frac{\left|j(\hat{\theta}_h(w))\right|^{1/2}}{\left|\partial w/\partial \hat{\theta} \right|_{\hat{\theta}=\hat{\theta}_h(w)}} \{1 + O(n^{-3/2})\}. \qquad (11.24)$$

The next step is to obtain an asymptotic expansion for the sum that appears on the right-hand side of (11.24). Derivatives of $l(\theta; \hat{\theta}, a)$ with respect to $\hat{\theta}$, that is, mixed derivatives $l_{m;n} = l_{m;n}(\theta; \hat{\theta}, a)$ defined by (11.21), will appear in this expansion. Furthermore, let

$$\hat{l}_{m;n} = l_{m;n}(\theta; \hat{\theta}, a)\Big|_{\theta=\hat{\theta}} .$$

Note that

$$\frac{\partial w}{\partial \hat{\theta}} = 2(\hat{l}_{1;} + \hat{l}_{;1} - l_{;1}) = 2(\hat{l}_{;1} - l_{;1}), \tag{11.25}$$

since, because of the likelihood equation, $\hat{l}_{1;} = 0$. Consider the expansion of $l_{;1}$, as a function of θ, around $\hat{\theta}$

$$l_{;1} = \hat{l}_{;1} + \hat{l}_{1;1}(\theta - \hat{\theta}) + \frac{1}{2}\hat{l}_{2;1}(\theta - \hat{\theta})^2 + \frac{1}{6}\hat{l}_{3;1}(\theta - \hat{\theta})^3 + \ldots. \tag{11.26}$$

Differentiation of the likelihood equation $\hat{l}_{1;} = 0$ with respect to $\hat{\theta}$, gives

$$\hat{l}_{2;} + \hat{l}_{1;1} = 0,$$

so that $\hat{l}_{1;1} = -\hat{l}_{2;} = \hat{j}$. Therefore (11.25) can be expanded as

$$\begin{aligned}
\frac{\partial w}{\partial \hat{\theta}} &= 2\left\{(\hat{\theta} - \theta)\hat{j} - \frac{1}{2}\hat{l}_{2;1}(\hat{\theta} - \theta)^2 + \frac{1}{6}\hat{l}_{3;1}(\hat{\theta} - \theta)^3 + \ldots\right\} \\
&= 2(\hat{\theta} - \theta)\hat{j}\left\{1 - \frac{\hat{l}_{2;1}}{2\hat{j}}(\hat{\theta} - \theta) + \frac{\hat{l}_{3;1}}{6\hat{j}}(\hat{\theta} - \theta)^2 + \ldots\right\}. \tag{11.27}
\end{aligned}$$

Expansion (11.24) can thus be rewritten in the form

$$p_{W|A=a}(w; \theta, a) = \frac{c(\theta, a)}{2}e^{-w/2}$$

$$\times \sum_{h=1}^{2} \frac{1}{\left|(\hat{\theta} - \theta)\hat{j}^{1/2}\left\{1 - \frac{\hat{l}_{2;1}}{2\hat{j}}(\hat{\theta} - \theta) + \frac{\hat{l}_{3;1}}{6\hat{j}}(\hat{\theta} - \theta)^2 + \ldots\right\}\right|_{\hat{\theta}=\hat{\theta}_h(w)}}$$

$$\times \left\{1 + O(n^{-3/2})\right\}. \tag{11.28}$$

Expanding $l(\theta; \hat{\theta}, a)$ around $\hat{\theta}$, we get

$$l(\theta; \hat{\theta}, a) = l(\hat{\theta}; \hat{\theta}, a) + \hat{l}_{1;}(\theta - \hat{\theta}) + \frac{1}{2}\hat{l}_{2;}(\theta - \hat{\theta})^2 + \frac{1}{6}\hat{l}_{3;}(\theta - \hat{\theta})^3 + \frac{1}{24}\hat{l}_{4;}(\theta - \hat{\theta})^4 + \ldots$$

from which we obtain the expansion for w

$$w = (\hat{\theta} - \theta)^2\hat{j}\left\{1 + \frac{1}{3\hat{j}}\hat{l}_{3;}(\hat{\theta} - \theta) - \frac{1}{12\hat{j}}\hat{l}_{4;}(\hat{\theta} - \theta)^2 + \ldots\right\}, \tag{11.29}$$

which gives

$$(\hat{\theta} - \theta)^2 \hat{j} = w \left\{ 1 + \frac{1}{3\hat{j}} \hat{l}_{3;}(\hat{\theta} - \theta) - \frac{1}{12\hat{j}} \hat{l}_{4;}(\hat{\theta} - \theta)^2 + \ldots \right\}^{-1},$$

and makes it possible to rewrite (11.28) as

$$p_{W|A=a}(w; \theta, a) = \frac{c(\theta, a)}{2\sqrt{w}} e^{-w/2}$$

$$\times \sum_{h=1}^{2} \frac{\left\{ 1 + \frac{1}{3\hat{j}} \hat{l}_{3;}(\hat{\theta} - \theta) - \frac{1}{12\hat{j}} \hat{l}_{4;}(\hat{\theta} - \theta)^2 + \ldots \right\}^{1/2}}{\left\{ 1 - \frac{\hat{l}_{2;1}}{2\hat{j}}(\hat{\theta} - \theta) + \frac{\hat{l}_{3;1}}{6\hat{j}}(\hat{\theta} - \theta)^2 + \ldots \right\}} \Bigg|_{\hat{\theta} = \hat{\theta}_h(w)}$$

$$\times \{ 1 + O(n^{-3/2}) \}. \tag{11.30}$$

Because $\sqrt{1+x} = 1 + x/2 - x^2/8 + O(x^3)$ and $(1-x)^{-1} = 1 + x + x^2 + O(x^3)$, (11.30) can be written as

$$p_{W|A=a}(w; \theta, a) = \frac{c(\theta, a)}{2\sqrt{w}} e^{-w/2}$$

$$\times \sum_{h=1}^{2} \left\{ 1 + (\hat{\theta} - \theta)\hat{c}_1 + (\hat{\theta} - \theta)^2 \hat{c}_2 + O(n^{-3/2}) \right\} \Bigg|_{\hat{\theta} = \hat{\theta}_h(w)},$$

where $\hat{c}_1$ and $\hat{c}_2$ are functions of $(\hat{\theta}, a)$ of order $O_p(1)$. Lastly, by expanding $\hat{c}_1$ and $\hat{c}_2$ around θ, we get

$$p_{W|A=a}(w; \theta, a) = \frac{c(\theta, a)}{2\sqrt{w}} e^{-w/2}$$

$$\times \sum_{h=1}^{2} \left\{ 1 + (\hat{\theta} - \theta)c_1 + (\hat{\theta} - \theta)^2 c_2' + O(n^{-3/2}) \right\} \Bigg|_{\hat{\theta} = \hat{\theta}_h(w)}, \tag{11.31}$$

where c_2' includes one term from the expansion of $\hat{c}_1$.

Now, $\hat{\theta} - \theta$ must be expressed in terms of w in the two regions $\hat{\theta} < \theta$ and $\hat{\theta} > \theta$. From (11.29),

$$\frac{\sqrt{w}}{\hat{j}^{1/2}} = |\hat{\theta} - \theta| \left\{ 1 + \frac{1}{6\hat{j}} \hat{l}_{3;}(\hat{\theta} - \theta) - \left(\frac{1}{24\hat{j}} \hat{l}_{4;} + \frac{1}{72\hat{j}^2} \hat{l}_{3;}^2 \right) (\hat{\theta} - \theta)^2 + \ldots \right\} \tag{11.32}$$

which, through inversion, and neglecting terms of order $O(n^{-3/2})$, gives

$$(\hat{\theta} - \theta) = -\frac{\sqrt{w}}{\hat{j}^{1/2}} - \frac{1}{6} \frac{\hat{l}_{3;}}{\hat{j}^2} w + O(n^{-3/2})$$

for $\hat{\theta} < \theta$ and

$$(\hat{\theta} - \theta) = \frac{\sqrt{w}}{\hat{j}^{1/2}} - \frac{1}{6}\frac{\hat{l}_{3;}}{\hat{j}^2}w + O(n^{-3/2})$$

for $\hat{\theta} > \theta$. When these expressions are used in (11.31) with $h = 1$ and $h = 2$, respectively, we obtain

$$p_{W|A=a}(w; \theta, a) = \frac{c(\theta, a)}{\sqrt{w}}e^{-w/2}\left\{1 + \left(\frac{1}{\hat{j}}c'_2 - \frac{1}{6}\frac{\hat{l}_{3;}}{\hat{j}^2}\right)w + O(n^{-3/2})\right\}.$$

One last expansion around θ of the coefficient of w in round brackets, yields

$$p_{W|A=a}(w; \theta, a) = \frac{c(\theta, a)}{\sqrt{w}}e^{-w/2}\left\{1 + D(\theta, a)w + O(n^{-3/2})\right\}, \qquad (11.33)$$

with $D(\theta, a)$ of order $O(n^{-1})$. The approximation (11.14) for $\bar{c}(\theta, a)$ gives the expansion

$$p_{W|A=a}(w; \theta, a) = \frac{1}{\sqrt{2\pi}\sqrt{w}}e^{-w/2}\left\{1 + B(\theta, a)w + O(n^{-3/2})\right\} \qquad (11.34)$$

for the density of W, with $B(\theta, a)$ a function both of $D(\theta, a)$ and of the term of order $O(n^{-1})$ in the expansion of $c(\theta, a)$. Thus, W has chi-squared density with one degree of freedom with an error of order $O(n^{-1})$. Furthermore, since a does not appear in the leading term, the result holds both conditionally on a, and marginally. Hence, inference based on W using its first-order asymptotic distribution is stable.

There is a relation between the constants $c(\theta, a)$ and $D(\theta, a)$. Indeed, because the approximation used for the density of $\hat{\theta}$ is normalized, except for terms of order $O(n^{-3/2})$, so too must the approximation obtained for the density of W be normalized to the same order. It follows that the integral of (11.33) in which the error of order $O(n^{-3/2})$ is neglected must be equal to $1 + O(n^{-3/2})$. Using (11.13), we deduce that

$$\bar{c}(\theta, a)\{1 + D(\theta, a)\} = 1 + O(n^{-3/2}). \qquad (11.35)$$

On the other hand, again using (11.33) and taking (11.35) into account, we obtain the following expansion for the conditional null expectation of W

$$\begin{aligned} E_\theta(W|a) &= \bar{c}(\theta, a)\{1 + 3D(\theta, a)\} + O(n^{-3/2}) \\ &= \{1 + D(\theta, a)\}^{-1}\{1 + 3D(\theta, a)\} + O(n^{-3/2}) \\ &= 1 + 2D(\theta, a) + O(n^{-3/2}). \end{aligned} \qquad (11.36)$$

From (11.35) and (11.36) the relation

$$\bar{c}(\theta, a) = \{1 + D(\theta, a)\}^{-1} + O(n^{-3/2}) = \frac{1}{(E_\theta(W|a))^{1/2}} + O(n^{-3/2}) \quad (11.37)$$

is obtained linking the normalizing constant of the p^* formula and the conditional Bartlett correction $E_\theta(W|a)$.

All the previous argument can equally well be applied when $p > 1$. However, the proof is more involved, in that it requires integrating the p^* formula on a $(p-1)$-dimensional region; see Barndorff-Nielsen and Cox (1994, pp. 190-194). The main result is the expansion for the conditional null density of W

$$p_{W|A=a}(w; \theta, a) = c(\theta, a)(2\pi)^{p/2} \frac{e^{-w/2} w^{p/2-1}}{2^{p/2}\Gamma\left(\frac{p}{2}\right)} \left\{1 + Dw + O(n^{-3/2})\right\},$$

$$(11.38)$$

where $D = D(\theta, a)$ is of order $O(n^{-1})$. It follows from normalization that the constants D and $\bar{c}(\theta, a) = c(\theta, a)(2\pi)^{p/2}$ are linked by the relation

$$\bar{c}(\theta, a)(1 + pD) = 1 + O(n^{-3/2}), \quad (11.39)$$

which generalizes (11.35). Hence, the expansion for the density of W can be written as

$$p_{W|A=a}(w; \theta, a) = \frac{e^{-w/2} w^{p/2-1}}{2^{p/2}\Gamma\left(\frac{p}{2}\right)} \left\{1 + D(w - p) + O(n^{-3/2})\right\}. \quad (11.40)$$

Furthermore, the null expectation of W conditionally on a has the expansion

$$\begin{aligned} E_\theta(W|a) &= \bar{c}(\theta, a) p \{1 + D(2 + p)\} + O(n^{-3/2}) \\ &= p(1 + 2D) + O(n^{-3/2}). \end{aligned} \quad (11.41)$$

11.4   Effect of the Bartlett Correction

Note that in (11.40) there is no term of order $O(n^{-1/2})$ in brackets and the term of order $O(n^{-1})$ is linear in w: this is the structural element which is important in justifying the improvement in the χ_p^2 approximation which can be obtained through the Bartlett correction.

From (11.40) it can immediately be shown that

$$W^\star = p\frac{W}{E_\theta(W|a)}$$

has a chi-squared null distribution with p degrees of freedom with error of order $O(n^{-3/2})$. Firstly, observe that from (11.41),

$$W^\star = \frac{pW}{E_\theta(W|a)} = \frac{W}{1+2D} + O(n^{-3/2}).$$

Hence from (11.40), $W^\star$ has density

$$p_{W^\star|A=a}(w;\theta,a) = \frac{e^{-w/2}w^{p/2-1}e^{-wD}(1+2D)^{p/2}}{2^{p/2}\Gamma\left(\frac{p}{2}\right)}\left\{1+D(w-p)+O(n^{-3/2})\right\}.$$

Lastly, the expansions $e^{-wD} = 1 - wD + O(n^{-2})$ and $(1+2D)^{p/2} = 1+pD+O(n^{-2})$ give

$$p_{W^\star|A=a}(w;\theta,a) = \frac{e^{-w/2}w^{p/2-1}}{2^{p/2}\Gamma\left(\frac{p}{2}\right)}\left\{1+O(n^{-3/2})\right\}. \tag{11.42}$$

Thus the conditional Bartlett correction reduces the order of the error in the chi-squared approximation for the density of W.

From expansion (11.40), under regularity conditions, the marginal density of W may be expanded as

$$p_W(w;\theta) = \frac{e^{-w/2}w^{p/2-1}}{2^{p/2}\Gamma\left(\frac{p}{2}\right)}\left\{1+\bar{D}(w-p)+O(n^{-3/2})\right\}, \tag{11.43}$$

with $\bar{D} = \bar{D}(\theta)$ equal to the expectation of $D = D(\theta,a)$ with respect to the density of A; note that $\bar{D}$ remains of order $O(n^{-1})$. Thus,

$$E_\theta(W) = p\left(1+2\bar{D}\right) + O(n^{-3/2}). \tag{11.44}$$

Comparison between (11.44) and (9.79) shows that, effectively, the error is of order $O(n^{-2})$. With identical steps to those used in the conditional case, the corrected combinant

$$W^\star = p\frac{W}{E_\theta(W)} = \frac{W}{1+2\bar{D}} + O(n^{-3/2})$$

is seen to have a null χ_p^2 distribution with error of order $O(n^{-3/2})$. Barndorff-Nielsen and Hall (1988) show that, in reality, the error is of order $O(n^{-2})$; this is because, under regularity conditions, W is the square (a quadratic form if $p > 1$) of a variable whose density admits an Edgeworth expansion, hence, in the expansions both for the density of W and for that of $W^\star$, there can be no odd powers of $n^{-1/2}$ (see expansion (10.15)).

Where possible, it is usually preferable to apply the conditional correction, because this is more in line with the conditionality principle. See Barndorff-Nielsen and Cox, 1994, formulae (5.56)-(5.61), for a general expression. For a study of the numerical accuracy of the two corrections, see Eriksen (1987).

The results on the null distribution of W and of W^* still hold even when nuisance parameters are present, with reference to W_P (see sections 9.4.2 and 6.5.2 in Barndorff-Nielsen and Cox, 1994).

It should be emphasized that the improvement obtained by means of a Bartlett correction concerns the case where the reference model is continuous. Whether a similar result would be valid for lattice variables has, so far, neither been proved nor disproved. However, Frydenberg and Jensen (1989), with a numerical study on a multinomial model with a small number of cells, showed that the correction did not seem to produce any improvement. On the contrary, whenever the model has both continuous and discrete components, for example as happens when observations are censored, the Bartlett correction has the same effect as in the continuous case (Jensen, 1993).

Example 11.1 *Logistic regression*
This example has been drawn from Jensen (1993). Let $(y_1, x_1), \ldots, (y_n, x_n)$ be independent realizations of a r.v. (Y_i, X_i), $i = 1, \ldots, n$, where, conditionally on $X_i = x_i$, Y_i has a binomial distribution, $Bi(\phi_i, 1)$, with $\phi_i = \exp\{\beta_0 + \beta_1 x_i\}/\{1 + \exp(\beta_0 + \beta_1 x_i)\}$, $\beta_0, \beta_1 \in \mathbb{R}$. Assume that the X_i are continuous r.v.'s with identical marginal density $p_X(x)$. The log-likelihood function is

$$l(\beta_0, \beta_1) = \beta_1 \sum_{i=1}^{n} x_i y_i + \beta_0 \sum_{i=1}^{n} y_i - \sum_{i=1}^{n} \log\{1 + \exp(\beta_0 + \beta_1 x_i)\} .$$

For inference about β_1, with known β_0, the combinant W^* is obtained by dividing the uncorrected combinant W by $1 + b/n$, with

$$b = \frac{1}{4}(4 - e^{\beta_0} - e^{-\beta_0})\left\{\frac{3s_2^2 - 4s_1 s_3 + s_4}{(s_2 - s_1^2)^2} - 1\right\}$$
$$+ \frac{5}{12}(e^{\beta_0} - e^{-\beta_0} - 2)\left\{\frac{4s_2^3 - 3s_1^2 s_2^2 + 4s_1^3 s_3 - 6s_1 s_2 s_3 + s_3^2}{(s_2 - s_1^2)^3} - 1\right\},$$

where $s_r = n^{-1}\sum x_i^r$. This quantity is obtained from an evaluation of the expectation of W conditionally on the distribution constant statistic $(x_1, \ldots, x_n)$, according to (9.81).

Table 11.1 gives the results of a simulation study presented in Jensen (1993). The values were obtained on the basis of 10^6 simulations for $n \leq 30$ and on

Table 11.1: *Probabilities on the tails of W and W*, in percentages; logistic regression example.*

		$\alpha \times 100$		
n		5.00	2.50	1.00
10	W	9.21	5.65	2.82
	$W^\star$	6.30	3.33	1.76
20	W	6.66	3.60	1.60
	$W^\star$	5.24	2.66	1.07
30	W	6.02	3.18	1.35
	$W^\star$	5.08	2.56	1.02
50	W	5.57	2.87	1.19
	$W^\star$	5.01	2.51	1.00

the basis of 3×10^6 simulations for $n = 50$. The value of β_0 is set equal to zero and the values $x_1, \ldots, x_n$ are i.i.d. realizations of a $N(0,1)$ distribution. The nominal significance levels $\alpha = 0.01, 0.025, 0.05$ are considered. The marginal distribution of $W^\star$ is very closely approximated by the χ_1^2 distribution, while the same is not true for W. $\triangle$

The chi-squared approximation for the null distribution of a likelihood ratio statistic may be particularly inaccurate if there is a nuisance parameter present whose dimension is large when compared with the sample size. Many simulation studies carried out with reference to the usual problems of testing hypotheses on multivariate normal distributions, have shown that the Bartlett correction can give very accurate results. The example below, taken from Porteous (1985), is representative (further references are given in section 11.7).

Example 11.2 *Test on a partial correlation*
On the basis of a random sample of size n from a $N_4(0, \Sigma)$ distribution, we wish to test $H_0 : \sigma^{14} = 0$, where, as usual, σ^{ij} is the generic element of Σ^{-1}. By I denote a generic subset of $\{1, 2, 3, 4\}$ and with $S_{(I)}$ the sample covariance matrix where the rows and columns with indices in I have been deleted. The test statistic W for H_0 is

$$W = n \log \left\{ \frac{|S_{(1,2,3)}|}{|S|} \frac{|S_{(2,3,4)}|}{|S_{(2,3)}|} \right\}.$$

Table 11.2: *Probabilities on the tails of W and W*; partial correlation example.*

	α	5	10	15	20	25
				n		
W	0.05	0.2633	0.1118	0.0862	0.0747	0.0714
	0.01	0.1429	0.0385	0.0256	0.0202	0.0178
$W^\star$	0.05	0.0492	0.0518	0.0536	0.0512	0.0504
	0.01	0.0093	0.0113	0.0105	0.0103	0.0108

For the expression of the Bartlett correction, obtained by directly approximating the null expectation of W, see Porteous (1985). Table 11.2 shows the results of a simulation study (with 10 000 trials) from this latter paper. Note that the distribution of W is quite different from the asymptotic distribution, even for moderate sample sizes. However, the improvement effected by the Bartlett correction is remarkable.

$\triangle$

11.5 Modified Versions of Z and Z_p

When the parameter of interest is a scalar, it is convenient to use Z or Z_P for inference. First consider a one-parameter model and the combinant $W = 2\{l(\hat{\theta}) - l(\theta)\}$ associated with it. Since W has a null asymptotic χ_1^2 distribution,

$$Z = \text{sgn}(\hat{\theta} - \theta)\sqrt{W} \qquad (11.45)$$

has a null asymptotic $N(0, 1)$ distribution. This can immediately be justified, because, from expansion (9.77),

$$2\{l(\hat{\theta}) - l(\theta)\} = \left(\frac{l_*}{\sqrt{i}}\right)^2 + O_p(n^{-1/2})$$

and, since $(\hat{\theta} - \theta) = i^{-1}l_* + \cdots$ then, asymptotically, $\text{sgn}(\hat{\theta} - \theta) = \text{sgn}(l_*)$, and

$$Z = \frac{l_*}{\sqrt{i}} + O_p(n^{-1/2}).$$

A more accurate approximation, conditionally on an ancillary statistic a, can be obtained by using the $p^\star$ formula. For simplicity, assume that the correspondence between $\hat{\theta}$ and $z = \text{sgn}(\hat{\theta} - \theta)\sqrt{w}$ is one-to-one and smooth on the support of $\hat{\theta}$ (for asymptotic validity it is enough that the correspondence is one-to-one and smooth in a neighbourhood of θ). Note that, because of (11.25),

$$\frac{dz}{d\hat{\theta}} = \operatorname{sgn}(\hat{\theta} - \theta)\frac{1}{2}\frac{dw}{d\hat{\theta}}w^{-1/2} = \frac{\hat{l}_{;1} - l_{;1}}{z} .$$

Hence,

$$p_{Z|A=a}(z;\theta,a) = \bar{c}(\theta,a)\phi(z)\frac{z}{u}\left\{1 + O(n^{-3/2})\right\} , \tag{11.46}$$

with $u = \hat{j}^{-1/2}(\hat{l}_{;1} - l_{;1})$. Expansion (11.27) makes it possible to check that $z/u = 1 + O(n^{-1/2})$, thus, taking (11.14) into account,

$$p_{Z|A=a}(z;\theta,a) = \phi(z) + O(n^{-1/2}). \tag{11.47}$$

The error here is of order $O(n^{-1/2})$ since from (11.22)

$$P_\theta(Z > 0|a) = P_\theta(\hat{\theta} > \theta|a) = \frac{1}{2} + \frac{t_3}{6\sqrt{2\pi}} + O(n^{-3/2}) ,$$

and t_3 is of order $O(n^{-1/2})$. Because a does not appear in the leading term of (11.47), the result holds both conditionally on a and marginally. Hence, inference based on the first-order asymptotic distribution of Z is stable.

Since Z converges more slowly to its own asymptotic distribution than does W, it is interesting to find a normalizing transformation of Z. In Example 10.6, we saw that, in an exponential family, the transformation $(Z - m)/v$, with m and v^2 approximations for the mean and variance of Z, reduces error in the normal approximation. Constructions such as $(Z - m)/v$ are, however, usually very difficult to generalize because it is difficult to determine m and v as they may depend on ancillaries (see Barndorff-Nielsen, 1986). Both for simplicity, and in order to obtain a more accurate approximation, it is better to adopt the modified directed deviance, $Z^\star$, proposed by Barndorff-Nielsen (1986, 1990a).

Note that Z coincides with $r_{\hat{\theta}}$ as defined in (11.17). From (11.19) and (11.20), we deduce that the combinant

$$Z^\star = Z + Z^{-1}\log\frac{U}{Z} , \tag{11.48}$$

where

$$U = \hat{j}^{-1/2}\left(\hat{l}_{;1} - l_{;1}\right) , \tag{11.49}$$

is such that

$$P_\theta(Z^\star \le z|a) = \Phi(z)\left\{1 + O(n^{-3/2})\right\} ,$$

hence, $Z^\star$ has a $N(0,1)$ distribution with error of order $O(n^{-3/2})$ both conditionally on a and marginally.

Because $l_{;1} = (\partial/\partial\hat{\theta})l(\theta; \hat{\theta}, a)$, U can be interpreted as a dual of the standardized score

$$\frac{l_*}{\sqrt{i}} = i^{-1/2}\left(l_{1;} - \hat{l}_{1;}\right),$$

where $\hat{l}_{1;} = 0$. Just like $l_*/\sqrt{i}$, so too U is parameterization invariant.

The asymptotic normality of $Z^\star$, with error of order $O(n^{-3/2})$, is a direct consequence of (11.46). Essentially, the argument is as follows: if $z^\star = z + z^{-1}\log(u/z)$, then

$$(z^\star)^2 = z^2 + z^{-2}\log^2\left(\frac{u}{z}\right) + 2\log\frac{u}{z},$$

hence

$$-\frac{z^2}{2} + \log\frac{z}{u} = -\frac{1}{2}\left\{(z^\star)^2 - z^{-2}\log^2\left(\frac{u}{z}\right)\right\}.$$

Furthermore,

$$z^{-2}\log^2\left(\frac{u}{z}\right) = (z^\star - z)^2$$

and

$$-\frac{z^2}{2} + \log\frac{z}{u} = -\frac{(z^\star)^2}{2} + \frac{1}{2}(z^\star - z)^2$$

which makes it possible to write (11.46) in the equivalent form

$$p_{Z^\star|A=a}(z^\star; \theta, a) = \bar{c}(\theta, a)H(z^\star)\phi(z^\star)\left\{1 + O(n^{-3/2})\right\},$$

with $H(z^\star) = \exp\{(z - z^\star)^2/2\}(dz/dz^\star)$, where z is expressed in terms of $z^\star$. Using Taylor expansions, Barndorff-Nielsen (1990a) shows that $\bar{c}(\theta, a)H(z^\star) = 1 + O(n^{-3/2})$ thus, a normal approximation with error of order $O(n^{-3/2})$ holds for the distribution of $Z^\star$.

Example 11.3 *One parameter exponential family*

Let $y_1, \ldots, y_n$ be independent observations of a r.v. with p.d.f. in an $\mathcal{F}^1_{ne}$,

$$p(y; \theta) = \exp\{\theta y - K(\theta)\}p_0(y),$$

with $\theta \in \tilde{\Theta} \subseteq \mathbb{R}$, and $\tilde{\Theta}$ the natural parameter space. The quantities required to calculate $Z^\star$ are: $l(\theta; \hat{\theta}) = n(\theta K'(\hat{\theta}) - K(\theta))$, $l_{;1} = n\theta K''(\hat{\theta}) = \theta\hat{j}$ and $\hat{l}_{;1} = n\hat{\theta}K''(\hat{\theta}) = \hat{\theta}\hat{j}$. Specifically, we obtain

$$U = \hat{j}^{-1/2}\left(\hat{l}_{;1} - l_{;1}\right) = \hat{j}^{1/2}(\hat{\theta} - \theta) = \text{sgn}(\hat{\theta} - \theta)\sqrt{\hat{j}(\hat{\theta} - \theta)^2}.$$

Table 11.3: *Exact and approximate distribution function of $Y \sim IG(1,1)$.*

y	2	4	8	16
$P\{Y \leq y\}$	0.885474	0.979075	0.998740	0.999990
$1 - \Phi(Z^\star(y))$	0.899819	0.982608	0.998974	0.999992

Note that U coincides with Z_e, that is, with the signed Wald statistic in the natural parameterization. △

Example 11.4 *Inverse Gaussian distribution*
Let y be one observation from an $IG(\phi, \lambda)$ distribution, with known λ. The maximum likelihood estimate of ϕ is a one-to-one function of y through the relation $\sqrt{\lambda/\hat\phi} = y$ and the log-likelihood function may be written as

$$l(\phi; \hat\phi) = \lambda^{1/2} \left(\phi^{1/2} - \frac{1}{2}\phi\hat\phi^{-1/2} \right)$$

from which we obtain

$$Z = \lambda^{1/4}\hat\phi^{-1/4} \left(\hat\phi^{1/2} - \phi^{1/2} \right) ,$$

$$U = \frac{1}{2}\left\{ 1 + \left(\frac{\phi}{\hat\phi}\right)^{1/2} \right\} Z$$

and

$$Z^\star = Z + Z^{-1} \log\left\{ \frac{1}{2}\left(1 + \left(\frac{\phi}{\hat\phi}\right)^{1/2} \right) \right\} .$$

Table 11.3 shows some exact and approximate probabilities on the upper tail of $Y \sim IG(1,1)$. The approximation is remarkably accurate, even for y values that are extreme in the upper tail. △

Example 11.5 *Extreme value distribution*
Let $Y \sim EV(\mu, 1)$, with density $p_Y(y; \mu) = \exp\{(y - \mu) - e^{y-\mu}\}$, $y, \mu \in \mathbb{R}$. Then

$$Z = \text{sgn}(y - \mu)\sqrt{2\left(e^{y-\mu} - (y - \mu) - 1\right)},$$

$$U = e^{y-\mu} - 1.$$

Table 11.4 shows exact and approximate values of the distribution function of $Y \sim EV(0, 1)$. The approximation is also extremely accurate in this case. △

Table 11.4: *Evaluation of the d.f. of* $Y \sim EV(0,1)$.

y	$F_Y(y)$	$\Phi(r^\star(y))$
-10	$0.0^4 45419$	$0.0^4 47564$
-9	$0.0^3 12338$	$0.0^3 12946$
-8	$0.0^3 33540$	$0.0^3 35054$
-7	$0.0^3 91147$	$0.0^3 94980$
-6	0.00247568	0.00256908
-5	0.00671530	0.00693059
-4	0.0181489	0.0185997
-3	0.0485680	0.0493582
-2	0.126577	0.127505
-1	0.307799	0.307870
0	0.632121	0.630559
1	0.934012	0.933075
2	0.999382	0.999364

In order to calculate $Z^\star$ an ancillary a must be specified. In most cases this is not straightforward. One alternative solution is offered by the combinant

$$Z^\dagger = Z + Z^{-1} \log\left(\frac{V}{Z}\right) \tag{11.50}$$

where

$$V = \sqrt{\frac{\hat{i}}{i}} \frac{1}{\sqrt{\hat{j}}} l_* .$$

V is parameterization invariant. Barndorff-Nielsen and Chamberlin (1991) showed that $Z^\dagger$ has a $N(0,1)$ null distribution with error of order $O(n^{-1})$ both conditionally, on any ancillary statistic a, and marginally.

In the multiparameter case with $\theta = (\tau, \zeta)$ and scalar τ, Barndorff-Nielsen (1991b) obtains a modified version of $Z_P = \text{sgn}(\hat{\tau} - \tau)\sqrt{W_P}$ which has a $N(0,1)$ null distribution with error of order $O(n^{-3/2})$. This modified version is a generalization of (10.71), which refers to an $\mathcal{F}_{ne}^p$. Once again, assume that the minimal sufficient statistic can be expressed as $(\hat{\theta}, a) = (\hat{\tau}, \hat{\zeta}, a)$, with a ancillary. The modified version of Z_P is

$$Z_P^\star = Z_P + \frac{1}{Z_P} \log \frac{C U_P}{Z_P}, \tag{11.51}$$

where

$$C = \frac{|\tilde{l}_{\zeta;\hat{\zeta}}|}{\{|\tilde{\jmath}_{\zeta\zeta}||\hat{\jmath}_{\zeta\zeta}|\}^{1/2}} \tag{11.52}$$

and

$$U_P = j_P(\hat{\tau})^{-1/2} \frac{\partial}{\partial \hat{\tau}} \{l_P(\tau) - l_P(\hat{\tau})\} . \tag{11.53}$$

The matrix $l_{\zeta;\hat{\zeta}}$ in (11.52) has the generic element

$$l_{a;b} = \frac{\partial^2 l(\tau, \zeta; \hat{\tau}, \hat{\zeta}, a)}{\partial \zeta^a \partial \hat{\zeta}^b} , \qquad a, b = 1, \ldots, p-1 . \tag{11.54}$$

The partial derivative which appears in (11.53) should be understood to be a derivative with respect to $\hat{\tau}$ of the normalized profile log-likelihood, when this is considered as a function of τ, $\hat{\tau}$, $\hat{\zeta}_\tau$ and of a. This derivative can be rewritten in the form

$$\frac{\partial}{\partial \hat{\tau}} \left\{ l_P(\tau; \hat{\tau}, \hat{\zeta}, a) - l_P(\hat{\tau}; \hat{\tau}, \hat{\zeta}, a) \right\}$$

$$= \tilde{l}_{;\hat{\tau}} - \hat{l}_{;\hat{\tau}} + \frac{\partial \hat{\zeta}}{\partial \hat{\tau}} \left(\tilde{l}_{;\hat{\zeta}} - \hat{l}_{;\hat{\zeta}} \right) \tag{11.55}$$

where the quantity $\partial \hat{\zeta}/\partial \hat{\tau}$ can be obtained by differentiating the likelihood equation for $\hat{\zeta}_\tau$ with respect to $\hat{\tau}$, which gives

$$\frac{\partial \hat{\zeta}}{\partial \hat{\tau}} = - \left(\tilde{l}_{\zeta;\hat{\zeta}} \right)^{-1} \tilde{l}_{\zeta;\hat{\tau}} .$$

Example 11.6 *Expression of Z_P^* in an $\mathcal{F}_{ne}^p$*
If the reference model has density in an $\mathcal{F}_{ne}^p$ as given by (10.55), then Z_P^* will coincide with the combinant r_P^* defined by (10.71). Indeed, the log-likelihood function is

$$l(\tau, \zeta; \hat{\tau}, \hat{\zeta}) = \tau t + \zeta \cdot u - nK(\tau, \zeta)$$
$$= n \left\{ \tau K_\tau(\hat{\tau}, \hat{\zeta}) + \zeta \cdot K_\zeta(\hat{\tau}, \hat{\zeta}) - K(\tau, \zeta) \right\} ,$$

where $K_\tau(\tau, \zeta)$, $K_\zeta(\tau, \zeta)$ and other analogous functions are defined as in section 5.7. Since $l_{\zeta;\hat{\zeta}} = nK_{\zeta\zeta}(\hat{\tau}, \hat{\zeta})$ and $j_{\zeta\zeta} = nK_{\zeta\zeta}(\tau, \zeta)$, we get

$$C = \frac{\left| K_{\zeta\zeta}(\hat{\tau}, \hat{\zeta}) \right|^{1/2}}{\left| K_{\zeta\zeta}(\tau, \hat{\zeta}_\tau) \right|^{1/2}}$$

which is the quantity given by (10.69). Furthermore,

$$\tilde{l}_{;\hat{\tau}} - \hat{l}_{;\hat{\tau}} = n\left\{(\tau - \hat{\tau})K_{\tau\tau}(\hat{\tau},\hat{\zeta}) + (\hat{\zeta}_\tau - \hat{\zeta})K_{\zeta\tau}(\hat{\tau},\hat{\zeta})\right\},$$

$$\frac{\partial\hat{\zeta}}{\partial\hat{\tau}} = -\left(K_{\zeta\zeta}(\hat{\tau},\hat{\zeta})\right)^{-1}K_{\zeta\tau}(\hat{\tau},\hat{\zeta}),$$

and

$$\tilde{l}_{;\hat{\zeta}} - \hat{l}_{;\hat{\zeta}} = n\left\{(\tau - \hat{\tau})K_{\tau\zeta}(\hat{\tau},\hat{\zeta}) + (\hat{\zeta}_\tau - \hat{\zeta})K_{\zeta\zeta}(\hat{\tau},\hat{\zeta})\right\}.$$

Therefore, taking into account the expression (5.53) of $j_p(\tau)$, it is immediate to check that U_p coincides with the quantity v_p defined by (10.70).　　△

Example 11.7 *Inference about the shape parameter of a Weibull distribution*
Let $t_1, \ldots, t_n$ be independent realizations of T with density (1.29) (see Problem 1.10). We wish to obtain the expression of $Z_p^\star$ when the shape parameter ν is of interest. It is convenient to make a logarithmic transformation of the data, hence, $y_i = \log t_i$ are independent realizations of $Y \sim EV(\mu, \sigma)$, with $\sigma = \nu^{-1}$ and $\mu = -\log\lambda$. This brings us back to a scale and location family, where the log-likelihood $l(\mu, \sigma; \hat{\mu}, \hat{\sigma}, a)$ is given by (7.33), with $q_0(y) = e^y - y$. Thus, we immediately obtain the quantities required for calculating $Z_p^\star$ as in (11.51). Specifically,

$$Z_P = \text{sgn}(\hat{\sigma} - \sigma)\left\{2\left(-n\log\frac{\hat{\sigma}}{\sigma} + \left(1 - \frac{\hat{\sigma}}{\sigma}\right)\sum a_i + n\log\left(n^{-1}\sum e^{\frac{\hat{\sigma}}{\sigma}a_i}\right)\right)\right\}^{1/2}$$

and

$$CU_P = \frac{\hat{\sigma}}{\sigma}\frac{(\hat{\sigma}/\sigma)\left\{\left(n\sum a_i e^{\frac{\hat{\sigma}}{\sigma}a_i}\right)\left(\sum e^{\frac{\hat{\sigma}}{\sigma}a_i}\right)^{-1} - \sum a_i\right\} - n}{\left\{\sum a_i^2 e^{a_i} - n^{-1}\left(\sum a_i\right)^2 - 2\sum a_i\right\}^{1/2}}.$$

In these expressions, $\hat{\sigma}$ is the solution to (7.39). The confidence intervals for ν with approximate level $1-\alpha$ are $(1/\sigma_{\alpha/2}^\star, 1/\sigma_{1-\alpha/2}^\star)$, with $\sigma_\alpha^\star$ the solution in σ of the equation $Z_P^\star = u_\alpha$, where $\Phi(u_\alpha) = \alpha$. Table 11.5 gives the results of a small simulation study, based on 2000 trials, which aims to assess the actual coverage levels. For comparison, the coverage probabilities of the usual intervals based on the one-sided Wald statistic with observed information, Z_{eP}, are also given. The values shown in brackets offer information about symmetry of coverage. They estimate the coverage probability associated with the intervals

(lower bound of the two-sided interval, estimator),

Table 11.5: *Estimated coverage levels (in %) of the confidence intervals for ν based on $Z_P^\star$ and Z_{e_P}, $\nu = 1$.*

n	nom. lev. (in %)	$Z_P^\star$	Z_{e_P}
5	90	90	90
		(48)	(25)
	95	95	96
		(50)	(27)
10	90	90	88
		(44)	(32)
	95	94	95
		(47)	(33)
20	90	92	91
		(44)	(35)
	95	96	97
		(46)	(37)
50	90	91	91
		(45)	(38)
	95	95	96
		(47)	(40)

where the estimator is the value of ν which renders the combinant concerned equal to zero. In the case of Z_{e_P}, the estimator is $\hat{\nu}$. Observe that while the overall coverage level is also satisfactory for intervals based on the first order asymptotic distribution of Z_{e_P}, adequate coverage symmetry is only obtained by $Z_P^\star$. △

Example 11.8 *Normal distribution (cont.)*
Under the same assumptions as in Examples 4.14 and 4.17, it is easy to calculate $Z_P^\star$. Indeed,

$$Z_P = \text{sgn}(\hat{\mu} - \mu)\sqrt{n \log \frac{\hat{\sigma}_\mu^2}{\hat{\sigma}^2}}$$

and

$$CU_P = \sqrt{n}\frac{\hat{\sigma}^2}{\hat{\sigma}_\mu^2}\left(\frac{\hat{\mu} - \mu}{\hat{\sigma}}\right).$$

A Taylor expansion shows that the confidence intervals for μ with level $1 - \alpha$

are of the form $\hat{\mu} \pm sc_{n,\alpha}$, where $s^2 = n\hat{\sigma}^2/(n-1)$ and

$$c_{n,\alpha} = \frac{u_{1-\alpha/2}}{\sqrt{n}}\left\{1 + \frac{1}{4(n-1)}\left(1 + u_{1-\alpha/2}^2\right) + O(n^{-2})\right\}. \qquad (11.56)$$

On the other hand, the $1 - \alpha/2$ quantile of a Student's t distribution with g degrees of freedom has the Cornish–Fisher expansion

$$t_{g,1-\alpha/2} = u_{1-\alpha/2} + \frac{h(u_{1-\alpha/2})}{4g} + O(g^{-2}),$$

with $h(x) = x^3 + x$. Unlike the intervals based on profile likelihood or approximate conditional likelihood, obtained in Example 4.17, the intervals based on Z_P^* agree with the exact intervals except for an error of order $O(n^{-5/2})$.   △

Example 11.9 *Behrens–Fisher problem*
Let $x = (x_1,\ldots,x_{n1})$ be a random sample of size n_1 from a $N(\mu_1,\sigma_1^2)$ distribution and let $y = (y_1,\ldots,y_{n2})$ be a random sample of size n_2 from a $N(\mu_2,\sigma_2^2)$ distribution, with x and y independent. Consider the problem of testing $H_0 : \mu_1 = \mu_2$ versus $H_1 : \mu_1 < \mu_2$. Let us denote by $\bar{x}$, $\bar{y}$, $\hat{\sigma}_1^2$ and $\hat{\sigma}_2^2$ the sample means and variances of the two samples. Under H_0, the m.l.e. $\hat{\mu}$ of the common mean $\mu = \mu_1 = \mu_2$ is a solution of the equation

$$\mu^3 - a\mu^2 + b\mu - c = 0,$$

where

$$a = \frac{(2n_1 + n_2)\bar{y} + (n_1 + 2n_2)\bar{x}}{n_1 + n_2}$$

$$b = \frac{n_1(\bar{y}^2 + \hat{\sigma}_2^2 + 2\bar{x}\bar{y}) + n_2(\bar{x}^2 + \hat{\sigma}_1^2 + 2\bar{x}\bar{y})}{n_1 + n_2}$$

$$c = \frac{n_1\bar{x}(\bar{y}^2 + \hat{\sigma}_2^2) + n_2\bar{y}(\bar{x}^2 + \hat{\sigma}_1^2)}{n_1 + n_2}.$$

The constrained m.l.e.'s of the two variances are

$$\tilde{\sigma}_1^2 = \hat{\sigma}_1^2 + (\hat{\mu} - \bar{x})^2 \quad \text{and} \quad \tilde{\sigma}_2^2 = \hat{\sigma}_2^2 + (\hat{\mu} - \bar{y})^2.$$

Let us denote by $l(\mu_1, \mu_2, \sigma_1^2, \sigma_2^2)$ the log-likelihood associated to the unconstrained model. The signed square root of the log-likelihood ratio statistic for H_0 is

$$Z_P = \text{sgn}(\bar{y} - \bar{x})\sqrt{W},$$

with

$$W = 2\left\{l(\bar{x}, \bar{y}, \hat{\sigma}_1^2, \hat{\sigma}_2^2) - l(\hat{\mu}, \hat{\mu}, \tilde{\sigma}_1^2, \tilde{\sigma}_2^2)\right\}$$
$$= 2\left(-\frac{n_1}{2}\log\hat{\sigma}_1^2 - \frac{n_2}{2}\log\hat{\sigma}_2^2 + \frac{n_1}{2}\log\tilde{\sigma}_1^2 + \frac{n_2}{2}\log\tilde{\sigma}_2^2\right).$$

To compute the modified version $Z_p^\star$, it is convenient to use the reparameterization $\psi(\mu_1, \mu_2, \sigma_1^2, \sigma_2^2) = (\mu_1, \tau, \sigma_1^2, \sigma_2^2)$, where $\tau = \mu_2 - \mu_1$. With some algebra, we obtain

$$CU_p = \frac{\sqrt{n_1 n_2 \hat{\sigma}_1^2 \hat{\sigma}_2^2}\left\{\hat{\sigma}_1^2(\hat{\mu} - \bar{y})^3 + \hat{\sigma}_1^2\hat{\sigma}_2^2(\bar{x} - \bar{y}) - \hat{\sigma}_2^2(\hat{\mu} - \bar{x})^3\right\}}{\hat{\sigma}_1^2\hat{\sigma}_2^2\sqrt{n_1\tilde{\sigma}_2^4(\hat{\sigma}_1^2 - (\hat{\mu} - \bar{x})^2) + n_2\tilde{\sigma}_1^4(\hat{\sigma}_2^2 - (\hat{\mu} - \bar{y})^2)}}.$$

Table 11.6 gives the results of a simulation study, based on 50 000 trials, which aims to assess the actual significance levels of $Z_p^\star$, compared with the significance levels of Z_p and of the Welch statistic V_W (cf. e.g. Stuart and Ord, 1991, section 20.33). The sample sizes and the values of the nuisance parameters are also shown in the table. The significance levels of $Z_p^\star$ are remarkably close to the nominal ones, for all sample sizes and values of the variances in the two samples. There is an improvement also in comparison with Welch's solution, which requires the use of special tables to compute the significance levels.

The results in this example have been obtained by Bellio (1996, personal communication) and are very similar to those in Jensen (1992).   △

11.6 Modified Profile Likelihood

Let $p > 1$ and $\theta = (\tau, \zeta)$ with τ a k-dimensional parameter of interest ($k \geq 1$). As was discussed in section 4.6, the profile likelihood $L_p(\tau) = L(\tau, \hat{\zeta}_\tau)$ offers a general, and fairly simple tool for inference about τ. If a pseudo-likelihood which is a genuine likelihood and which does not entail loss of information is not available, asymptotic methods could suggest improvements for $L_p(\tau)$, through the introduction of modifying factors.

One initial example of this is the Cox and Reid approximate conditional likelihood, introduced in section 4.7. However, this does have some drawbacks: firstly, an orthogonal parameterization must be explicitly available, and, secondly, it is not invariant under interest-respecting reparameterizations. In recent years, a variety of proposals have been put forward as suitable modifications to $L_p(\tau)$ (see section 11.7). Here, we briefly describe the modified profile likelihood introduced by Barndorff-Nielsen (1983, 1988, 1994). This is

Table 11.6: *Significance levels (×100) for Behrens–Fisher problem.*

	$\alpha \times 100$	$Z_P^{\star}$	Z_P	V_W
$n_1 = 5$, $n_2 = 5$	1.0	1.1	2.5	1.3
$\sigma_1 = 1$, $\sigma_2 = 20$	5.0	5.2	8.3	4.8
	10.0	10.1	14.0	9.3
$n_1 = 9$, $n_2 = 5$	1.0	1.1	2.4	1.2
$\sigma_1 = 1$, $\sigma_2 = 20$	5.0	5.1	8.2	4.8
	10.0	10.1	13.8	9.3
$n_1 = 16$, $n_2 = 6$	1.0	1.1	2.1	1.1
$\sigma_1 = 1$, $\sigma_2 = 20$	5.0	5.2	7.7	4.5
	10.0	10.2	13.3	9.1
$n_1 = 20$, $n_2 = 10$	1.0	1.0	1.6	0.8
$\sigma_1 = 1$, $\sigma_2 = 20$	5.0	5.1	6.5	4.5
	10.0	10.1	11.7	9.3
$n_1 = 50$, $n_2 = 40$	1.0	1.0	1.1	0.9
$\sigma_1 = 1$, $\sigma_2 = 40$	5.0	5.1	5.4	4.8
	10.0	10.1	10.6	9.6

based on the p^* formula and gives a pseudo-likelihood which does not have either of the drawbacks mentioned above for approximate conditional likelihood. However, an ancillary statistic must be available, unless, as happens when going from Z^* to $Z^\dagger$, stable approximations of the modified profile likelihood with error of order $O(n^{-1})$ are used.

The modified profile likelihood is defined as

$$L_{MP}(\tau) = L_P(\tau)\, M(\tau)\,, \tag{11.57}$$

where the modification factor $M(\tau)$ is

$$M(\tau) = \left| l_{\zeta;\hat{\zeta}}\left(\tau,\hat{\zeta}_\tau;\hat{\tau},\hat{\zeta},a\right)\right|^{-1}\left| j_{\zeta\zeta}\left(\tau,\hat{\zeta}_\tau;\hat{\tau},\hat{\zeta},a\right)\right|^{1/2}.$$

Note that the ancillary has to be specified only for the calculation of the first determinant. An alternative expression for (11.57) is

$$L_{MP}(\tau) = L_P(\tau)\, D(\tau)\left| j_{\zeta\zeta}\left(\tau,\hat{\zeta}_\tau;\hat{\tau},\hat{\zeta},a\right)\right|^{-1/2}, \tag{11.58}$$

where

$$D(\tau) = \frac{\left| j_{\zeta\zeta}\left(\tau,\hat{\zeta}_\tau;\hat{\tau},\hat{\zeta},a\right)\right|}{\left| l_{\zeta;\hat{\zeta}}\left(\tau,\hat{\zeta}_\tau;\hat{\tau},\hat{\zeta},a\right)\right|}.$$

Differentiating the likelihood equation for $\hat{\zeta}_\tau$ with respect to $\hat{\zeta}$, immediately shows that

$$D(\tau) = \left|\frac{\partial \hat{\zeta}_\tau}{\partial \hat{\zeta}}\right|.$$

It is easy to check that $L_{MP}(\tau)$ is invariant under interest-respecting reparameterizations. Furthermore, the modification factor $M(\tau)$ is of order $O_p(1)$ hence, $L_P(\tau)$ and $L_{MP}(\tau)$ are asymptotically equivalent to the first order.

Expression (11.57) can be obtained as an approximation of a marginal likelihood for τ, if the marginal likelihood exists, or of a conditional likelihood for τ, if that exists. More precisely, suppose that one of these two factorizations does hold: either

$$p_{\hat{\tau},\hat{\zeta}|a}(\hat{\tau},\hat{\zeta};\tau,\zeta,a) = p_{\hat{\tau}|a}(\hat{\tau};\tau,a)p_{\hat{\zeta}|\hat{\tau},a}(\hat{\zeta};\tau,\zeta,\hat{\tau},a) \tag{11.59}$$

or

$$p_{\hat{\tau},\hat{\zeta}|a}(\hat{\tau},\hat{\zeta};\tau,\zeta,a) = p_{\hat{\zeta}|a}(\hat{\zeta};\tau,\zeta,a)p_{\hat{\tau}|\hat{\zeta},a}(\hat{\tau};\tau,\hat{\zeta},a)\,. \tag{11.60}$$

If (11.59) holds, the marginal density of $\hat{\tau}$ gives a marginal likelihood for τ. An approximation for such a function can be obtained as the ratio between the p^* formula for the density of $(\hat{\tau}, \hat{\zeta})$ given a and the p^* formula for the density of $\hat{\zeta}$ given $(\hat{\tau}, a)$. If (11.60) holds, the conditional density of $\hat{\tau}$ given $\hat{\zeta}$ gives a conditional likelihood for τ. This function could be approximated by the ratio between the p^* formula for the density of $(\hat{\tau}, \hat{\zeta})$ given a and the p^* formula for the density of $\hat{\zeta}$ given a. The result of the two approximations is the same and is $L_{MP}(\tau)$ (Barndorff-Nielsen and Cox, 1994, section 8.2).

For a multiparameter exponential family with density (10.55), where a factorization such as (11.60) holds, the modified profile likelihood is

$$L_{MP}(\tau) = L_P(\tau) \left| K_{\zeta\zeta}(\tau, \hat{\zeta}_\tau) \right|^{1/2},$$

which coincides with the approximation given by (10.57) for the conditional likelihood, based on the double saddlepoint expansion.

Under random sampling from an element of a scale, location and shape family with density (7.42), a factorization such as (11.59) holds and the function $L_{MP}(\tau)$ coincides with the approximation, with error of order $O(n^{-3/2})$, of the marginal likelihood obtained by applying the Laplace expansion in Example 9.9.

In general, the ancillary a must be specified if (11.57) is to be made explicit. However, if $\hat{\zeta}_\tau = \hat{\zeta}$, the matrix $(\partial \hat{\zeta}_\tau / \partial \hat{\zeta})$ is the identity matrix, hence, using the representation (11.58) of the modified profile likelihood, we get

$$L_{MP}(\tau) = L_{AC}(\tau), \tag{11.61}$$

where $L_{AC}(\tau)$ is the Cox and Reid (1987) approximate conditional likelihood defined by (4.42). Linear regression offers important examples of this coincidence, see Example 4.16, as does the Neyman–Scott problem, see Example 4.7 and Problem 4.23. If the parameters τ and ζ are orthogonal, because of (4.38), (11.61) holds with error of order $O_p(n^{-1})$. Even when there is no orthogonality, (11.61) could stand with error of order $O_p(n^{-1})$. This is true for example if τ is the shape parameter of a composite group family, such that the right invariant measure on the group $\mathcal{G}$ has a constant density with respect to Lebesgue measure (Barndorff-Nielsen and McCullagh, 1993).

Barndorff-Nielsen (1994) proposed two approximations of $L_{MP}(\tau)$ with error of order $O(n^{-1})$ for a scalar parameter of interest. The ancillary statistic does not have to be specified for either of these two approximations. One of them is

$$L_B^\dagger(\tau) = L_P(\tau) \exp\{b_0(\tau)\}, \tag{11.62}$$

with

$$b_0(\tau) = \int_{\hat{\tau}}^{\tau} b(t, \hat{\zeta}_t) dt , \tag{11.63}$$

where $b(\tau, \zeta) = -E_{\tau,\zeta}\{l_\tau(\tau, \hat{\zeta}_\tau)\}$, that is, it is the term of order $O(1)$ of the null expectation of the profile score given in (9.89). From the tensorial behaviour of the summands of (9.89) shown in section 9.5.3 we deduce that $L^\dagger_B(\tau)$ is invariant under interest-respecting reparameterizations. By construction, the score function of this pseudo-likelihood has zero null expectation with error of order $O(n^{-1})$. Since $L^\dagger_B(\tau)$ can be obtained as an approximation of $L_{MP}(\tau)$, it follows that, for this latter too, the score function has zero null expectation with error of order $O(n^{-1})$.

Example 11.10 *Equicorrelated multivariate normal observations and application to the analysis of variance with random effects*
Let $y_1, \ldots, y_n$ be independent realizations of $Y \sim N_p([\kappa^j], [\kappa^{j,l}])$, $j, l, \ldots = 1, \ldots, p$, with $\kappa^j = \mu$ and

$$\kappa^{j,l} = \sigma^2(1-\rho)\left(\delta_{jl} + \frac{\rho}{1-\rho}u_{jl}\right) ,$$

where $\delta_{jl} = 1$ if $j = l$ and $\delta_{jl} = 0$ if $j \neq l$, $u_{jl} = 1$. Note that the components of Y have a common mean μ, common variance σ^2 and that the correlation coefficient between each pair of distinct components is equal to ρ. Using the result of Problem 9.3, we have $[\kappa^{j,l}]^{-1} = [\kappa_{jl}]$, with

$$\kappa_{jl} = \frac{1}{\sigma^2(1-\rho)}\left(\delta_{jl} - \frac{\rho}{1+\rho(p-1)}u_{jl}\right) .$$

Furthermore,

$$\det[\kappa^{j,l}] = (\sigma^2)^p(1-\rho)^{p-1}\{1 + \rho(p-1)\} .$$

Observe that $\det[\kappa^{j,l}] > 0$ if and only if $-1/(p-1) < \rho < 1$. Thus, the density of Y is

$$p_Y(y; \mu, \sigma^2, \rho) = \frac{1}{(2\pi\sigma^2)^{p/2}(1-\rho)^{(p-1)/2}\{1 + \rho(p-1)\}^{1/2}}$$

$$\times \exp\left\{-\frac{1}{2\sigma^2(1-\rho)}\left(\sum_{j=1}^{p}(y_j - \mu)^2 - \frac{\rho p^2}{1+\rho(p-1)}(\bar{y} - \mu)^2\right)\right\} .$$

For observations $y_i = (y_{i1}, \ldots, y_{ip})$, $i = 1, \ldots, n$, the log-likelihood is

$$
\begin{aligned}
l(\mu, \sigma^2, \rho) &= -\frac{1}{2\sigma^2(1-\rho)} SSW - \frac{p}{2\sigma^2\{1 + \rho(p-1)\}}(SSB + n\bar{y}^2) \\
&\quad + \frac{np\mu}{\sigma^2\{1 + \rho(p-1)\}}\bar{y} - \frac{np\mu^2}{2\sigma^2\{1 + \rho(p-1)\}} \\
&\quad - \frac{np}{2}\log\sigma^2 - \frac{n(p-1)}{2}\log(1-\rho) \\
&\quad - \frac{n}{2}\log(1 + \rho(p-1)),
\end{aligned}
\tag{11.64}
$$

with $SSW = \sum_{i=1}^{n}\sum_{j=1}^{p}(y_{ij} - \bar{y}_i)^2$, $SSB = \sum_{i=1}^{n}(\bar{y}_i - \bar{y})^2$, $\bar{y}_i = \sum_{j=1}^{p} y_{ij}/p$, $\bar{y} = \sum_{i=1}^{n}\sum_{j=1}^{p} y_{ij}/(np)$. Note that the reference model is a full exponential family of order 3.

Let us consider ρ as the parameter of interest and (μ, σ^2) as a nuisance parameter. It is easy to check that $\hat{\mu}_\rho = \bar{y}$ and that

$$
\hat{\sigma}_\rho^2 = \frac{SSW}{np(1-\rho)} + \frac{SSB}{n\{1 + \rho(p-1)\}}.
$$

The profile log-likelihood is

$$
\begin{aligned}
l_p(\rho) &= \frac{n(p-1)}{2}\log\left(1 + p\frac{\rho}{1-\rho}\right) \\
&\quad - \frac{np}{2}\log\left\{SSW\left(1 + p\frac{\rho}{1-\rho}\right) + pSSB\right\}.
\end{aligned}
$$

In order to calculate the modified profile likelihood, we need to express the sufficient statistic $(\bar{y}, SSW, SSB)$ as a function of the overall maximum likelihood estimate $(\hat{\mu}, \hat{\sigma}^2, \hat{\rho})$, itself a sufficient statistic. With easy calculations, we get

$$
\begin{aligned}
\hat{\mu} &= \bar{y}, \\
\hat{\sigma}^2 &= \frac{SSW + pSSB}{np}, \\
\hat{\rho} &= 1 - \frac{p}{p-1}\frac{SSW}{SSW + pSSB}.
\end{aligned}
$$

Hence, $SSW = n(p-1)\hat{\sigma}^2(1-\hat{\rho})$, $SSB = n\hat{\sigma}^2\{1 + \hat{\rho}(p-1)\}/p$. By inserting these expressions in (11.64), we obtain

$$
l(\mu, \sigma^2, \rho; \hat{\mu}, \hat{\sigma}^2, \hat{\rho}) = -\frac{n(p-1)\hat{\sigma}^2(1-\hat{\rho})}{2\sigma^2(1-\rho)}
$$

$$-\frac{p}{2\sigma^2\{1+\rho(p-1)\}}\left\{\frac{n\hat{\sigma}^2}{p}\{1+\hat{\rho}(p-1)\}+n\hat{\mu}^2\right\}$$

$$+\frac{np\mu}{\sigma^2\{1+\rho(p-1)\}}\bar{y}-\frac{np\mu^2}{2\sigma^2\{1+\rho(p-1)\}}$$

$$-\frac{np}{2}\log\sigma^2-\frac{n(p-1)}{2}\log(1-\rho)-\frac{n}{2}\log\{1+\rho(p-1)\}.$$

With $\zeta=(\mu,\sigma^2)$, the ingredients of the modification factor of the profile likelihood are

$$|j_{\zeta\zeta}(\rho,\hat{\zeta}_\rho)|=\frac{n^5p^5\{1+\rho(p-1)\}^2}{\left\{pSSB+\left(1+p\frac{\rho}{1-\rho}\right)SSW\right\}^3},$$

$$|l_{\zeta;\hat{\zeta}}(\rho,\hat{\zeta}_\rho)|=\frac{n^5p^5\{1+\rho(p-1)\}}{2(SSW+pSSB)\left\{pSSB+\left(1+p\frac{\rho}{1-\rho}\right)SSW\right\}^2}.$$

Finally,

$$L_{MP}(\rho)=\left(1+p\frac{\rho}{1-\rho}\right)^{n(p-1)/2}\left\{SSW\left(1+p\frac{\rho}{1-\rho}\right)+pSSB\right\}^{-(np-1)/2}.$$

The statistics SSW and SSB are independent, with marginal distribution

$$SSW\sim\sigma^2(1-\rho)\chi^2_{n(p-1)},$$

$$pSSB\sim\sigma^2(1-\rho)\left(1+p\frac{\rho}{1-\rho}\right)\chi^2_{n-1}.$$

The distribution of SSB/SSW does not depend on (μ,σ^2) and is an F distribution with $n-1$ and $n(p-1)$ degrees of freedom, multiplied by the scale factor $(n-1)\{1+p\rho/(1-\rho)\}/\{np(p-1)\}$. It is not hard to check that the marginal likelihood based on SSB/SSW coincides with $L_{MP}(\rho)$.

One important application of this is in the analysis of variance with random effects where, usually, one assumes that $y_{ij}=\mu+a_i+e_{ij}$ ($j=1,\ldots,p$, $i=1,\ldots,n$), with the effect a_i realization of a $N(0,\sigma_a^2)$, the error e_{ij} realization of a $N(0,\sigma_e^2)$ and independence between these realizations. The corresponding statistical model is of the type considered above, with $\sigma^2=\sigma_e^2+\sigma_a^2$ and $\rho=\sigma_a^2/(\sigma_e^2+\sigma_a^2)$. However, the family is not full, because the parameter space must satisfy the constraint $\rho>0$. Often the parameter of interest is the ratio between the variance of the effect and the error variance, $\rho/(1-\rho)$. Because this is a one-to-one function of ρ, inference could be based on the likelihood functions $L_P(\rho)$ and $L_{MP}(\rho)$, restricted to positive values of ρ. $\triangle$

It is important to stress that when combinants such as W or Z are defined starting from the modified profile likelihood they will have normal or chi-squared null asymptotic distributions with errors of the same order as the corresponding combinants based on the unmodified profile likelihood. However, it is true that in practice one could expect to obtain more precise evaluations, especially if $p - k$ is large (see for example DiCiccio and Stern, 1994a, section 4, for some simulation results). In order to obtain an improvement also in terms of the error order, it is necessary to make further corrections (Mukerjee and Chandra, 1991; DiCiccio and Stern, 1994a, 1994b; Ghosh and Mukerjee, 1994; Reid, 1995a, section 4).

11.7 Bibliographic Note

In-depth discussions of the p^* formula can be found in Barndorff-Nielsen (1980, 1983, 1984, 1988, 1990c), Skovgaard (1990), McCullagh (1987, section 8.6), Fraser (1988), Fraser and Reid (1988). Further references are to be found in the bibliographic note to Chapter 6 in Barndorff-Nielsen and Cox (1994). For a study of the relation with the Laplace expansion, and also for obtaining approximations for the marginal density of components of $\hat{\theta}$, see Barndorff-Nielsen (1990b). A proof of the validity of the p^* formula in curved exponential families is given in Barndorff-Nielsen and Cox (1994, section 7.4).

The idea of improving the asymptotic approximation of the distribution of W and W_p through correction of the expectation appears in Bartlett (1937), see Problem 11.12. An early collection of formulae for the correction factor is to be found in Bartlett (1954). For a historical survey of associated theory, see Jensen (1993). Barndorff-Nielsen and Cox (1984b) derived the results on the distribution of W conditionally on ancillary statistics and the link between the Bartlett correction and the normalizing constant in the p^* formula. Evaluations through simulation of the efficacy of the correction in problems on multivariate normal parameters can be found, among others, in Porteous (1985), Møller (1986), Eriksen (1987) and Jensen (1991). Cordeiro (1983, 1987) and Cordeiro and Paula (1989) obtained specifications for generalized linear models of formula (9.83). Cox (1984) considered the Bartlett correction both in a model of analysis of variance with random effects and for the Behrens–Fisher problem. DiCiccio, Hall and Romano (1991) have shown that (11.43), and the consequent improvement in the approximation obtained by means of the Bartlett correction, are also valid in a nonparametric setting for the empirical log-likelihood ratio (4.54). However, corrections similar to that of Bartlett do not exist for the score test; see Cordeiro and Ferrari (1991) and Rao and Mukerjee (1995) for some polynomial modifications, and Cordeiro, Ferrari and Paula (1993) and Cribari-Neto and Ferrari (1995) for applications to generalized linear models.

The material in section 11.5 regarding modified versions of Z and Z_P was developed by Barndorff-Nielsen (1986, 1990a, 1991b, 1994) and by Barndorff-Nielsen and Chamberlin (1991, 1994). In the latter paper an extension of $Z^\dagger$ to the case $p > 1$ is proposed. Further modified versions of Z and Z_P have been developed more recently, with the aim either of finding a way to avoid having to specify the ancillary (DiCiccio and Martin, 1993; Barndorff-Nielsen, 1994, 1995b; Skovgaard, 1996) or of obtaining valid approximations in a large deviation region (Jensen, 1992, 1995b; Skovgaard, 1996; Barndorff-Nielsen and Wood, 1996). Usually, such modified versions have an expression that is analogous to that of $Z^\star$, with a different definition of the combinant U.

The effects of the sampling rule on the asymptotic behaviour of W, Z and their modified versions are analyzed in Barndorff-Nielsen and Cox (1984a) and in Pierce and Peters (1994).

The modified profile likelihood was first introduced by Barndorff-Nielsen (1983); see Barndorff-Nielsen and Cox (1994, sections 8.2, 8.3 and 8.5) for further study, also as regards the behaviour of $L_{MP}(\tau)$ in models with incidental parameters. The stable version $L_B^\dagger(\tau)$ defined by (11.62) is equivalent to order $O(n^{-1})$ to the modification proposed by McCullagh and Tibshirani (1990), which entails modifying the profile score function so as the first two Bartlett relations, (9.20) and (9.21), are satisfied, at least approximately. Barndorff-Nielsen (1994, 1995b) proposed other stable versions of the modified profile likelihood. DiCiccio, Martin, Stern and Young (1996) show that the modified profile likelihood satisfies the information identity (1.9) with error of order $O(n^{-1})$, while this is not true, in general, for other modification of the profile likelihood, which, typically, satisfy (1.9) with error of order $O(1)$.

Recent applications of higher-order asymptotic methods to predictive inference are in Vidoni (1995) and in Barndorff-Nielsen and Cox (1995). Lastly, we mention the recent contributions by Reid (1995b) and Sweeting (1995a, 1995b) who examine the relations, highlighted by asymptotic considerations, between frequentist inference and the non-personalistic Bayesian approach.

11.8 Problems

11.1 It is known that if $Y_1, \dots, Y_n$ are i.i.d. with an $IG(\phi, \lambda)$ distribution, then

$\bar{Y}_n$ and $\sum(Y_i^{-1} - \bar{Y}_n^{-1})$ are independent and have distribution $IG(n\phi, n\lambda)$ and $\lambda^{-1}\chi_{n-1}^2$, respectively (Tweedie, 1957). Exploit this result in order to show that the p^* formula for the density of $(\hat{\phi}, \hat{\lambda})$ is exact. Show that the p^* formula is also exact for the density of $\hat{\phi}$ when λ is known.

[Section 11.2]

11.2 Show that the p^* formula satisfies the properties *i)–iii)* described on page 436.

[Section 11.2]

11.3 Under the same assumptions as in Example 2.8, obtain the p^* formula for the conditional density of $\hat{\sigma}$ given a.

[Section 11.2]

11.4 With reference to the model $(\mathcal{Y}, p_Y(y; \theta), \Theta)$, with $\Theta \subseteq \mathbb{R}^p$, let

$$l_{R_m;S_n}(\theta; \hat{\theta}, a) = \frac{\partial^{m+n}}{\partial\theta^{r_1}\cdots\partial\theta^{r_m}\partial\hat{\theta}^{s_1}\cdots\partial\hat{\theta}^{s_n}} l(\theta; \hat{\theta}, a)$$

be a generic mixed derivative. By expanding around θ the likelihood equation $l_{r;}(\hat{\theta}; \hat{\theta}, a) = 0$, obtain the asymptotic expansion

$$
\begin{aligned}
(\hat{\theta} - \theta)^r &= j^{rs}(\theta; \theta, a)l_s; \\
&\overset{\bullet}{-} \frac{1}{2}j^{rs}(\theta; \theta, a)j^{tu}(\theta; \theta, a)j^{vw}(\theta; \theta, a)l_{s;tv}(\theta; \theta, a)l_{u;}l_{w;} \\
&\overset{\bullet}{+} O_p(n^{-3/2})
\end{aligned}
\tag{11.65}
$$

which represents the observed version of expansion (9.61). Use expansion (11.65) to obtain

$$E_\theta\left\{(\hat{\theta} - \theta)^r \,\middle|\, a\right\} = -\frac{1}{2}j^{rs}(\theta; \theta, a)j^{tu}(\theta; \theta, a)l_{s;tu}(\theta; \theta, a) + O(n^{-3/2}). \tag{11.66}$$

This expansion allows us to check (11.10). Also, obtain the following expansion:

$$\text{Cov}_\theta\left((\hat{\theta} - \theta)^r(\hat{\theta} - \theta)^s \,\middle|\, a\right) = -j^{rs}(\theta; \theta, a) + O(n^{-2}). \tag{11.67}$$

[Section 11.2; Barndorff-Nielsen and Cox, 1994, section 5.4]

11.5 Let $Y_1, \ldots, Y_n$ be a random sample of size n from an exponential r.v. with density $p_Y(y; \theta) = \theta e^{-\theta y}$, $y > 0$, $\theta > 0$. Check directly that (11.15) gives an approximation for the density of $\hat{\theta}$ with error of order $O(n^{-1})$.

[Section 11.2]

11.6 Obtain the p^* and $p^\dagger$ formulae for the density of the score vector.

[Section 11.2; Barndorff-Nielsen and Cox, 1994, p. 179]

11.7 Show that the standardized normalizing constant $\bar{c}(\theta, a)$ of the p^* formula (11.7) and the expectation $E_\theta(W|a)$ are linked by

$$\bar{c}(\theta, a) = \left(\frac{p}{E_\theta(W|a)} \right)^{p/2} + O(n^{-3/2}). \tag{11.68}$$

[Section 11.3]

11.8 Consider the profile log-likelihood ratio W_P, as defined, for example, in (9.84). Assume that the Bartlett correction is calculated starting from $W_P = 2\{l(\hat{\tau}, \hat{\zeta}) - l(\tau, \zeta)\} - 2\{l(\tau, \hat{\zeta}_\tau) - l(\tau, \zeta)\}$. Specifically, calculate the expected value of the first summand conditionally on an ancillary statistic a and the expected value of the second summand conditionally on an ancillary statistic a_0 for the submodel with fixed τ. Indicate the correction thus obtained with kb. Show that b and the standardized normalizing constants $\bar{c}$ and $\bar{c}_0$ of the p^* formulae for, respectively, $\hat{\theta} = (\hat{\tau}, \hat{\zeta})$ in the model with parameter $\theta = (\tau, \zeta)$ and for $\hat{\zeta}_\tau$ in the submodel with fixed τ, are linked by

$$b = \left(\frac{\bar{c}}{\bar{c}_0} \right)^{-2/k} + O(n^{-3/2}). \tag{11.69}$$

[Section 11.3; Barndorff-Nielsen and Cox, 1994, section 6.5.2]

11.9 Let V_n be a sequence of random variables with density

$$p_{V_n}(v) = \left(1 - \frac{\delta}{n} \right) p_{\chi_p^2}(v) + \frac{\delta}{n} p_{\chi_{p+2}^2}(v),$$

where $p_{\chi_g^2}(v)$ is the χ_g^2 density and $\delta/n \in (0, 1)$. Show that $pV_n/E(V_n)$ has density χ_p^2 with an error of order $O(n^{-2})$.

[Section 11.4]

11.10 Consider p independent random samples $y_1, \ldots, y_p$, of size $n_1, \ldots, n_p$, respectively, from p gamma distributions, $Ga(1, \theta_j)$, $j = 1, \ldots, p$. Obtain the null expectation of the statistic W_P for testing $H_0 : \theta_1 = \ldots = \theta_p$ versus the alternative that the p distributions do not all have the same mean. Compare the Bartlett correction obtained with the approximation suggested in (11.69) and show, directly, that expansion (11.69) holds.

[Section 11.4; Hollas, 1991]

11.11 Let $y_1, \ldots, y_n$ be independent realizations of a gamma distribution, $Ga(\nu, \phi)$. Obtain the combinants W_P and $W_P^\star$ for inference about the mean.
[Section 11.4; Jensen and Kristensen, 1991]

11.12 Consider m independent random samples $\boldsymbol{y}_1, \ldots, \boldsymbol{y}_m$, of size $n_1, \ldots, n_m$, respectively, from m normal distributions, $N(\mu_j, \sigma_j^2)$, $j = 1, \ldots, m$. Obtain the null expectation of the combinant W_P for testing $H_0 : \sigma_1^2 = \ldots = \sigma_m^2$ versus the alternative that not all the m distributions have the same variance. Compare the Bartlett correction obtained with the approximation suggested by (11.69) and show, directly, that (11.69) holds.
[Section 11.4; Barndorff-Nielsen and Cox, 1994, Example 6.16]

11.13 Show that the quantity U defined by (11.49) is parameterization invariant.
[Section 11.5]

11.14 Show that in a location family the combinant U defined by (11.49) coincides with Z_u, the signed version of W_u, defined by (3.45).
[Section 11.5]

11.15 Carry out detailed calculations in order to obtain $Z^\star$ in Example 11.4.
[Section 11.5]

11.16 Show that in a location family $Z^\dagger = Z^\star$.
[Section 11.5]

11.17 Let $y_1, \ldots, y_n$ be independent realizations of a von Mises distribution with density (2.7). Obtain the combinants Z and $Z^\star$ for inference about θ with known λ.
[Section 11.5; Barndorff-Nielsen and Cox, 1994, Example 6.29]

11.18 Obtain the general expression of $Z_P^\star$ in an $\mathcal{F}_{n_e}^p$ with density

$$p_s(s; \theta) = e^{\theta \cdot s - nK(\theta)} p_0(s) ,$$

when the parameter of interest is a non-linear scalar function of the natural parameter θ.
[Section 11.5; Barndorff-Nielsen and Cox, 1994, Example 6.24]

11.19 Obtain expansion (11.56). Under the same assumptions as in Example 11.8, show that the confidence intervals for μ, deduced from the asymptotic distribution of the log-likelihood ratio based on the modified profile likelihood, coincide, up to and including terms of order $O(n^{-3/2})$, with the analogous intervals obtained from the approximate conditional likelihood in Example 4.17.

[Sections 11.5 and 11.6]

11.20 Under the same assumptions as in Example 11.10, obtain the expressions of Z_P and Z_P^* for inference about ρ. Adapt these quantities in order to obtain tests for $\rho = 0$ against $\rho > 0$. Discuss the possibility of using these tests under a model of analysis of variance with random effects.

[Sections 11.5 and 11.6]

11.21 Under the same assumptions as in Example 4.8, obtain the modified profile likelihood for ρ and check, directly, that this gives an approximation with error of order $O(n^{-3/2})$ of the marginal likelihood based on (4.16).

[Section 11.6]

11.22 Under the same assumptions as in Example 11.7, show that the modified profile likelihood $L_{MP}(\sigma)$ coincides with the approximate conditional likelihood $L_{AC}(\sigma)$.

[Section 11.6; Barndorff-Nielsen and McCullagh, 1993, section 3]

APPENDIX A

LAWS OF LARGE NUMBERS AND CENTRAL LIMIT THEOREMS

First, the case of identically distributed random variables is considered; then some results for independent random variables (which are not necessarily identically distributed) are given. Lastly, useful results that are related to the asymptotic behaviour of regular functions of converging sequences of random variables are surveyed.

In the following, $\{Y_i\}$ is used to indicate a sequence of random variables $i = 1, 2, \ldots$, with $\{S_n\}$ the corresponding sequence of sums, $S_n = \sum_{i=1}^{n} Y_i$, and $\{\bar{Y}_n\}$ the sequence of sample means, $\bar{Y}_n = S_n/n$.

A.1 Sums of i.i.d. Random Variables

a) Laws of large numbers

Theorem A.1 (*Khintchine's weak law.*) Let $\{Y_i\}$ be a sequence of independent and identically distributed random variables with finite expectation $E(Y_i) = \mu$. Then, $\bar{Y}_n \overset{p}{\to} \mu$.

Theorem A.2 (*Kolmogorov's strong law (I).*) Let $\{Y_i\}$ be a sequence of independent and identically distributed random variables. Then, $\bar{Y}_n \overset{as}{\to} \mu$ if and only if the expectation $E(Y_i)$ is finite and equal to μ.

b) Central limit theorems

Theorem A.3 (*The finite variance case: Lindberg-Lévy's theorem.*) Let $\{Y_i\}$ be a sequence of independent and identically distributed variables with expectation μ and finite variance $\sigma^2 > 0$. Then

$$\frac{\sqrt{n}(\bar{Y}_n - \mu)}{\sigma} \overset{d}{\to} N(0, 1).$$

469

Theorem A.4 (*Multivariate central limit theorem.*) Let $\{\mathbf{Y}_i\}$ be a sequence of d-dimensional independent and identically distributed random variables with mean vector $\mu = [\kappa^r]$ and finite covariance matrix $\Sigma = [\kappa^{r,s}]$, $r, s = 1, \ldots, d$, $\kappa^{r,r} < +\infty$ for $r = 1, \ldots, d$. Then, denoting by $\bar{\mathbf{Y}}_n$ the vector of the sample means,

$$\sqrt{n}(\bar{\mathbf{Y}}_n - \mu) \xrightarrow{d} N_d(0, \Sigma).$$

Theorem A.5 (*The infinite variance case: convergence to stable distributions.*) Let $\{Y_i\}$ be a sequence of independent and identically distributed random variables with infinite variance. If two sequences of constants exist, $\{a_n\}$ and $\{b_n\}$, such that $S_n/b_n - a_n$ converges in distribution to a non-degenerate limit, then the limiting distribution must be stable with characteristic exponent $0 < \theta < 2$ and Y_i is said to be in the **domain of attraction** of this stable distribution. The sequence b_n must have the form $b_n = n^{1/\theta} b_0(n)$, where $b_0(n)$ is a slowly varying function (see section 3.2.6).

Theorem A.6 (*A local limit theorem.*) Let $\{Y_i\}$ be a sequence of independent and identically distributed random variables with expectation μ and finite variance $\sigma^2 > 0$; furthermore assume
i) (continuous case) that their characteristic function $C_{Y_i}(t)$ satisfies the condition

$$\int_{-\infty}^{+\infty} |C_{Y_i}(t)|^m \, dt < +\infty, \tag{A.1}$$

for some integer $m \geq 1$; or
ii) (discrete case) that Y_i is a **lattice variable**, that is, with support $\mathcal{Y}_i \subseteq \{a, a \pm h, a \pm 2h, \ldots\}$, $a \in \mathbb{R}$, $h > 0$.
Then, the density function of

$$S_n^* = \frac{\sqrt{n}(\bar{Y}_n - \mu)}{\sigma}$$

(or $(\sigma\sqrt{n}/h)P(S_n^* = s_n)$ in the discrete case) converges uniformly to $\phi(\cdot)$ for converging sequences of points in the support of S_n^*.

c) Law of iterated logarithm

Theorem A.7 Let $\{Y_i\}$ be a sequence of independent and identically distributed random variables with expectation μ and finite variance $\sigma^2 > 0$. Then

$$P\left\{\limsup_{n \to +\infty} \frac{\sum_{i=1}^n Y_i - n\mu}{\sigma\sqrt{2n\log\log n}} = 1\right\} = 1.$$

A.2 Sums of Independent Random Variables

a) Laws of large numbers

Theorem A.8 (*Chebyshev's weak law.*) Let $\{Y_i\}$ be a sequence of independent random variables with $E(Y_i) = \mu_i$ and $\mathrm{Var}(Y_i) = \sigma_i^2$, finite for every i. If

$$\lim_{n \to +\infty} \frac{1}{n^2} \sum_{i=1}^{n} \sigma_i^2 = 0$$

then

$$\bar{Y}_n - \bar{\mu}_n \xrightarrow{P} 0 ,$$

where $\bar{\mu}_n = n^{-1} \sum_{i=1}^{n} \mu_i$.
The theorem is also valid for sequences of uncorrelated random variables.

Theorem A.9 (*Kolmogorov's strong law (II).*) Let $\{Y_i\}$ be a sequence of random variables with $E(Y_i) = \mu_i$ and $\mathrm{Var}(Y_i) = \sigma_i^2$, finite for every i. If

$$\sum_{i=1}^{+\infty} \frac{\sigma_i^2}{i^2} < +\infty$$

then

$$\bar{Y}_n - \bar{\mu}_n \xrightarrow{as} 0 ,$$

where $\bar{\mu}_n = n^{-1} \sum_{i=1}^{n} \mu_i$.
The result is also valid for uncorrelated variables, so long as the more restrictive condition

$$\sum_{i=1}^{+\infty} \frac{\sigma_i^2 (\log i)^2}{i^2} < +\infty$$

is satisfied.

b) Central limit theorems

Theorem A.10 (*A sufficient condition: Liapunov's theorem.*) Let $\{Y_i\}$ be a sequence of independent random variables with $E(Y_i) = \mu_i$, $\mathrm{Var}(Y_i) = \sigma_i^2 > 0$ and $E|Y_i - \mu_i|^3 = \beta_i$, finite for every i. Furthermore, let

$$B_n = \left(\sum_{i=1}^{n} \beta_i \right)^{1/3} \quad \text{and} \quad C_n = \left(\sum_{i=1}^{n} \sigma_i^2 \right)^{1/2} .$$

If

$$\lim_{n \to +\infty} \frac{B_n}{C_n} = 0$$

then

$$\frac{\sum_{i=1}^{n}(Y_i - \mu_i)}{C_n} \xrightarrow{d} N(0,1) \, .$$

Theorem A.11 (*A necessary and sufficient condition: Lindberg-Feller's theorem.*) Let $\{Y_i\}$ be a sequence of independent random variables with distribution function $F_{Y_i}(y)$, $E(Y_i) = \mu_i$ and $\text{Var}(Y_i) = \sigma_i^2 > 0$, finite for every i. Furthermore let

$$C_n = \left(\sum_{i=1}^{n} \sigma_i^2 \right)^{1/2} \, .$$

Then

$$\frac{\sum_{i=1}^{n}(Y_i - \mu_i)}{C_n} \xrightarrow{d} N(0,1)$$

and

$$\lim_{n \to +\infty} \max_{1 \le i \le n} \frac{\sigma_i}{C_n} = 0$$

hold if, and only if, for every $\varepsilon > 0$,

$$\lim_{n \to +\infty} \frac{1}{C_n^2} \sum_{i=1}^{n} \int_{|y - \mu_i| > \varepsilon C_n} (y - \mu_i)^2 dF_{Y_i}(y) = 0 \, .$$

A.3 Smooth Functions of Converging Sequences

a) Convergence in probability

Theorem A.12 Let $\{Y_n\}$, $n = 1, 2, \ldots$, be a sequence of random variables which converges in probability to a constant c. If $g(\cdot)$ is a continuous function defined on the support of Y_n, then $g(Y_n) \xrightarrow{p} g(c)$.

b) Convergence in distribution

Theorem A.13 Let $\{Y_n\}$, $n = 1, 2, \ldots$, be a sequence of random variables such that

$$\sqrt{n}(Y_n - \theta) \xrightarrow{d} U \, ,$$

with $\theta \in \mathbb{R}$ and U a non-degenerate random variable. Furthermore, let $g(\theta)$ be a twice differentiable real function with $g'(\theta) \neq 0$. Then

$$\sqrt{n}(g(Y_n) - g(\theta)) \xrightarrow{d} g'(\theta)U .$$

Under the same assumptions, the asymptotic expansion

$$\sqrt{n}(g(Y_n) - g(\theta)) = \sqrt{n}g'(\theta)(Y_n - \theta) + o_p(1) = g'(\theta)U + o_p(1)$$

holds, so that, asymptotically, $E\{g(Y_n)\} = g(\theta)$ and

$$\mathrm{Var}\{g(Y_n)\} = (g'(\theta))^2 \, \mathrm{Var}(U)/n .$$

The local linearization of $g(\cdot)$ in the above theorem and, more generally, local approximations of $g(Y_n)$ based on Taylor expansions, are usually referred to as the **delta method** (see also Theorems A.14 and A.15). The delta method is widely used to obtain approximations for moments of $g(Y_n)$ in terms of moments of Y_n (cf. section 8.5).

c) Convergence to the normal distribution

Theorem A.14 (*Univariate case.*)   Let $\{Y_n\}$, $n = 1, 2, \ldots$, be a sequence of random variables such that

$$\sqrt{n}(Y_n - \theta) \xrightarrow{d} N(0, \sigma^2) ,$$

with $\theta \in \mathbb{R}$. Furthermore let $g(\cdot)$ be a twice differentiable real function defined on $\mathbb{R}$ such that $g'(\theta) \neq 0$. Then

$$\sqrt{n}(g(Y_n) - g(\theta)) \xrightarrow{d} N(0, (g'(\theta))^2 \sigma^2) .$$

Theorem A.15 (*Multivariate case.*)   Let $\{\mathbf{Y}_n\}$, $n = 1, 2, \ldots$, be a sequence of d-dimensional random variables such that

$$\sqrt{n}(\mathbf{Y}_n - \theta) \xrightarrow{d} N_d(0, \Sigma) ,$$

with $\theta \in \mathbb{R}^d$ and covariance matrix Σ. Also, let $\mathbf{g}(\cdot)$ be a function defined on $\mathbb{R}^d$ with values in $\mathbb{R}^k$, $k \leq d$, with twice differentiable components $g_i(\cdot)$, $i = 1, \ldots, k$. Then

$$\sqrt{n}(\mathbf{g}(\mathbf{Y}_n) - \mathbf{g}(\theta)) \xrightarrow{d} N_k(0, D\Sigma D^T) ,$$

where $D = [d_{ij}]$ is a $k \times d$ matrix with $d_{ij} = \partial g_i(\theta)/\partial \theta^j$, $i = 1, \ldots, k$ and $j = 1, \ldots, d$.

A.4   Bibliographic Note

Any intermediate probability textbook deals, in detail, with the laws of large numbers and central limit theorems; see e.g., Billingsley (1986) and Grimmet and Stirzaker (1992). For in depth explanation of limit theorems and stable laws, see Laha and Rohatgi (1979, Chapter 5).

Laws of large numbers and central limit theorems are also extended to sequences of dependent random variables. The extensions that are most important for the study of the asymptotic properties of likelihood quantities are those related to martingales; for the main results and further references, see e.g. Hall (1985) and Andersen, Borgan, Gill and Keiding (1993, Chapter 2).

APPENDIX B

ASYMPTOTIC DISTRIBUTION OF EXTREMES

B.1   Basic Results

Let $Y_1, Y_2, \ldots$ be a sequence of independent and identically distributed one-dimensional random variables, each with distribution function $F_Y(y)$. Then $Y_{(n)} = \max(Y_1, \ldots, Y_n)$ has distribution function $F_{Y_{(n)}}(y) = \{F_Y(y)\}^n$. We can immediately see that, as n diverges, $F_{Y_{(n)}}(y)$ converges to a degenerate distribution with unit mass in the upper extreme of the support of $F_Y(y)$. Since this does not give any useful approximation, one could ask whether it is possible to identify suitable sequences of constants, $\{a_n\}$ and $\{b_n\}$, with $a_n \in \mathbb{R}$ and $b_n > 0$, such that

$$P\{b_n(Y_{(n)} - a_n) \leq y\} = \{F_Y(y/b_n + a_n)\}^n \to \Lambda(y) \qquad (B.1)$$

as $n \to +\infty$, with $\Lambda(\cdot)$ the distribution function of a non-degenerate random variable. Gnedenko (1943) has shown that if such a limiting distribution exists, then it must be one of the following three types (called, respectively, Fréchet, Weibull on $\mathbb{R}^-$ and Gumbel):

$$\Lambda_1(y) = \begin{cases} 0 & \text{if } y \leq 0 \\ \exp(-y^{-\theta}) & \text{if } y > 0, \text{ with } \theta > 0, \end{cases} \qquad (B.2)$$

$$\Lambda_2(y) = \begin{cases} \exp\{-(-y)^\theta\} & \text{if } y \leq 0, \text{ with } \theta > 0 \\ 1 & \text{if } y > 0, \end{cases} \qquad (B.3)$$

$$\Lambda_3(y) = \exp(-e^{-y}), \quad y \in \mathbb{R}. \qquad (B.4)$$

Note that Λ_2 is the distribution function of $-Z$, with Z distributed according to a Weibull law, with density

$$p_Z(z; \theta) = \theta z^{\theta-1} \exp(-z^\theta), \quad z > 0, \theta > 0. \qquad (B.5)$$

The previous results extend immediately to the sample minimum

$$Y_{(1)} = \min(Y_1, \ldots, Y_n)$$

since $Y_{(1)} = -\max(-Y_1, \ldots, -Y_n)$.

If there exist sequences $\{a_n\}$ and $\{b_n\}$ such that (B.1) holds, $F_Y(y)$ is said to be in the domain of attraction of $\Lambda(y)$. To see whether $F_Y(y)$ is in the domain of attraction of Λ_j, $j = 1, 2, 3$, we only have to know how $F_Y(y)$ behaves on the upper tail of the distribution.

In particular, $F_Y(y)$ is in the domain of attraction of Λ_1 if the support of Y is unbounded and if

$$1 - F_Y(y) \sim cy^{-\theta}, \qquad \text{as } y \to +\infty, \ \theta > 0, \qquad (\text{B.6})$$

where $\sim$ indicates that as y diverges the limit of the ratio between the two quantities is equal to 1. If (B.6) holds, the standardization required is $(cn)^{-1/\theta} Y_{(n)}$.

The distribution function $F_Y(y)$ is in the domain of attraction of Λ_2 if the support of Y has $y_0 < +\infty$ as its upper bound and if

$$1 - F_Y(y) \sim c(y_0 - y)^{\theta}, \qquad \text{as } y \to y_0 \ (y < y_0), \ \theta > 0. \qquad (\text{B.7})$$

To obtain convergence, the standardization needed is $(Y_{(n)} - y_0)(nc)^{1/\theta}$.

Lastly, $F_Y(y)$ is in the domain of attraction of Λ_3 if the support of Y has no finite upper bound and if

$$1 - F_Y(y) \sim ce^{-y^h}, \qquad \text{as } y \to +\infty, \ h > 0.$$

The constants a_n and b_n are determined, for example, by the relations $a_n = F_Y^{-1}(1 - 1/n)$ and $b_n = np_Y(a_n)$.

On the basis of these results we can immediately state that, for example, the uniform distribution on $[0, 1]$ is in the domain of attraction of Λ_2. More specifically, $c = \theta = 1$ and convergence to Λ_2 can be obtained by considering the sequence $n(Y_{(n)} - 1)$. The exponential distribution with mean equal to one is in the domain of attraction of Λ_3; in particular, the standardization that should be considered is $Y_{(n)} - \log n$.

B.2 Bibliographic Note

For an introduction to the theory of extremes and its application in statistics, Gumbel (1958) remains the basic reference; see also, David (1981, Chapter 9). More recent developments concern the multivariate theory of extremes with dependent observations. For this, see the books by Leadbetter, Lindgren and Rootzén (1983), Resnick (1987) and Galambos (1987).

B.3   Problems

B.1 Show that if Y has p.d.f. $p_Y(y) = e^{-y}$, $y > 0$, the r.v. $Z = Y^{1/\theta}$, $\theta > 0$, has p.d.f. (B.5). Show that $\log Z = -(1/\theta)W$, with W distributed according to Λ_3.

B.2 Verify that if Y satisfies relation (B.7) then the limiting distribution of the sample maximum is Λ_2.

B.3 Check the statements at the end of section B.1 relating, respectively, to convergence of the sample maximum from a uniform distribution on $[0, 1]$ and from an exponential distribution with mean equal to one.

B.4 Show that the Cauchy distribution is in the domain of attraction of Λ_1.

B.5 Show that the normal distribution is in the domain of attraction of Λ_3.

B.6 Show that the parametric family with distribution function

$$F_Y(y; \xi) = \exp\left\{ -(1 + \xi y)^{-1/\xi} \right\},\qquad\qquad (B.8)$$

on the support $1 + \xi y > 0$, with $\xi \in \mathbb{R}$, includes, as special cases, the three distributions Λ_j, $j = 1, 2, 3$, which can be obtained, respectively, with $\xi < 0$, with $\xi > 0$ and as a limit when $\xi \to 0$.

APPENDIX C

PARAMETRIC INFERENCE: BASIC TERMINOLOGY

Let $\mathcal{F} = \{p_Y(y; \theta), \theta \in \Theta \subseteq \mathbb{R}^p\}$ be a parametric statistical model for data y in the sample space $\mathcal{Y}$. The data y are a realization of the random variable Y with density in $\mathcal{F}$.

C.1 Point Estimation

An **estimate** is defined by a mapping $\tilde{\theta} : \mathcal{Y} \to \Theta$ and is denoted by $\tilde{\theta}(y)$.

The random variable $\tilde{\theta}(Y)$ is called an **estimator**.

An estimator is **unbiased** for θ if $E_\theta\left\{\tilde{\theta}(Y)\right\} = \theta$.

The **bias** of an estimator is $E_\theta\left\{\tilde{\theta}(Y) - \theta\right\}$.

The **mean squared error** of an estimator is $E_\theta\left\{\left(\tilde{\theta}(Y) - \theta\right)^2\right\}$.

An estimator based on $Y = (Y_1, \ldots, Y_n) \sim p_Y(y; \theta)$ is **weakly consistent** for θ if $\tilde{\theta}(Y) \xrightarrow{P} \theta$, i.e. if $\tilde{\theta}(Y)$ converges in probability to θ as $n \to +\infty$, for every $\theta \in \Theta$.

An estimator based on $Y = (Y_1, \ldots, Y_n) \sim p_Y(y; \theta)$ is **strongly consistent** for θ if $\tilde{\theta}(Y) \xrightarrow{as} \theta$, i.e. if $\tilde{\theta}(Y)$ converges almost surely to θ as $n \to +\infty$, for every $\theta \in \Theta$.

C.2 Hypothesis Testing

A **null hypothesis**, denoted by H_0, states that θ belongs to a subset Θ_0 of Θ.

479

A test or test statistic identifies two regions in $\mathcal{Y}$: the acceptance region A_{Θ_0}, where the data do not indicate evidence against H_0, and the rejection region, $\bar{A}_{\Theta_0} = \mathcal{Y} \setminus A_{\Theta_0}$, where the data indicate evidence against H_0.

A test has level or size α if $\sup_{\theta \in \Theta_0} P_\theta \left\{ Y \in \bar{A}_{\Theta_0} \right\} \leq \alpha$.

The power function of a test is $\beta(\theta) = P_\theta \left\{ Y \in \bar{A}_{\Theta_0} \right\}$.

A scalar test statistic $t(y)$ is said to be one-sided if the rejection region is $\bar{A}_{\Theta_0} = \{ y \in \mathcal{Y} : t(y) \geq t_\alpha \}$ (large values significant) or $\bar{A}_{\Theta_0} = \{ y \in \mathcal{Y} : t(y) \leq t_\alpha \}$ (small values significant).

For a one-sided scalar test statistic with large values significant, the observed significance level or P-value for observation y^{obs} is given by

$$\alpha^{obs} = \sup_{\theta \in \Theta_0} P_\theta \left(t(Y) \geq t(y^{obs}) \right) .$$

An analogous definition of the P-value is given when small values are significant.

For a one-sided test statistic with a discrete distribution and with large values significant, the mid-P value for observation y^{obs} is

$$\text{mid-}P = \sup_{\theta \in \Theta_0} \left\{ P_\theta \left(t(Y) > t(y^{obs}) \right) + \frac{1}{2} P_\theta \left(t(Y) = t(y^{obs}) \right) \right\} .$$

An alternative hypothesis, denoted by H_1, states that θ belongs to a subset Θ_1 of Θ, with $\Theta_0 \bigcap \Theta_1 = \emptyset$.

A test for $H_0 : \theta \in \Theta_0$, based on $Y = (Y_1, \ldots, Y_n) \sim p_Y(y; \theta)$, is consistent if $\lim_{n \to +\infty} \beta(\theta) = 1$, for every $\theta \in \Theta \setminus \Theta_0$.

C.3   Confidence Regions

A confidence region for θ with confidence level $1 - \alpha$ is a subset of Θ, defined as a function of y and denoted by $C(y)$, such that

$$P_\theta \left\{ \theta \in C(Y) \right\} \geq 1 - \alpha, \qquad \text{for every } \theta \in \Theta .$$

The (null) coverage probability of $C(y)$ is

$$P_\theta \left\{ \theta \in C(Y) \right\} , \qquad \theta \in \Theta .$$

A confidence region with level $1 - \alpha$ can be obtained from the acceptance regions A_{θ_0} of a family of level-α tests for $H_0 : \theta = \theta_0$, $\theta_0 \in \Theta$. For each $y \in \mathcal{Y}$, let

$$C(y) = \{\theta \in \Theta : y \in A_\theta\} .$$

Then $C(y)$ is a confidence region with level $1 - \alpha$.

APPENDIX D

RELATIONS BETWEEN THE FREQUENCY-DECISION AND FISHERIAN PARADIGMS

The decision paradigm outlined in section 3.5 was developed primarily in order to meet the need to formalize Fisher's ideas, with particular reference to the second of the three problems of the theory of statistics, (see section 1.3), that is, inference problems. These are problems of choosing an appropriate statistical procedure for a given inferential problem. Over the years, because of the lively debate between Fisher and Neyman, general opinion gradually began to accept that the two paradigms, Fisherian and decisional, were essentially opposites. This conflict is also clear in the terminology. For example, the expression *testing hypotheses* is typical of the decision paradigm, whereas one speaks of *significance tests* in the Fisherian school. On the other hand, it is also clear that terminological divergence must be a symptom of deeper conceptual divergence. Thus, one must ask precisely which are the key points where the two paradigms do not agree and, whether it is possible to reconcile the two positions. Lehmann (1993) proposed an incisive analysis of the question with reference to testing hypotheses; many of his considerations could be generalized. Here, no exhaustive comparison will be given; for specialized discussions, see Lehmann (1993), Cox (1958, 1977, 1986) and Skovgaard (1989).

i) **Loss functions and constraints.** Clearly, the discussion must be limited to inference contexts, where the possible introduction of a loss function serves only to define the problem in mathematical terms and has nothing whatever to do with real, tangible losses. Section 3.5 was devoted to the three main classes of parametric inference procedures: point estimation, testing hypotheses and confidence regions. Since there is a close relationship between optimality in tests and optimality in confidence regions, comments can be limited to the first two classes of procedures. First, we should note that both the loss functions and the criteria for constraint considered in hypothesis testing obey the principle of parameterization invariance, while this is not always so in the case of the decision approach to point estimation. In particular, the theory of unbiased estimators with a quadratic loss function does not obey the principle of parameterization invariance. The choice of loss functions and centring criteria would seem to be more convincing when both of them represent intrinsic qualities, such as, at least in the one-parameter case, median unbiasedness. The restriction of equivariance agrees both with the principle of parameterization invariance and, at least with reference to specific classes of transformations, with the requirement of invariance under one-to-one transformations of data, see section 2.2.

ii) **Point estimation and confidence regions.** A point estimate should always be accompanied by an evaluation of error and should thus be seen as a summarizing quantity for a confidence region. Hence, we could argue that point estimation should not be taken in isolation, but should rather be dealt with as a limiting case of confidence regions. The criterion of median unbiasedness would fit this view. A possible multiparameter extension is suggested by Skovgaard (1989): a natural point estimate is the intersection of all the confidence regions at all positive levels. Whenever these confidence regions are uniformly most accurate, the corresponding point estimates are maximally concentrated around the true value of the parameter (see formula (3.59) in the one-parameter case). If this is agreed, then all efforts to trace inference procedures can concentrate on testing hypotheses. A family of tests with nested acceptance regions, automatically provides a family of confidence regions and a point estimate.

iii) **Fixed significance levels and P-values.** As an example of the conflict between the Fisherian and Neyman–Pearson paradigms one often refers to the fact that the result of a test is expressed in the first case, as an observed significance level or P-value, and, in the second as a comparison with a pre-assigned level α. It should be remembered (see Lehmann, 1993) that the initial reason for resorting to pre-assigned levels was that there were no tools for fast calculation available at the time, nor were there sufficiently detailed tables. Clearly an observed P-value offers much fuller information and can always be compared with a fixed level. Furthermore, there are no counter-indications for evaluating the P-value of an optimal test.

iv) Conditioning. In the decision paradigm, one assumes that the choice between competing decisions takes place in the pre-experimental phase. Thus it follows, quite naturally, that the expectation (3.56) is calculated by considering the entire space $\mathcal{Y}$ as the reference sample space. This would seem to exclude the conditionality principle, which would require that the risk is evaluated conditionally on an available distribution constant, or ancillary, statistic. This is one of the main differences between the Fisherian and decision paradigms: according to the latter, conditioning is seen as an unjustifiable abandonment of the search for global, i.e. unconditional, optimality. It should be stressed, however, that *any* application of the principle of repeated sampling requires the specification of a reference set. The probability statements relating to uncertainties in inference depend on the choice of this set. This has two possible interpretations. We could say that, from the Neyman–Pearson point of view, what matters is that inference procedures should be optimal: whether they are optimal with regard to all the $\mathcal{Y}$ space or with regard to the conditional space given $A = a$ depends on the aims of the inference. On the other hand, what is important from the Fisherian point of view, is that inference should be relevant, this means that, when a test has been assigned, evaluation of the observed significance level is carried out with respect to the conditional space given $A = a$. Thus if, as usually happens e.g. in quality control, repetitions of the experiment are really carried out and are not merely hypothetical, it is important that the test should be unconditionally optimal; the observed significance level could, however, be evaluated for the optimal unconditional test on the conditional space. See Barnard (1976, 1977b, 1979) for a fuller discussion.

v) Maximum power. According to Fisher, the problem of choosing between competing tests for the same null hypothesis does not even arise in the context of scientific research. In his view, an experienced researcher, does not resort to optimality theories in order to choose the test to apply to a specific problem; experience offers enough of a basis on which to decide about the most appropriate test, that is, about which aspects of the data could detect deviation from the null hypothesis. Consequently, the Fisherian vision, only requires a null hypothesis and indeed strongly criticizes any references to alternative hypotheses. This raises two points. Firstly, the increasing complexity of problems and of statistical models suitable for dealing with them makes basing the choice of a test only on the researcher's experience and commonsense somewhat discouraging: obviously experience must not be ignored but a few general and theoretical guidelines would not come amiss. Secondly, it can be difficult to judge some aspects of the data as being anomalous without knowing the direction where the anomaly will be found. Also, in order to be useful, a test must reflect deviations from the null hypothesis: it is hard to see why, when choosing between two competing statistics, that with higher resolution should not be chosen. Furthermore, the evaluation of power can be very important in experimental design, for example when determining the sample size required in order to attain an acceptable level of discrimination: in this case one is concerned with pre-experimental evaluations which cannot be reasoned out according to the conditionality principle.

vi) Generality of application. The existence of an optimal procedure depends both on the structure of the statistical model and on the complexity of the problem. Such a possibility is essentially limited to cases where the interest parameter is a scalar and where the reference parametric family has either an exponential or a group structure. In this framework, procedures obtained with the aim of maximizing power sometimes match the solutions proposed by Fisher on the basis of arguments based on sufficiency and conditionality (this is true in the case of similar tests in multiparameter exponential families, see section 5.8.2). It is clear that the problems approached within the theory of statistics cover a very broad spectrum and that optimality alone does not propose techniques with sufficiently general applicability. In this respect, likelihood methods are, without doubt, attractive. They represent a sort of regular generalization of optimal techniques: in many cases where there is an optimal procedure this matches a procedure based on the likelihood for the same problem (see, for example, section 5.8.1). Usually, likelihood methods have optimality properties, at least in an approximate sense (see sections 3.4.2, 3.5.2, 9.4.3); see Skovgaard (1989) for a survey.

vii) Distribution problems. Optimality theory does not usually also provide solutions to distribution problems: often an inference problem is considered solved as soon as, for example, the structure of the optimal test has been determined. Finding critical points or calculating P-values is generally illustrated with reference to particular cases where the exact distribution of the statistic is known. Obviously, this is a further restraint on the applicability of optimal procedures: even in cases where these exist there are no general tools for resolving distribution problems. On the other hand, in most cases, resorting to approximate distributions renders these procedures equivalent to those based on the likelihood. On the contrary, distribution problems dominate within the Fisherian view. Priority is given to the search for exact solutions and there is ever present concern that, in small samples, approximations based on the usual first-order asymptotic theory may not be adequate: the distribution problem of the sample correlation coefficient in a normal bivariate model is typical of this (Fisher, 1921; see example 8.9). The pioneering work done by Fisher in this area has been continued by the recent development of higher-order asymptotic methods (Chapters 8-11). This theory offers simple, effective and general approximations to distribution problems arising in optimality theory as well (see section 10.10.2).

REFERENCES

Abramowitz, M. and Stegun, I.A. (1965). *Handbook of Mathematical Functions*. Dover, New York.

Adimari, G. (1996). On empirical likelihood ratio for smooth functions of M-functionals. *Scand. J. Statist.*, to appear.

Adimari, G. (1997). Empirical likelihood type confidence intervals under random censorship. *Ann. Inst. Statist. Math.*, to appear.

Agresti, A. (1990). *Categorical Data Analysis*. Wiley, New York.

Agresti, A. (1992). A survey of exact inference for contingency tables (with discussion). *Statist. Sci.*, **7**, 131-177.

Akahira, M. and Takeuchi, K. (1981). *Asymptotic Efficiency of Estimators: Concepts and Higher Order Asymptotic Efficiency*. Lecture Notes in Statistics, **7**, Springer-Verlag, New York.

Albert, A. and Anderson, J.A. (1984). On the existence of maximum likelihood estimates in logistic regression models. *Biometrika*, **71**, 1-10.

Amari, S.-I. (1982). Geometrical theory of asymptotic ancillarity and conditional inference. *Biometrika*, **69**, 1-17.

Amari, S.-I. (1985). *Differential Geometrical Methods in Statistics*. Lecture Notes in Statistics, **28**, Springer-Verlag, Berlin.

Amari, S.-I. (1987). Differential-geometrical theory of statistics – towards new developments. In *Differential Geometry in Statistical Inference*, S.-I. Amari, O.E. Barndorff-Nielsen, R.E. Kass, S.L. Lauritzen, C.R. Rao (eds.), IMS Lecture Notes - Monograph Series, **10**, Institute of Mathematical Statistics, Hayward, California, 19-94.

Amari, S.-I. (1988). Statistical curvature. In *Encyclopedia of Statistical Sciences*, S. Kotz, N.L. Johnson, C.B. Read (eds.), Wiley, New York, vol. **8**, 642-646.

Andersen, E.B. (1970). Asymptotic properties of conditional maximum likeli-

489

hood estimators. *J. Roy. Statist. Soc.* B, **32**, 283-301.

Andersen, E.B. (1971). Asymptotic properties of conditional likelihood ratio tests. *J. Amer. Statist. Assoc.*, **66**, 630-633.

Andersen, E.B. (1973). *Conditional Inference and Models for Measuring*. Mentalhygiejnisk Forlag, Copenhagen.

Andersen, E.B. (1980). *Discrete Statistical Models with Social Sciences Applications*. North-Holland, Amsterdam.

Andersen, P.K., Borgan, Ø., Gill, R.D. and Keiding, N. (1993). *Statistical Models based on Counting Processes*. Springer-Verlag, New York.

Andrews, D.F. and Stafford, J.E. (1993). Tools for the symbolic computation of asymptotic expansions. *J. Roy. Statist. Soc.* B, **55**, 613-627.

Anscombe, F.J. (1948). The transformation of Poisson, binomial and negative binomial data. *Biometrika*, **35**, 246-254.

Anscombe, F.J. (1964). Normal likelihood functions. *Ann. Inst. Statist. Math.*, **16**, 1-19.

Atkinson, A.C. and Cox, D.R. (1988). Transformations. In *Encyclopedia of Statistical Sciences*, S. Kotz, N.L. Johnson, C.B. Read (eds.), Wiley, New York, vol. **9**, 312-318.

Azzalini, A. (1983). Maximum likelihood estimation of order m for stationary stochastic processes. *Biometrika*, **70**, 381-387.

Azzalini, A. (1996). *Statistical Inference Based on the Likelihood*. Chapman and Hall, London.

Bai, Z.D. and Rao, C.R. (1991). Edgeworth expansion of a function of sample means. *Ann. Statist.*, **19**, 1295-1315.

Bar-Lev, S.K. and Reiser, B. (1982). An exponential subfamily which admits UMPU tests based on a single test statistic. *Ann. Statist.*, **10**, 979-989.

Barnard, G.A. (1963). Some logical aspects of the fiducial argument. *J. Roy. Statist. Soc.* B, **25**, 111-114.

Barnard, G.A. (1974a). The foundations of statistics 1964-1974. *Bull. Inst. Math. Applic.*, **10**, 344-347.

Barnard, G.A. (1974b). Conditionality, pivotals and robust estimation. In *Proc. Conf. Foundational Questions in Statistical Inference. Memoir 1*, O.E. Barndorff-Nielsen, P. Blæsild and G. Schou (eds.), Dept. Theor. Statistics, Aarhus University, 61-80.

Barnard, G.A. (1976). Conditional inference is not inefficient. *Scand. J. Statist.*, **3**, 132-134.

Barnard, G.A. (1977a). Pivotal inference and the Bayesian controversy. *Bull. Int. Statist. Inst.*, **47**, 543-551.

Barnard, G.A. (1977b). On ridge regression and the general principles of estimation. *Utilitas Math.*, **11**, 299-311.

Barnard, G.A. (1979). In contradiction to J. Berkson's dispraise: conditional tests can be more efficient. *J. Statist. Plann. Inf.*, **3**, 181-187.

Barndorff-Nielsen, O.E. (1978). *Information and Exponential Families*. Wiley, Chichester.

Barndorff-Nielsen, O.E. (1980). Conditionality resolutions. *Biometrika*, **67**, 293-310.

Barndorff-Nielsen, O.E. (1983). On a formula for the distribution of the maximum likelihood estimator. *Biometrika*, **70**, 343-365.

Barndorff-Nielsen, O.E. (1984). On conditionality resolution and the likelihood ratio for curved exponential models. *Scand. J. Statist.*, **11**, 157-170. Correction: (1985) **12**, 191.

Barndorff-Nielsen, O.E. (1986). Inference on full and partial parameters based on the standardized signed log likelihood ratio. *Biometrika*, **73**, 307-322.

Barndorff-Nielsen, O.E. (1988). *Parametric Statistical Models and Likelihood*. Lecture Notes in Statistics, **50**, Springer-Verlag, Heidelberg.

Barndorff-Nielsen, O.E. (1990a). A note on the standardized signed log likelihood ratio. *Scand. J. Statist.*, **17**, 157-160.

Barndorff-Nielsen, O.E. (1990b). p^* and Laplace's method *REBRAPE*, **4**, 89-103.

Barndorff-Nielsen, O.E. (1990c). Approximate interval probabilities. *J. Roy. Statist. Soc.*, **52**, 485-496.

Barndorff-Nielsen, O.E. (1991a). Likelihood theory. In *Statistical Theory and Modelling*, D.V. Hinkley, N. Reid and E.J. Snell (eds.), Chapman and Hall, London, 232-264.

Barndorff-Nielsen, O.E. (1991b). Modified signed log likelihood ratio. *Biometrika*, **78**, 557-563.

Barndorff-Nielsen, O.E. (1994). Adjusted versions of profile likelihood and directed likelihood, and extended likelihood. *J. Roy. Statist. Soc. B*, **56**, 125-140.

Barndorff-Nielsen, O.E. (1995a). Diversity of evidence and Birnbaum's theorem (with discussion). *Scand. J. Statist.*, **22**, 513-522.

Barndorff-Nielsen, O.E. (1995b). Stable and invariant adjusted profile likelihood and directed likelihood for curved exponential models. *Biometrika*, **82**, 489-499.

Barndorff-Nielsen, O.E. (1995c). Quasi profile and directed likelihoods from estimating functions. *Ann. Inst. Statist. Math.*, **47**, 461-464.

Barndorff-Nielsen, O.E. and Blæsild, P. (1975). S-ancillarity in exponential families. *Sankhyā*, **37**, 354-385.

Barndorff-Nielsen, O.E., Blæsild, P. and Eriksen, P.S. (1989). *Decomposition and Invariance of Measures, with a View to Statistical Transformation Models*. Lecture Notes in Statistics **58**, Springer-Verlag, Heidelberg.

Barndorff-Nielsen, O.E., Blæsild, P., Jensen, J.L. and Sørensen, M. (1985). The fascination of sand. In *A celebration of Statistics*, A.C. Atkinson and S. Fienberg (eds.), Springer-Verlag, New York.

Barndorff-Nielsen, O.E., Blæsild, P., Pace, L. and Salvan, A. (1991). Formulas for asymptotic statistical calculations. *Reserch Report n. 207, Dept. Theor. Statist., Univ. of Aarhus*.

Barndorff-Nielsen, O.E. and Chamberlin, S.R. (1991). An ancillary invariant modification of the signed log likelihood ratio. *Scand. J. Statist.*, **18**, 341-352.

Barndorff-Nielsen, O.E. and Chamberlin, S.R. (1994). Stable and invariant adjusted directed likelihood. *Biometrika*, **81**, 485-499.

Barndorff-Nielsen, O.E. and Cox, D.R. (1979). Edgeworth and saddlepoint approximations with statistical applications (with discussion). *J. Roy. Statist. Soc.* B, **41**, 279-312.

Barndorff-Nielsen, O.E. and Cox, D.R. (1984a). The effect of sampling rules on likelihood statistics. *Int. Statist. Rev.*, **52**, 309-326.

Barndorff-Nielsen, O.E. and Cox, D.R. (1984b). Bartlett adjustments to the likelihood ratio statistic and the distribution of the maximum likelihood estimator. *J. Roy. Statist. Soc.* B, **46**, 483-495.

Barndorff-Nielsen, O.E. and Cox, D.R. (1989). *Asymptotic Techniques for Use in Statistics*. Chapman and Hall, London.

Barndorff-Nielsen, O.E. and Cox, D.R. (1994). *Inference and Asymptotics*. Chapman and Hall, London.

Barndorff-Nielsen, O.E. and Cox, D.R. (1995). Prediction and asymptotics. Manuscript.

Barndorff-Nielsen, O.E., Cox, D.R. and Reid, N. (1986). The role of differential geometry in statistical theory. *Int. Statist. Rev.*, **54**, 83-96.

Barndorff-Nielsen, O.E. and Hall, P. (1988). On the level-error after Bartlett adjustment of the likelihood ratio statistic. *Biometrika*, **75**, 374-378.

Barndorff-Nielsen, O.E., Jensen, J.L. and Sørensen, M. (1995). Some stationary processes in discrete and continuous time. *Research Report n. 241, Department of Theoretical Statistics, Univ. of Aarhus.*

Barndorff-Nielsen, O.E. and Jupp, P.E. (1987). Differential geometry, profile likelihood, L-sufficiency and composite transformation models. *Ann. Statist.*, **16**, 1009-1043.

Barndorff-Nielsen, O.E. and Jupp, P.E. (1989). Approximating exponential models. *Ann. Inst. Statist. Math.*, **41**, 247-267.

Barndorff-Nielsen, O.E. and McCullagh, P. (1993). A note on the relation between modified profile likelihood and the Cox-Reid adjusted profile likelihood. *Biometrika*, **80**, 321-328.

Barndorff-Nielsen, O.E. and Sørensen, M. (1994). A review of some aspects of asymptotic likelihood theory for stochastic processes. *Int. Statist. Rev.*, **62**, 133-165.

Barndorff-Nielsen, O.E. and Wood, A.T.A. (1996). On large deviations and choice of ancillary for p^* and the modified directed likelihood. *Bernoulli*, to appear.

Barnett, V. (1982). *Comparative Statistical Inference*, 2-nd ed., Wiley, Chichester.

Bartlett, M.S. (1936a). Statistical information and properties of sufficiency. *Proc. Roy. Soc.* A, **154**, 124-137.

Bartlett, M.S. (1936b). The square root transformation in analysis of variance. *J. Roy. Statist. Soc.*, Suppl. 3, 68-78.

Bartlett, M.S. (1937). Properties of sufficiency and statistical tests. *Proc. Roy. Soc.* A, **160**, 268-282.

Bartlett, M.S. (1953). Approximate confidence intervals. II. More than one unknown parameter. *Biometrika*, **40**, 306-317.

Bartlett, M.S. (1954). A note on the multiplying factors for various χ^2 approximations. *J. Roy. Statist. Soc.* B, **16**, 296-298.

Bartlett, M.S. (1990). Chance or chaos? *J. Roy. Statist. Soc.* A, **153**, 321-347.

Basu, D. (1955). On statistics independent of a complete sufficient statistic. *Sankhyā*, **15**, 377-380.

Basu, D. (1964). Recovery of ancillary information. *Sankhyā* A, **26**, 3-16.

Basu, D. (1977). On the elimination of nuisance parameters. *J. Amer. Statist. Assoc.*, **72**, 355-366.

Basu, D. (1978). On partial sufficiency: a review. *J. Statist. Plann. Inf.*, **2**, 1-13.

Beaumont, G.P. (1980). *Intermediate Mathematical Statistics*. Chapman and Hall, London.

Becker, N. and Gordon, I. (1983). On Cox's criterion for discriminating between alternative ancillary statistics. *Int. Statist. Rev.*, **51**, 89-92.

Bellhouse, D.R. (1990). On the equivalence of marginal and approximate conditional likelihoods for correlation parameters under a normal model. *Biometrika*, **77**, 743-746.

Benzécri, J.P. (1973). *L'Analyse des Données. 1 La Taxonomie*. Dunod, Paris.

Berger, J.O. (1985). *Statistical Decision Theory and Bayesian Analysis*, 2-nd ed.. Springer-Verlag, New York.

Berliner, L.M. (1992). Statistics, probability and chaos. *Statist. Sci.*, **7**, 69-122.

Bernardo, J.M. and Smith, A.F.M. (1994). *Bayesian Theory*. Wiley, Chichester.

Besag, J. (1974). Spatial interaction and the statistical analysis of lattice systems. *J. Roy. Statist. Soc.* B, **36**, 192-236.

Besag, J. (1977). Efficiency of pseudo-likelihood estimation for simple Gaussian fields. *Biometrika*, **64**, 616-618.

Besag, J. (1978). Some methods of statistical analysis of spatial data. *Bull. Int. Statist. Inst.*, **47**, 77-92.

Besag, J. (1986). On the statistical analysis of dirty pictures (with discussion). *J. Roy. Statist. Soc.* B, **48**, 259-302.

Bhat, U.N. (1985). Markov processes. In *Encyclopedia of Statistical Sciences*, S. Kotz, N.L. Johnson and C.B. Read (eds.), Wiley, New York, vol. 5, 256-270.

Bhattacharya, R. and Denker, M. (1990). *Asymptotic Statistics*. Birkhäuser, Basel.

Bickel, P.J. (1982). On adaptive estimation. *Ann. Statist.*, **10**, 647-671.

Bickel, P.J. and Doksum, K.A. (1977). *Mathematical Statistics, Basic Ideas and Selected Topics*. Holden Day, Oakland, California.

Bickel, P.J., Klaassen, C.A.J., Ritov, Y. and Wellner, J.A. (1993). *Efficient and Adaptive Estimation in Semiparametric Models*. John Hopkins University Press, Baltimore.

Billingsley, P. (1986). *Probability and Measure*. Wiley, New York.

Bishop, Y.M.M., Fienberg, S.E. and Holland, P.W. (1975). *Discrete Multivariate Analysis*. MIT Press, Cambridge, Mass..

Bjørnstad, J.F. (1990). Predictive likelihood: a review (with discussion). *Statist. Sci.*, **5**, 242-265.

Blæsild, P. and Jensen, J.L. (1985). Saddlepoint formulas for reproductive exponential models. *Scand. J. Statist.*, **12**, 193-202.

Box, G.E.P. and Cox, D.R. (1964). An analysis of transformations. *J. Roy. Statist. Soc.* B, **26**, 211-252.

Breslow, N.E. and Day, N.E. (1980). *Statistical Methods in Cancer Research. 1: The Analysis of Case-Control Studies*. I.A.R.C., Lyon.

Brown, L.D. (1967). The conditional level of Student's t test. *Ann. Math. Statist.*, **38**, 1068-1071.

Brown, L.D. (1986). *Fundamentals of Statistical Exponential Families*. IMS Lecture Notes - Monograph Series, **9**, Institute of Mathematical Statistics, Hayward, California.

Burridge, J. (1981). A note on maximum likelihood estimation for regression models using grouped data. *J. Roy. Statist. Soc.* B, **43**, 41-45.

Butler, R.W. (1986). Predictive likelihood inference with applications. *J. Roy. Statist. Soc.* B, **48**, 1-38.

Butler, R.W. (1989). Approximate predictive pivots and densities. *Biometrika*, **76**, 489-501.

Butler, R.W., Huzurbazar, S. and Booth, J.G. (1992a). Saddlepoint approximations for the generalized variance and Wilks' statistic. *Biometrika*, **79**, 157-169.

Butler, R.W., Huzurbazar, S. and Booth, J.G. (1992b). Saddlepoint approximations for the Bartlett-Nanda-Pillai trace statistic in multivariate analysis. *Biometrika*, **79**, 705-715.

Butler, R.W., Huzurbazar, S. and Booth, J.G. (1993). Saddlepoint approximations for tests of block independence, sphericity and equal variances and covariances. *J. Roy. Statist. Soc.* B, **55**, 171-183.

Casella, G. (1992). Conditional inference from confidence sets. In *Current issues in Statistical Inference: Essays in Honor of D. Basu*, M. Ghosh, P.K. Pathak (eds.), IMS Lecture Notes - Monograph Series, **17**, Institute of Mathematical Statistics, Hayward, California, 1-12.

Chamberlin, S.R. (1989). The foundation and application of inferential estimation. Ph.D. thesis, Department of Statistics and Actuarial Science, University of Waterloo, Canada.

Chamberlin, S.R. and Sprott, D.A. (1989). Linear systems of pivotals and associated pivotal likelihoods with applications. *Biometrika*, **76**, 685-691.

Chatfield, C. (1985). The initial examination of data (with discussion). *J. Roy. Statist. Soc.* A, **148**, 214-253.

Chatfield, C. (1995). Model uncertainty, data mining and statistical inference (with discussion). *J. Roy. Statist. Soc.* A, **158**, 419-466.

Chatterjee, S. and Yilmaz, M.R. (1992). Chaos, fractals and statistics. *Statist. Sci.*, **7**, 49-68.

Cheng, K.F. and Wu, J.W. (1994). Testing goodness of fit for a parametric family of link functions. *J. Amer. Statist. Assoc.*, **89**, 657-664.

Cheng, R.C.H. and Traylor, L. (1995). Non-regular maximum likelihood problems (with discussion). *J. Roy. Statist. Soc.* B, **57**, 3-44.

Chhikara, R.S. and Folks, J.L. (1989). *The Inverse Gaussian Distribution*. Marcel Dekker, New York.

Coles, S.G. and Tawn, J.A. (1994). Statistical methods for multivariate extremes: an application to structural design. *Appl. Statist.*, **43**, 1-48.

Cook, R.D. and Weisberg, S. (1982). *Residuals and Influence in Regression*. Chapman and Hall, London.

Cordeiro, G.M. (1983). Improved likelihood ratio statistics for generalized linear models. *J. Roy. Statist. Soc.* B, **45**, 404-413.

Cordeiro, G.M. (1987). On the corrections to the likelihood ratio statistics. *Biometrika*, **74**, 265-274.

Cordeiro, G.M. and Ferrari, S.L.P. (1991). A modified score test having chi-squared distribution to order $O(n^{-1})$. *Biometrika*, **78**, 573-582.

Cordeiro, G.M., Ferrari, S.L.P. and Paula, G.A. (1993). Improved score tests for generalized linear models. *J. Roy. Statist. Soc.* B, **55**, 661-674.

Cordeiro, G.M. and McCullagh, P. (1991). Bias corrections in generalized linear models. *J. Roy. Statist. Soc.* B, **53**, 629-643.

Cordeiro, G.M. and Paula, G.A. (1989). Improved likelihood ratio statistics for exponential family nonlinear models. *Biometrika*, **76**, 93-100.

Cornish, E.A. and Fisher, R.A. (1937). Moments and cumulants in the specification of distributions. *Int. Statist. Rev.*, **5**, 307-322.

Cox, D.R. (1958). Some problems connected with statistical inference. *Ann. Math. Statist.*, **29**, 357-372.

Cox, D.R. (1971). The choice between alternative ancillary statistics. *J. Roy. Statist. Soc.* B, **33**, 251-255.

Cox, D.R. (1972). Regression models and life tables (with discussion). *J. Roy. Statist. Soc.*, **34**, 187-220.

Cox, D.R. (1975). Partial likelihood. *Biometrika*, **62**, 269-276.

Cox, D.R. (1977). The role of significance tests (with discussion). *Scand. J. Statist.*, **4**, 49-70.

Cox, D.R. (1980). Local ancillarity. *Biometrika*, **67**, 279-286.

Cox, D.R. (1983). Some remarks on overdispersion. *Biometrika*, **70**, 269-274.

Cox, D.R. (1984). Effective degrees of freedom and the likelihood ratio test. *Biometrika*, **71**, 487-493.

Cox, D.R. (1986). Some general aspects of the theory of statistics. *Int. Statist. Rev.*, **54**, 117-126.

Cox, D.R. (1990). Role of models in statistical analysis. *Statist. Sci.*, **5**, 169-174.

Cox, D.R. (1993). Likelihood and asymptotics. In *Tre lauree honoris causa*, Univ. of Padova, SGE, Padova.

Cox, D.R. (1995). Some recent developments in statistical theory. *Scand. Actuarial J.*, 29-34.

Cox, D.R. and Hinkley, D.V. (1974). *Theoretical Statistics*. Chapman and Hall, London.

Cox, D.R. and Reid, N. (1987). Parameter orthogonality and approximate conditional inference (with discussion). *J. Roy. Statist. Soc.* B, **49**, 1-39.

Cox, D.R. and Reid, N. (1993). A note on the calculation of adjusted profile likelihood. *J. Roy. Statist. Soc.* B, **55**, 467-471.

Cox, D.R. and Snell, E.J. (1989). *Analysis of Binary Data*, 2-nd ed.. Chapman and Hall, London.

Cribari-Neto, F. and Ferrari, S.L.P. (1995). Second order asymptotics for score tests in generalised linear models. *Biometrika*, **82**, 426-432.

Critchley, F., Marriot, P. and Salmon, M. (1993). Preferred point geometry and statistical manifolds. *Ann. Statist.*, **21**, 1197-1224.

Critchley, F., Marriot, P. and Salmon, M. (1994). Preferred point geometry and the local differential geometry of the Kullback–Leibler divergence. *Ann. Statist.*, **22**, 1587-1602.

Cruddas, A.M., Reid, N. and Cox, D.R. (1989). A time series illustration of approximate conditional likelihood. *Biometrika*, **76**, 231-237.

D'Agostino, R.B. and Stephens, M.A. (1986). *Handbook of Goodness-of-Fit Techniques*. Marcel Dekker, New York.

Daniels, H.E. (1954). Saddlepoint approximations in statistics. *Ann. Math. Statist.*, **25**, 631-650.

Daniels, H.E. (1983). Saddlepoint approximations for estimating equations. *Biometrika*, **70**, 89-96.

Daniels, H.E. (1987). Tail probability approximations. *Int. Statist. Rev.*, **55**, 37-48.

David, H.A. (1981). *Order Statistics*. Wiley, New York.

Davison, A.C. (1988). Approximate conditional inference in generalized linear models. *J. Roy. Statist. Soc.* B, **50**, 445-461.

Davison, A.C. (1990). Bootstrap methods. *Handbook of Applicable Mathematics: Supplement*, 409-436.

Davison, A.C. (1992). Treatment effect heterogeneity in paired data. *Biometrika*, **79**, 463-474.

Davison, A.C. and Hinkley, D.V. (1988). Saddlepoint approximations in re-sampling methods. *Biometrika*, **75**, 417-431.

Davison, A.C., Hinkley, D.V. and Worton, B.J. (1992). Bootstrap likelihoods. *Biometrika*, **79**, 113-130.

Davison, A.C. and Tsai, C.-L. (1992). Regression model diagnostics. *Int. Statist. Rev.*, **60**, 337-353.

Dawid, A.P. (1975). On the concepts of sufficiency and ancillarity in the presence of nuisance parameters. *J. Roy. Statist. Soc.* B, **37**, 248-258.

Dawid, A.P. (1983a). Inference, statistical: I. In *Encyclopedia of Statistical Sciences*, S. Kotz, N.L. Johnson, C.B. Read (eds.), Wiley, New York, vol. **4**, 89-105.

Dawid, A.P. (1983b). Invariant prior distributions. In *Encyclopedia of Statistical Sciences*, S. Kotz, N.L. Johnson, C.B. Read (eds.), Wiley, New York, vol. **4**, 228-236.

De Bruijn, N.G. (1961). *Asymptotic Methods in Analysis*. Dover, New York.

de Finetti, B. (1974). *Theory of Probability. A Critical Introductory Treatment*, vol. 1. Wiley, London.

de Finetti, B. (1975). *Theory of Probability. A Critical Introductory Treatment*, vol. 2. Wiley, London.

Desmond, A.F. and Chapman, G.R. (1993). Modelling task completion data with inverse Gaussian mixtures. *Appl. Statist.*, **42**, 603-613.

DiCiccio, T.J. (1984). On parameter transformations and interval estimation. *Biometrika*, **71**, 477-485.

DiCiccio, T.J., Hall, P. and Romano, J.P. (1991). Empirical likelihood is Bartlett-correctable. *Ann. Statist.*, **19**, 1053-1061.

DiCiccio, T.J. and Martin, M.A. (1993). Simple modifications for signed roots of likelihood ratio statistics. *J. Roy. Statist. Soc.* B, **55**, 305-316.

DiCiccio, T.J., Martin, M.A., Stern, S.E. and Young, G.A. (1996). Information bias and adjusted profile likelihoods. *J. Roy. Statist. Soc.* B, **58**, 189-203.

DiCiccio, T.J., Martin, M.A. and Young, G.A. (1994). Analytical approximations to bootstrap distribution functions using saddlepoint methods. *Statistica Sinica*, **4**, 281-295.

DiCiccio, T.J. and Romano, J.P. (1988). A review of bootstrap confidence intervals (with discussion). *J. Roy. Statist. Soc.* B, **50**, 338-354.

DiCiccio, T.J. and Stern, S.E. (1994a). Constructing approximately standard normal pivots from signed roots of adjusted likelihood ratio statistics. *Scand. J. Statist.*, **21**, 447-460.

DiCiccio, T.J. and Stern, S.E. (1994b). Frequentist and Bayesian Bartlett corrections of test statistics based on adjusted profile likelihoods. *J. Roy. Statist. Soc.* B, **56**, 397-408.

Dobson, A.J. (1990). *An Introduction to Statistical Modelling*, 2-nd ed.. Chapman and Hall, London.

Draper, D. (1995). Assessment and propagation of model uncertainty (with discussion). *J. Roy. Statist. Soc.* B, **57**, 45-97.

Draper, N.R. and Tierney, D.E. (1972). Regions of positive and unimodal series expansion of the Edgeworth and Gram-Charlier approximations. *Biometrika*, **59**, 463-465.

Du Mouchel, W.H. (1983). Estimating the stable index α in order to measure tail thickness: a critique. *Ann. Statist.*, **11**, 1019-1031.

Durbin, J. (1961). Some methods of constructing exact tests. *Biometrika*, **48**, 41-55.

Durbin, J. (1980). Approximations for densities of sufficient estimators. *Biometrika*, **67**, 311-333.

Dynkin, E.B. (1951). Necessary and sufficient statistics for a family of probability distributions. *Select. Trans. Math. Statist. Prob.*, **1**, 23-41.

Eaton, M.L. (1989). *Group Invariance Applications in Statistics*. NSF-CBMS Regional Conference Series, **1**, Institute of Mathematical Statistics, Hayward, California.

Edgeworth, F.Y. (1905). The law of error. *Cambridge Phil. Trans.*, **20**, 36-66, 113-141.

Edwards, A.W.F. (1972). *Likelihood*. Cambridge University Press, Cambridge.

Edwards, A.W.F. (1974). The history of likelihood. *Int. Statist. Rev.*, **42**, 9-15.

Efron, B. (1975). Defining the curvature of a statistical problem (with applications to second order efficiency) (with discussion). *Ann. Statist.*, **3**, 1189-1242.

Efron, B. (1978). The geometry of exponential families. *Ann. Statist.*, **6**, 362-376.

Efron, B. (1979). Bootstrap methods: another look at the jackknife. *Ann. Statist.*, **7**, 1-26.

Efron, B. (1981). Nonparametric standard errors and confidence intervals (with discussion). *Canad. J. Statist.*, **9**, 139-172.

Efron, B. (1982). *The Jackknife, the Bootstrap and Other Resampling Plans*. SIAM, Philadelphia.

Efron, B. (1986a). Why isn't everyone a Bayesian? *Amer. Statistician*, **40**, 1-11.

Efron, B. (1986b). Double exponential families and their use in generalized linear models. *J. Amer. Statist. Assoc.*, **87**, 98-107.

Efron, B. and Hinkley, D.V. (1978). Assessing the accuracy of the maximum likelihood estimator: observed versus expected Fisher information (with discussion). *Biometrika*, **65**, 457-487.

Efron, B. and Tibshirani, R. (1986). Bootstrap methods for standard errors, confidence intervals, and other measures of statistical accuracy. *Statist. Sci.*, **1**, 54-77.

Efron, B. and Tibshirani, R. (1993). *An Introduction to the Bootstrap*. Chapman and Hall, London.

Eriksen, P.S. (1987). Proportionality of covariance matrices. *Ann. Statist.*, **15**, 732-748.

Esscher, F. (1932). On the probability function in the collective theory of risk. *Skand. Akt. Tidsskr.*, **15**, 175-195.

Esséen, C.G. (1956). A moment inequality with an application to the central limit theorem. *Skand. Aktuar.*, **39**, 160-170.

Evans, M. and Swartz, T. (1995). Methods for approximating integrals in statistics with special emphasis on Bayesian integration problems. *Statist. Sci.*, **10**, 254-272.

Farrington, C.P. (1996). On assessing goodness of fit of generalized linear models to sparse data. *J. Roy. Statist. Soc. B*, **58**, 349-360.

Feller, W. (1971). *An Introduction to Probability Theory*. Vol. 2, 2-nd ed.. Wiley, New York.

Ferguson, H., Reid, N. and Cox, D.R. (1991). Estimating equations from modified profile likelihood. In *Estimating Functions*, V.P. Godambe (ed.), Clarendon Press, Oxford, 279-293.

Ferguson, T.S. (1962). Location and scale parameters in exponential families of distributions. *Ann. Math. Statist.*, **33**, 986-1001. Correction: (1963) **34**, 1603.

Ferguson, T.S. (1967). *Mathematical Statistics: a Decision Theoretic Approach*. Academic Press, New York.

Fernholz, L.T. (1983). *Von Mises Calculus for Statistical Functionals*. Lecture Notes in Statistics, **19**, Springer-Verlag, New York.

Feuerverger, A. (1989). On the empirical saddlepoint approximation. *Biometrika*, **76**, 457-464.

Field, C.A. and Ronchetti, E.M. (1990). *Small Sample Asymptotics*. IMS Lecture Notes - Monograph Series, **13**, Institute of Mathematical Statistics, Hayward, California.

Fienberg, S.E. and Hinkley, D.V. (eds.) (1980). *Fisher: an Appreciation.* Lecture Notes in Statistics, 1, Springer-Verlag, New York.

Firth, D. (1991). Generalized linear models. In *Statistical Theory and Modelling*, D.V. Hinkley, N. Reid and E.J. Snell (eds.), Chapman and Hall, London, 55-82.

Firth, D. (1993). Recent developments in quasi-likelihood methods. *Bull. Int. Statist. Inst.*, 55, 341-358.

Fisher, R.A. (1912). On an absolute criterion for fitting frequency curves. *Mess. Math.*, 41, 155-160.

Fisher, R.A. (1915). Frequency distribution of the values of the correlation coefficient in samples from an indefinitely large population. *Biometrika*, 10, 507-521.

Fisher, R.A. (1920). A mathematical examination of the methods of determining the accuracy of an observation by the mean error, and by the mean square error. *Monthly Not. Roy. Astr. Soc.*, 80, 758-770.

Fisher, R.A. (1921). On the 'probable error' of a coefficient of correlation deduced from a small sample. *Metron*, 1, 3-32.

Fisher, R.A. (1922a). On the mathematical foundations of theoretical statistics. *Phil. Trans. Roy. Soc.* A, 222, 309-368.

Fisher, R.A. (1922b). On the interpretation of χ^2 from contingency tables and the calculation of P. *J. Roy. Statist. Soc.*, 85, 87-94.

Fisher, R.A. (1925). Theory of statistical estimation. *Proc. Camb. Phil. Soc.*, 22, 700-725.

Fisher, R.A. (1934). Two new properties of mathematical likelihood. *Proc. Roy. Soc.* A, 144, 285-307.

Fisher, R.A. (1935). The logic of inductive inference. *J. Roy. Statist. Soc.*, 98, 39-54.

Fisher, R.A. (1950). *Contributions to Mathematical Statistics.* Wiley, New York.

Fisher, R.A. (1956). *Statistical Methods and Scientific Inference.* Oliver and Boyd, Edinburgh. (3-rd ed. 1973, Collier Macmillan, London.)

Fisher, R.A. (1958). *Statistical Methods for Research Workers*, 13-th ed.. Oliver and Boyd, Edinburgh.

Fisher, R.A. (1971). *Collected Papers, vol. 1, (1912-1924).* University of Adelaide, Adelaide.

Forster, J.J., McDonald, J.W. and Smith, P.W.F. (1996). Monte Carlo exact conditional tests for log-linear and logistic models. *J. Roy. Statist. Soc. B*, **58**, 445-453.

Fraser, D.A.S. (1956). Sufficient statistics with nuisance parameters. *Ann. Math. Statist.*, **27**, 838-842.

Fraser, D.A.S. (1957). *Nonparametric Methods in Statistics.* Wiley, New York.

Fraser, D.A.S. (1968). *The Structure of Inference.* Wiley, New York.

Fraser, D.A.S. (1976). Necessary analysis and adaptive inference. *J. Amer. Statist. Assoc.*, **71**, 99-110.

Fraser, D.A.S. (1979). *Inference and Linear Models.* McGraw Hill, New York.

Fraser, D.A.S. (1983). Inference, statistical: II. In *Encyclopedia of Statistical Sciences*, S. Kotz, N.L. Johnson, C.B. Read (eds.), Wiley, New York, vol. **4**, 105-114.

Fraser, D.A.S. (1988). Normed likelihood as saddlepoint approximation. *J. Mult. Anal.*, **27**, 181-193.

Fraser, D.A.S. and Reid, N. (1988). On conditional inference for a real parameter: a differential approach on the sample space. *Biometrika*, **75**, 251-274.

Frydenberg, M. and Jensen, J.L. (1989). Is the 'improved likelihood ratio statistic' really improved in the discrete case? *Biometrika*, **76**, 655-661.

Galambos, J. (1987). *The Asymptotic Theory of Extreme Order Statistics.* Robert Krieger Publishing Co., Malabar, Florida.

Gatto, R. and Ronchetti, E.M. (1996). General saddlepoint approximations of marginal densities and tail probabilities. *J. Amer. Statist. Assoc.*, **91**, 666-673.

Gayen, A.K. (1951). The frequency distribution of the product-moment correlation coefficient in random samples of any size drawn from non-normal universes. *Biometrika*, **38**, 219-247.

Gelfand, A.E. and Dalal, S.R. (1990). A note on overdispersed exponential families. *Biometrika*, **77**, 55-64.

Gelman, A., Carlin, J.B., Stern, H.S. and Rubin, D.B. (1995). *Bayesian Data Analysis.* Chapman and Hall, London.

Ghosh, J.K. (1994). *Higher Order Asymptotics.* NSF-CBMS Regional Conference Series, **4**, Institute of Mathematical Statistics, Hayward, California.

Ghosh, J.K. and Mukerjee, R. (1994). Adjusted versus conditional likelihood:

power properties and Bartlett-type adjustments. *J. Roy. Statist. Soc.* B, **56**, 185-188.

Ghosh, M. and Sen, P.K. (1989). Median unbiasedness and Pitman closeness. *J. Amer. Statist. Assoc.*, **84**, 1089-1091.

Gibbons, J.D. (1988). Sign tests. In *Encyclopedia of Statistical Sciences*, S. Kotz, N.L. Johnson, C.B. Read (eds.), Wiley, New York, vol. **8**, 471-475.

Gill, R.D. (1989). Non- and semi-parametric maximum likelihood estimators and the von Mises method (part I). *Scand. J. Statist.*, **16**, 97-128.

Gill, R.D. and Johansen, S. (1990). A survey of product-integration with a view towards applications in survival analysis. *Ann. Statist.*, **18**, 1501-1555.

Gill, R.D. and van der Vaart, A.W. (1993). Non- and semi-parametric maximum likelihood estimators and the von Mises method (part II). *Scand. J. Statist.*, **20**, 271-288.

Gnedenko, B.V. (1943). Sur la distribution limite du terme maximum d'une série aléatoire. *Ann. Math.*, **44**, 423-453.

Godambe, V.P. (1960). An optimum property of regular maximum likelihood estimation. *Ann. Math. Statist.*, **31**, 1208-1212.

Godambe, V.P. (ed.) (1991). *Estimating Functions*. Clarendon Press, Oxford.

Godambe, V.P. and Kale, B.K. (1991). Estimating functions: an overview. In *Estimating Functions*, V.P. Godambe (ed.), Clarendon Press, Oxford, 3-20.

Gong, G. and Samaniego, F.J. (1981). Pseudo maximum likelihood estimation: theory and applications. *Ann. Statist.*, **89**, 861-869.

Good, I.J. and Gaskins, R.A. (1971). Nonparametric roughness penalties for probability densities. *Biometrika*, **58**, 255-277.

Gosset, W.S. ('Student') (1908). On the probable error of a mean. *Biometrika*, **6**, 1-25.

Green, P.J. (1987). Penalized likelihood for general semiparametric regression models. *Int. Statist. Rev.*, **55**, 245-260.

Grimmet, G.R. and Stirzaker, D.R. (1992). *Probability and Random Processes*. Oxford Science Publications, Oxford.

Gumbel, E.J. (1958). *Statistics of Extremes*. Columbia University Press, New York.

Haldane, J.B.S. (1938). The approximate normalization of a class of frequency distributions. *Biometrika*, **29**, 392-404.

Hall, P. (1985). Martingales. In *Encyclopedia of Statistical Sciences*, S. Kotz, N.L. Johnson, C.B. Read (eds.), Wiley, New York, vol. **5**, 278-285.

Hall, P. (1992). *The Bootstrap and Edgeworth Expansion*. Springer-Verlag, New York.

Hall, P. and La Scala, B. (1990). Methodology and algorithms of empirical likelihood. *Int. Statist. Rev.*, **58**, 109-127.

Halmos, P.R. and Savage, L.J. (1949). Application of the Radon-Nikodym theorem to the theory of sufficient statistics. *Ann. Math. Statist.*, **20**, 225-241.

Hampel, F.R. (1974). The influence curve and its role in robust estimation. *J. Amer. Statist. Assoc.*, **69**, 383-393.

Hampel, F.R., Ronchetti, E.M., Rousseeuw, P.J. and Stahel, W.A. (1986). *Robust Statistics: the Approach based on Influence Functions*. Wiley, New York.

Hand, D.J. (1994). Deconstructing statistical questions (with discussion). *J. Roy. Statist. Soc.* A, **157**, 317-356.

Harvill, J.L. and Newton, H.J. (1995). Saddlepoint approximations for the difference of order statistics. *Biometrika*, **82**, 226-231.

Helland, I.S. (1995). Simple counterexamples against the conditionality principle. *Amer. Statistician*, **49**, 351-356.

Hinkley, D.V. (1988). Bootstrap methods (with discussion). *J. Roy. Statist. Soc.* B, **50**, 321-337.

Holland, P.W. (1973). Covariance stabilizing transformations. *Ann. Statist.*, **1**, 84-92.

Hollas, E.A. (1991). Performance of Bartlett adjustment for certain likelihood ratio tests. *Commun. Statist.-Simula.*, **20**, 449-462.

Holst, L. (1981). Some conditional limit theorems in exponential families. *Ann. Prob.*, **9**, 818-830.

Hougaard, P. (1982). Parametrizations of non-linear models. *J. Roy. Statist. Soc.* B, **44**, 244-252.

Hougaard, P. (1986). Survival models for heterogeneous populations derived from stable distributions. *Biometrika*, **73**, 387-396.

Huber, P.J. (1964). Robust estimation of a location parameter. *Ann. Math. Statist.*, **35**, 73-101.

Huber, P.J. (1967). The behavior of maximum likelihood estimates under nonstandard conditions. In *Proc. 5-th Berk. Symp. Math. Statist. Prob.*, vol. 1, 221-233.

Huzurbazar, V.S. (1950). Probability distributions and orthogonal parameters. *Proc. Camb. Phil. Soc.*, **46**, 281-284.

Jackson, O.A.Y. (1968). Some results on tests of separate families of hypotheses. *Biometrika*, **55**, 315-323.

Janicki, A. and Weron, A. (1994). Can one see α-stable variables and processes? *Statist. Sci.*, **9**, 109-126.

Jeffreys, H. (1946). An invariant form for the prior probability in estimation problems. *Proc. Roy. Soc. London* A, **196**, 453-461.

Jeffreys, H. (1961). *Theory of Probability*, 3-rd ed.. Clarendon Press, Oxford.

Jensen, J.L. (1988). Uniform saddlepoint approximations. *Adv. Appl. Prob.*, **20**, 622-634.

Jensen, J.L. (1991). A large deviation-type approximation for the "Box class" of likelihood ratio criteria. *J. Amer. Statist. Assoc.*, **86**, 437-440.

Jensen, J.L. (1992). The modified signed likelihood statistic and saddlepoint approximations. *Biometrika*, **79**, 693-703.

Jensen, J.L. (1993). A historical sketch and some new results on the improved log likelihood ratio statistic. *Scand. J. Statist.*, **20**, 1-15.

Jensen, J.L. (1995a). *Saddlepoint Approximations*. Clarendon Press, Oxford.

Jensen, J.L. (1995b). A simple derivation of a natural large deviation modified likelihood ratio statistic. *Research Report n. 328, Dept. Theor. Statist., Univ. of Aarhus*.

Jensen, J.L. (1995c). Asymptotic expansions at work. *Scand. Actuarial J.*, 143-152.

Jensen, J.L. and Kristensen, L.B. (1991). Saddlepoint approximations to exact tests and improved likelihood ratio tests for the gamma distribution. *Commun. Statist.-Theor. Meth.*, **20**, 1515-1532.

Jørgensen, B. (1982). *Statistical Properties of the Generalized Inverse Gaussian Distribution*. Lecture Notes in Statistics, 9, Springer-Verlag, Berlin.

Jørgensen, B. (1983). Maximum likelihood estimation and large-sample inference for generalized linear and nonlinear regression models. *Biometrika*,

70, 19-28.

Jørgensen, B. (1986). Some properties of exponential dispersion models. *Scand. J. Statist.*, **13**, 187-197.

Jørgensen, B. (1987). Exponential dispersion models. *J. Roy. Statist. Soc.* B, **49**, 127-162.

Jørgensen, B. (1992). Exponential dispersion models and extensions: a review. *Int. Statist. Rev.*, **60**, 5-20.

Jørgensen, B. (1993). A review of conditional inference: is there a universal definition of nonformation? *Bull. Int. Statist. Inst.*, **55**, vol. 2, 323-340.

Kalbfleisch, J.D. (1975). Sufficiency and conditionality (with discussion). *Biometrika*, **62**, 251-268.

Kalbfleisch, J.D. and Prentice, R.L. (1973). Marginal likelihoods based on Cox's regression and life model. *Biometrika*, **60**, 267-278.

Kalbfleisch, J.D. and Sprott, D.A. (1970). Applications of likelihood methods to models involving a large number of parameters (with discussion). *J. Roy. Statist. Soc.* B, **32**, 175-208.

Kalbfleisch, J.D. and Sprott, D.A. (1974). Marginal and conditional likelihoods. *Sankhyā* A, **35**, 311-328.

Kaplan, E.L. and Meier, P. (1958). Non-parametric estimation from incomplete observations. *J. Amer. Statist. Assoc.*, **53**, 457-481.

Kappenman, R.F. (1975). Conditional confidence intervals for the double exponential distribution parameters. *Technometrics*, **17**, 233-235.

Karlin, S. (1952). Pólya type distributions. II. *Ann. Math. Statist.*, **28**, 281-308.

Kass, R.E. (1989). The geometry of asymptotic inference. *Statist. Sci.*, **4**, 188-234.

Kaufmann, A. and Gupta, M.M. (1991). *Introduction to Fuzzy Arithmetic. Theory and Applications.* Van Nostrand Reinhold, New York.

Keating, J.P., Glaser, R.E. and Ketchum, N.S. (1990). Testing hypotheses about the shape parameter of a gamma distribution. *Technometrics*, **32**, 67-82.

Kendall, W.S. (1993). Computer algebra in probability and statistics. *Statist. Neerlandica*, **47**, 2-25.

Kent, J.T. (1982). Robust properties of likelihood ratio tests. *Biometrika*, **69**, 19-27.

Kiefer, J. and Wolfowitz, J. (1956). Consistency of the maximum likelihood estimator in the presence of infinitely many nuisance parameters. *Ann. Math. Statist.*, **27**, 887-906.

Kim, J.S. and Proschan, F. (1988). Total positivity. In *Encyclopedia of Statistical Sciences*, S. Kotz, N.L. Johnson, C.B. Read (eds.), Wiley, New York, vol. **9**, 289-297.

Koehn, U. and Thomas, D.L. (1975). On statistics independent of a sufficient statistic: Basu's lemma. *Amer. Statistician*, **29**, 40-42.

Kolassa, J.E. (1994). *Series Approximation Methods in Statistics*. Lecture Notes in Statistics, **88**, Springer-Verlag, New York.

Kolassa, J.E. (1996). Higher-order approximations to conditional distributions. *Ann. Statist.*, **24**, 353-364.

Kolassa, J.E and McCullagh, P. (1990). Edgeworth series for lattice distributions. *Ann. Statist.*, **18**, 981-985.

Kolassa, J.E. and Tanner, M.A. (1994). Approximate conditional inference in exponential families via the Gibbs sampler. *J. Amer. Statist. Assoc.*, **89**, 697-702.

Konishi, S. (1978). An approximation to the distribution of the sample correlation coefficient. *Biometrika*, **65**, 654-656.

Konishi, S. (1981). Normalizing transformations of some statistics in multivariate analysis. *Biometrika*, **68**, 647-651.

Koul, H.L. (1992). *Weighted Empiricals and Linear Models*. IMS Lecture Notes - Monograph Series, **21**, Institute of Mathematical Statistics, Hayward, California.

Kumon, M. and Amari, S.-I. (1984). Estimation of a structural parameter in the presence of a large number of nuisance parameters. *Biometrika*, **71**, 445-459.

Laha, R.G. and Rohatgi, V.K. (1979). *Probability Theory*. Wiley, New York.

Lawless, J.F. (1972). Conditional confidence interval procedures for the location and scale parameters of the Cauchy and logistic distributions. *Biometrika*, **59**, 377-386.

Lawless, J.F. (1973). Conditional versus unconditional confidence intervals for the parameters of the Weibull distribution. *J. Amer. Statist. Assoc.*, **68**, 655-669.

Lawless, J.F. (1978). Confidence interval estimation for the Weibull and extreme value distributions. *Technometrics*, **20**, 355-368.

Lawless, J.F. (1982). *Statistical Models and Methods for Lifetime Data*. Wiley, New York.

Lawley, D.N. (1956). A general method for approximating to the distribution of the likelihood ratio criteria. *Biometrika*, **43**, 295-303.

Leadbetter, M.R., Lindgren, G. and Rootzén, H. (1983). *Extremes and Related Properties of Random Sequences and Series*. Springer-Verlag, New York.

Le Cam, L. (1986). *Asymptotic Methods in Statistical Decision Theory*. Springer-Verlag, New York.

Le Cam, L. and Yang, G.L. (1990). *Asymptotics in Statistics*. Springer-Verlag, New York.

Lehmann, E.L. (1959). *Testing Statistical Hypotheses*, 1-st ed.. Wiley, New York.

Lehmann, E.L. (1981). An interpretation of completeness and Basu's theorem. *J. Amer. Statist. Assoc.*, **76**, 335-340.

Lehmann, E.L. (1983). *Theory of Point Estimation*. Wiley, New York.

Lehmann, E.L. (1986). *Testing Statistical Hypotheses*, 2-nd ed.. Wiley, New York.

Lehmann, E.L. (1990). Model specification: the views of Fisher, Neyman and later developments. *Statist. Sci.*, **5**, 160-168.

Lehmann, E.L. (1993). The Fisher, Neyman-Pearson theories of testing hypotheses: one theory or two? *J. Amer. Statist. Assoc.*, **88**, 1242-1249.

Lehmann, E.L. and Scheffé, H. (1950). Completeness, similar regions and unbiased estimation. *Sankhyā* A, **10**, 305-340.

Lehmann, E.L. and Scholz, F.W. (1992). Ancillarity. In *Current issues in Statistical Inference: Essays in Honor of D. Basu*, M. Ghosh, P.K. Pathak (eds.), IMS Lecture Notes - Monograph Series, **17**, Institute of Mathematical Statistics, Hayward, California, 32-51.

Letac, G. (1990). *Familles exponentielles*. Lecture notes, Laboratoire de Statistique et Probabilité, Univ. Paul Sabatier, Toulouse.

Letac, G. and Mora, M. (1990). Natural real exponential families with cubic variance functions. *Ann. Statist.*, **18**, 1-37.

Li, G. (1995). Nonparametric likelihood ratio estimation of probabilities for truncated data. *J. Amer. Statist. Assoc.*, **90**, 997-1003.

Li, G., Hollander, M., McKeague, I.W. and Yang, J. (1996). Nonparametric likelihood ratio confidence bands for quantile functions from incomplete survival data. *Ann. Statist.*, **24**, 628-640.

Liang, K.-Y. (1987). Estimating functions and approximate conditional likelihood. *Biometrika*, **74**, 695-702.

Liang, K.-Y. and Self, S.G. (1996). On the asymptotic behaviour of the pseudolikelihood ratio test statistic. *J. Roy. Statist. Soc.* B, **58**, 785-796.

Liang, K.-Y. and Zeger, S.L. (1995). Inference based on estimating functions in the presence of nuisance parameters (with discussion). *Statist. Sci.*, **10**, 158-199.

Lieberman, O. (1994a). Saddlepoint approximation for the least squares estimator in first order autoregression. *Biometrika*, **81**, 807-811. Correction: (1996) **83**, 247.

Lieberman, O. (1994b). Saddlepoint approximation for the distribution of a ratio of quadratic forms in normal variables. *J. Amer. Statist. Assoc.*, **89**, 924-928.

Lindsay, B.G. (1980). Nuisance parameters, mixture models, and the efficiency of partial likelihood estimators. *Phil. Trans. Roy. Soc.*, **296**, 639-665.

Lindsay, B.G. (1983a). The geometry of mixture likelihoods: a general theory. *Ann. Statist.*, **11**, 86-94.

Lindsay, B.G. (1983b). Efficiency of the conditional score in a mixture setting. *Ann. Statist.*, **11**, 486-497.

Lindsay, B.G. (1983c). The geometry of mixture likelihoods II: the exponential family. *Ann. Statist.*, **11**, 783-792.

Lindsay, B.G. (1988). Composite likelihood methods. *Contemporary Mathematics*, **80**, 221-239.

Lindsay, B.G. (1995). *Mixture Models: Theory, Geometry and Applications.* NSF-CBMS Regional Conference Series, 5, Institute of Mathematical Statistics, Hayward, California.

Lindsey, J.K. (1996). *Parametric Statistical Inference.* Clarendon Press, Oxford.

Linhart, H. and Zucchini, W. (1986). *Model Selection.* Wiley, New York.

Lloyd, C.J. (1992). Effective conditioning. *Austral. J. Statist.*, **34**, 241-260.

Lugannani, R. and Rice, S. (1980). Saddlepoint approximations for the distribution of the sum of independent random variables. *Adv. Appl. Prob.*, **12**, 475-490.

Lukács, E. (1970). *Characteristic Functions*, 2-nd ed.. Griffin, London.

Mallick, B.K. and Gelfand, A.E. (1994). Generalized linear models with unknown link functions. *Biometrika*, **81**, 237-245.

Mann, H.B. and Wald, A. (1943). On stochastic limits and order relationships. *Ann. Math. Statist.*, **14**, 217-226.

Mardia, K.V. (1972). *Statistics of Directional Data.* Academic Press, London.

Mardia, K.V., Kent, J.T. and Bibby, J.M. (1979). *Multivariate Analysis.* Academic Press, London.

McCullagh, P. (1983). Quasi-likelihood functions. *Ann. Statist.*, **11**, 59-67.

McCullagh, P. (1984a). Local sufficiency. *Biometrika*, **71**, 233-244.

McCullagh, P. (1984b). Tensor notation and cumulants of polynomials. *Biometrika*, **71**, 461-476.

McCullagh, P. (1985). On the asymptotic distribution of Pearson's statistic in linear exponential families. *Int. Statist. Rev.*, **53**, 61-67.

McCullagh, P. (1986a). The conditional distribution of goodness-of-fit statistics for discrete data. *J. Amer. Statist. Assoc.*, **81**, 104-107.

McCullagh, P. (1986b). Quasi-likelihood functions. In *Encyclopedia of Statistical Sciences*, S. Kotz, N.L. Johnson, C.B. Read (eds.), Wiley, New York, vol. **7**, 464-467.

McCullagh, P. (1987). *Tensor Methods in Statistics.* Chapman and Hall, London.

McCullagh, P. (1991). Quasi-likelihood and estimating functions. In *Statistical Theory and Modelling*, D.V. Hinkley, N. Reid and E.J. Snell (eds.), Chapman and Hall, London, 265-286.

McCullagh, P. (1992). Conditional inference and Cauchy models. *Biometrika*, **79**, 247-259.

McCullagh, P. and Nelder, J.A. (1989). *Generalized Linear Models*, 2-nd ed.. Chapman and Hall, London.

McCullagh, P. and Tibshirani, R. (1990). A simple method for the adjustment of profile likelihoods. *J. Roy. Statist. Soc.* B, **52**, 325-344.

McLeish, D.L. and Small, C.G. (1988). *The Theory and Applications of Statistical Inference Functions.* Lecture Notes in Statistics, **44**, Springer-Verlag, New York.

McLeish, D.L. and Small, C.G. (1992). A projected likelihood function for semiparametric models. *Biometrika*, **79**, 93-102.

McPherson, G. (1989). The scientists' view of statistics — a neglected area (with discussion). *J. Roy. Statist. Soc.* A, **152**, 221-240.

Monrad, D. and Stout, W. (1988). Stable distributions. In *Encyclopedia of Statistical Sciences*, S. Kotz, N.L. Johnson, C.B. Read (eds.), Wiley, New York, vol. **8**, 617-621.

Monti, A.C. and Ronchetti, E.M. (1993). On the relationship between empirical likelihood and empirical saddlepoint approximation for multivariate M-estimators. *Biometrika*, **80**, 329-338.

Moore, D.F. (1986). Asymptotic properties of moment estimators for overdispersed counts and proportions. *Biometrika*, **73**, 583-588.

Morris, C.N. (1982). Natural exponential families with quadratic variance functions. *Ann. Statist.*, **10**, 65-80.

Møller, J. (1986). Bartlett adjustments for structured covariances. *Scand. J. Statist.*, **13**, 1-15.

Mudholkar, G.S. (1983). Fisher's z-transformation. In *Encyclopedia of Statistical Sciences*, S. Kotz, N.L. Johnson, C.B. Read (eds.), Wiley, New York, vol. **3**, 130-135.

Mukerjee, R. and Chandra, T.K. (1991). Bartlett-type adjustment for the conditional likelihood ratio statistic of Cox and Reid. *Biometrika*, **78**, 365-372.

Murphy, S.A. (1995a). Asymptotic theory for the frailty model. *Ann. Statist.*, **23**, 182-198.

Murphy, S.A. (1995b). Likelihood ratio-based confidence intervals in survival analysis. *J. Amer. Statist. Assoc.*, **90**, 1399-1405.

Murray, M.K. and Rice, J.W. (1993). *Differential Geometry and Statistics.* Chapman and Hall, London.

Mykland, P.A. (1994). Bartlett type identities for martingales. *Ann. Statist.*, **22**, 21-38.

Mykland, P.A. (1995a). Dual likelihood. *Ann. Statist.*, **23**, 396-421.

Mykland, P.A. (1995b). Martingale expansions and second order inference. *Ann. Statist.*, **23**, 703-731.

Nelder, J.A. and Lee, Y. (1992). Likelihood, quasi-likelihood and pseudo-likelihood: some comparisons. *J. Roy. Statist. Soc.* B, **54**, 273-284.

Nelder, J.A. and Wedderburn, R.W.M. (1972). Generalized linear models. *J. Roy. Statist. Soc.* A, **135**, 370-384.

Neyman, J. and Scott, E.L. (1948). Consistent estimates based on partially consistent observations. *Econometrica*, **16**, 1-32.

Owen, A.B. (1988). Empirical likelihood ratio confidence intervals for a single functional. *Biometrika*, **75**, 237-249.

Owen, A.B. (1990). Empirical likelihood ratio confidence regions. *Ann. Statist.*, **18**, 90-120.

Pace, L. and Salvan, A. (1990). Best conditional tests for separate families of hypotheses. *J. Roy. Statist. Soc.* B, **52**, 125-134.

Pace, L. and Salvan, A. (1992). A note on conditional cumulants in canonical exponential families. *Scand. J. Statist.*, **19**, 185-191.

Pace, L. and Salvan, A. (1994a). The geometric structure of the expected/observed likelihood expansions. *Ann. Inst. Statist. Math.*, **46**, 649-666.

Pace, L. and Salvan, A. (1994b). A note on invariance of the Laplace approximation under change of variable. *Working paper n. 1994.7, Dip. di Scienze Statistiche, Univ. di Padova.*

Padmanabhan, A.R. (1977). Ancillary statistics which are not invariant. *Amer. Statistician*, **31**, 124.

Pearson, K. (1900). On a criterion that a given system of deviations from the probable in the case of a correlated system of variables is such that it can be reasonably supposed to have arisen from random sampling. *Phil. Mag. Series 5*, **50**, 157-175.

Peers, H.W. and Iqbal, M. (1985). Asymptotic expansions for confidence limits in the presence of nuisance parameters with applications. *J. Roy. Statist. Soc.* B, **47**, 547-554.

Pfanzagl, J. (1970a). On asymptotic efficiency of median unbiased estimates. *Ann. Math. Statist.*, **41**, 1500-1509.

Pfanzagl, J. (1970b). Median unbiased estimates for M.L.R. families. *Metrika*, **17**, 30-39.

Pfanzagl, J. (1971). On median unbiased estimates. *Metrika*, **18**, 154-173.

Pfanzagl, J. (1979). On optimal median unbiased estimators in the presence of nuisance parameters. *Ann. Statist.*, **7**, 187-193.

Pfanzagl, J. (1982). *Contributions to a General Asymptotic Statistical Theory.* Lecture Notes in Statistics, **13**, Springer-Verlag, Berlin.

Pfanzagl, J. (1985). *Asymptotic Expansions for General Statistical Models.* Lecture Notes in Statistics, **31**, Springer-Verlag, Berlin.

Pfanzagl, J. (1990). *Estimation in Semiparametric Models*. Lecture Notes in Statistics, **63**, Springer-Verlag, Berlin.

Pfanzagl, J. (1993). Incidental versus random nuisance parameters. *Ann. Statist.*, **21**, 1663-1691.

Pierce, D.A. (1973). On some difficulties with a frequency theory of inference. *Ann. Statist.*, **1**, 241-250.

Pierce, D.A. (1982). The asymptotic effect of substituting estimators for parameters in certain types of statistics. *Ann. Statist.*, **10**, 475-478.

Pierce, D.A. and Peters, D. (1992). Practical use of higher-order asymptotics for multiparameter exponential families (with discussion). *J. Roy. Statist. Soc.* B, **54**, 701-737.

Pierce, D.A. and Peters, D. (1994). Higher-order asymptotics and the likelihood principle: one-parameter models. *Biometrika*, **81**, 1-10.

Pierce, D.A. and Schafer, D.W. (1986). Residuals in generalized linear models. *J. Amer. Statist. Assoc.*, **81**, 977-986.

Pitman, E.J.G. (1937). The closest estimates of statistical parameters. *Proc. Cambridge Phil. Soc.*, **33**, 212-222.

Pitman, E.J.G. (1938). The estimation of the location and scale parameters of a continuous population of any given form. *Biometrika*, **31**, 9-12.

Porteous, B.T. (1985). Improved likelihood ratio statistics for covariance selection models. *Biometrika*, **72**, 97-101.

Pregibon, D. (1980). Goodness of link tests for generalized linear models. *Appl. Statist.*, **29**, 15-24.

Qin, J. and Lawless, J.F. (1994). Empirical likelihood and general estimating equations. *Ann. Statist.*, **22**, 300-325.

Quesenberry, C.P. and Starbuck, R.R. (1976). On optimal tests for separate hypotheses and conditional probability integral transformations. *Commun. Statist.-Theor. Meth.*, **5**, 507-524.

Randles, R.M. and Wolfe, D.A. (1979). *Introduction to the Theory of Nonparametric Statistics*. Wiley, New York.

Rao, C.R. (1945). Information and accuracy attainable in the estimation of statistical parameters. *Bull. Calcutta Math. Soc.*, **37**, 81-91.

Rao, C.R. (1973). *Linear Statistical Inference and its Applications*, 2-nd ed.. Wiley, New York.

Rao, C.R. (1992). R.A. Fisher: the founder of modern statistics. *Statist. Sci.*, **7**, 34-48.

Rao, C.R. and Mukerjee, R. (1995). Comparison of Bartlett-type adjustments for the efficient score statistic. *J. Statist. Plann. Inf.*, **46**, 137-146.

Rasch, G. (1960). *Probabilistic Models for some Intelligence and Attainment Tests.* Studies in Mathematical Psychology I. Danish Inst. Educational Research, Copenhagen.

Read, C.B. (1985). Median unbiased estimators. In *Encyclopedia of Statistical Sciences*, S. Kotz, N.L. Johnson, C.B. Read (eds.), Wiley, New York, vol. 5, 424-426.

Reid, N. (1988). Saddlepoint methods and statistical inference (with discussion). *Statist. Sci.*, **3**, 213-238.

Reid, N. (1991). Approximations and asymptotics. In *Statistical Theory and Modelling*, D.V. Hinkley, N. Reid and E.J. Snell (eds.), Chapman and Hall, London, 287-305.

Reid, N. (1995a). The roles of conditioning in inference (with discussion). *Statist. Sci.*, **10**, 138-157, 173-199.

Reid, N. (1995b). Likelihood and Bayesian approximation methods. In *Bayesian Statistics 5*, J.M. Bernardo, J.O. Berger, A.P. Dawid, A.F.M. Smith (eds.), Oxford University Press, Oxford, 1-18.

Rémon, M. (1984). On a concept of partial sufficiency: L-sufficiency. *Int. Statist. Rev.*, **52**, 127-136.

Resnick, S.I. (1987). *Extreme Values, Point Processes and Regular Variation.* Springer-Verlag, New York.

Ripley, B.D. (1988). *Statistical Inference for Spatial Processes.* Cambridge University Press, Cambridge.

Robinson, J. (1982). Saddlepoint approximations for permutation tests and confidence intervals. *J. Roy. Statist. Soc.* B, **44**, 91-101.

Rockafellar, R.T. (1970). *Convex Analysis.* Princeton University Press, Princeton.

Ronchetti, E.M. (1990). Small sample asymptotics: a review with applications to robust statistics. *Comp. Statist. Data Anal.*, **10**, 207-223.

Ronchetti, E.M. and Welsh, A.H. (1994). Empirical saddlepoint approximations for multivariate M-estimators. *J. Roy. Statist. Soc.* B, **56**, 313-326.

Rousseeuw, P.J. and Leroy, A. (1987). *Robust Regression and Outlier Detec-*

tion. Wiley, New York.

Royall, R.M. (1986). Model robust confidence intervals using maximum likelihood estimators. *Int. Statist. Rev.*, **54**, 221-226.

Sakia, R.M. (1992). The Box-Cox transformation technique: a review. *The Statistician*, **41**, 169-178.

Sandved, E. (1965). A principle for conditioning on an ancillary statistic. *Skand. Aktuar. Tidskr.*, **49**, 39-47.

Santner, T.J. and Duffy, D.E. (1986). A note on A. Albert and J.A. Anderson's conditions for the existence of maximum likelihood estimates in logistic regression models. *Biometrika*, **73**, 755-758.

Sen, P.K. (1992). The Pitman closeness of statistical estimators: latent years and the renaissance. In *Current issues in Statistical Inference: Essays in Honor of D. Basu*, M. Ghosh, P.K. Pathak (eds.), IMS Lecture Notes - Monograph Series, **17**, Institute of Mathematical Statistics, Hayward, California, 52-74.

Serfling, R.J. (1980). *Approximation Theorems of Mathematical Statistics*. Wiley, New York.

Seshadri, V. (1993). *The Inverse Gaussian Distribution*. Clarendon Press, Oxford.

Severini, T.A. (1993). Local ancillarity in the presence of a nuisance parameter. *Biometrika*, **80**, 305-320.

Severini, T.A. (1994). On approximate elimination of nuisance parameters by conditioning. *Biometrika*, **81**, 649-661.

Severini, T.A. (1995). Information and conditional inference. *J. Amer. Statist. Assoc.*, **90**, 1341-1346.

Shun, Z. and McCullagh, P. (1995). Laplace approximation of high dimensional integrals. *J. Roy. Statist. Soc. B*, **57**, 749-760.

Silvapulle, M.J. (1981). On the existence of maximum likelihood estimators for the binomial response model. *J. Roy. Statist. Soc. B*, **43**, 310-313.

Silvapulle, M.J. and Burridge, J. (1986). Existence of maximum likelihood estimates in regression models for grouped and ungrouped data. *J. Roy. Statist. Soc. B*, **48**, 100-106.

Skovgaard, I.M. (1986). A note on differentiation of cumulants of log likelihood derivatives. *Int. Statist. Rev.*, **54**, 29-32.

Skovgaard, I.M. (1987). Saddlepoint expansions for conditional distributions.

J. Appl. Prob., **24**, 875-887.

Skovgaard, I.M. (1989). A review of higher order likelihood methods. *Bull. Int. Statist. Inst.*, **53**, vol. 3, 331-351.

Skovgaard, I.M. (1990). On the density of minimum contrast estimators. *Ann. Statist.*, **18**, 779-789.

Skovgaard, I.M. (1996). An explicit large-deviation approximation to one-parameter tests. *Bernoulli*, **2**, 145-165.

Small, C.G. and Murdoch, D.J. (1993). Nonparametric Neyman-Scott problems: telescoping product methods. *Biometrika*, **80**, 763-779.

Smith, R.L. (1989). Extreme value analysis. *Statist. Sci.*, **4**, 367-393.

Smith, W.L. (1983). Generating functions. In *Encyclopedia of Statistical Sciences*, S. Kotz, N.L. Johnson, C.B. Read (eds.), Wiley, New York, vol. **3**, 372-376.

Smyth, G.K. (1994). A note on modelling cross-correlations: hyperbolic secant regression. *Biometrika*, **81**, 396-402.

Sørensen, M. (1983). On maximum likelihood estimation in randomly stopped diffusion-type processes. *Int. Statist. Rev.*, **51**, 93-110.

Sprott, D.A. (1975). Marginal and conditional sufficiency. *Biometrika*, **62**, 599-605.

Sprott, D.A. (1990). Inferential estimation, likelihood, and linear pivotals. *Can. J. Statist.*, **18**, 1-15.

Stafford, J.E. (1996). A robust adjustment of the profile likelihood. *Ann. Statist.*, **24**, 336-352.

Strawderman, R.L., Casella, G. and Wells, M.T. (1996). Practical small sample asymptotics for regression problems. *J. Amer. Statist. Assoc.*, **91**, 643-654.

Struik, D.J. (1966). *A Concise History of Mathematics*. Dover, New York.

Stuart, A. and Ord, J.K. (1991). *Kendall's Advanced Theory of Statistics*, vol. **2**, 5-th ed.. Edward Arnold, London.

Sverdrup, E. (1966). The present state of decision theory and the Neyman-Pearson theory. *Rev. Int. Statist. Inst.*, **34**, 309-333.

Sweeting, T.J.A. (1995a). A framework for Bayesian and likelihood approximations in statistics. *Biometrika*, **82**, 1-23.

Sweeting, T.J.A. (1995b). A Bayesian approach to approximate conditional inference. *Biometrika*, **82**, 25-36.

Taniguchi, M. (1991). *Higher Order Asymptotic Theory for Time Series Analysis*. Lecture Notes in Statistics, **68**, Springer-Verlag, New York.

Tarter, M. and Lock, M. (1994). *Model Free Curve Estimation*. Chapman and Hall, London.

Tawn, J.A. (1990). Modelling multivariate extreme value distributions. *Biometrika*, **77**, 245-253.

Thisted, R.A. (1988). *Elements of Statistical Computing*. Chapman and Hall, New York.

Tierney, L. and Kadane, J.B. (1986). Accurate approximations for posterior moments and marginal densities. *J. Amer. Statist. Assoc.*, **81**, 82-86.

Tierney, L., Kass, R.E. and Kadane, J.B. (1989a). Fully exponential Laplace approximations to expectations and variances of nonpositive functions. *J. Amer. Statist. Assoc.*, **84**, 710-716.

Tierney, L., Kass, R.E. and Kadane, J.B. (1989b). Approximate marginal densities of nonlinear functions. *Biometrika*, **76**, 425-433. Correction: (1991) **78**, 233-234.

Tsou, T.-S. and Royall, R.M. (1995). Robust likelihoods. *J. Amer. Statist. Assoc.*, **90**, 316-320.

Tukey, J.W. (1977). *Exploratory Data Analysis*. Addison-Wesley, Reading, Mass.

Tukey, J.W. (1980). We need both exploratory and confirmatory. *Amer. Statistician*, **34**, 23-25.

Tunnicliffe-Wilson, G. (1989). On the use of marginal likelihood in time series problems. *J. Roy. Statist. Soc.* B, **51**, 15-28.

Tweedie, M.C.K. (1957). Statistical properties of inverse Gaussian distributions. *Ann. Math. Statist.*, **28**, 362-377, 696-705.

Upton, G.J.G. (1992). Fisher's exact test. *J. Roy. Statist. Soc.* A, **155**, 395-402.

Uthoff, V.A. (1970). An optimum test property of two well known statistics. *J. Amer. Statist. Assoc.*, **65**, 1597-1600.

van Beek, P. (1972). An application of the Fourier method to the problem of sharpening the Berry-Esséen inequality. *Z. Wahrsch. Verw. Geb.*, **23**, 187-197.

van der Vaart, A.W. (1996). Efficient maximum likelihood estimation in semiparametric mixture models. *Ann. Statist.*, **24**, 862-878.

Vardi, Y. and Lee, D. (1993). From image deblurring to optimal investments: maximum likelihood solutions for positive linear inverse problems (with discussion). *J. Roy. Statist. Soc.* B, **55**, 569-612.

Ventura, L. (1994). Approximate marginal likelihoods and approximate MPI tests for location-scale models. *Working Paper n. 1994.5, Dip. Scienze Statist., Univ. di Padova.*

Vidoni, P. (1995). A simple predictive density based on the p^*-formula. *Biometrika*, **82**, 855-863.

von Mises, R. (1947). On the asymptotic distribution of differentiable statistical functions. *Ann. Math. Statist.*, **18**, 309-348.

Wald, A. (1949). Note on the consistency of maximum likelihood estimate. *Ann. Math. Statist.*, **20**, 595-601.

Wald, A. and Wolfowitz, J. (1948). Optimum character of the sequential probability ratio test. *Ann. Math. Statist.*, **19**, 326-339.

Wang, S. (1990). Saddlepoint approximations in resampling analysis. *Ann. Inst. Statist. Math.*, **42**, 115-131.

Wang, S. (1992a). Tail probability approximation in the first order noncircular autoregression. *Biometrika*, **79**, 431-434.

Wang, S. (1992b). General saddlepoint approximations in the bootstrap. *Statist. Prob. Letters*, **13**, 61-66.

Wang, S. (1993). Saddlepoint expansions in finite population problems. *Biometrika*, **80**, 583-590.

Waterman, R.P. and Lindsay, B.G. (1996a). A simple and accurate method for approximate conditional inference applied to exponential family models. *J. Roy. Statist. Soc.* B, **58**, 177-188.

Waterman, R.P. and Lindsay, B.G. (1996b). Projected score methods for approximating conditional scores. *Biometrika*, **83**, 1-13.

Wedderburn, R.W.M. (1974). Quasi-likelihood functions, generalised linear models, and the Gauss-Newton method. *Biometrika*, **61**, 439-447.

Wedderburn, R.W.M. (1976). On the existence and uniqueness of the maximum likelihood estimates for certain generalized linear models. *Biometrika*, **63**, 27-32.

Welch, B.L. (1947). The generalization of 'Student's' problem when several different population variances are involved. *Biometrika*, **34**, 28-35.

White, H. (1982). Maximum likelihood estimation in misspecified models.

Econometrica, **50**, 1-25.

Wilson, E.B. and Hilferty, M.M. (1931). The distribution of chi-square. *Proc. Nat. Acad.*, **17**, 684-688.

Withers, C.S. (1983). Expansions for the distribution and quantiles of a regular functional of the empirical distribution with applications to nonparametric confidence intervals. *Ann. Statist.*, **11**, 577-587.

Wong, W.H. (1986). Theory of partial likelihood. *Ann. Statist.*, **14**, 88-123.

Wong, W.H. and Li, B. (1992). Laplace expansions for posterior densities of nonlinear functions of parameters. *Biometrika*, **79**, 393-398.

Wong, W.H. and Severini, T.A. (1991). On maximum likelihood estimation in infinite dimensional parameter spaces. *Ann. Statist.*, **19**, 603-632.

Yamada, S. and Morimoto, H. (1992). Sufficiency. In *Current issues in Statistical Inference: Essays in Honor of D. Basu*, M. Ghosh, P.K. Pathak (eds.), IMS Lecture Notes - Monograph Series, **17**, Institute of Mathematical Statistics, Hayward, California, 86-98.

Yanez, N.D. and Wilson, J.R. (1995). Comparison of quasi-likelihood models for overdispersion. *Austral. J. Statist.*, **37**, 217-231.

Yates, F. (1984). Tests of significance for 2×2 contingency tables (with discussion). *J. Roy. Statist. Soc.* A, **147**, 426-463.

Ying, Z. (1993). Maximum likelihood estimation of parameters under a spatial sampling scheme. *Ann. Statist.*, **21**, 1567-1590.

Young, G.A. (1994). Bootstrap: more than a stab in the dark? *Statist. Sci.*, **9**, 382-415.

Zacks, S. (1981). *Parametric Statistical Inference*. Pergamon Press, Oxford.

Zhu, Y. and Reid, N. (1994). Information, ancillarity, and sufficiency in the presence of nuisance parameters. *Can. J. Statist.*, **22**, 111-123.

AUTHOR INDEX

SUBJECT INDEX

www.ingramcontent.com/pod-product-compliance
Lightning Source LLC
Chambersburg PA
CBHW050632190326
41458CB00008B/2242